ADVANCED TECHNOLOGY FOR COMMAND AND CONTROL SYSTEMS ENGINEERING

Edited by
Stephen J. Andriole

AFCEA International Press
Fairfax, Virginia

Book Design and Production by Brenda N. Jackie

November 1990

Published by AFCEA International Press (AIP)
A Division of the Armed Forces Communications
and Electronics Association (AFCEA)
4400 Fair Lakes Court
Fairfax, Virginia 22033-3899 USA
(703)631-6190 or (800)336-4583 ext. 6190

Library of Congress Cataloging-in-Publication Data

Advanced technology for command and control systems engineering/edited by Stephen J. Andriole.
p. cm.
Includes index.
ISBN 0-916159-22-1 : $29.95
1. Command and control systems. I. Andriole, Stephen J.
UB212.A34 1990
355.3'3041--dc20 90-19757
CIP

Contents

Acknowledgments

First and foremost, I'd like to thank the contributors to this volume. On behalf of the command and control research and development community, I appreciate the time stolen from over-committed schedules to produce a set of superb papers. The field is better off now than before because of the willingness of analysts and practitioners to share their expertise with a wide audience.

I'd also like to thank the professionals at AFCEA International Press for making the process—once again—smooth and rewarding. As always, Brenda Jackie, Al Campen and Judy Shreve deserve very special mention.

Stephen J. Andriole
Marshall, Virginia
September 1990

Preface

The importance of command and control systems to our national security is widely recognized. From the National Command Authorities to an Army platoon leader, C^2 systems provide the means by which a commander directs his forces to accomplish a mission. These C^2 systems consist of the personnel, equipment, communications, facilities, and procedures that the commander utilizes to control his forces. History provides many examples of cases where an excellent commander supported by a suitable C^2 system has successfully accomplished his mission. Unfortunately, history also provides many examples, usually more publicized, where the C^2 system has had a significant negative impact on the outcome. Examples ranging from Pearl Harbor, the Pueblo, the Liberty, to the Vincennes incident come quickly to mind.

The importance of command and control systems is reflected in the portion of the defense budget devoted to this area. In recent years, the C^3 budget has exceeded seven percent of the defense budget. The intelligence function which is an integral element of command and control has a separate classified budget. As the defense budget tightens, it becomes increasingly important that the C^3I dollars be employed effectively.

In view of the recognized importance of C^2 systems from both the mission effectiveness and cost-effectiveness standpoint, it is vital that we develop an intellectual basis for C^2. The intellectual basis must provide us with the necessary theory, tools, and techniques to design and evaluate command and control systems.

One of the objectives of the new Center of Excellence in Command, Control, Communications, and Intelligence (C^3I) at George Mason University is to help establish this intellectual basis for C^2. The research program, which is currently sponsored by the Virginia Center of Innovative Technology, the Defense Communications Agency, CECOM, ASD(C^3I), MITRE, PRC/ATI, Lockheed, AFCEA International, and AFCEA NOVA, conducts a broad spectrum research and development program. The research program includes projects in requirements engineering, modeling and simulation, automated planning, decision support systems, advanced communications technologies, and cognitive engineering. A quick survey of the table of contents will show that all of these topics are represented in this book. A second objective of the Center is to develop an educational program which provides a unique curriculum leading to an masters of science degree in C^3I. One of the results of the curriculum development will be a series of books on various aspects of C^3I. It is appropriate that the first C^2 book to be published by a faculty member associated with the Center is edited by Professor Stephen J. Andriole, who has played a key role in the creation of the Center.

This book should be of interest to a widespread audience, including people who specify systems requirements, system engineers, software system developers, and potential users. The advanced technology for C^2 system engineering discussed in this book should enable the community to create more effective command and control systems.

Harry L. Van Trees
Distinguished Professor of Information Technology,
Electrical and System Engineering and
Director, Center of Excellence in
Command, Control, Communications and Intelligence
George Mason University

Introduction

This book tries to synthesize an enormous amount of disparate material. Such is the nature of multidisciplinary systems engineering (MSE). Command and control (C^2) and its supporting cast of characters — communications systems, intellegence collection and processing systems, platforms and weapons — is complex, dynamic and to a great extent conceptually, analytically and operationally elusive. There are countless "models" of command and control, perennially controversial measures of effectiveness, and raging debates about whether neural networks, expert systems, or linear programs make more sense for the same application. In may respects C^2 approaches, models, tools, methods and techniques constitute a proverbial Tower of Babel.

This edited volume tries to lend some order. The organizing concept is systems engineering, though but a few of the contributors even use the term. The book also tries to paint a technology picture at a particular point in time, a picture that should fade over time, yielding to the inevitable progress of technology and the means by which we apply it. The book reflects the breadth and depth of the field, the richness of the methodological persepctives that inform C^2 systems design and development, and the issues and challenges that remain.

Systems Engineering in Perspective

What is systems engineering? Why is it essential to the design, development and deployment of command and control systems?

Systems engineering is a problem-solving process with roots in the behavioral, computer, engineering, management, and mathematical sciences. According to at least one source (DOD, 1974 as quoted in Chambers, 1986), the systems engineering process is a:

"... logical sequence of activities and decisions transforming an operational need into a description of system performance parameters and a preferred system configuration ..."

The same source recognizes the importance of the following nine phases of the systems engineering process:

1. Mission requirements analysis;
2. Functional analysis;
3. Allocation;
4. Synthesis;
5. Logistic engineering;
6. Life cycle cost analysis;
7. Optimization;
8. Production engineering analysis; and
9. Generation of system specifications.

Other sources (Eisner, 1988) regard the systems engineering process as more elaborate, consisting of the following "elements":

1. Requirements analysis;
2. Requirements allocations;
3. Functional analysis;
4. Functional allocation;
5. Specification analysis;
6. Specification allocation;
7. Specification development;
8. Preliminary design:
    A. System level;
    B. Sub-system level;
9. Interface definition;
10. Schedule development;
11. Preliminary cost-analysis;
12. Technical performance measurement;
13. Trade-off/alternative analysis;
14. Pre-planned product improvement;
15. Final design:
    A. System level;
    B. Sub-system level;
16. Schedule update;
17. Cost update;
18. Fabrication;
19. Coding;
20. Preliminary testing;
21. De-bugging and reconfiguration;

22. Testing and integration;
23. Updates:
    A. Schedule;
    B. Cost;
    C. Technical performance measurement;
24. Documentation;
25. Training; and
26. Production.

As the above implies, the systems engineering process—when properly implemented—consists of a number of steps that when taken together solve a specific problem. The systems engineering process is by nature and purpose generic; specific substantive application areas—like C^2—borrow from the generic process and tailor elements from the generic process to the substantive problem at hand.

The MSE State-of-Mind

It is important to understand the differences in the perspectives of systems analysts, systems engineers, "other" engineers, programmers, for example, and students of general systems theory. Let's start with programmers, since they are most often—rightly or wrongly—associated with the design and development of (C^2) information systems. Programmers are instinctively "bottom-up" when it comes to their software vis-a-vis the entire information system. They see their code as accomplishing some aspect or aspects of the system's operation. They take a "micro-view" of the system, but change perspectives when it comes to the design and development of their individual software systems. Good programmers consciously or unconsciously adopt a systems approach to the design and development of their pieces of the overall systems puzzle. Such analytical schizophrenia is at first blush difficult to fathom, but upon reflection makes perfect sense. The programmer is not repsonsible for the entire system, only a piece of it; use of a "top-down" systems approach makes the most sense when you are responsible for identifying a set of interrelated parts that together must perform some specific function (like a computer program).

The contribution of computer programmers in the systems design and development process is measured several levels down from the contribution of the systems analyst, who worries about how all the pieces will fit together. Mills (1985) goes even farther by drawing a distinction between the systems designer and the "systems architect," who has "control" over the entire design, at least during the initial stages of the project.

"Other," that is, non-systems engineers, also see systems from the bottom-up, except vis-à-vis their own responsibilities. The mechanical engineer might worry about the hydraulic *system* or the structural integrity of the materials he chooses

The Systems Engineering Process

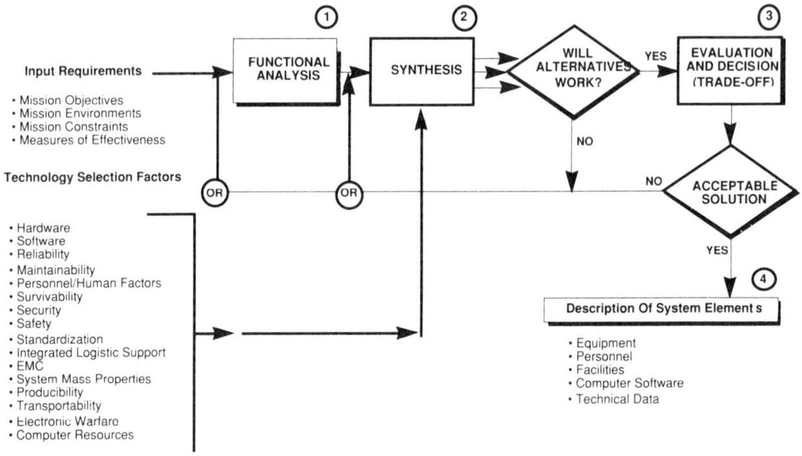

Figure 1.

to use. The electrical engineer might worry about circuitry that supports larger operations, while the chemical engineer will probably focus on certain reactions that drive subsequent processes. All of these engineers worry about their "little" systems; the systems architect worries about how they all fit together.

The systems engineer worries about everything. The systems engineer is the quintessential top-down problem-solver. The systems engineer applies the systems approach to the design and development of complex systems. The systems engineer is the architect, the analyst, and a whole lot more.

Systems engineers lead, integrate, synthesize, prioritize, measure, test, prototype, and monitor. They adhere to a set of procedures that are simultaneously explicit and imprecise. Systems engineering is as much an art as it is an applied science. Believe it or not, there is a systems engineering "standard" (MIL-STD-499A) definition:

> "Systems engineering is the application of scientific and engineering efforts to (a) transform operational need into a description of system performance parameters and a system configuration through the use of an iterative process of definition, synthesis, analysis, design, test, and evaluation; (b) integrate related technical parameters and ensure compatibility of all physical, functional, and program interfaces in a manner that optimizes the total system definition and design; (c) integrate reliability, maintainability, safety, survivability, human, and other such factors into the total engineering effort to meet cost, schedule, and technical performance objectives."

The government's Systems Engineering Management Guide (DSMC, 1986) describes the process graphically in Figure 1. The steps in the process — which is deliberately generic—suggest how systems engineers proceed to design and develop complex systems. *More to the point, the figure suggests how they design designs.*

Systems engineers apply this life cycle to a variety of problems. There are urban systems engineers, transportation systems engineers, sewer systems engineers, space station systems engineers, and C^2 systems engineers. Each of these applications calls for a detailed interpretation of the generic life cycle presented in Figure 1, though they all have several important things in common.

First and foremost, is the nature of the tools and techniques necessary to implement a generic or problem-specific systems engi neering life cycle: *they are always multidisciplinary.* There is simply no way one can design or develop a system today without the benefit of a multidisciplinary perspective. All phases of the life cycle call for multiple methods and approaches that cannot be found under a single (or even a few) disciplinary umbrella(s). It is interesting that many of the systems engineering texts fail to emphasize this aspect of systems engineering. *The second common denominator is iteration.* Complex systems are never designed right the first time. Another important similarity is *the need for benefit and cost-benefit analyses, while a fourth is the need to coordinate all design decisions with validated requirements.*

All of these common aspects exist in every single systems engineering project. We have already established the relevance of the behavioral, management and computer sciences for requirements analysis. Prototyping means iteration, and good systems engineers are constantly looking for cost-effective solutions to complicated design alternatives. But much more importantly, *the systems engineer is driven by a top-down specification of requirements*, as suggested in Figure 2 (DSMC, 1986). The systems engineer uses the methods, tools and techniques from a variety of fields and disciplines to design and develop the system.

The multidisciplinary systems engineering (MSE) state-of-mind is critical to successful systems design and development. *It seeks information about the purpose of the system, the motivation behind its development, user expectations about performance, and mangerial constraints.* MSE never specifies requirements, never addresses testing or evaluation questions, and never develops maintenance schedules *independent of requirements. The ultimate reason why a system is to exist is the filter through which all design decisions must successfully pass.*

Key MSE Leveraging Principles

What are the MSE "watchwords," the key leveraging principles? The first is *functional hierarchical decomposition of requirements.* MSE calls for the identi-

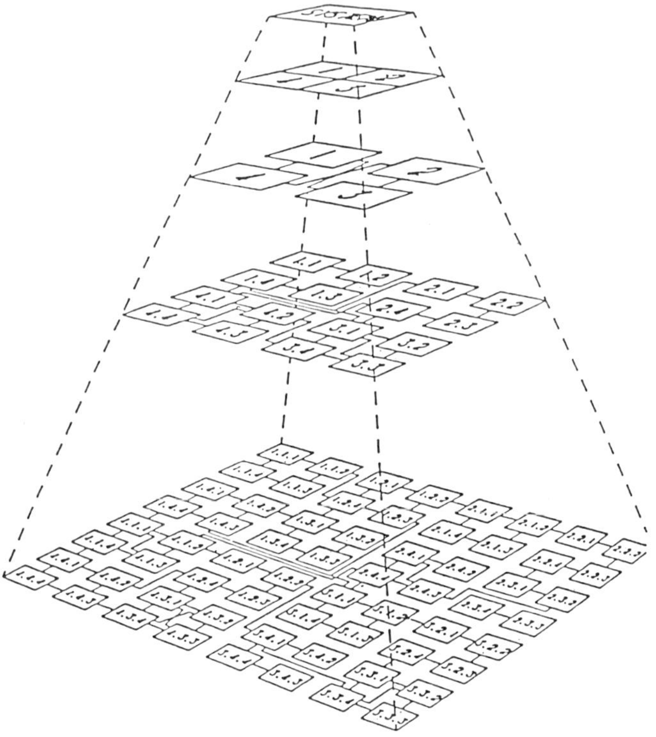

Figure 2. Top Down, Hierarchical Decomposition of System Requirements.

fication, definition and validation of requirements on multiple levels that suggest how requirements interrelate and how they might be satisfied through design.

Another key principle is iteration, or the recognition that it is impossible to capture requirements the first time through a system concept. MSE assumes the need for iteration—in all phases of the design and development process. Rapid and not-so-rapid prototyping assumes the need to iterate on user requirements, alternative system concepts, software requirements, testing and evaluation and even documentation.

Yet another key concept is synthesis, or the ability to integrate disparate components into a coherent, cost-effective whole greater than the sum of the components. Here is where systems engineering generally and MSE specifically becomes an art. The process of creative synthesis has been described (Smith, 1988), but not formally represented. While there are steps that suggest *what* and *where* to synthesize elements of the design, there are far fewer guidelines for instructing the information systems engineer about *how* to proceed.

Nevertheless, synthesis remains an important part of the MSE process and one that is well suited to a multidisciplinary perspective.

Trade-off analysis is always part of the MSE process. The mix among requirements, constraints, and alternative designs yields a continuing need to evaluate, prioritize, and test. Every time we build a prototype we engage in trade-off analysis. Every time we decide upon one course of design action versus another, we trade off. The more complex the design challenge, the more trade-offs we have to make. The more constraints we face, the more important the trade-offs become.

Last, but certainly not least, is the essential need to remain as *multidisciplinary* as possible. This calls for expertsie that cuts across the social, behavioral, computer, mathematical, engineering, management, and physical sciences. But note that systems engineers cannot be expected to know all there is to know about these fields and disciplines, but only—at a minimum—that they indeed even exist! An attitude that accepts contributions—regardless of their disciplinary race, creed or religion—is what keeps the door open to creative problem-solving. Once we close our minds to someone else's methods, tools and techniques we dramatically reduce the probability of success.

Advanced Technology for Command and Control Systems Engineering

This edited volume assumes the value of multidisciplinary systems engineering. It also recognizes the need for new ideas, concepts and creative applied technologies. Part I focuses on C^2 information systems engineering. System, user and software requirements analysis is examined; new approaches and methods are proposed. A new information systems engineering life cycle is even introduced.

Part II focuses on C^2 modeling and simulation. Modeling and simulation are necessary systems engineering tools. Models of command processes and simulations of courses of actions provide valuable insight into the systems design process. Part II deals with war gaming, human performance modeling, and simulation for requirements analysis, systems design and training.

Part III examines the range of intelligent systems technology available to the multidisciplinary C^2 systems engineer. Some challenges are extended to the "conventional" artificial intelligence (AI) community, and some applied case studies are described. Evaluation is also an important theme of Part III: how do we know if an expert advisory system enhances or degrades problem-solving?

Part IV looks at a particular kind of C^2 systems and the means by which they are engineered. This part of the book examines some case studies, acknowledges the importance of evaluation, and presents some advanced technology for next generation decision support systems engineering.

Part V focuses on the requirements analysis and modeling of substantive C^2 processes and functions. Part V deals with some controversial systems, some tricky allocation problems, and how technology can shape our approach to tactical planning systems engineering.

Part VI describes some advanced communications technology and how it has been—and most likely will be—leveraged to support situation assessment and command decision-making. The emphasis is on technology that will help revolutionize C^2 communications systems engineering.

Part VII turns to some issues. Computer-aided systems engineering (CASE) tools are described and a call for "a change in system design emphasis" (from machine to human) is sounded. The "prospects for cognitive systems engineering" are explored and the nature of multidisciplinary systems engineering education and training is scrutinized.

The book tries to leverage technology in the larger C^2 systems engineering process. Hopefully the chapters will expand your conceptual and methodological toolboxes.

References

Chambers, GJ (1986) "The Systems Engineering Process: A Technical Bibliography." *IEEE Transactions on Systems, Man and Cybernetics*, Volume 16, Number 5, September - October 1986, pp. 712-721

Defense Systems Management College, *Systems Engineering Management Guide*, US Government Printing Office, 1986

Eisner H, *Computer-Aided Systems Engineering*, Prentice-Hall, 1988

Mills, JA (1985) "A Pragmatic View of the Systems Architect," *Communications of the ACM*, Volume 28, Number 7, pp. 708-717.

Smith, CU (1988) "Applying Synthesis Principles to Create Responsive Software Systems," *IEEE Transactions on Software Engineering*, Volume 14, Number 10, pp. 1394-1408.

United States Department of Defense, *Engineering Management, MIL-STD-499A (USAF)*, May 1, 1974, pp. 11-18.

PART 1

C^2 Information Systems Engineering

Software Reuse in Command and Control Systems

Christine L. Braun and Alan B. Salisbury

Introduction

The "software crisis" within the Department of Defense (in particular) has been well-documented. Billions of dollars are spent annually on the development of new software and the maintenance of existing software. While the most visible and vexing problems are with what can be described as the virtually routine escalation of costs and slippages of schedules associated with essentially all new software, the continually escalating costs of maintaining ever-increasing quantities of operational software are of at least equal importance. Software for command and control (C^2) systems is no exception, and may be at the forefront of areas of concern.

Software "reuse"—the use of the same software in multiple systems—offers the very real potential of dramatically reducing the development cost of software for new systems, and significantly lowering maintenance costs as well. There are several contributing factors that support this conclusion. First and foremost is the high degree of functional commonality among C^2 systems, ranging from high level application commonality (e.g., Order of Battle), through common support functions (e.g., automated message handling), to more generic functions (e.g., man-machine interface). These are rich areas with an even larger base of common subfunctions that assure a high potential for reuse of software across multiple systems. A second factor is the increasing use of Ada as the implementation language for C^2 systems. The wide availability of Ada compilers for diverse operating environments greatly facilitates the transport of reusable designs and programs for reuse.

In addition to savings in development cost and time, other benefits accrue from software reuse in C^2 systems. Interoperability, as an example, can be achieved more readily between systems sharing common software. Typically a large amount of interoperability software is given over to the requirement for multiple conversions, including such items as data elements and codes, and information formats that are different between two systems. Such conversions can be eliminated in part or all depending on the degree of shared common software. Similar benefits occur in the areas of reliability, maintainability and usability.

The Army Tactical Command and Control System (ATCCS) presents an excellent case study for the high potential of software reuse in command and control systems. ATCCS is a family of five systems, each controlling one of the five major battlefield functional areas (maneuver, fire support, intelligence/electronic warfare, air defense and combat service support), plus an overlaid distributed integrating system providing "force level" control for the commander. Each system has been developed independently, several employing different languages on different processors, many implementing the same functions differently with little existing commonality. Under the direction of an integrating program executive officer, a major effort is being made to identify common functions to be implemented with reusable software for all five systems. ATCCS can thus be expected to evolve over time to employ increasing levels of common (reusable) software, affording a highly visible opportunity to demonstrate the advantages of reusable software.

In many ways, software reuse is not a new phenomenon. The use of common operating systems and DBMS, for example, are well-established forms of reuse. Working actively to increase reuse through the disciplined processes of functional (domain) analysis, software cataloging and retrieval, and program synthesis from libraries of software components is, however, a relatively new direction. The potential payoff for developers and maintainers of systems is exceptionally high.

What is Software Reuse?

Software reuse can be simply defined as "the use of a software component more than once." This is a very broad definition. A *software component* can be any product of the software development process—a unit of code, a design specification, a test case, etc. Use of a component more than once can mean anything from informal reuse of a design by its designer to the widespread use of a large software package such as a Data Base Management System. The component can be used unchanged, or it can be modified to fit the new application. Technologies that help automate the production of programs, e.g., applications generators, 4GLs, can also be thought of as approaches to reuse.

Much software reuse occurs today. At a low level, use of standard packages such as math routines is everyday practice. At a higher level, applications typically include commercial operating systems, DBMSs, etc. We don't usually call these standard practices "software reuse," but that is exactly what they are. The objective of software reuse researchers is to extend this practice to many more applications. By understanding how reuse works in these established cases, we can find ways to institutionalize the practice.

A Reuse Spectrum

There are many approaches to identifying and reusing software components. One consideration is what we might call bottom-up versus top-down approaches.

Bottom-up reuse looks for small, widely-usable bottom-level components such as math packages, sorting and searching routines, etc. There are many publicly-available libraries that offer such components. Books of general-purpose C routines are available; many authors include the source code on a disk. Ada software components are available free from the government's SIMTEL20 repository, and may be purchased from various vendors. The payoff with bottom-up reuse comes from widespread use of the components.

Top-down reuse depends on an overall view of system structure to identify larger-scale components that can be reused in a particular class of systems. It looks for *generic architectures* that can serve as design frameworks and permit large-scale reuse. An example is the Data Base Management System. The original developers of the DBMS concept recognized that many systems included a large group of data manipulation operations that could be extracted and packaged. Providing these services in an off-the-shelf package saves the developer a lot of work; however, it requires the constraint of the design to adapt to the functions and interfaces provided by the DBMS. Effectively, the invention of the DBMS has dictated a generic architecture for a class of systems. The payoff with top-down reuse comes from the large-scale of reuse; typically the instances of reuse are fewer.

The greatest benefits of reuse are realized when we can combine these two approaches, offering a generic design architecture as a starting point, with a menu of lower-level components available to fill in the specific functions in the architecture. Clearly, such an approach only makes sense when we constrain our focus to a particular application domain.

Domain-Specific Reuse

Reuse is much more widespread in some application domains than in others. The compiler development field is widely referenced as a model of successful reuse in practice, and is worth examining as an example. A generic compiler architecture (a front end that checks program syntax and semantics and a back end that produces object code, listings, etc.) is almost universally accepted. The existence of such a generic compiler architecture has led to significant reuse in building compilers:

- Compiler designers never have to start from scratch. Design reuse is standard; all designs build from the same top-level structure.
- Because all compilers have the same structure, interfaces between parts of the compilers have tended to standardize. For example, there is a standard interface between Ada front ends and back ends called a Diana-tree.
- The generic architecture establishes a common vocabulary for talking about compiler parts. If a compiler designer says "I need a global optimizer for Ada that operates on a Diana-tree representation," other compiler designers will all understand what this means.
- Because all compilers are built from the same functional parts with standard interfaces, and because designers refer to them with a common vocabulary, reuse of parts is greatly simplified. For example, compilers for the same language routinely reuse the entire compiler front end.

Major advances in software reuse are possible through establishing such generic architectures for other application domains. In particular, the command and control domain is exceptionally well-suited to such an approach, as we will illustrate later.

Reuse-Directed Design

Given a generic architecture like that described, and a set of components that implement major portions of the architecture, it is possible to *direct* the design activity so that it makes maximum reuse of the available components. An analogy to the hardware world is highly applicable. Today major subfunctions of electronic systems are available on integrated circuits, or "chips." Hardware designers don't design systems down to the gate logic level, then look to see what chips might implement some of their design. Instead, they know what chips are available, and direct their designs toward these chips. Clearly, the latter approach leads to much greater reuse, with consequent savings in development cost. However, software design today typically follows the former model. We design down to the detailed level and then may try to match our design to existing software components. Not surprisingly, we find few perfect fits. A goal in the reuse field is to move toward the directed approach in order to maximize opportunities for reuse; we believe that such an approach is possible for command and control systems.

Why Reuse in Command and Control Applications?

C^2 is a particularly fertile area for software reuse. This is true both because the nature and structure of such systems offer the opportunity for substantial reuse, and because the potential benefits of reuse are so significant. The following subsections briefly discuss some of these opportunities and benefits.

Opportunities for Reuse

System Size and Cost

C^2 systems are large and expensive, and becoming more so every day. As hardware becomes cheaper and more powerful, the incentive to build more and more software to take advantage of it is very great. However, building software is labor-intensive; it just gets more expensive as the price of labor goes up. It is essential to find ways to control these costs through finding less labor-intensive ways to build software, and software reuse appears to offer a partial solution. The payoff of software reuse is greatest when it can be applied to large systems; hence the C^2 domain is a major focus for the reuse community.

Functional Commonality

A major reason to focus on reuse in a particular application domain is simply the recognition that systems in that domain have many functions in common. While there are many kinds of C^2 systems, they all have common functions. Furthermore, there is much functional commonality within a single C^2 system, leading to reuse opportunities even within a single project. Some of the major common high-level C^2 functions include system management, database interaction, message processing, man-machine interface, decision support, and communications processing. At lower levels, there is significant opportunity to reuse software for manipulating basic data structures and screen interface objects, for performing common functions, and for producing reports. Table 1, from [Quanrud] indicates the reuse potential of the components of a typical C^2 system architecture.

Interface Standardization

Systems that conform to established standards for interfaces between functions provide an excellent foundation for reuse. As noted earlier, common interfaces are one aspect of a generic architecture for an application domain. One of the key concepts in reuse is that of using an *object-oriented* design approach to building components around invariant objects. Basically, object-oriented design means that a system's design is expressed in terms of objects and the set of operations that can be performed on them. Each object and its operations are grouped into a single package. Any system that needs to deal with that object can then incorporate that package. An excellent example of such an object is a communications interface standard (e.g., TADIL-B, JINTACCS). Packages that implement such interfaces have great reuse potential; they can potentially be reused by any application that needs to support the particular interface.

Component	Likelihood of Reuse Across Segment	Application	Hardware	Software
System Management	H	H	L	L
Data Base	H	H	H	M
Information Processing	H	H	H	H
Message Validation	H	H	H	M
Message Processing	H	H	H	M
Message Routing	H	H	H	M
Message Recording	H	H	H	M
Overload Processing	H	H	M	M
Data Base Information Handling	H	H	H	L
Statistical Analysis	H	H	H	H
Man-Machine Interface				
Operator Entry	H	H	M	M
Operator Display	H	H	M	M
Decision Support				
Query	H	H	H	M
Standing Request for Information	H	H	H	L
Word Processing	H	H	H	M
Information Generation	H	H	H	H
Graphics	H	H	M	M
Decision Implementation Support	H	H	H	H
Situation Analysis	L	L	H	H
Communications Processing	H	H	L	M

Table 1. C^2 Component Reuse Potential.

Architecture Standardization

As noted previously, a common application architecture is a key to high-payoff reuse of larger scale components. There is already a substantial degree of architecture standardization in the C^2 domain, providing a natural framework for reuse. This is particularly true within classes of systems serving a particular DOD community or a particular functional area. For example, NATO Command, Control, and Information systems tend to have similar top-level architectures. Some of this standardization is imposed by the DOD/MOD level organizations specifying the systems, either from a natural desire to stick to a proven approach, or from a recognition that such common structures will facilitate interoperability and will also make systems easier to maintain once turned over to appropriate maintenance personnel. Other standardization is imposed by contractors. A given contractor, even if not consciously focusing on reuse, tends to reuse design—to adopt a proven design approach from past programs.

Number of New Systems

Some application areas will not benefit greatly from reuse because they are so highly specialized that only a few systems are ever built. Command and control is a broader domain. Several new system developments are ongoing at any given point in time, and old systems are continually being upgraded. This, like the large size of the systems, makes this a good area to invest in developing a reuse-based approach.

Development Through Evolution

A non-trivial level of reuse already occurs in C^2 applications, simply because these systems are typically built by evolving existing systems. Thus, designers and programmers are comfortable with not starting from scratch on every project. Reuse researchers hope to build on this practice in adopting a more structured and comprehensive approach. In particular, this system evolution tends to occur only within an individual contractor's organization; the goal is to spread it across contractors and across services.

Language and Processor Commonality

In many application areas reuse has been hampered by the fact that a wide variety of languages and processors are used. Language differences make code reuse impossible; processor differences make it difficult by creating a dependence on software portability. This has traditionally been true of the C^2 community, but the picture is changing. With the transition to Ada, language will cease to be an issue. Because Ada is designed to allow the development of software that is portable across processors, the impact of using different processors is also decreased. Furthermore, there is a trend toward processor standardization within particular segments of the community, for example, the Army's ATCCS Common Hardware/Software. The relatively low use in this domain of special-purpose processors, which typically require programming in a unique assembly language, is also an advantage.

Benefits of C^2 Reuse

Development Cost Savings

Reuse clearly has the potential to reduce software development cost, and not just in the coding phase. Every line of code reused is one less that must be designed, coded, and debugged. Top-level design of the system is, of course, required. However, because it is directed to an established structure and a library of available components, the process is simplified. Furthermore, system testing proceeds much more smoothly because many parts of the system are substructures that have already been thoroughly debugged.

Reliability

A less immediately obvious benefit of reuse is improved reliability. Systems that are made out of already-proven components have the potential for much higher reliability. As the basic components in the reuse library are used in systems, fielded, and subjected to day-to-day use, they will evolve to a highly reliable state. Any changes will be reflected in the library version of the component, so that new systems can take advantage of this field shake-down and start with a much more reliable baseline.

Interoperability

Interoperability of C^2 systems is both critical and difficult. When two implementations simply follow a written interface definition to implement the interface, two different interpretations frequently arise. Significant DOD resources are devoted to establishing and testing for interoperability. One benefit of reuse is greatly simplified interoperability. Reuse of the same software to implement an interface standard ensures that implementations of the standard are compatible with one another.

Usability

A major benefit of using the same software to implement C^2 functions is that of uniformity from the user's perspective. Today, most C^2 systems have completely different soldier-machine interfaces, report formats, etc. The soldier who must interact simultaneously with several systems to obtain the information needed to carry out a mission has a very difficult job. He must switch among interactive formats (e.g., from function-key oriented to mouse-oriented), read tables that contain related information in different formats and units, and view maps drawn in different scales and with different notations. To the extent that the software that implements these user interfaces is reused, however, these interfaces will be much more standardized. The soldier, with less training, can perform the job more simply and quickly and with fewer opportunities for error.

Maintainability

Eventually the maintenance of C^2 systems becomes the responsibility of the DOD. Typically, maintenance organizations—for example the Army's Life-Cycle Support Center—must maintain a number of different systems. These systems add up to millions of lines of source code, written in different styles and using completely different design and implementation approaches to solve the same problems. Software reuse will simplify the job of the maintainers by making the systems much more similar overall, and by allowing single-point maintenance of common software.

A Generic C^2 Architecture

Earlier we noted that high-payoff application area reuse comes from use of a common architecture model, or *generic architecture*. A generic architecture is simply a decomposition of the system into functions, with established interfaces between the functions. Figure 1 is an example top-level generic architecture. Such an architecture can serve as the top-level framework for reuse, providing a common view of system functions and structure.

Ideally, there will be standards (at least within the organization) for the interfaces between the components, for example, for the way the Man/Machine Interface calls and is called by the Application Support Routines. Reuse opportunities are greatly increased by the existence of such standards.

The elements of this architecture can then be decomposed into subelements. For example, the message handling element might be subdivided into several subelements as shown in Figure 2.

The designer then uses this generic architecture as a top-level design structure, and as a frame of reference for searching for reusable components. For example, the designer might decide to see if any reusable components are available to implement a message validation requirement. The reuse system, because it is based on this standard architecture model, recognizes the term "Message Validation" and offers additional choices to refine the request, for instance, a menu of available message formats. After refining the request, the designer will be offered one or more candidate message validation components for reuse and may find it possible to reuse one as is; this will be possible only if it conforms to the interface requirements of the system, e.g. the same interface to Message Processing. It may be possible to reuse it with modifications, for example, adaptation of the interfaces. Perhaps only the design can be reused. Even here, the gains are many.

This concept of common architectures has been adopted by the Army Tactical Command and Control System (ATCCS) program, which unites a family of five systems into a single integrated capability. The potential for reuse in

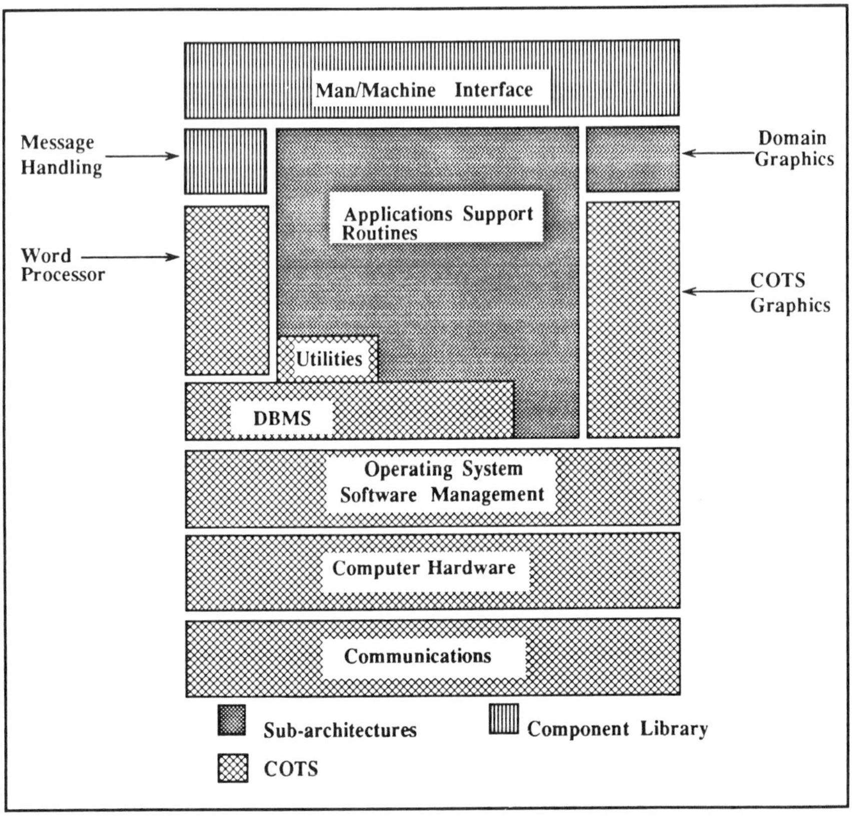

Figure 1. Generic Command and Control Architecture.

ATCCS is great, but the diversity of functions and contractors makes it difficult to achieve. The Army has recognized the importance of a common architecture in implementing ATCCS reuse. Figure 3, from the U.S. Army CECOM's SOFTCON '89 conference [Giordano], shows the top-level architecture envisioned for ATCCS. Table 1, showing potential functional commonality, is also from a study in support of the ATCCS program.

Implementing Reuse in Command and Control Applications

The preceding sections have described the benefits of reuse, have introduced basic concepts, and described some of the changes that must be made to make reuse a reality. But how do we really make these changes? Where do generic

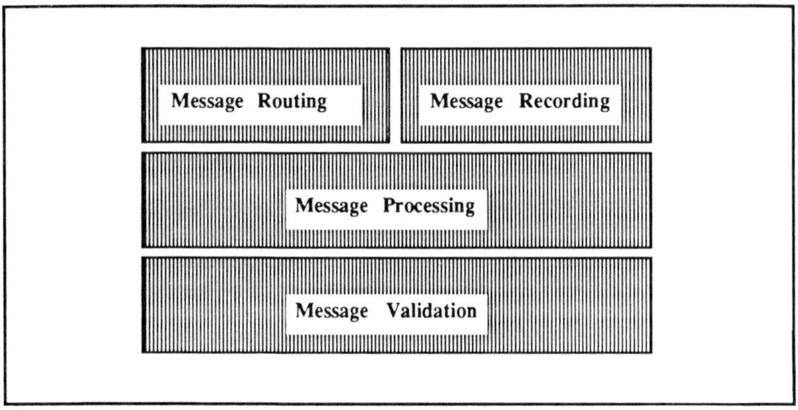

Figure 2. Message Handling Functions.

architectures come from? What is a reuse library and how does it work? What makes some software more reusable than other software? How do we overcome cultural stumbling blocks? These are the focuses of researchers in software reuse. The following subsections briefly describe some major research thrusts; [Tracz] is a good reference for the reader who would like to follow up on these ideas.

Domain Analysis

Domain analysis simply means analyzing the application domain to identify common functions and applications architectures. It results in a generic archi-

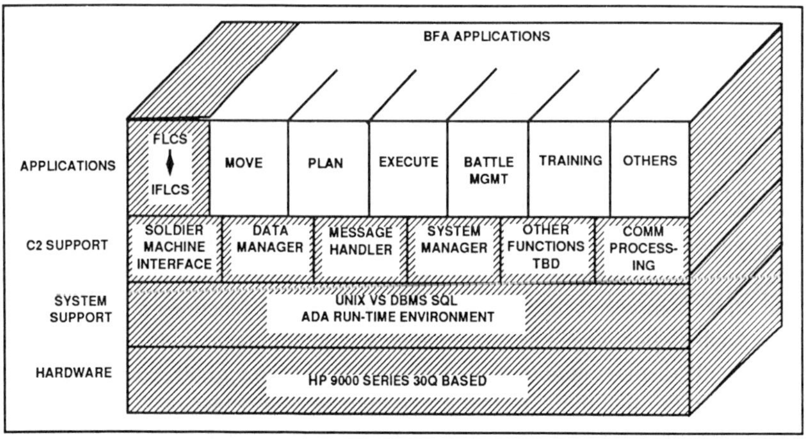

Figure 3. ATCCS Common Software Architecture.

jobs. We must work toward a view in which the programmer perceives the craft as finding the most cost-effective way to put together a system from available pieces, rather than creating elegant, highly-specific implementations.

A separate class of issues relates to government procurement practices. The current system can create disincentives for contractors to reuse their own software, and even more to provide software that can be reused by others. Also, the government is not well-equipped to become a software supplier on a large scale; even handling the small amount of GFE software in use today is complex and time-consuming. Liability for reused software is another problem. The government, recognizing the potential payoff of software reuse, is beginning to address these issues.

Summary

The DOD stands to benefit significantly from reuse in the C^2 domain, particularly as systems transition to Ada. There are a number of reasons why this is a particularly fruitful domain for software reuse, including significant functional commonality and a high degree of standardization. A reuse approach that incorporates an understanding of C^2 applications, using a generic architecture model as a framework for reuse, provides the most promise for large-scale reuse. If the promise of reuse is to become a reality, however, much work needs to be done. Development of a generic architecture recognized throughout the C^2 community, with interface standards connecting the components, will not be easy. There will be many who have a strong interest in perpetuating particular viewpoints. One of the goals of the Software Technology for Adaptable, Reliable Systems (STARS) program [Greene] is to develop such industry-wide application models. Classification and retrieval technology, particularly with respect to the implementation of domain-specific systems, requires further R&D investment to achieve the full potential of domain-based reuse. Finally, many cultural and procurement issues must be overcome.

It is important to understand, though, that reuse is not an all-or-nothing proposition. While we are working on the big problems we can solve smaller ones, and gradually increase our practice of reuse. Early prototype reuse libraries can help develop our understanding of how to most effectively support users. Reuse on an organization level is a lot better than no reuse at all. By starting small, we can gradually shift the way we develop software toward an industry in which reuse is standard practice.

References

Braun, C.L., Goodenough, J.B., Eanes, R.S., "Ada Reusability Guidelines," Technical Report 3285-2-208/2, SofTech, Inc., April 1985.

Giordano, R.F., "Army Command and Control System Common Software," Proceedings of SOFTCON '89, January 1989.

Greene, J.S., "Consolidated Technical Development Plan for the Software Technology for Adaptable, Reliable Systems (STARS) Competing Prime Contractors," STARS Joint Program Office, November 1988.

Prieto-Diaz, R., Freeman, P., "Classifying Software for Reusability," *IEEE Software*, January 1987, pp. 6-17.

Ruegsegger, T., "A Center of Excellence for Software Reuse," VG 1200, SofTech, Inc., February 1988.

Fitzgerald, J., Mathis, R., "Use of an Expert System in Software Component Reuse," Proceedings of AIDA-88, November 1988.

Quanrud, R.B., "Generic Architecture Study," Technical Report 3451-4-14/2, SofTech, Inc., January, 1988.

Swanson, M., Curry, S., "Results of an Asset Engineering Program: Predicting the Impact of Software Reuse," Proceedings of the National Conference on Software Reusability and Portability, September, 1987.

Tracz, W., *Software Reuse: Emerging Technology*, IEEE Computer Society Press, 1988.

Software System Requirements Engineering for Command and Control

James D. Palmer

Introduction

Significant progress has been made in recent years in addressing the requirements phase for software systems in general and in support of C^2 particularly. Most of this activity has been centered on organizing the *form* of the requirements specification to assure that the specifications are complete, consistent, and logically correct. More recently, research and development efforts have been aimed at addressing the *substance* as well as the form of the requirement.

Recognition that errors in software requirements lead to 50 percent to 75 percent of the problems in delivered software has been the essential cause behind the increased effort to address software requirements substance. One of the primary sources of errors stems from lack of involvement of the user at critical points in the software requirements specification process. Software systems requirements engineering processes require the interaction of the user and the systems developer throughout the specification process.

The goal of the requirements engineering process is to formulate accurate and correct statements of user needs that convey a high level of understanding of the software product necessary to produce the desired outcome. These requirements must realistically be based upon both the objectives of the user group as well as constraints and alterables that affect the solution that can be synthesized. In this way, system level C^2 software requirements are no different from general C^2 system requirements.

Software requirements processes generally follow a two-phase procedure; 1) identification of system requirements; and 2) development of software requirements. The first activity relates to the identification and development of user requirements. They must be presented in a consistent, understandable, logical, and integrated way so that both user and developer understand the problem that should and will be solved. The second activity relates to the technical document that is prepared for the internal use of the software systems development organization. The software specifications represent a translation of user requirements to technical specifications for software development.

In this paper we will examine two kinds of software requirements, functional and purposeful (non-functional), derived from system level requirements. We will review the notion of who is the "user." We will then provide a brief overview of systems level requirements, examine knowledge acquisition approaches, and describe the contents of the software requirements specifications document. Finally, we will look at the characteristics of the software specification, examine risk factors related to the quality of the information provided, review the development processes, and then look at support tools to assist the software developer.

Functional and Purposeful (Non-Functional) Software Requirements

Usually, two documents are developed in the process of preparing requirements specifications. The first of these is the user system requirements, a top-level document concerned with the interpretation of the user's perceived needs which are to be served by the system. This document will cover all of the alterables, constraints, and needs to be met by the system under development [Palmer, 1989].

The second needed document contains the software requirements specification. It is concerned with the detailed breakdown of the software product to be developed. This document is used internally for the identification of tasks in the production of software and for assignment of personnel to various tasks to be accomplished in the evolution of the software product.

Two difficulties impede progress in identifying requirements:

- *Knowledge acquisition difficulties.* This relates to lack of definitive information concerning the user needs and the translation of these into software system requirements, lack of an experiential base from which to comprehend the requirements, and other cognitive factors that must be resolved to produce the desired software product. These are basic difficulties that lie in the interface between the developer and the user in the identification of system level requirements.
- *Technical difficulties.* These relate to hardware, algorithms, interfaces, performance, capacity, and other issues that have the potential for technical resolution. These are technical difficulties in the development of a set of software requirements specifications from the user requirements specifications.

System and software requirements specifications may generally be grouped in two categories: functional and purposeful (non-functional) [Yeh, 1983]. Functional requirements represent the way in which the internal system components interact with their environment. These functional requirements specifications present a precise set of properties that the system must satisfy. Purposeful requirements represent the restrictions or constraints that are placed on the types of solutions that we may consider for the system design.

There will also need to be systems management requirements specifications that include: 1) operating environment concerns; 2) overall design concept or objectives; 3) trustworthiness requirements, including quality factors; 4) maintenance concerns, including the need for software evolution over time; and 5) economic resources, and time allocations, for software development.

On the basis of these, we prepare *a software development plan*, or *systems management plan*, or set of *systems management requirements* that are incorporated as an inherent part of the system requirements. The two items: 1) *systems management requirements*, and 2) *systems technical requirements* comprise the system requirements.

With respect to the *systems management requirements* portion of the *system requirements*, we must:

- *evaluate the system level functional requirements* for completeness and feasibility;
- *evaluate the system level requirements for compatibility* with other systems that are to be operational in the client's environment;
- *explore relevant aspects of the client's environment*, including existing systems, to determine how the new system might best be employed;
- *identify other organizations and units that need to have interfaces* with the system to be developed;
- *evaluate the client stated available resources*, including economic and time available resources, and
- *develop a systems management strategy* that incorporates these considerations and which serves, together with the *system technical requirements*, as input to the software requirements phase of development effort.

There are four very major objectives that are to be fulfilled by following the requirements set forth in the *systems management plan*. These include:

- to further define and develop the *scope of the development effort*,
- to explicate and specify the *resources needed for development*, including people, time, hardware, software, and client cooperation;
- to explicate and develop *detailed development schedules*, including important deliverables and milestones; and
- to explicate and *prepare cost estimates* across the phased development effort to follow.

These must realistically be based upon both the objectives of the user group as well as constraints and alterables that affect the solution that can be synthesized. Based on this it is important that we have a common understanding of the term user.

Definition of the User

For our purposes we will utilize a broad definition of the term "user" or "client." User is the person or group associated with the conceptualization, operation, and maintenance of large scale complex systems. In this definition we see the user first in the role of system conceptualization, and as an associate in the development of requirements specifications. We may find the user in the role of maintenance of the software to satisfy the evolving requirements objectives as these change after installation. Hence, we will generally consider the user as the person or group with responsibilities that extend beyond only "user" as conventionally understood. Thus, we define the "user" as that individual or group responsible for the articulation of the problem to be solved through implementation of the software product, for the problem definition statement, for approving the installation of the system following design, and potentially for some aspects of maintenance of the system over its useful lifetime.

It is clear that this definition may include many separate groups in the single name of user. For large-scale software systems development, it is not unusual to have one group defining the goals and objectives of the system, another working with the software developer in turning these into requirements specifications, still another group may have the responsibility of working with the developer in installation, and yet another group is likely to have the responsibility, with or without support from the developer, for life cycle maintenance.

System Requirements Specifications

A close examination of development case histories shows that inadequate resources, both time and money, have generally been allocated to the initial phases of the software development life cycle. What has become abundantly clear from experiences gained in the development of large-scale software systems is that *skimping on the front end of a software development project will cost the user orders of magnitude more time and money during the life of the software product than would otherwise be the case. It will often result in an inferior product, or no product at all.*

Understanding problems to be resolved and the acquisition of knowledge concerning these problems are difficult tasks. But, we must understand issues before we attempt to resolve them through producing software products that meet user expectations. Otherwise, we will generally produce software products that fall short of user expectations and needs.

Knowledge Acquisition Approaches

There have been many studies of information requirements determination perspectives. The results of some of this work have led to the increased use of prototyping approaches for the analysis of both system and software requirements specifications. There has been an adoption of systems engineering principles for software systems development, and with this has come a substantive improvement in knowledge acquisition approaches.

The primary methods used to elicit information about user needs and systems requirements include talking with users, or reviewing existing documents, or use of prototypes to explore various alternatives. The common processes are:

- ask people
- the traditional interview process
- structured interview techniques
- responses to written questionnaires
- employ nominal group approaches when interactive approaches fail
- review existing plans and documents
- study the existing system
- study the evolving system

These constitute the ways in which knowledge acquisition is conducted. Few software developers are professionally trained in these fields of expertise.

Knowledge acquisition commences with interaction between the user or domain expert and the software developer or knowledge engineer. The software developer utilizes knowledge acquisition techniques noted above to discover constraints, alterables, and interfaces for the proposed system. This information is then converted to system level definitions. The process is an iterative one and continues until the user and the developer are satisfied that the system level requirements specifications will meet user system needs.

The most commonly used techniques for requirements specification may be taken from unstructured interviews, protocol analysis, or structured interview approaches. In many cases, elicitation is accomplished by the software systems design engineers, or individuals in comparable positions, who formulate questions and pose these to the user.

If the design is relatively well-formulated and many of the parameters are known, the structured interview may prove better, faster, and more economical of both time and resources than the unstructured approach. Such an interview is pre-structured by the software systems developers to reflect the experience of that group in addressing similarly structured software development problems.

The information thus developed, involves subjective interpretation. Therefore, considerable interaction between user and elicitation group must take place before the design characteristics are truly credible.

Much of human expertise operates on intuition, which is experientially based wisdom. Intuitive skill based knowledge is most difficult to code into requirements statements, rules or even heuristics that might lead to interpretation of the process of decision making. Using intuition, an expert is frequently able to scan a situation, size up the alternatives, and propose a course of action without resorting to detailed rules. In cases such as this, the expert is able to aggregate information into larger packages, review these, assess alternatives, and propose courses of action without resorting to any form of formal knowledge or rule based knowledge.

One of the most difficult problems to confront the software systems engineering design team in the elicitation of requirements information is characterized by the worn-out phrase: *communication.* Even if the user is fully aware of all the requirements, constraints, and needs of the system to be built, there comes the time when this information must be communicated to the software development team. In a typical situation, the software development team is not familiar with the system that is to be constructed. In addition the user lacks familiarity with the design methods to be used for software development. Thus, any attention to development of communication skills to alleviate these problem areas, is of great benefit.

The processes and procedures of knowledge acquisition and the methods utilized to gather this information will have a profound effect on the developed software and the ability of that product to meet the needs of the user. It is important for the user to be aware of rules, procedures, and skills that may be used to express user requirements. It is also important for the software systems engineering design team to be aware of the potential for misinterpretation, failure to obtain all the necessary information, and the need for clear concise communication with the user. This is a need, and there is a treachery associated with it. Experts do not think in terms of precise rules and procedures. They use skill based holistic reasoning. At this time, computers operate on the basis of rules and procedures. There is an inherent danger in asking people to express ideas in a (rule based) format that they do not use.

In a large complicated software development project, it is usually necessary that the development of requirements specifications take place over much of the development cycle to overcome the requirements specification articulation difficulties noted. It is extremely useful for this process to continue until the user is satisfied that the system requirements identification and software requirements specifications have captured the actual requirements identified by the user.

In the event that both user and developer are novices at knowledge acquisition methods, it is likely that the resulting system requirements identification and software requirements specifications will prove to be unsatisfactory. This is because the user will be unable to provide proper guidance and inputs to the software developers and the software developers will not understand that they are not obtaining the desired system identification needs. Provisions must be made to handle these problems. Some of the possible approaches would be to acquire formal training and preparation in knowledge acquisition methods on the part of user and software developer, perform an apprenticeship to learn by observation, or employ expert facilitators to assist in the knowledge acquisition processes.

Software Requirements Specification Documentation

The requirements specification process results in a deliverable, the requirements definition document. The requirements definition document will provide information on such procedures as the preliminary interface requirements specification, operational conceptualization, and how the software is to be used by the user.

The software requirements specifications document [ANSI/IEEE Std 830-1984] is intended to:

- establish a basis for agreement between customer and supplier on what the software product should do;
- reduce the development effort;
- provide a basis for estimating costs and schedule
- provide a baseline for validation and transfer;
- facilitate transfer of the software product to new users or new machines; and
- serve as a basis for enhancement.

A prototype software requirements specifications outline is shown below.

Table of Contents

1. Introduction
    - 1.1 Purpose
    - 1.2 Scope
    - 1.3 Definitions, Acronyms, and Abbreviations
    - 1.4 References
    - 1.5 Overview
2. General Description
    - 2.1 Product Perspective
    - 2.2 Product Functions
    - 2.3 User Characteristics
    - 2.4 General Constraints
    - 2.5 Assumptions and Dependencies
3. Specific Requirements (Prototype Outline 2 is shown)
    - 3.1 Functional Requirements (for interface requirements)
        - 3.1.1 Functional Requirement 1
            - 3.1.1.1 Specification
                - 3.1.1.1.1 Introduction
                - 3.1.1.1.2 Inputs
                - 3.1.1.1.3 Processing
                - 3.1.1.1.4 Outputs
            - 3.1.1.2 External Interfaces

3.1.1.2.1 User Interfaces
3.1.1.2.2 Hardware Interfaces
3.1.1.2.3 Software Interfaces
3.1.1.2.4 Communication Interfaces
3.1.2 Functional Requirement 2
...
3.1.*n* Functional Requirement *n*
3.2 Performance Requirements
3.3 Design Constraints
3.4 Attributes
3.4.1 Security
3.4.2 Maintainability
....
3.5 Other Requirements
3.5.1 Database
3.5.2 Operations
3.5.3 Site Adaption

The system level requirements specifications document is a top-level identification of the user's perceived needs to be ultimately served by the software system, and a translation of these to system level requirements. The software requirements specification document represents a translation of these user requirements specifications into software specifications to be used for the next phase for the software product to be developed.

For new system needs, prototyping is most useful for the development of requirements. System level requirements specifications form the input for the prototyping model and are taken through a series of steps. These include writing the system level requirements specifications into a proper format for input to the prototype, preliminary evaluation by the software developer, performing a rapid prototype, evaluation and refinement of these specifications by the software developer, and then to a "final" product of software requirements specifications. The process, called evolutionary development, is iterated as many times as necessary to assure that the software requirements specifications represent the system level requirements specifications as provided by the user. When modifying existing systems, it is also valuable to experiment with the present system to modify or add to the current requirements specifications for a new or different application. A similar approach to that described above for a new set of system level requirements would be applicable.

Information Risk

As we have seen, the knowledge acquisition process leading to system, and thence software requirements specifications, is anything but perfect. It frequently leads to a high level of uncertain, imprecise and ambiguous information, and interpretations of the system requirements. These factors also complicate the software requirements specification process.

If inadequate information is used, it will almost always later lead to a functionally inadequate system. Particularly troublesome are: 1) imprecise, conflicting, ambiguous, information and overstatement of requirements; 2) changes in the system requirements specifications that occur during the development phases due to poorly derived specifications; and 3) changes in the system requirements specifications that occur during development due to user modification of needs.

These problem areas must be addressed as early as possible. As the possibility for changes exists throughout the design process, the system and software requirements specifications must be reviewed and refined as the system requirements change. As we continue to monitor changes in system level requirements throughout the requirements phase, we will modify the software requirements to assure that changes at the system level are reflected at the software requirements specifications level.

The need to accommodate to changes in requirements specifications must be tempered by the thought that, if things are changing too rapidly, it may be necessary to defer development of the project. In this event, the volatility of the requirements specifications is such that it is not possible to stabilize them. More investigation on the part of the user is required to better define the system level requirements.

Errors reduce the quality of the software product and decrease productivity. Some of the errors may go undetected until discovered during operation and a greater investment of time and effort is required to overcome the problems. Imprecision occurs because the user has incomplete information, is poorly informed as to specific constraints, is unaware of changing aspects of the mission of the software, or lacks the ability to state the problem in complete and concise terms. Conflict occurs because the user is generally unaware of the consequences of goal selection on software design and may choose software functionality factor combinations that are in conflict with each other. Conflict may be due also to specifications in domains such as operating environments or existing hardware that do not take pre-existing conditions into account. Specification overstatement is either an outcome of user inexperience or exuberance to have the system satisfy all possible requirements or ignorance of the potential consequences of over-specification. Poorly derived specifications will continue to plague a project long after development. Changing requirements need to be stabilized or the project is at considerable risk due to volatility of requirements.

Software Requirements Development Processes

Upon receipt of the information that has been developed from software requirements specifications, conceptual and logical design parameters are established and software developers initiate the following: 1) review the project scope; 2) establish the software development plan; 3) examine the design parameters for problems; and 4) provide procedures for problem definition and solution. This is needed to assure that the requirements specifications are as complete as possible. High level information flows and structures are examined, potential system interfaces are proposed, and incorporation of any constraints and alterables that have been enumerated during the software requirements specifications process are taken into account. Following this activity, validation and verification criteria are established and metrics to determine success (or failure) of the project are defined. The principles following are desirable: 1) accurate representation of problem areas; 2) problem partitioning for subproblem review; and 3) logical and physical representation of the system for ease of establishing the software design.

Following these principles will help us insure that an adequate software specifications document has been prepared for the next phases in the development of the software product. Agreement at this time between the user and the software development team on these specifications is mandatory in order that the software specifications documents reflect the needs of the user and the design implications of the software design group.

Specific Methods and Techniques

In this section we will consider several of the automated procedures and associated tools that assist in the development of requirements specifications. A recently developed approach administered contractually by the U.S. Air Force Rome Air Development Center [Rzepka, 1988] proposes a requirements engineering model. The model envisions a system and software requirements specifications engineering test bed designed to incorporate engineering methods and tools in a host environment. The objectives of the host environment are to assist in the development of system and software requirements specifications analysis, documentation, and validation. The process model used by RADC is depicted in Figure 1. System requirements specifications conceptualization is first accomplished by the user. These ideas are refined and documented. Specific software system level goals are defined, and an iterative process commences that results in final software requirements specifications and partial design.

This model incorporates several tools. These tools include the requirements apprentice component of the programmers apprentice developed by MIT, an object-oriented design tool developed by Computer Sciences Corporation, and a

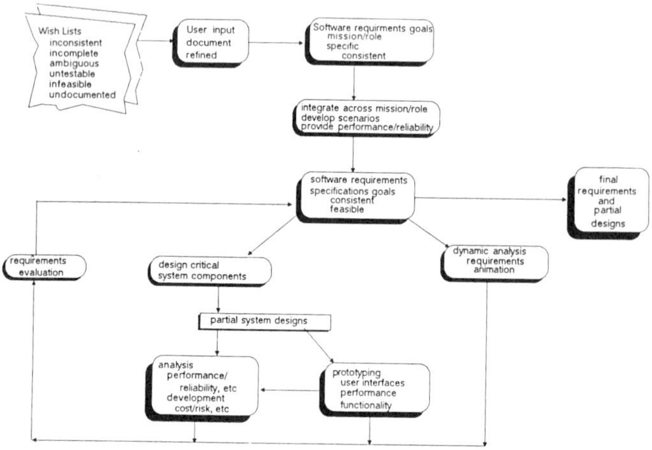

Figure 1. Requirements Engineering Process Model.

knowledge-based requirements assistant developed by Sanders Associates, Incorporated. The inputs for this model are documented system level requirements statements. These statements are analyzed for consistency between various system objectives, and for conflict between requirements specifications.

An example of the kind of operations performed by one of the tools in the test bed is from the approach taken by Computer Sciences Corporation. Requirements specifications are decomposed into data flow diagrams using an object-oriented language, Smalltalk 80. In this system, requirements are manually transformed into structured data flows using DeMarco's rules. The outputs are: 1) a data flow diagram for all requirements; 2) a database; and 3) a hierarchical tree structure showing the relationships between all objects. These resulting software specifications are reviewed and the process iterated until the user representative and software developer are satisfied that the software requirements specifications represent the system level requirements.

Following this, a partial conceptual and logical design is completed and becomes the output of the model. Requirements specifications thus prepared are feasible, testable, and traceable to the system level requirements statements that served as input to the model.

This model is representative of new initiatives to gain control over the software requirements specifications engineering process and to introduce formal methods, tools and validation procedures to assist in providing software requirements specifications that reflect user needs and are correct. This model provides little support for direct interactive identification of user needs, because the input is user-generated system-level statements that are accepted without challenge.

Several Computer Aided Software Engineering (CASE) tools have been developed in recent years that are intended to address the software requirements specifications issues. The current custom of handling software requirements specifications issues has generally been limited to provisions for software engineer support such as interactive graphics, program management, configuration control, consistency checking through database construction, and prototyping. Some of the CASE tools that have explicitly addressed the requirements area include: EPOS, CASE 2000, TAGS, PSL/PSA, SADT, and SREM/SYREM.

Two knowledge-based automated assistance tools have been developed that begin to address the substance of software requirements specifications: GLITTER-S and WATSON. GLITTER-S [Fickas, 1987] incorporates purposeful requirements and permits constraint alterables, such as development resources and organizational goals, to become factors in the definition process. WATSON [Kelly, 1987] uses domain knowledge to build software requirements specifications from scenarios. It examines and corrects inconsistencies, finishes incomplete software requirements specification statements by adding rules, eliminates unreachable states, and ensures all pertinent data are handled in all states.

One additional knowledge-based assistant has been developed to handle information risk related to incomplete and imprecise information from system requirements specifications. The architecture for this assistant examines qualitative information, covers a large search space and manages changes at the level of individual words.

In this system, the user and the software systems design team interact with the knowledge base management system to examine the user generated requirements. The knowledge base management system provides for access to the knowledge bases for functional and purposeful requirements. The outputs are system level requirements that have reduced conflicts, managed volatility and managed over-specification.

A prototype system has been constructed to demonstrate the validity of the conceptual architecture using a blackboard architecture [Samson, 1988]. The blackboard architecture has three components: 1) knowledge sources in which the knowledge bases are maintained as separate modules; 2) a blackboard data structure that keeps current state information in a global database; and 3) a control structure that manages contributions from various knowledge sources. All knowledge sources are independent and are able to communicate only through the central control structure. Four knowledge bases have been defined to implement the blackboard architecture. These are shown in Figure 2 [Samson, 1988] and include an imprecision specialist, a quality goals specialist, a conflict specialist, and a domain specialist. The imprecision expert looks for imprecise words from the requirements and requests a quantitative value for each of these on a semantic differential scale of 1 (least) to 7 (most) important to the system level needs. The conflict expert possesses pre-programmed knowledge as to which of the software characteristics conflict and which ones do not. The qual-

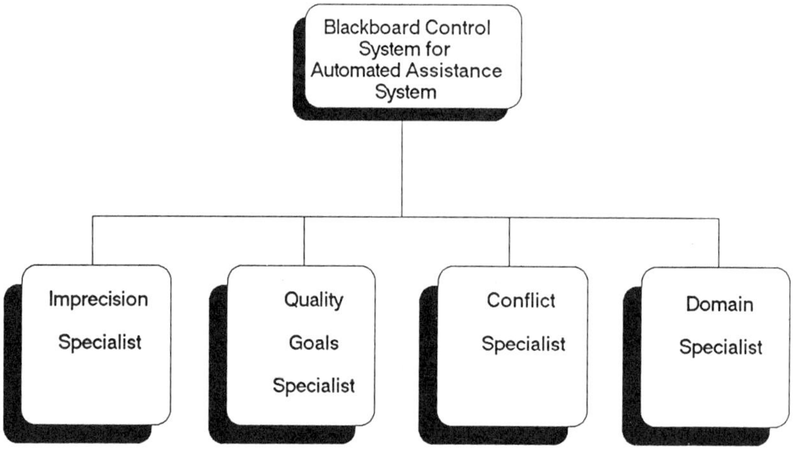

Blackboard Control System with Knowledge Bases

Figure 2

ity goals expert identifies non-functional quality goals and requests more information as to how these goals are to be attained. The domain expert possesses domain-specific knowledge and is a changeable module in accordance with the requirements definition under consideration.

The prototype was constructed in Smalltalk/V [Goldberg and Robson, 1983]. The use of an object oriented language provided system extensibility to be able to: 1) include information as it is made available; 2) utilize inheritance characteristics which are important to link requirements development; and 3) employ information hiding which permits individualization of specific instances of an object. It also promotes high reusability and provides a means for rapid prototyping.

The prototype automated assistance system provides the user and software design team an interactive environment in which to identify system level requirements issues and concerns related to imprecision, conflict and over-specification problems. In using this assistant, the order and number of times of access of any of the problem areas is at the choice of the users.

References

ANSI/IEEE Std 830-1984

Fickas, S., "Automating the Specification Process," TR 87-05, Computer Science Department, University of Oregon, 1987.

Goldberg, A. and Robson, D., *Smalltalk-80: The Language and its Implementation*, Reading, MA, Addison-Wesley, 1983.

Kelley, V.E. and Nonnenmann, U., "Inferring Formal Software Specifications from Episodic Descriptions," in *6th International Conference of the AAAI*, Seattle, WA, 1987.

Rzepka, W.E., "A Requirements Engineering Testbed: Concept, Status and First Results," Rome Air Development Center, Griffis Air Force Base, NY 1988.

Samson, Dolly, *Automated Assistance System for Software Requirements Definition*, Ph.D. Dissertation, George Mason University, Fairfax, VA, 1988.

Yeh, R.T., "Requirements Analysis-A Management Perspective," *Proceedings COMSPAC '82*, pp 410-416, 1982.

Specification in a World of Ever-Changing Requirements

Alan M. Davis

Introduction to the C^2 System Requirements Problem

The engineering processes associated with systems and software development are similar to those of any engineering discipline. At the beginning of the process, only a fuzzy, relatively abstract understanding of the system to be built is available. As time progresses, the product becomes less and less abstract, the engineer's view of the final system gains more focus, and we get closer and closer to a working system. Although in practice and in theory this process is continuous, we usually talk about it in terms of discrete steps, or *stages*. The primary reason for establishing these discrete steps is to provide visibility, track progress, and enable engineering management to better allocate and control resources. A series of these discrete steps is usually called a *development life cycle model*. Figure 1 shows a typical life cycle model for a system composed of both hardware and software. Each discrete stage is indicated by a box. Arrows show the order in which progression from stage to stage may occur; bold arrows show the primary product synthesis path. At the completion of each stage, some type of review of the product is performed by peers, management, and/or customers, and the results of that stage, if approved, are baselined and usually put under some type of configuration control.

It is hard to argue with the demands made by a life cycle model such as that shown in Figure 1. All the model really says is:

- that you should know what problem you are solving before you attempt to solve it

- that you decide what the external behavior of your system is going to be before you build the system
- that you should decide on what its overall structure will be before you decide what each particular component will look like
- that you should decide what each component will look like before you build it
- that you will decide *how* you are going to test the system before you test it, and
- that you will test the product iteratively, a little at a time, until the final system is checked out thoroughly before delivering it to the customer.

These six "demands" are in fact achievable for *most* applications. This fact helps to explain why some variation of the life cycle model shown in Figure 1 has been in place on most systems development jobs during the past 15 or so years. However, there exist applications in which these demands are just not achievable. One such application is C^2 systems. Since this paper is addressing the problems of specification, let us examine the attributes of an application that would make the first two demands unachievable. The requirements for command and control systems, by their very nature, are very hard to define up front and constantly evolve as new threats are identified and new counters to old threats are realized. C^2 systems built using the traditional way of developing large-scale systems as shown in Figure 1 often [BER87]:

- don't satisfy user needs because the real problem was not understood prior to the freezing of the requirements,
- cost much more than necessary to develop because of constant redirection during development,
- don't satisfy user needs because it is impossible in some cases to understand what is needed without an operational system to play with, and
- don't satisfy user needs because they changed during development.

Requirements Specification Strategies

As previously explained, one of the biggest problems facing developers of C^2 systems is coming to grips with the real user needs. User needs are generally expressed in a document called a requirements specification. This is a brief summary of state-of-the-art and state-of-the-practice requirements specification techniques. It also explains the characteristics of C^2 systems that make various techniques more or less applicable.

In any system development life cycle, the system requirements stage includes two activities: problem analysis and system specification [DAV88, DAV90]. During the problem analysis activity, an analyst tries to uncover all aspects of the problem to be solved. The analyst's domain of discourse is purely the problem domain. During system specification, the system requirements specification

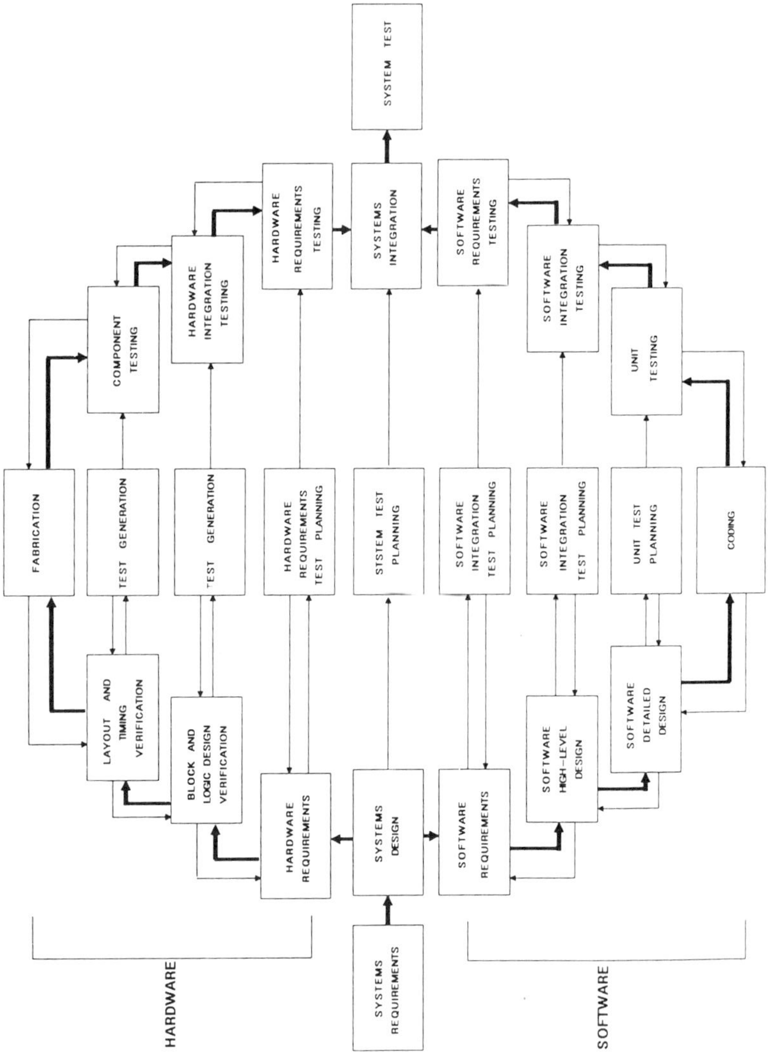

Figure 1. Typical System Development Lifecycle.

is written. It documents the external behavior and external characteristics of the system to be built. The document's domain of discourse is purely the solution domain, although care must be taken to specify the solution system only from an external viewpoint; the internal structure of the system belongs in a system design document.

Traditionally, requirements specifications have been written in English. Although this practice results in documents that are relatively easy to understand by customers and users who are applications experts (but not necessarily computer experts), the documents suffer from a number of major problems, all of which compound the problems of C^2 system development:

- English (or any natural language) is inherently ambiguous resulting in requirements that have multiple interpretations,
- there is no way to insure internal consistency among various requirements stated within any document,
- due to the ambiguity, the documents serve as poor starting points for system design, software or hardware requirements, and for testing, and
- modifications to the documents are difficult due to the inability to cross-reference between related requirements or to check for inconsistencies after a modification is made.

The most obvious, and least expensive, improvement is the use of databases to store the requirements and word processors to produce the requirements documents. These two improvements solve the problems of cross-referencing between related requirements and the purely mechanical difficulty of making changes to documents. However, most of the above problems are not solved by just using databases and word processors. A more promising avenue is the incorporation of formal models into the requirements specification. These models greatly reduce the ambiguity, but because they might be more difficult to understand by the non-computer expert, they are best handled as an adjunct to, not a replacement for, the natural language [DAV88a].

Some of the more promising formal models for the specification of external system behavior for C^2 systems are decision tables and trees, finite-state machines, statecharts, Structured Analysis/Real-Time, and Object-Oriented Analysis:

- Decision tables and trees [MOR82] capture, in a non-ambiguous fashion, the relationships that exist between a set of conditions and the action that one wants either a human or the system to take as a result of those conditions. They are most helpful in those aspects of a C^2 system definition that are decision-intensive. Examples of decision tables and trees may be found in [DAV88a].
- Finite-state machines are hypothetical machines that can be in any of a given number of states, and that (1) produce outputs that are purely a function of the current state and the triggering stimulus received from the environment, and (2) change state purely as a function of the current state and the triggering stimulus. They are of most assistance in describing the real-time behavioral aspects of C^2 systems. Figure 2 (adapted from [OFF87]) shows an example of a state

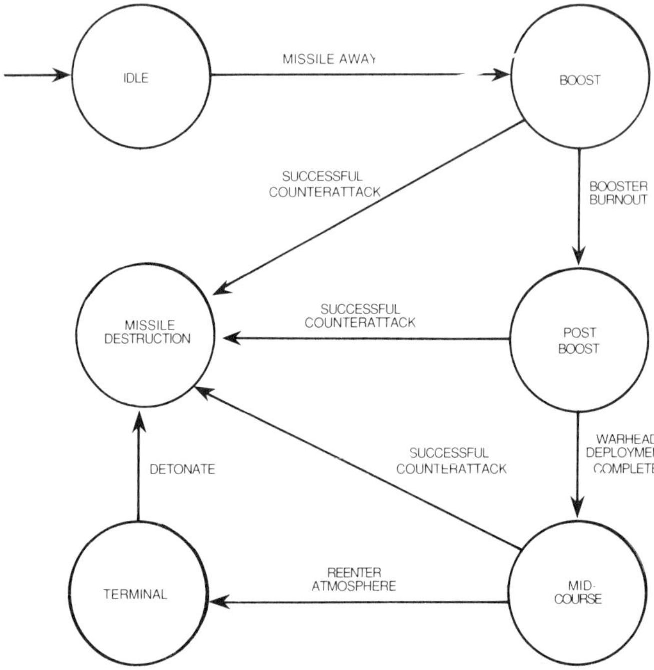

Figure 2. Sample State Transition Diagram: Life of a Ballistic Missile.

transition diagram showing the life of a ballistic missile. Because finite-state machines can grow quite large and unwieldy (resulting in difficulty in both creating and understanding) in complex applications, at least two major efforts have been undertaken to simplify their presentation. In particular, as part of the Distributed Computing Design System (DCDS) [ALF85], TRW has defined the R-net as a way of defining and presenting finite-state machines in smaller palatable chunks [ALF76]. Each R-net captures all system behaviors (i.e., state changes and external system responses) resulting from a single stimulus. DCDS keeps the R-nets separate for presentation in the specification to increase readability, but merges them together internally to detect inconsistencies. As part of the Requirements Language Processor (RLP) [DAV79], GTE has defined the stimulus-response sequence, which replaces a single finite-state machine with a set of user-machine interaction scenarios, each of which describes one system feature from the user's perspective. RLP keeps the stimulus-response sequences separate for presentation in the specification to increase readability, but merges them together internally to detect inconsistencies.

• Statecharts [HAR87] add considerable semantic richness to the finite-state model. In particular, Harel has added the ability to specify transitions based not only upon stimuli and current state, but also upon conditions that exist in the system or its environment. Statecharts also add the ability to define abstract states which can later be refined into more detailed state machines in a variety of ways. They are most helpful in system-level requirements of complex systems that can be most readily modeled as a set of independently operating processes. Numerous examples of statecharts used for requirements specification can be found in [DAV88a].

• The term "Structured Analysis/Real-Time" is used to describe both Hatley's [HAT84, HAT87] and Ward's [WAR85, WAR86] extensions to the DeMarco's Structured Analysis (SA) [DEM79]. The original DeMarco SA was primarily a problem decomposition and system/software design approach, with little or no capability to describe external system behavior. The real-time (RT) extensions add that capability among other things. In Ward's SA/RT, a variety of control signals and state machine descriptions can be used along with a methodology that keeps the model builder in the requirements domain. In Hatley's SA/RT, state transition diagrams and process activation tables precisely and non-ambiguously define the order in which processes in the SA model are activated, deactivated, etc. Although both are aimed by their respective authors toward real-time systems, their reliance on data-flow diagrams as the primary building block (as opposed to, say, statecharts which rely on state machines as the primary building block) make them most suitable for environments that already are familiar with structured analysis.

• Object-Oriented Analysis (OOA) [COA89] provides a notation and methodology to describe a problem in terms of the real-world entities comprising the system and its environment, their attributes, their interrelationships, and the services each provides to its immediate surroundings. Once defined, this model can serve as an effective structure in which to embed external behavioral requirements. The advantage of this approach is that (1) the requirements directly map into the real world, (2) if you are following Object-Oriented Design (OOD) [BOO86], then there is a high correlation between requirements and design which should improve maintainability and modifiability, and (3) experience has shown that the evolution of user needs tends to make radical changes and additions to the functionality of the system, but not to the real-world entities. For example, in a C^3I application, what you want the system to do with the ships and aircraft being tracked will change, but the existence of entities called ships and aircraft will not. Figure 3 shows an OOA model of a particular aspect of a C^2 system (adapted from [HER87]).

Appropriate use of any of these techniques within a system requirements specification can greatly reduce ambiguity in the document, can lower the probability of disagreements between developers and customers due to misinterpretation, and can reduce the cost of changes to the document because automated

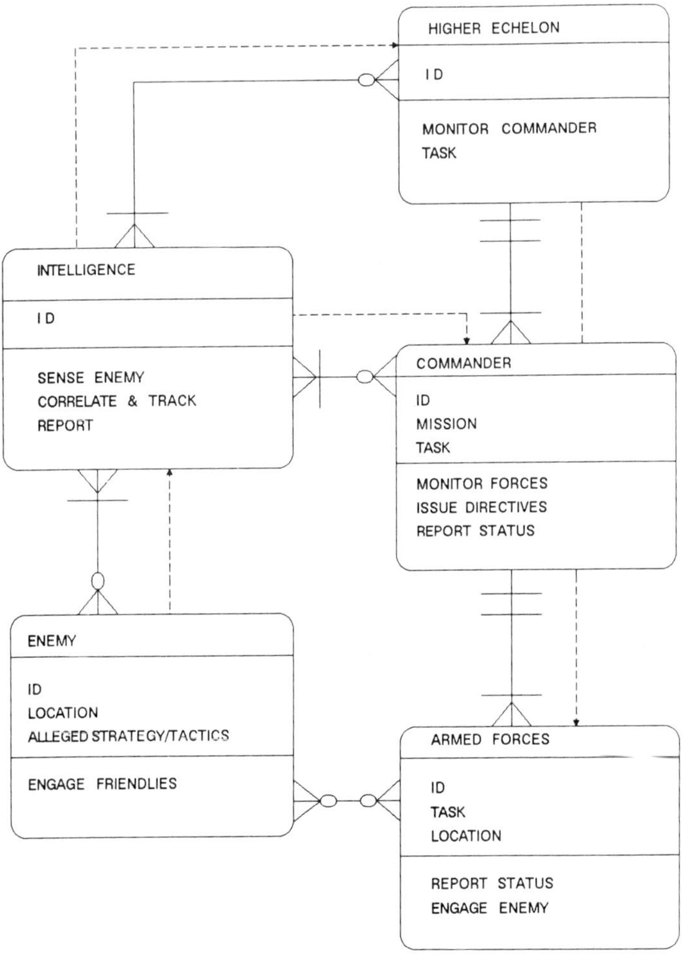

Figure 3. Sample OOA Model of Command and Control.

tools associated with each technique can check for both ambiguity and consistency. Thus, after any change is made to the document, a quick pass through the appropriate tool easily uncovers any new inconsistencies that were added as a result of the change. This will reduce the negative implications of the first two problems listed at the first section. More details on all these techniques, their associated tools, and their impacts on the quality of the requirements specification can be found in [DAV90].

Prototyping

This section provides a brief tutorial on prototyping and explains how various prototyping strategies can be applied effectively to reduce development risk in C^2 systems, reduce full life cycle costs, and reduce the impacts of ill-defined requirements and requirements that cannot be understood until after the user sees a working system.

A prototype is a partial implementation built for the express purpose of learning more about a problem to be solved, or a solution to that problem. There are at least two entirely different approaches to building systems that are called "prototyping" [GOM86]. The first approach results in a *throwaway prototype* [GOM81]. The throwaway prototype is constructed in a quick-and-dirty manner using any available method to reduce its development cost. After construction, it is used by customers, potential system users, or developers in order to learn something about the problem being solved, or about possible solutions to the problem. Once we learn what we want to learn, we discard the prototype. By the way, we absolutely *must* throw it away because it was built without attention to quality; it is impossible to retrofit quality. Figure 4 shows how the throwaway prototype fits into the overall life cycle (adapted from [DAV90]). Notice that one can construct throwaway prototypes during any of the synthesis stages of the development life cycle. The second approach results in an *evolutionary prototype* [BAL82]. The evolutionary prototype is constructed in a quality manner, with few if any shortcuts. After being used by customers, potential users, or developers to learn whatever we want to learn, it is repeatedly modified and reused, until eventually it evolves into the actual product. Figure 5 shows how evolutionary prototypes fit into the overall development life cycle (adapted from [DAV90]).

It is possible to use throwaway or evolutionary prototyping techniques for either the software or hardware aspects of a system, or both. It is also possible to mix and match. For example, you can procure or fabricate the production hardware and use it as a base for either throwaway or evolutionary software prototypes. Or, you could construct a throwaway prototype of the hardware (usually called a "breadboard"), on which you can construct either throwaway or evolutionary software prototypes. Or, you can plan to evolve the hardware (for example, building it first out of off-the-shelf components, and slowly migrating toward custom integrated circuits to replace selected subsystems), while it houses either throwaway or evolutionary software prototypes. It is also possible to construct throwaway software (or hardware) prototypes on top of an evolving, stable, evolutionary software (or hardware) prototype. In all these cases, the software and hardware can in fact each evolve separately. The important thing to remember is that the final disposition of any software or hardware prototype must be decided beforehand because the disposition will determine how you construct it. That is, a throwaway prototype is created in a rapid way disregard-

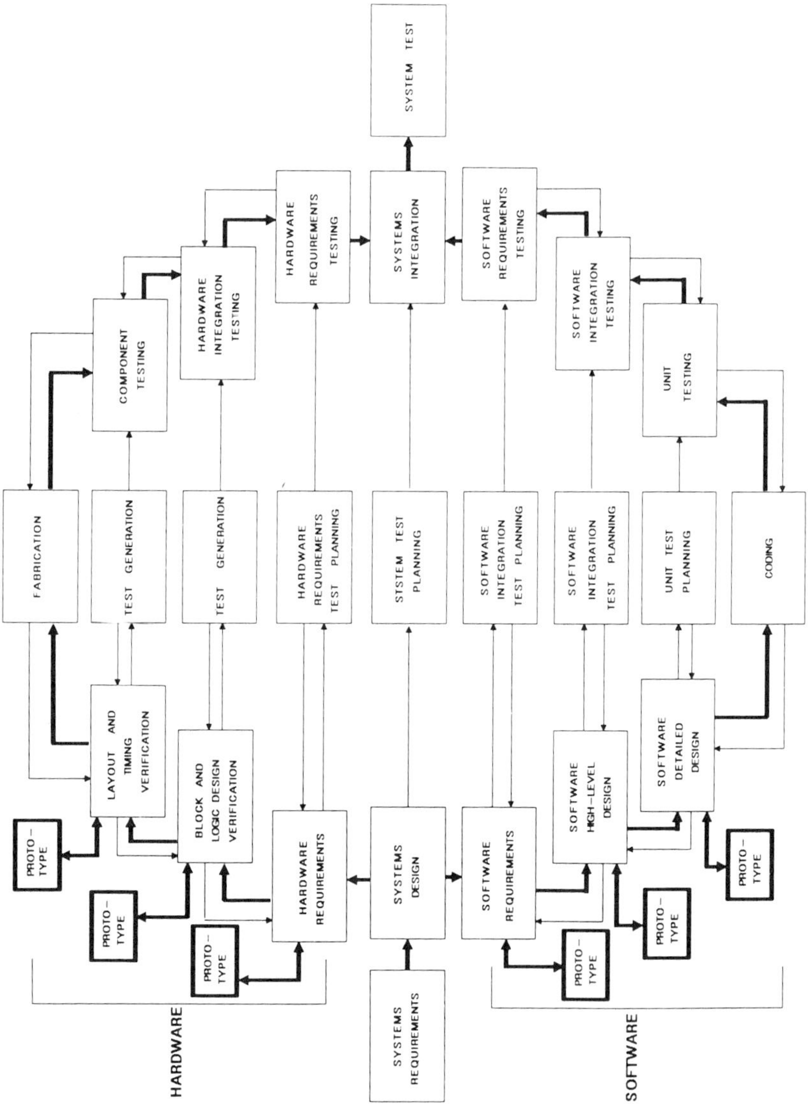

Figure 4. Throwaway Prototypes and the System Development Life Cycle.

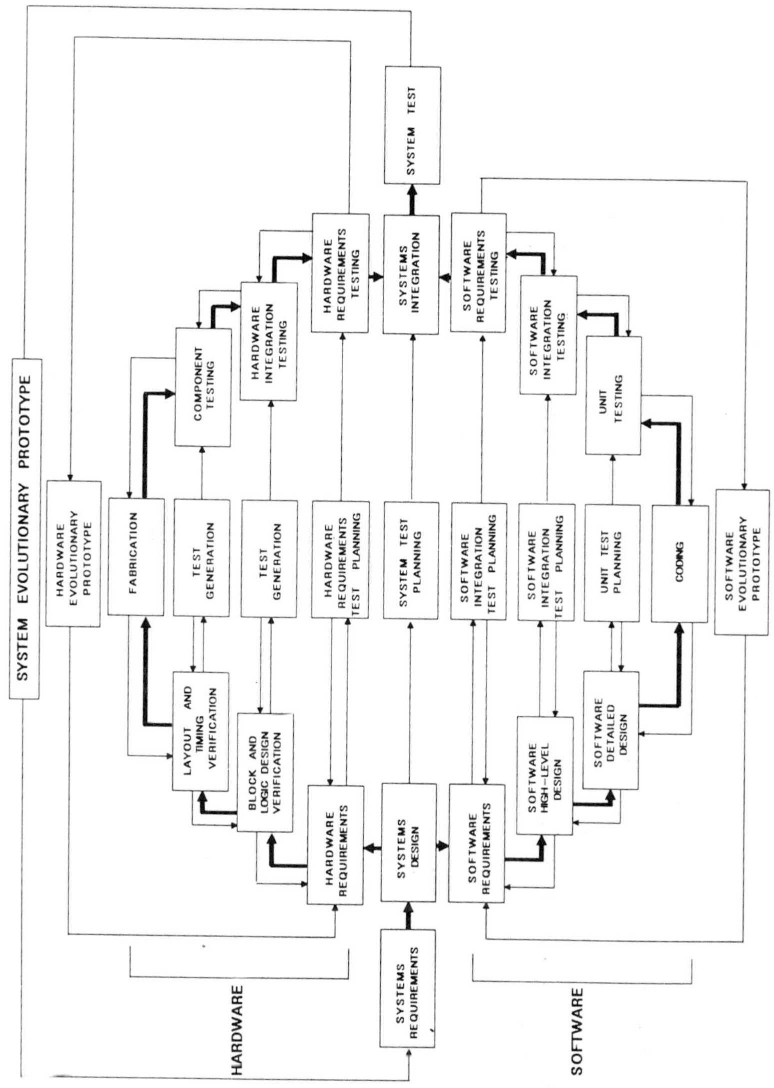

Figure 5. Evolutionary Prototypes and the System Development Life Cycle.

ing most of the "ilities"; while an evolutionary prototype is constructed following a rigorous quality-producing method. Note also that in the two cases you probably construct a completely different set of functions. For example, in a throwaway prototype, you would implement only those functions that you do *not* understand (obviously, it makes no sense to build functions that you already understand since you are going to throw them away after construction), while in an evolutionary prototype, you would start by implementing only those functions that you *do* understand (that way you evolve your system from a solid foundation).

Selective use of prototypes can help to alleviate the problems of C^2 system development that were listed at the end of the first section. For example, construction of a throwaway prototype during any of the requirements stages helps to solve the first three problems by providing users with a system up-front to validate understandings of and to learn the real user needs. Construction of an evolutionary prototype helps to solve the last three problems by providing an environment that facilitates the inclusion of new requirements easily, and by constantly providing a working system in which users can learn of new needs. It also reduces costs by taking one step at a time, thus reducing the risks associated with building the wrong system in one large increment. Using both types of prototypes, namely an evolutionary prototype as a stable repository for known system functions and throwaway prototypes to validate new concepts prior to their incorporation into the evolutionary prototype, solves all four of the problems. This combined approach, called *field prototyping*, is fully described in [DAV89].

Configuration Management Strategies

Configuration management (CM) has the responsibility to manage changes that occur to a system over the course of its life. Included within CM are the practices of identification (to uniquely identify any component or any system composed of components), configuration control (to establish baselines and identify and enforce procedures for the orderly approval and dissemination of changes to those baselines), and configuration status accounting (to record the status of all baselines and requests to alter those baselines). This section explores the unique implications on CM of the ever-changing requirements and the use of prototypes in C^2 systems.

All systems undergo change. For systems undergoing traditional developments, change comes in two flavors: evolutionary and revolutionary. Evolutionary change occurs when engineers move forward in the engineering process. Adding a circuit layout to a project repository that heretofore contained only a logic diagram and adding code to a repository that heretofore contained only design are examples of evolutionary changes. These are positive, inevitable, and predictable. Revolutionary change occurs when engineers or customers or users detect that something is wrong and needs correction. Discovering a previously unstated requirement after the requirements specification is frozen and uncovering a race condition in the implementation after construction is completed are examples of revolutionary changes. These are positive (in that the change is for the better) and inevitable, but unfortunately unpredictable.

C^2 systems are unique only in the relative distributions of revolutionary and evolutionary changes occurring during their development and use. Because user needs are often not isolated until after deployment, there will obviously be a considerably larger number of revolutionary changes in a typical development. The implication of this is simply that the CM planners must recognize this fact and streamline the process of managing such changes as much as possible. Using throwaway prototypes on C^2 systems does not alter CM practices. Using evolutionary prototypes, however, will introduce a considerably larger set of evolutionary changes into the system. The result is that C^2 system developments using evolutionary prototyping will have a much higher load of both types of changes. Configuration management can become *the* single biggest contributor to project failure unless careful attention is paid to its planning and streamlining. A number of books have been written that address configuration management of source code only (e.g., [BAB86, BUC82]), but more comprehensive discussions of the planning, streamlining, and automation of configuration management exist [BER80, BRY88]. A more thorough discussion of the impact of prototyping on configuration management appears in [DAV89].

Acquisition Management Strategies

C^2 systems are usually procured using a traditional acquisition life-cycle of concept exploration, demonstration and validation, full-scale development, and production and deployment. However, in many cases the requirements are really not fully understood until the end of the full-scale development phase. This section explores alternatives to this traditional acquisition strategy for the procurement of systems. In particular, alternatives are presented to take into account the build-a-little/learn-a-little approach, and to take into account the use of prototyping.

A useful alternative to this approach has been put forth by Bersoff and Davis [BER89]. In this paper, a three-stage procurement process replaces the previous full-scale development stage: Rapid Prototyping, Prototype Deployment, and Full-Scale Engineering Development. During the Rapid Prototyping stage, the *least* understood parts of the system are constructed rapidly. These are typically security, user interface, or perhaps complex computational models. This contract should be incentive-based (that is CPIF or CPAF) to encourage the developers to reuse existing components and create new components in as rapid a manner as possible. This stage should be competed, with three to five contracts awarded.

Approximately nine to 12 months into the first stage, the Prototype Deployment stage begins. Here, the *best* understood parts of the system are constructed. This understanding may have been derived from the Rapid Prototyping stage or from the experience base of the users or developers. This should be built as a relatively quality product and expected to stay viable in a fully deployed and supported, operational setting, under considerable change, for about four or five years. Prototype Deployment stage contractors should be limited to one or two contractors only, and ideally should be limited to contractors who participated in the first stage. It is expected that first stage activities would continue throughout this phase to serve as constant low-risk testbeds for new ideas, prior to their incorporation into this stage two prototype.

About one or two years after the beginning of stage two, the Full-Scale Engineering Development stage should begin. In this stage, a full-scale system is built following all standard documentation procedures to insure a quality product. Only those requirements already validated in the stage two prototype are acceptable to include in the system. This activity, performed by one contractor determined by an open competition, can thus be awarded on a fixed-price (i.e., FFP) basis to reduce risk for the customer. It is expected that the stage one and two activities would continue throughout stage three. Figure 6 (adapted from [BER89]) shows how the three stages interplay to produce a quality product for the customer. It shows two complete cycles, each of which includes all three stages of procurement. The heights of the shaded areas represent the amount of activity.

An approach to procurement as described here can make effective use of the various technologies described in this paper. In particular, throwaway prototypes are used during the first stage, evolutionary prototyping and configuration management are used during the second stage, and formal specification techniques and configuration management techniques are used during the last stage. This careful orchestration of the technologies and selection of appropriate acquisition strategies can insure that all of the problems associated with ever-changing requirements for C^2 systems can be kept under control.

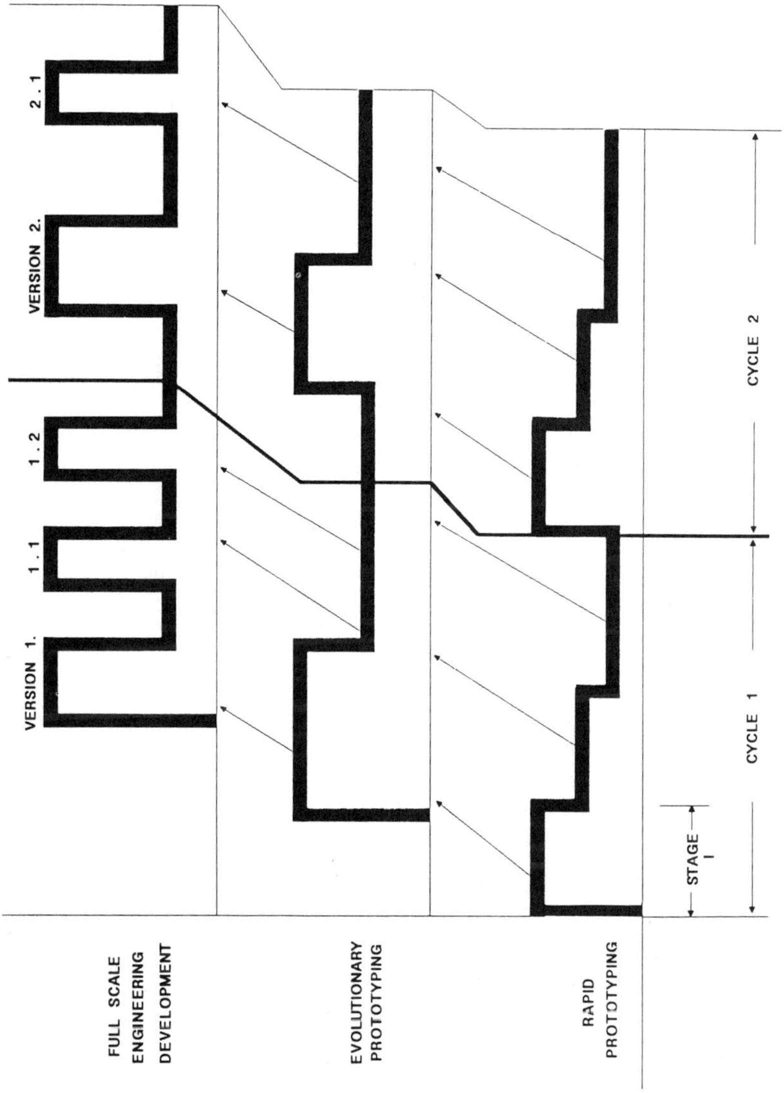

Figure 6. An Integrated Three-Stage Acquisition Process.

References

[ALF76] Alford, M., and I. Burns, "R-Nets: A Graph Model for Real-Time Software Requirements," *Symposium on Computer Software Engineering*, New York: Polytechnic Press, 1976.

[ALF85] Alford, M., "SREM at the Age of Eight, The Distributed Design System," *IEEE Computer*, 18, 4 (April 1985), pp. 36-46.

[BAB86] Babich, W., *Software Configuration Management: Coordination for Team Productivity*, Reading, Massachusetts: Addison Wesley, 1986.

[BAL82] Balzer, R., et al., "Operational Specifications as the Basis for Rapid Prototyping," *ACM Software Engineering Notes*, 7, 5 (December 1982), pp. 3-16.

[BER80] Bersoff, E., et al., *Software Configuration Management: An Investment of Product Integrity*, Englewood Cliffs, New Jersey: Prentice Hall, 1980.

[BER87] Bersoff, E., et al., "A New Look at the C3I Software Life Cycle,"*SIGNAL*, 41, 8 (April 1987), pp. 85-93.

[BER89] Bersoff, E., and A. Davis, "Some Thoughts on Acquisition Alternatives For C^3I Systems," Unpublished.

[BOO86] Booch, G., "Object-Oriented Development," *IEEE Transactions on Software Engineering*, 12, 2 (February 1986), pp. 211-221.

[BRY88] Bryan, W., and S. Siegel, *Software Product Assurance: Techniques for Reducing Software Risk*, New York: Elsevier Sciences, 1988.

[BUC82] Buckle, J., *Software Configuration Management*, London: MacMillan Press Ltd., 1982

[COA89] Coad, P., and E. Yourdon, *OOA: Object-Oriented Analysis*, Englewood Cliffs, New Jersey: Prentice Hall, 1989.

[DAV79] Davis, A., et al., "RLP: An Automated Tool for the Processing of Requirements," *IEEE COMPSAC '79*, Washington, D.C.: Institute of Electrical and Electronics Engineers, October 1979, pp. 289-299.

[DAV88] Davis, A., "A Taxonomy for the Early Stages of the Software Development Life Cycle," *Journal of Systems and Software*, 8, 4 (September 1988), pp. 297-311.

[DAV88a] Davis, A., "A Comparison of Techniques for the Specification of External System Behavior," *Communications of the ACM*, 14, 9 (September 1988), pp. 1098-1115.

[DAV88b] Davis, A., et al., "A Strategy for Comparing Alternate Software Development Life Cycle Models," *IEEE Transactions on Software Engineering*, 14, 10 (October 1988), pp. 1453-1461.

[DAV89] Davis, A., and E. Bersoff, "Impacts of New Life Cycle Models on Software Configuration Management," submitted to *Communications of the ACM*, June 1989.

[DAV90] Davis, A., *Software Requirements: Analysis and Specification*, Englewood Cliffs, New Jersey: Prentice Hall, 1990.

[DEM79] DeMarco, T., *Structured Analysis and System Specification*, Englewood Cliffs, New Jersey: Prentice Hall, 1979.

[GOM81] Gomaa, H., and P. Scott, "Prototyping as a Tool in the Specification of User Requirements," *Fifth IEEE International Conference on Software Engineering*, Washington, D.C.: Institute of Electrical and Electronics Engineers, 1981, pp. 333-342.

[GOM86] Gomaa, H., "Prototypes—Keep Them or Throw Them Away?," *InfoTech State of the Art Report on Software Prototyping*, 1986.

[HAR87] Harel, D., "Statecharts: A Visual Formalism for Complex Systems," *Science of Computer Programming*, 8 (1987), pp. 231-274.

[HAT84] Hatley, D., "The Use of Structured Methods in The Development of a Large Software-Based Avionics System," *AIAA/IEEE Sixth Digital Avionics Systems Conference*, Washington, D.C.: Institute of Electrical and Electronics Engineers, 1984, pp. 6-15.

[HAT87] Hatley, D., and I. Pirbhai, *Strategies for Real-Time System Specification*, New York: Dorset House, 1987.

[HER87] Herres, R., "Equipment, Personnel and Procedures— Foundations for Future C^2 Architecture," in S. Andriole, ed., *Principles of Command and Control*, Washington, D.C.: AFCEA International Press, 1987, pp. 413-426.

[MOR82] Moret, B., "Decision Trees and Diagrams," *ACM Computing Surveys*, 14, 4 (December 1984), pp. 593-623.

[OFF87] Offutt, J., and J. Danese, "Communications for SDI," in S. Andriole, ed., *Principles of Command and Control*, Washington, D.C.: AFCEA International Press, 1987, pp. 263-276.

[WAR85] Ward, P., and S. Mellor, *Structured Development for Real-Time Systems*, Englewood Cliffs, New Jersey: Prentice Hall, 1985.

[WAR86] Ward, P., "The Transformation Schema: An Extension of the Data Flow Diagram to Represent Control and Timing," *IEEE Transactions on Software Engineering*, 12, 2 (February 1986), pp. 198-210.

Hypermedia-based Requirements Engineering

Peter Aiken

The Conceptual Origins of Hypermedia

The conceptual basis for the hypermedia workstation is largely influenced by the work of three researchers. This section describes the contributions of Vannevar Bush, Ted Nelson, and Douglas Engelbart who collectively supplied concepts contributing to the rudiments of hypermedia technology. Three concepts—associative indexing, hypertext, and augmentation technology—initiated a chain of research culminating in current interest in hypermedia technology.

The Importance of "Selection" in Information Retrieval

Hugh G. J. Aitken has said, "Scientists discover new knowledge ... largely in response to problems posed by scientists, and the new knowledge so generated is consumed largely by scientists themselves" [Aitk76, p. 315]. In the July 1945 issue of *Atlantic Monthly*, Vannevar Bush posed one such problem described at the time as having the potential to alter the "relationship between thinking man and the sum of our knowledge." As Director of the Office of Scientific Research and Development, Bush was concerned with setting a post-war research agenda for the thousands of scientists who were currently applying science to warfare. The area in which he saw the most pressing need was support for the investigator who "is staggered by the findings and conclusions of thousands of other workers—conclusions which he cannot find time to grasp, much less remember, as they appear" [Bush45, p. 101].

In "As We May Think," Bush noted the inadequacy of current means of transmitting and disseminating research results because the published record exceeded the ability of an individual to make use of it. In order to be useful, a scientific record must be consultable. One means of making records more accessible was to reduce the cost of information storage and distribution. Data compression was Bush's solution and, extrapolating current trends, he postulated a time when the cost of reproducing the *Encyclopedia Britannica* on microfiche would be a nickel. Further, the fiche version could be mailed anywhere for the cost of a first class postage stamp.

In Bush's vision of the future, scientists would use an integrated recording system, combining photography and voice recording techniques to create their records. He wrote, "One can picture a future investigator in his laboratory. His hands are free, and he is not anchored. As he moves about and observes, he photographs and comments. Time is automatically recorded to tie the two records together. If he goes into the field, he may be connected by radio to his recorder. As he ponders over his notes in the evening, he again talks his comments into the record" [Bush45, p. 104].

After describing the means of coping with the increasing rate of growth of the scientific record, Bush turned his thoughts to the process of utilizing the information. He identified the "prime action of use" for information as *selection.* Then current means of automating the selection process involved examining every potential piece of information and selecting those meeting the specified criteria. Bush wanted a system more like the telephone exchange, capable of making direct connections between circuits without employing search mechanisms. He criticized the artificiality of indexing systems as being contrary to the way the human mind works.

"It operates by association. With one item in its grasp, it snaps instantly to the next that is suggested by the association of thoughts, in accordance with some intricate web of trails carried by the cells of the brain. It has other characteristics, of course; trails that are not frequently followed are prone to fade, items are not fully permanent, memory is transitory. Yet the speed of action, the intricacy of trails, the detail of mental pictures, is awe-inspiring beyond all else in nature."[Bush45, p. 106]

Noting the futility of attempting to artificially duplicate this process, Bush suggested a memex as an alternate implementation vehicle. He described memex as a private mechanized system for accessing and filing communications and records. Printed material would be distributed on microfilm, accessible through a reader. Traditional index-based means of accessing the material would be available. More importantly, memex systems would also support automated associative indexing. Users would be able to create unobtrusive buttons in and around the data being viewed. The buttons would represent associative links to other information which could be accessed by *tapping a button.* Creation of buttons would be as natural as turning the pages of a book. While the human brain is faster and more flexible, the associative indices created and stored in memex would not fade as a function of the passage of time and/or fre-

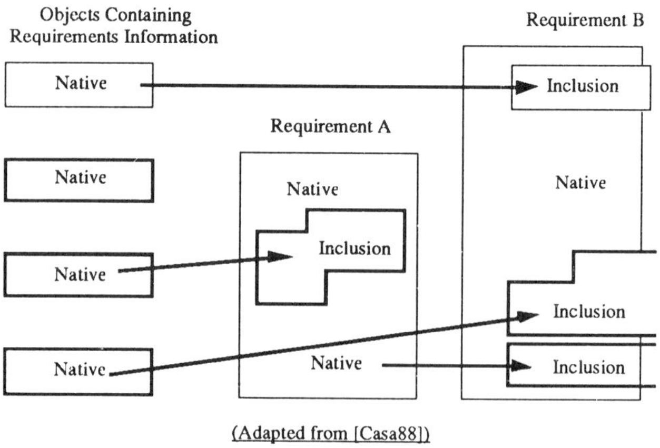

Figure 1. Hypertext Document Creation.

quency of access. In addition, humans would benefit from significant increases in the clarity of information items recalled by using the system. Bush predicted more efficient and effective scientific progress using memex and other systems supporting this associative linking process. One of the primary motivations behind this research is the belief that the ability to create these associative indices will positively effect requirement engineering efforts.

Hypertext

In the fall of 1960, Ted Nelson was a graduate student at Harvard University taking a course on computers in the social sciences. He was quoted as being astounded that automated support for organizing notes did not exist in forms that he considered useful. This useful form evolved into what he termed *hypertext*, meaning non-sequential reading and writing [Conk87]. Since the 1960's, Ted Nelson has been advocating use of a hypertext-based approach to creating and preserving knowledge called Project Xanadu.

Documents in hypertext systems consist of original text, created specifically for the document in question and *inclusions*, information originally created or included in other documents but now part of a new document. (See Figure 1.) Based on this model, Nelson's original concept has evolved into a grand vision for hypertext as "a repository publishing network for anybody's documents and contents, which users may combine and link to freely" [Casa88, p. 24]. To fulfill his vision, Nelson has also created

- methods for keeping track of the immense quantity of links generated by authors as they create and include text objects, and

- a document numbering system suitable for tracking the very large and growing collection of documents existing on distributed systems and for tracking the origin of all inclusions [Nels88, p. 225].

Nelson's vision has had a significant impact on the development of hypermedia technology. As hypertext has evolved into the primary vehicle for implementing the associative indices described by Bush, his status, originally that of a maverick, has changed to that of a visionary. Nelson has stated that products based on Xanadu concepts and running on Sun workstations should ship during 1989 [Casa88].

Augmentation Technology

A 1962 summary of Doug Engelbart's research to date, *Augmenting Human Intellect: A Conceptual Framework*, completed the conceptual basis for hypermedia technology. Inspired by J. C. R. Licklider's concept of human-computer symbiosis [Lick60] and by the concepts developed by Bush and Nelson, Engelbart saw a potential usefulness for man-computer synergism [Casa88]. He defined augmentation technology as interactive computerized methods or devices capable of extending native human sensory, mental, and motor capabilities [Enge62].

In 1968, Engelbart implemented some of his ideas in a system called NLS (oN Line System) which has since evolved into its present form, Augment, the information repository for ARPAnet [Casa88]. NLS was the first implementation of hypertext and was quickly extended to include a number of other features such as integrated computer to computer communications and conferencing facilities, electronic mail and other support for group collaborative processes, integrated text and video imaging, the mouse, and the first windowing system, and the first context-dependent Help system. Engelbart collectively referred to these efforts aimed at providing consistent support for knowledge workers as augmentation technology.

"By augmenting man's intellect we mean increasing the capability of a man to approach a complex problem situation, gain comprehension to suit his personal needs, and to derive solutions to problems. Increased capacity in this respect is taken to mean a mixture of the following: that comprehension can be gained more quickly; that better comprehension can be gained; that a useful degree of comprehension can be gained where previously the situation was too complex; that solutions can be produced more quickly; that better solutions can be produced; that solutions can be found where previously the human could find none." [Enge62]

Augmentation technology helped to define the thrust of modern support-based technology. In general terms, it defines the nature of support required by the knowledge worker [Engl62]. Hypermedia support for requirements engineering is a direct extension of augmentation technology. Functions supported

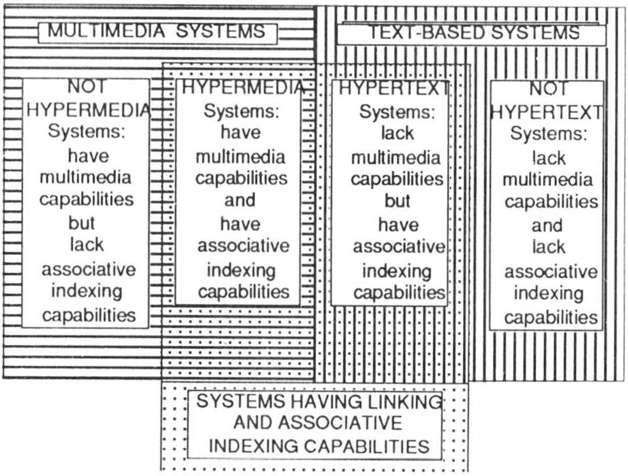

Figure 2. The Relationship Between Three Often Confused Terms.

by the hypermedia workstation are memex-like functions tailored to provide support for requirements engineering tasks. Much of that support is focused on providing support for the requirements engineer's ability to create associative indices while working with requirements information.

Hypermedia Defined and Differentiated

Currently, the literature presents a bewildering number of often conflicting definitions for the terms "hypermedia," "hypertext," and "multimedia." (For examples see [Nels67], [Yank87], [Deli86], [Garr86], [Conk87], [Garg87], [Aksc88], [Casa88], and [Smit88].) This section presents a set of definitions in hopes of clarifying the terms used in this chapter. Figure 2 presents one means of conceptualizing the distinctions between hypermedia, multimedia, and hypertext systems.

Hypermedia

As shown above, hypermedia systems can be described as a subset of multimedia systems but provide support for the creation and maintenance of associative indices. The most appropriate definition for hypermedia is one by Frank Halasz who describes hypermedia systems as: "information representation and management systems that organize information into networks of multimedia nodes interconnected by links. Each node generally contains a large chunk of 'content' such as a document, a drawing, or a voice annotation. The links are used to represent interrelations among these nodes" [Hala87, p. 350]. Hypermedia systems contain the following four components: 1) a computer-based data analysis and management system with 2) a high bandwidth human-computer interface and 3) capabilities permitting users to quickly and easily navigate, manipulate, and maintain multimedia data by 4) linking hypermedia objects containing chunks of unique, identifiable information into personalized associative indices.

Data are stored in hypermedia systems as objects, each containing a variable amount of information stored on a variety of media. The media can be classified with four general categories: text, graphics, motion, and sound. The contents of a hypermedia object can be as small as a single bit of information or it can be quite large (i.e., a digitized two-hour movie occupying 550 megabytes on CD ROM disc). Each hypermedia object is stored in a separate window. Using system utilities, users navigate, manipulate and maintain structures of objects linked into associative indices.

Associative indices are used to maintain data in an essentially unconstrained form. Unconstrained implies that data models created using the workstation are not restricted to customary patterns of data organization. Data models can be created by users as dictated by personal taste. This permits organization of data from different perspectives or in different contexts. Robert Akscyn and Jeff Conklin have highlighted additional differences between hypermedia and other types of systems to help establish boundaries differentiating hypermedia systems [Aksc88] and [Conk87]. They note:

- Multimedia Document Systems lack user ability to create, manipulate, and maintain links;
- Windowing Systems lack database capabilities;
- File Systems are databases but generally lack linking capabilities and navigational support;
- Outliners have limited support for references occurring between outline entries and provide only hierarchical links between entries;
- Text Formatting Systems permit only hierarchical links between chunks of text; and
- Data Base Management Systems lack the ability to link multimedia items together and a unified user interface across media types.

Multimedia

Hypermedia systems are often mislabeled as multimedia systems and vice versa. Consequently, a whole array of developmental and commercial, multimedia and hypermedia products share the same assortment of names [see Aksc88]. Multimedia is properly defined as: "Including or involving the use of several media" [Amer82, p. 821]. Multimedia-based systems generally seek to create a more effective presentation by synergistically combining various media (i.e., text, graphics, video images, etc.) to enhance user interaction with screen events [Brew86].

Hypertext

Hypertext has been variously defined as:

- " ... a combination of natural language text with the computer's capacity for interactive branching, or dynamic display ... of a nonlinear text ... which can not be printed conveniently on a conventional page" [Nels67];
- A system permitting "authors or groups of authors to link information together, create paths through a corpus of related material, annotate existing texts, and create notes that point readers to either bibliographic data or the body of the referenced text" [Yank85];
- A facility for machine support of arbitrary cross-linking between (textual) items [Conk86]; and
- Non-sequential reading and writing [Casa88].

However, the most useful definition is to define hypertext systems as similar to hypermedia systems but supporting only text-based objects.

Features of Hypermedia-based Systems

Eight noteworthy features of hypermedia systems are described below. Not all features are present in all hypermedia systems. However, many of them are present in many hypermedia systems.

Higher Potential Information Bandwidth in Human-Computer Dialogue

In both directions, hypermedia's multimedia capabilities offer the highest potential amount of information exchange because hypermedia excludes no media as potential input or output. In fact, Alan Kay has noted "the computer, as a medium itself, can be all other media if the embedding and viewing methods are sufficiently well provided" [Kay88, p. 254].

Better Information Access and Exchange Capabilities Through a Unified Software Interface

Hypermedia has been envisioned as a "cassette player for information." This implies anyone is capable of retrieving the information, provided they have access to the appropriate hypermedia technology. Consider the analogy of hypermedia providing an information infrastructure similar to the transportation infrastructure existing as railroads, highways, and airports [Atki87].

Non-traditional Linking of Presentation Media

Information stored on CD ROM, video tapes, audio recordings, and other non-traditional presentation and storage media can be incorporated within a collection of data. Individual data items can be represented with multiple complementary media types. The key is to store and present the user with information using the most appropriate format.

Support for Reporting Beyond Traditional Methods

Associative indexing capabilities permit users to create buttons and active screen objects. With these users can branch to different data objects within a given context without having to access an index or otherwise employ a search mechanism.

Non-traditional Linking of Facts

The direct branching capability is not restricted to jumping around within the same data collection. Branching across conventional subject boundaries is an important capability permitting users to focus on the content of the data and ignore its organization.

Availability of Concepts Unique to Object-oriented Systems

Object-oriented concepts such as encapsulation, classes, inheritance, and polymorphism provide a number of advantages over conventional systems. Encapsulation of data and methods in objects permits the data itself to know which type of access procedures to invoke to accomplish specific objectives. Classes permit data structures and methods to be shared among related groups. Inheritance simplifies aspects of organizing and classifying data items. Polymorphism enables the implementation of a single set of access methods for all types of objects.

Toolkit Availability

Hypermedia technology offers the ability to create and access a whole toolkit of utilities in a fashion similar to the UNIX Programmer's Workbench concept, but with expanded information handling capabilities.

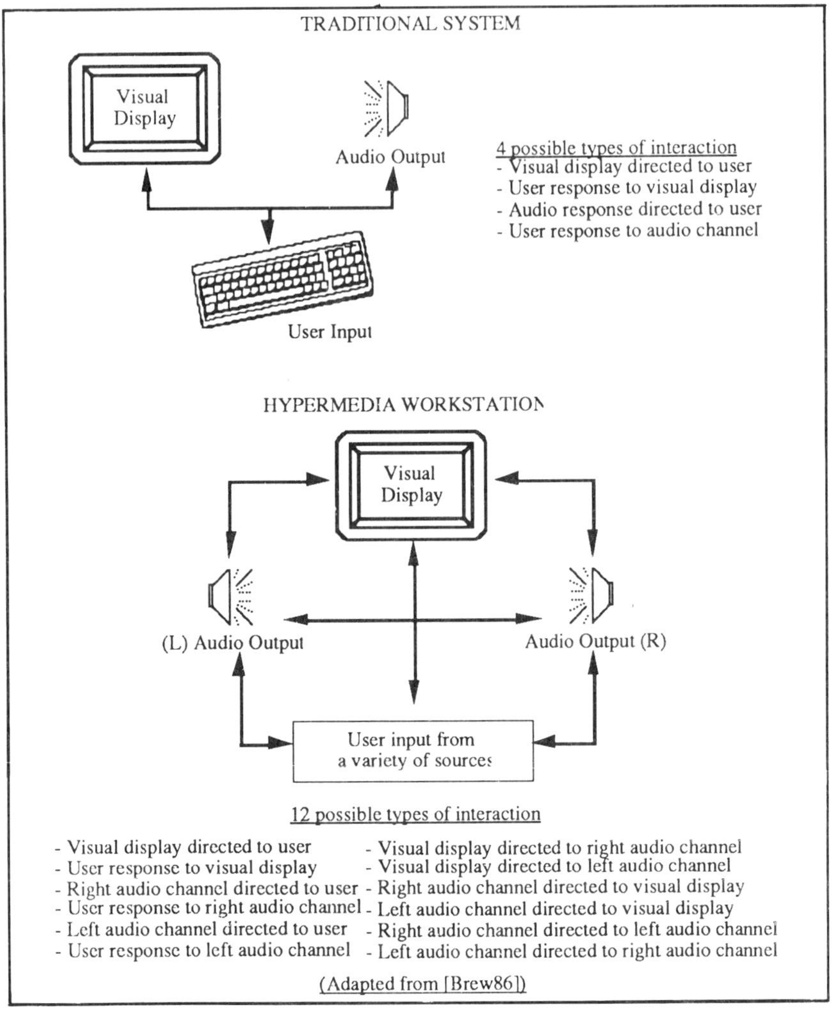

Figure 3. Possible Hypermedia Interaction Modes.

Increased Interaction Modes

Traditional user-computer interaction consists of a user interacting with a visual display, often supplemented with audio tones. Hypermedia interaction permits more total interactions. (See Figure 3. This may be particularly important because recent research indicates that the source of a event may be particularly important [Gave88].

Classes of Systems Able to Benefit From Hypermedia-based Requirements Engineering

Applying hypermedia-based technologies to a given requirements engineering application typically requires relatively expensive hardware, often including: more expensive storage technologies, audio/video playback and recording equipment, and color display technologies. Until the cost of this hardware decreases substantially it will not be cost-effective to use hypermedia in situations where the requirements are straightforward. Thus, hypermedia-based requirements engineering for systems where inputs and outputs are known and able to be represented in text-based format, and processing requirements are able to be specified procedurally will not be cost-effective for a number of years. On the other hand, there are several classes of situations where the use of hypermedia will be worth considering. These include:

- software systems where the problem solution cannot be described procedurally;
- software systems developed in an environment where current, text and static graphic-based requirements engineering tools and techniques cannot adequately capture and represent the issues involved (for example, the positions of the various stakeholders, political constraints, or lengthy background information);
- software systems where the design issues are too complex to be represented with static diagrams;
- software systems specifically concerned with sound-based, and/or motion-based requirements; and
- multimedia or hypermedia-based software systems.

Requirements Engineering Functions

The requirements engineering portion of the software development life cycle refers to all development activities occurring prior to system construction. This portion encompasses the most important phases of the software development life cycle: system requirements and system design. These phases incorporate the requirements engineering tasks of elicitation, task/user/environmental profiling, analysis, modeling, design, and prototyping and others. An examination of these tasks resulted in the identification of four basic requirements engineering functions. (See Figure 4.) The application of these functions is shown in the context of a prototyping development methodology in Figure 5.

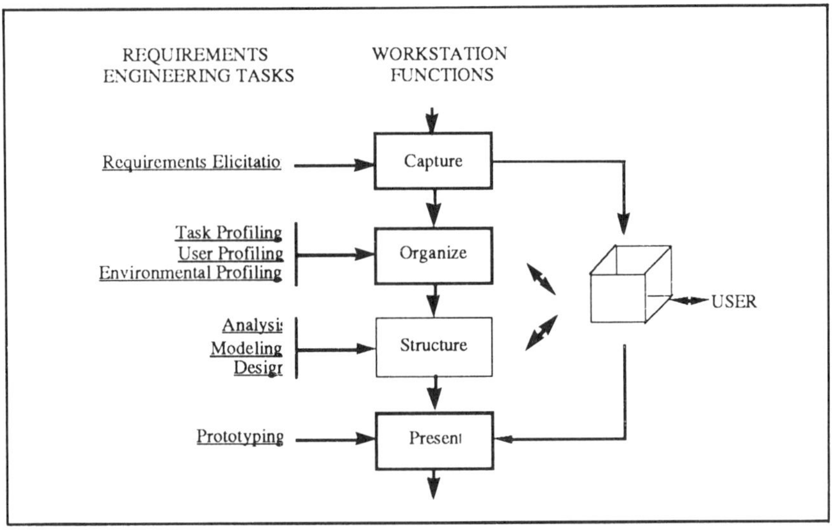

Figure 4. Mapping Between Conventional Requirements Engineering Tasks and Requirements Engineering Functions.

Mapping Between Conventional Requirements Engineering Tasks and Requirements Engineering Functions

The first two functions are associated with the requirements analysis phase of the software development life cycle. They are 1) capturing information about the task to be performed, the system users, and any relevant organizational/situational characteristics, and 2) organizing the information into an easily accessible form for use in subsequent phases. The other two functions, associated with system design are 3) structuring the tasks identified during the requirements analysis into a model of the computer-based solution, and 4) presenting the model to the user in order to obtain corrective feedback. Each function is described in more detail below.

Capture

The capture function is concerned with acquiring information about three categories of requirements data.

- *User profiles* portray the users of the package along a number of dimensions including job function, level of computing experience, frequency of access, conditions under which the package will be used, and technical background. The goal is to identify the shape and size of the package's users.
- *Task profiles* describe the qualitative and quantitative aspects of the pro-

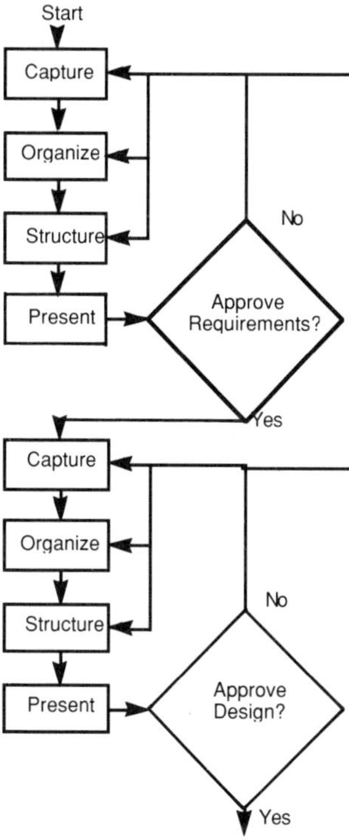

Figure 5. Recommended Requirements Engineering Process.

cesses, functions, behaviors, and performance of the package. It is important to describe all aspects of the task including both automated and manual sub-tasks to gain the most complete picture possible.

- *Environmental profiles* concern specific, relevant or other factors impacting the situation. This dimension of the matrix captures the flavor with which things are done in the host organization. (For example, in the military this translates directly into support for established doctrine.) [Andr86].

Organize

The organize function is concerned with arranging the captured requirements information into an easily accessible form so specific objects can be accessed throughout subsequent phases of the life cycle. Dallas E. Webster has noted " ... the way in which information is represented is frequently the critical factor in determining whether problems are easy or difficult to resolve, or whether a (package) does or doesn't work. So we must pay careful attention to the choice of representation media and methods to make sure they are sufficient or, more realistically, extensible or adaptable" [Webs88, p. 10]. Hypermedia technologies offer the ability to represent information using the most appropriate media. Once organized, the requirements information forms the basis for the software system design. Coders reference this information to supplement the system design when developing code. And when maintenance personnel approach a problem, they access the requirements in order to familiarize themselves with system's function(s).

Structure

The purpose of the structuring function is to increase the connectivity of the requirements information while designing a model of the computer-based solution for the host environment. Analysis of requirements information identifies constraints which then become design issues guiding model development. Webster has also stated, " ... the design process can be viewed as successive elaborations of representations, such as adding more information or even backtracking and exploring alternatives" [Webs88, p. 8]. Additionally, an effective environment for the design processes provides support for the required visual momentum and cognitive processes [Wood84] associated with these tasks.

Thus, the primary requirement of the structuring function is the ability to rapidly and easily navigate, manipulate, and maintain individual requirements data items. This reduces delays in system operation and cognitive processes that can cause the requirements engineer to lose the current train of thought, wasting valuable time, and potentially introducing other mistakes to the process. Requirements engineers require integrated tools permitting them to quickly manipulate multimedia chunks into associative indices forming the basis for the package model. They should be able to *directly* combine, divide, link, and redefine the chunks created during the organize function.

Present

The goal of the presentation function is to prevent disconnects from occurring between the representation of the package presented to the user by the requirements engineers and the user's/sponsor's concept of the package. This disconnect occurs when the user gains a different mental interpretation of the package prototype presented by the designer.

Presentation of the model to the user also can take a number of forms ranging from written reports to well-developed interactive prototypes. Written prototypes are easiest to produce but provide the least valuable information in terms of user feedback. The more a prototype is based on images, functions, and sounds familiar to the user, the smaller the chance of a disconnect occurring. Another rule of thumb is that the closer a system approximates the actual look

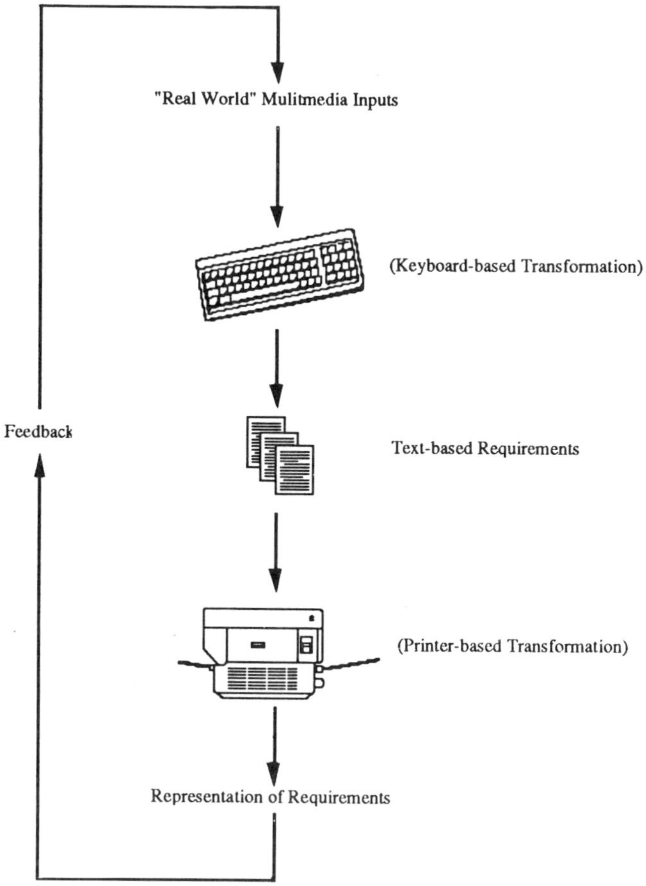

Figure 6. Data Transformation During Requirements Engineering.

and feel [Samu89] of the proposed system—the better the user will be able to correctly assess the system's ability to meet the required needs. As a result of previous functions, many sounds and images are already contained as objects in the workstation. These should be incorporated into the prototype. Ideally, the requirements engineering process repeats until the sponsor and the requirements engineers are certain of the correctness of the requirements data, its organization, the model of the system and the prototype.

Hypermedia Support for Requirements Engineering Functions

The information flow during the development of a software package can also be viewed as the result of a series of transformations on real world inputs. The goal of the first transformation is to obtain a text-based representation of the problem at hand. Typically input data is transformed from its representation in the multimedia, real world into a two-dimensional, serial document, although it is often supplemented with a table of contents, indices, and static graphs (i.e., various flow diagrams). The system design is generally made to fit a paper-based media. Thus requirements data enter the system through a keyboard and representations of the design leave the development system through a printer of some sort. (See Figure 6.)

Support For The Capture Function

Hypermedia support for the capture function is concerned with supporting processes used to obtain requirements information. The goal is to increase the amount and quality of information available to the requirements engineer. This is accomplished by increasing the volume of inputs and reducing the transformations required to get the requirements information into manipulable format. Augmented data capture facilitates the user's ability to capture requirements information from a variety of external sources and to create new data internally by transforming objects from one media to another. Sounds, pictures, and motion-based input data can supplement traditional text and static graphics-based representations of requirements information.

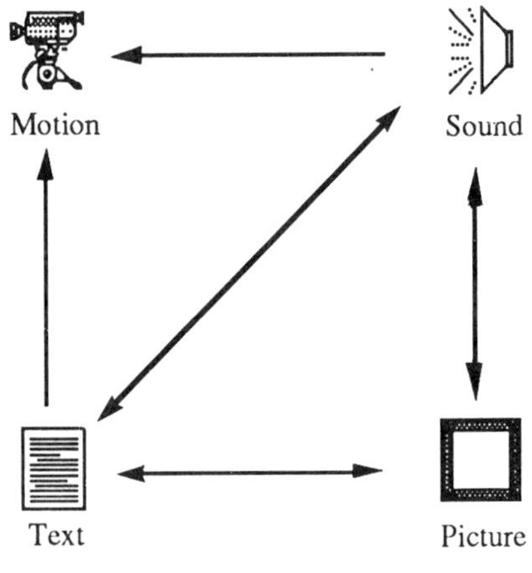

Figure 7. Potential Object Combinations.

Support For The Organize Function

Hypermedia support for the organize function is concerned with arranging the captured information, within dimension, into easily accessible task, user, and situational profiles. The goal is to assist the engineer to effectively and efficiently group requirements information into object collections. A partial list of useful object collections include objects grouped by:

- subject area within media classification,
- subject area irrespective of media classification,
- requirement,
- design constraints,
- design issues,
- design features,
- prototype features.

This process can be supported by

1. augmenting the requirements engineer's ability to transform unstructured objects into objects that can be identified by the search operations executed by the user, and

2. providing the ability to quickly and easily manipulate objects. Manipulations supported include: combine, divide, re-define, and attach.

Hypermedia can support the use of an unconstrained data model as defined above. Support for the organization function permits requirements engineers to quickly and easily access multimedia objects. Objects can be manipulated and redefined as additional objects. Users can combine objects to create the most appropriate representation of the requirements information. (See Figure 7) The system supports the use of many to many relationships between objects and augments user efforts to create and maintain those links. In addition, all requirements information is maintained on-line and is accessible thorough a single interface.

Support for the Structure Function

Hypermedia support for the structure function is concerned with increasing the connectivity of the information by establishing linkages across subject boundaries. The goal is to provide increased situation comprehension and augmented solution development. Two subfunctions supporting the requirements engineer's efforts to structure requirements information are: attach and link. The attach subfunction is used to create composite objects—objects made up of multiple media. The link function supports the process of creating associative indices.

Support for the Present Function

Hypermedia support for the present function is concerned with ensuring that the requirements engineer and the package sponsor have the same picture of the proposed package. The goal is to increase the communication between the engineer and the package sponsor through the use of prototyping strategies. Throughout the prototype evaluation users are presented with familiar sounds, images, and functions. This is accomplished by providing support for:

1. Human-computer Interface Design—The workstation facilitates the presentation of prototype interfaces using high bandwidth multimedia human-computer interface. This simplifies the use of realistic interface metaphors.

2. Program Flow and Program Control—The workstation supports simulation program flow and simulation of the host environment so the package can be observed in its intended environment.

3. Presentation Sequencing—This workstation provides interactive prototyping facilities for construction of prototypes closely approximating the look, feel, and operation of the package to be developed.

A Hypermedia Workstation for Requirements Engineering

Research efforts at the Center for Software Systems Engineering at George Mason University have resulted in the development of a hypermedia workstation supporting requirements functions as described in this chapter. The workstation was assembled using off-the-shelf hardware components and a specially developed software shell. The software shell is capable of navigating among, manipulating and maintaining requirements information existing on various media. The shell enables users to create and modify hypermedia objects and links between objects. The workstation is based on a Macintosh IIx computer. On-going research is evaluating the potential contribution hypermedia-based tools and techniques embodied in the workstation can make to the requirements engineering process.

References

Aitk76 Hugh G. J. Aitken, *Syntony and Spark: The Origins of Radio*, New York: John Wiley & Sons. 1976.

Aksc88 Robert Akscyn, *Tutorial: Design and Use of Hypermedia Systems*, ACM/SIGCHI Conference on Human Factors in Computing Systems, May 15, 1988.

Amer82 *American Heritage Dictionary*, Second Collage Edition, Boston: Houghton Mifflin Company, 1982.

Andr86 Stephen J. Andriole, *Handbook for the Design, Development, Evaluation, and Application of Interactive Military Decision Support Systems*, Marshall VA: International Information Systems, 1986.

Brew86 Bryan Brewer, "Compact Disc Interactive Audio," in (Steve Lambert and Suzanne Ropiequet, editors) *CD ROM: The New Papyrus*, Redmond, WA: Microsoft Press, 1986, pp. 273-294.

Bush45 Vannevar Bush, "As We May Think," *Atlantic Monthly*, July 1945, pp. 101-108.

Casa88 Lou Casabianca, "Ted Nelson Hypermedia Magician," in *Hypermedia*, Summer 1988, p. 23-25.

Conk86 Jeff Conklin, "A Survey of Hypertext," MCC Technical Report No. STP-356-86.

Deli86 Norman Delisle and Mayer Schwartz "Contexts—A Partitioning Concept for Hypertext" Conference of Computer-Supported Cooperative Work—Proceedings 1986, Austin, TX, 1986, pp. 147-152.

Enge62 Douglas Engelbart, *Augmenting Human Intellect: A Conceptual Framework*, Summary Report, Stanford Research Institute, on Contract AF 49(638)-1024, October 1962, 134 pp.

Garg87 Pankaj K. Garg and Walt Scacchi, "On Designing Intelligent Hypertext Systems for Information Management in Software Engineering," Hypertext '87 Papers, pp. 409-433.

Garr86 L. Nancy Garrett, Karen Smith and Norman Meyrowitz "Intermedia: Issues, Strategies and Tactics in the Design of a Hypermedia Document System" Conference of Computer-Supported Cooperative Work—Proceedings 1986, Austin, TX, 1986, pp. 163-174.

Gave88 William W. Gaver, personal communication 8/26/88.

Hala87 Frank G. Halasz, "Reflections on Notecards: Seven Issues for the Next Generation of Hypermedia Systems," Hypertext '87 Papers, pp. 345-366.

Kay88 Alan Kay and Adele Goldberg, "Personal Dynamic Media" in (Adele Goldberg, editor) *A History of Personal Workstations*, Reading, MA: Addison-Wesley Publishing Company, 1988, pp. 254-263.

Lick60 J. C. R. Licklider, "Man-Compute Symbiosis," *IRE* (now *IEEE*) *Transactions on Human Factors In Electronics*, March 1960, pp. 4-11.

Nels88 Theodor H. Nelson, "Managing Immense Storage: Project Xanadu Provides a Model for the Possible Future of Mass Storage," *BYTE*, January 1988, 13(1):225-238.

Samu89 Pamela Samuelson, "Why The Look and Feel of Software User Interfaces Should Not Be Protected By Copyright Law," *Communications of the ACM*, May 1989, 32(5)563-572.

Sher89 Chris Sherman, *The CD ROM Handbook*, New York: McGraw-Hill, 1989.

Smit88 John B. Smith and Stephen F. Weiss, "Hypertext," *Communications of the ACM*, July 1988, 31(7):816-819.

Webs88 Dallas E. Webster, "Mapping the Design Information Representation Terrain,"*IEEE Computer*, December 1988, 21(12):8-23.

Yank87 Nicole Yankelovich, Bernard J. Haan and Steven M. Drucker, "Connections in Context: The intermedia system," *IRIS* Technical Report 87-3, 1987.

System Level Requirements Specifications for Command and Control: An Approach to Risk Management and Volatility Reduction

James D. Palmer

Introduction

Generation of requirements for C^2 is often considered to be a deterministic activity in which needs of the several users, for whom the system design is intended, are either known or assumed. This has been especially true in system designs in which there have been less than explicit definition of user requirements and these have been left largely unanalyzed as to substance of the system requirements. This situation is characteristic of the current state of system level requirements specifications.

Conventional project support tools provide little help for the user in dealing with dynamic plans and requirements. Many problems appear when least expected: requirements change, critical factors vary in unanticipated ways, resources are not available when needed, and time and schedules stretch beyond reasonable bounds.

Systems support technology in use today is not capable of assisting in a review of the substance of requirements specifications. Generally, only the form of the requirements specification may be verified using these tools. As a consequence of this inability to analyze substance early on, we find that substance may be validated against the system level requirements specifications only at the time when the product is fielded. This is much too late in the development process to determine that fatal flaws may exist in the system, as conceived and designed.

For our discussion on system level requirements specifications for command and control, we will use the Joint Chiefs of Staff definition for C^2. This states that command and control is: "The exercise of authority and direction by a properly designated commander over assigned forces in the accomplishment of the mission."

This definition has been interpreted [Boyes and Andriole, 1987] to mean that commanders must also be given certain specific capabilities that include the ability to communicate to higher and lower command levels, obtain, process, analyze, synthesize, and display information.

Users are most often interested in determining alternative courses of action, or determining what countermeasures to take in a hostile jamming environment, or determining what to do if a portion of the system malfunctions. These purposes for the system are generally not able to be resolved exactly and measured precisely at the time of system level requirements decisions. This is one of the primary reasons behind the introduction of risk and volatility in system level requirements specifications. It leads to the question:

How do we come to select system level requirements specifications from inputs that may be less than concise, not always accurate, and frequently in conflict?

Experience in the requirements elicitation process clearly demonstrates that these factors are present during most of the project design life cycle.

Indeed, requirements elicitation experience shows that user needs tend to be stated in terms that are ambiguous, imprecise, incorrect, and/or infeasible, in some sense, for most situations. As a consequence, the information contained within the stated characteristics remains unavailable for use by system developers. In most cases when the system level requirements specifications are not made precise, complete, and correct, it is the system developer, and not the user, who will clarify these requirements and prepare the system design. In this way, the developers, and not the users, will proceed to make precise that which was imprecise without regard to the concerns of the user. This in no way is intended to cast aspersions at the system developer, rather it is intended to cast light on the problems of not clarifying system level requirements specifications prior to the design process.

The observation that user determination of user need reflects uncertainty is not due to ineptness or lack of interest on the part of the user, but rather to the situation extant in describing user requirements for C^2. Leading experts [c.f. Kroening, 1987] proclaim that the foremost problem in C^2 lies in the inability to explicitly state requirements definitions for the total system design, and most specifically for software requirements. Another severe constraint in the development of C^2 system requirements is the extremely long gestation period and considerable number of changes of personnel that may occur during the development of the design of a single system. This extensive length of time to prepare a set of system level requirements adds to the problems of uncertainty due to changes in system priorities, changes in system performance needs, and, perhaps, changes in the perceived threat situation for which the initial design was intended.

Military command and control requirements focus on the hard realities that surround human decision making in the face of severe environmental and time constraints. The environment frequently is rich with an overabundance of signals to process and understand. Yet, validation of even the most basic information needs may be lacking.

System goals that are to be satisfied by the design are generally compounded by multiple conflicting purposes. The field situation in which the system is to operate is characterized as being fluid, dynamic, and defying description. Most often the field situation is one of dynamic, high risk decision making with responsibility for human life hanging in the balance with little time to reflect and play "what if" scenarios. The goal for decision support systems is to assist human information processing in this volatile environment. This means that systems support needs are to strengthen the human decision making function, and not supplant the human in the loop.

We will look at the characteristics of C^2 system requirements in terms of the purposes of the system and then examine ways of managing risk and reducing volatility in system level requirements. Various facets of this problem have been addressed by others [Boyes and Andriole, 1987, Palmer, 1988]. Here, we will discuss one such aspect of this problem, a requirements engineering approach to deal with risk management and volatility reduction.

Characteristics of C^2 System Requirements

We have noted that there are severe environmental constraints coupled with time factors, human decision making activities and long gestation periods that plague the development of fieldable C^2 systems. In this section we will explore the nature of purposeful C^2 system requirements in order to provide guidance as to how to proceed to make them better understood. Two other types of requirements are present, functional and structural. These will not be discussed here except to provide a definition, as they are mainly concerned with the ways in which the system is realized. This occurs after we have determined the purposes of the system.

Purposeful system level requirements are those properties of the system that indicate *what* the system must perform in order to meet the needs that the user has defined as the mission of the system. Functional system level requirements are those properties of the system that indicate *how* the system will be designed to meet the performance requirements necessary to realize its purposes. Structural system requirements relate to the *means* of implementation of the design. The relationships between the three levels of requirements leads us to define a taxonomy for identification of total system level requirements for the

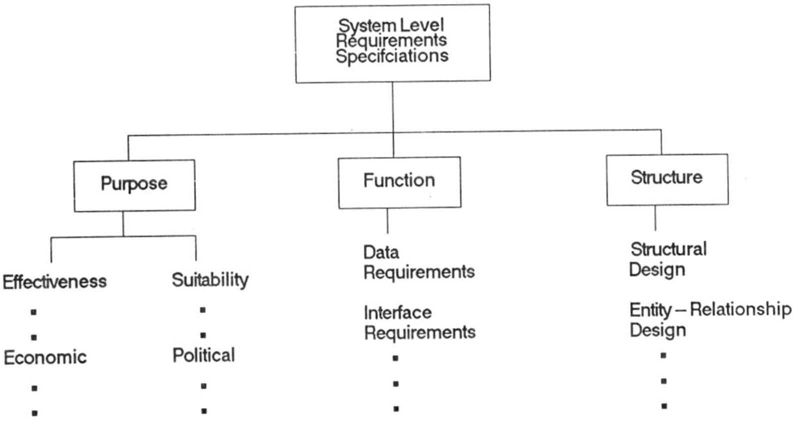

Figure 1. System Level Requirements Taxonomy.

mission under consideration. The three components for this are purposeful, functional, and structural. The relationship between these elements of system level requirements is shown in Figure 1.

We will probe system level characteristics of C^2 requirements through examination of the most commonly used set of terms applied to describe performance goals. These are system level requirements that state some of the *major purposes* of the system which relate to performance characteristics such as:

- *quality factors* such as reliability, maintainability, usability, interoperability, and portability;
- *interfaces* between the system and its users that effect the way in which the system interacts with the intended users;
- *performance characteristics* that affect factors such as response time, workload, information throughput, user productivity, and required storage space;
- *operating constraints* that relate to the external environment such as size, weight, power needs, cooling needs, personnel training and availability, and spatial configuration;
- *economic constraints* that are concerned with both immediate and long-term costs of the total program; and
- *political constraints* that deal with political realities and legal issues and concerns.

Included within these purposeful system level requirements specifications is a set of quality factors that are generally part of any system level requirements document. These quality factors relate to the purposes of the system and are what in the longer term makes the system usable. They are as follows:

- *Efficiency*: the amount of resource required to perform a function
- *Flexibility:* the effort required to modify operational systems and equipment

- *Integrity*: the extent to which access by unauthorized persons can be controlled
- *Interoperability*: the effort required to couple one system with another
- *Maintainability:* the effort required to locate and fix software errors
- *Portability*: the effort required to transfer from one hardware configuration or system environment to another
- *Reliability:* the extent to which the system can be expected to perform its intended function with required precision
- *Responsiveness:* the extent to which the system provides timely results
- *Reusability:* the extent to which system components and/or design can be used in other applications
- *Testability:* the effort required to test the system to insure that it performs its intended function
- *Usability:* the effort required to learn, operate, prepare input, and interpret output

Quality factors represent a much used subset of purposeful requirements. Quality factors are used in most large scale system acquisition projects to describe features desired/required in the product. We have divided these quality factors into two separate groups:

- *Effectiveness.* Those quality factors that relate to how proficient the system product is in accomplishing the assigned tasks.
- *Suitability.* Those quality factors that relate to how appropriate the system product is to perform the assigned tasks.

System effectiveness is concerned with system capability, system security, overall survivability, and adaptability to changing situations. System suitability is concerned with reliability, testability, usability, portability, maintainability, and flexibility. These quality factors and their relationships are shown in Figure 2.

While admirable system characteristics, these factors generally defy precise quantitative measurement, and consequently, make the development of requirements a formidable task. For example, some of the metrics used to determine the adequacy of these characteristics include: completeness, operability, decomposability, lack of redundancy, and size of the system. Another example relates to metric determination for system implementation. Implementation is dependent upon cost, risk, technical factors, and schedule. These are dependent upon factors such as the development life cycle, cost of design trade-offs, manning implications, and risk related to technical factors, cost factors, and schedule factors. As we can see, none of these metrics provides for precise quantitative observation.

In summary, the requirements elicitation process for C^2 system requirements is a difficult one. It is characterized by many unknowns and a high degree of risk for the overall system need. In the next sections we will look at a typical set of requirements statements for a C^2 system proposal and then follow this by an examination of a Requirements Engineering Process Model. This model should enable user and developer to come together to help resolve purposeful

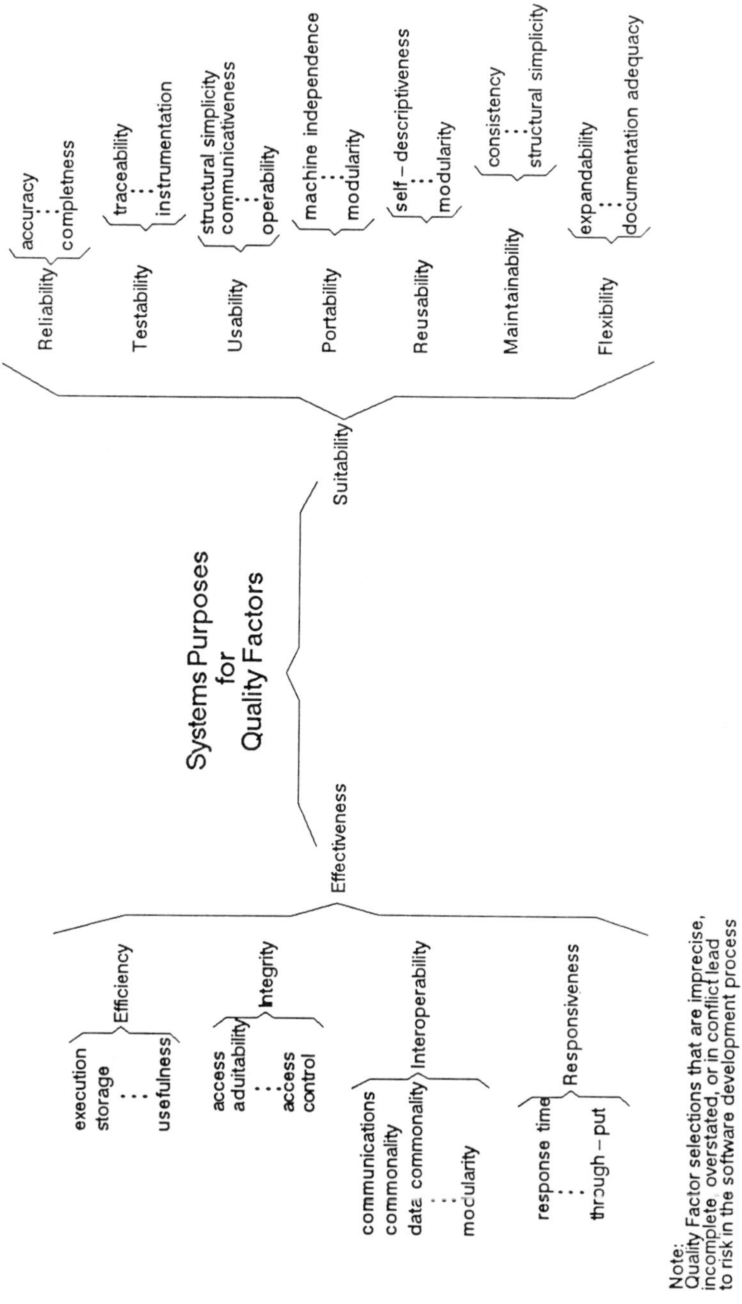

Figure 2. Quality Factors for System Level Requirements Specifications.

issues related to risk management and volatility reduction early in a development program.

An Example of System Level Requirements Statements for a C² System

We have noted that C² systems are intended to function in a myriad of ways. They are intended to track and report on activities that include:

- location of units in the field;
- coping with an overabundance of signals;
- being able to distinguish valid from specious signals; and
- doing all of this while surviving the severity of battle.

In addition, the system must provide support for human decision making within this dynamic confusing matrix of alternatives.

These functions and activities must be made explicit by the user and the developer during the requirements definition phase of the development. If these tasks are not accomplished at this time, they will be made explicit by the developer during the design phase.

To see the nature and form of the situation that must be handled during the system conceptualization phase of development, it is useful to consider a typical scenario. We will look at a hypothetical situation of being asked to provide system level requirements specifications to meet a particular threat. This threat is characterized by problems that have not been solved by the previous system; thus, a new system, rather than an upgrade, is required.

A program manager is given the task of preparing a concept paper for the new system and defining the system level requirements. The general statement of the requirement is as follows:

> The design is to be for a *system level* defense against a ballistic missile threat for theater level operations. The system must be *effective* against military targets in a conventional war with *minimal* leakage and have *some capability* for *general* targets such as *protection* of cities or military installations. The system will have a *standard* set of objectives and will be *user friendly*. The *various* databases, analytic functions, and graphs will *depend upon preset values* as determined by the particular theater operations guide. The system will have a *short response time* for a given scenario so that the user will not have *long wait states* for *complex* scenarios. *Sensitivity analysis* will be such that the user will be able to turn knobs to view *instant* scenes of the effect on performance on a display screen for any parameter *determined to be of interest*. The system will consist of four principal subsystems: sensors, weapons, communications, and command and control.

From this we see that the basic elements of a typical command and control decision support system includes requirements for several systems and subsystems. These systems include the following major subsystems:

- sensors;
- navigation systems;
- communications systems;
- facilities;
- equipment; and
- personnel procedures for planning, directing and controlling operations relevant to the missions assigned [JCS Pub 1-02].

These systems are expected to perform detection, classification, localization of the activity and provide prosecution information for decision making. The overall situation is one that at best may be considered fluid and system requirements must reflect this flexibility for the decision making process.

From our example, we are able to see that risk management and volatility reduction are highly desirable. These factors are manifested through the impact of ambiguity, imprecision, untestability, infeasibility, and conflict in system level requirements. Certain stipulations represent risk to the system design and volatility in requirements specifications. They are terms for which specific agreement must be reached in order to affect the functional and structural design of the system. They are as follows: system level; effective; minimal; some capability; general; protection; standard; user friendly; various; depend upon preset values; short response time; long wait states; complex; sensitivity analysis; instant; and interest.

We see that the apparently very simple guidance statement that has cited purposes of this system incorporates some sixteen words, terms, or phrases that are not well-defined, and therefore introduce risk and volatility.

These are terms that are typically utilized to describe the system level purposes. When they are not given specific quantitative values or ranges by the user (or the user in conjunction with the developer), the system developer will surely select the values or ranges. These selections will be those that fit within the experience of the developer's design expertise, rather than those that the user may have intended.

Based on these characteristics of C^2 system operating conditions, it is of little wonder that the determination of requirements is a difficult set of tasks. At the onset of requirements definition, we are faced with multiple unknowns, an uncertain environment and time constraints that make the likelihood of ferreting out the necessary and sufficient responses to these complexities nearly impossible. C^2 system requirements must take into account the nature of the threat situation, the deployment of forces at any given instant, commitments that have been previously made, environmental conditions through which deployment and support must be made and the limitations of our ability to respond in a particular situation.

The next section presents a requirements engineering approach that is intended to deal with management of risk and reduction of volatility seen to be present in requirements concepts. The goal of this approach is to provide a means of reducing volatility during the requirements conceptualization process and managing risk in accordance with the desires of the user, not the developer.

A Requirements Engineering Approach

The single most important activity in the design and development of a C^2 system is the preparation of an accurate set of requirements that is at once correct and will satisfy the needs of the user. As we have seen in the section on system characteristics, no other set of factors impacts C^2 system utility and validation more than the absence of correct and accurate requirements. Quality, usability, and ultimate fieldability of the product are enhanced when requirements meet user needs.

From need conceptualization through requirements definition to a complete set of requirements specifications to transition to the design process, it is necessary for the user and the developer to work together to reduce volatility and manage risk, while retaining a true sense of the C^2 system needs. Volatility and risk in requirements engineering may be attributed in part to incorrect or imprecise statements, conflict among selected criteria, incomplete needs statements, or over-specification of constraints.

The goal of the requirements engineering process is to formulate accurate and correct statements of user needs that convey a high level of understanding of the system product necessary to achieve a design that produces the desired outcome. These requirements must realistically be based upon both the objectives of the user group as well as constraints and alterables that affect the solution that can be synthesized. For this to occur it is important that all constituents participate in the requirements process in a meaningful and appropriate way. It is essential that user needs be embodied in the requirements, as well as those related to design preferences fashioned by the design team, as well as any constraints that apply to the application domain. In addition, requirements are expected to state constraints related to such factors as performance, reliability, availability, maintainability, safety, cost, and schedule. Alterables that impinge upon the solution; such as, choice of equipment or operating system, system level environmental concerns, or design preferences should come to light during this process. Requirements that accurately reflect the needs of the user, the capability of the designers and possess minimal levels of volatility and conflict will enhance the quality of the end product and increase productivity.

The requirements engineering activities, that we propose to assist the user and the system developer in the management of risk and reduction of volatility, provides a road map to follow in the conceptualization process. A matrix of these activities is shown in Figure 3. These activities are typically interactive and iterative. They include the range of actions that take place as requirements are identi-

Activity \ Function	Acquisition	Development	Documentation
1. make initial contact	x		
2. prepare system parameters	x	x	
3. review ASIS* system *current system, if any	x	x	
4. examine constraints		x	x
5. develop new requirements		x	x
6. review performance characteristics		x	x
7. propose evaluation methods	x	x	x
8. compare alternatives	x	x	x
9. produce system requirements		x	x

Figure 3. User and Developer Group Activities Matrix for Systems Level Requirements Specifications.

fied and a system level requirements specification document is prepared.

These activities—acquisition, development, and documentation of systems level requirements specifications—follow directly from systems engineering approaches and are described in greater detail as follows:

Formulate system level requirements issues:

- Identify needs, constraints and alterables, such as budget and any operational and legal or political requirements, through initial contact with users. Developers responsible for systems level requirements specifications interact with users to determine what the requirements should be, how the requirements should be stated, and how the requirements should be derived.
- The parameters of the system to be developed are derived from the interactions begun in step 1. The purpose of this activity is to identify objectives of user groups, and the ways to determine how system level objectives can be met.

This activity is applicable to the acquisition and development of system level requirements specifications.

- Identify alternatives to the existing or "ASIS" system, if one exists. In this step we examine the existing system, if any, in terms of requirements specifications options. The purpose of this step is to determine if any of the current system level specifications, and by inference the existing system, may be reused in the current application. We also look at alternative ways to determine the extent to which the requirements specifications are achieved. This step involves the acquisition and development of system level requirements specifications.

Analyze Issues and Select Approach(es)

- Once we have completed steps 1 through 3 above, we begin the process of reviewing and analyzing any constraints or issues related to system level requirements specifications. We then select an approach(es) to meet the needs stated in the system level requirements specifications. We identify the technical, operational, and economic aspects of the new (or "ASIS") system level requirements options, including preliminary estimates of limits on system level performance characteristics such as timing, size, errors and other operational features. In this step we are engaged in development of acquired requirements and documenting our outcomes.
- Next we develop new system level requirements from the information prepared and documented in step 4. We define the criteria for evaluation and assessment of these new system level requirements specifications. This activity involves continuing development and documentation of system level requirements specifications.

Interpretation

- At this point in our development of system level requirements specifications, we review performance requirements stated in these documents. We accomplish this by evaluating the system level requirements specifications in terms of user objectives. This usually involves the continuing development and documentation of system level requirements specifications.
- Next we propose methods of evaluating the requirements specifications to determine if the system product as delivered will meet the system level requirements specifications. A validation plan for requirements is necessary to assure that we have established feasible and testable requirements. In this step we are involved with additional acquisition activities, development, and documentation of system level requirements specifications.
- The next step is to rank order the system level requirements specifications in terms of their costs and benefits or effectiveness, and select the set of requirements specifications that best satisfy user needs within allocated resources. Alternative

approaches should be reviewed if there are any questions as to the system level requirements meeting user needs. This step involves interaction with the user to assure that the requirements so established meet user needs. Development and documentation of system level requirements specifications continues.

- Identify the final system level requirements and prepare system level requirements specifications. This final step may involve additional resolution of any remaining problems and issues. The entire system level requirements specifications are documented and form the input for system design and development.

The basic goal of these activities is to explore overall user needs for the development of the system product in terms of the technical, strategic, economic and other aspects of the system to be designed. The entire set of purposes of the system must be considered during this process.

It is important to establish as many of the constraints and system alterables as is feasible. For example, it may be essential that the system be able to be easily modified over the life of the product as the user is aware of substantial needs or changes that are likely to occur, but have not been conceptualized. The early establishment of constraints and alterables permits the best possible set of system level requirements specifications to be identified. The system level requirements selected during these activities are the ones that are placed in the systems level requirements specification document and subjected to analysis later on during this phase to determine the system level requirements specifications.

What is required is a tool or set of tools that assist the user in resolving system design goal issues during the requirements specifications phase of development. Tools that provide visibility into these dynamically changing environments are required. A comprehensive set of tools to assist the user in resolving or managing all of these issues is beyond our capability at this time.

At this point of time, the best we do is apply a set of tools that will track the form of the requirements and make certain that we have complete and consistent statements. Commercial tools are not available that will assist in the analysis and review of the substance of requirements. If we follow the nine steps that we have outlined in this section, we will go a long way toward risk management and volatility reduction in the requirements specifications process.

References

Boyes, J. L., and Andriole, S. J. (eds.), *Principles of Command and Control*, AFCEA International Press, Washington, D.C., 1987.

Kroening, D. W., "Army Command and Control Information Systems Requirements," in Boyes, J. L., and Andriole, S. J. (eds.), *Principles of Command and Control*, AFCEA International Press, Washington, D.C., pp 84-94, 1987.

Palmer, J. D., "Impact of Requirements Uncertainty on Software Productivity," *Proceedings, Twenty-seventh Annual Technical Symposium*, Washington D.C. Chapter Association for Computing Machinery, June 1988, pp. 15-85.

Flexible Life Cycling for Multidisciplinary C² Information Systems Engineering

Stephen J. Andriole

One of the several working premises of this paper is that the nature of the C^2 information systems engineering challenge is changing dramatically and fundamentally. A new era of "analytical computing" is upon us; the 1990s will see this segment of the market rise by several orders of magnitude. What does this really mean? Above all, it means that users will depend upon computers for more and different things than they have in the past. They will demand analytical support to complex problems; they will expect computers to execute data-intensive functions on their own. They will insist upon analytical freedom. They will not tolerate operator/system task misallocation, and they will assume that analytical algorithms will help them plan, assess, decide, monitor and allocate, not merely calculate, sort and display. The coming era of analytical computing also requires our design methodologies to evolve, to keep pace with changing user needs and expectations. The way we elicit, model and test requirements will change, just as the role of software—as we know it today—certainly will change.

This chapter proposes yet another life cycle (Davis, et al., 1988; Pressman, 1987; Royce, 1970). It was designed to be as modern and as comprehensive as possible. As you will see, it is really several life cycles, many life cycles, as many as you want. An attempt was made to design a life cycle that is flexible enough to accommodate a variety of multidisciplinary information systems engineering (MISE) challenges, yet specific enough to be measurable. It is not a standard, nor does it aspire to become one. It will need to be modified and/or replaced at regular intervals, because we are changing, the nature of the MISE world is expanding, and our methods are maturing.

This chapter also suggests how to organize, staff and manage the MISE project, given the proposed life cycle(s). Over the years, I have come to the conclusion that a large percentage of our problems can be traced directly to the management of individual projects and organizations. Section C of this chapter presents some ideas for project management.

A. A Flexible Multidisciplinary C^2 Information Systems Engineering Life Cycle

Figure 1 presents the new life cycle. It is "different" from "normal" ones because it not only suggests the steps but also the methods and tools that can be used to implement the steps. It also suggests what the "output" along the way should look like.

It is long, primarily because of the detail, but also because a number of "new" steps have been added. Its most notable aspect is its explicitness regarding how to perform the steps that comprise the life cycle. It is designed to be prescriptive and proactive; it is also designed to be flexible (see Section B below).

The life cycle is also steeped in iteration and prototyping. Each step in the life cycle is a mini-life cycle unto itself. The implication is clear: If you can't do it right the first time, then do it until you get it right, *and* you can't move on to the next phase of the cycle until you get it right . . .

The life cycle should be regarded as a living model. New steps may be added, new methods can be inserted, and more stringent requirements for advance might be specified. Over time we will learn from our design and development experiences and feed the important lessons back into the next iteration of the living life cycle. In other words, the life cycling business itself should be regarded as iterative.

The life cycle in Figure 1 has 18 specific, highly interrelated, steps, though the number of steps may grow if the prototyping and requirements conferencing steps get tedious. Each of these steps in turn requires that certain tasks and subtasks be performed. Each step has a gate at the end of its implementation process that determines if the next step should be taken.

The first step calls for the staffing of the requirements team. The range of expertise—given a particular problem—may be great; the expertise options are many and varied. Select the ones you need carefully.

Step two requires architects to acquire the necessary amount of domain expertise. Studies suggest (Curtis, et al., 1988) that high levels of domain expertise predict success. Systems and requirements analysts that do not have or seek domain expertise are asking for trouble. How can you fix a leaking faucet if you don't know what a washer is?

The third step calls for decisions about the elicitation of *user* requirements.

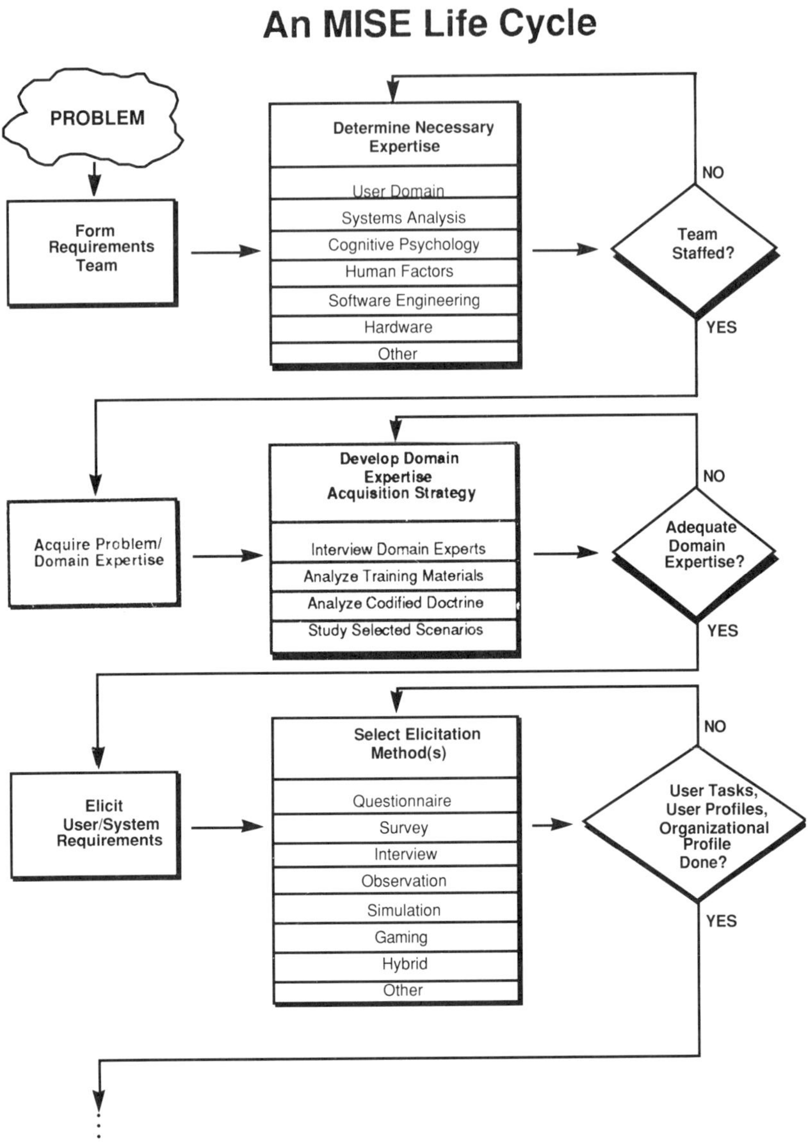
An MISE Life Cycle
PROBLEM
Form Requirements Team
Determine Necessary Expertise
User Domain
Systems Analysis
Cognitive Psychology
Human Factors
Software Engineering
Hardware
Other
Team Staffed?
NO
YES
Acquire Problem/ Domain Expertise
Develop Domain Expertise Acquisition Strategy
Interview Domain Experts
Analyze Training Materials
Analyze Codified Doctrine
Study Selected Scenarios
Adequate Domain Expertise?
NO
YES
Elicit User/System Requirements
Select Elicitation Method(s)
Questionnaire
Survey
Interview
Observation
Simulation
Gaming
Hybrid
Other
User Tasks, User Profiles, Organizational Profile Done?
NO
YES

Figure 1a.

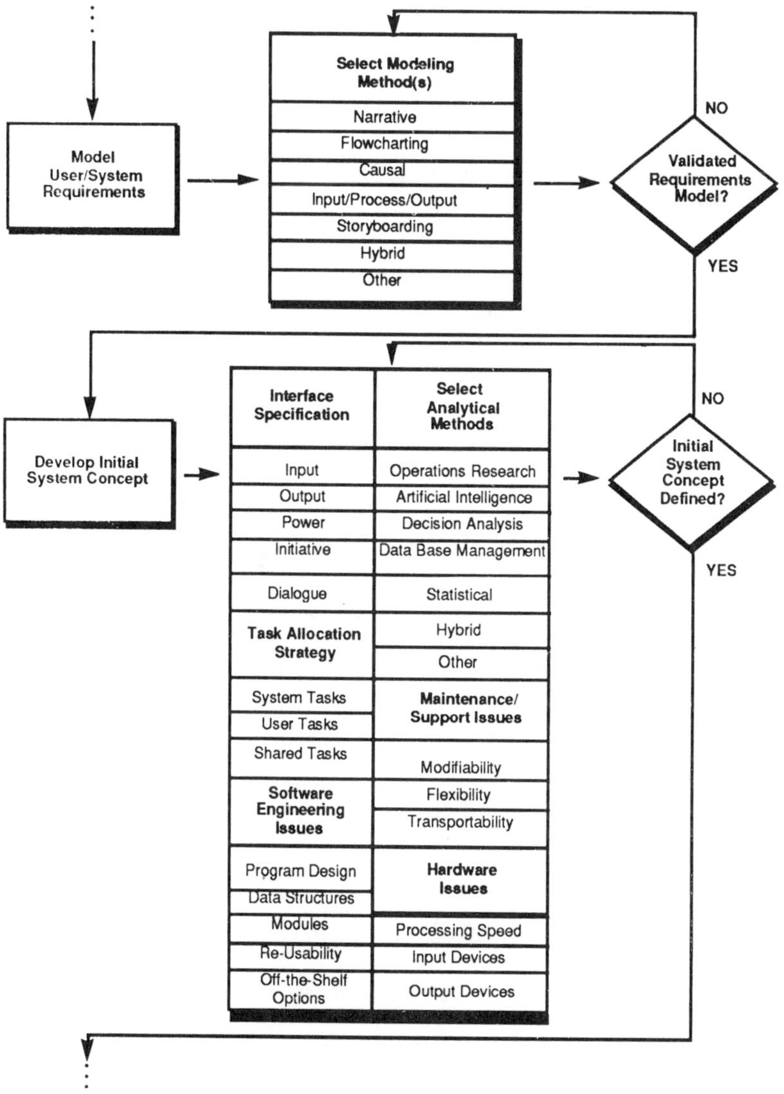
Model
User/System
Requirements
Select Modeling
Method(s)
Narrative
Flowcharting
Causal
Input/Process/Output
Storyboarding
Hybrid
Other
Validated
Requirements
Model?
NO
YES
Develop Initial
System Concept
Interface
Specification
Input
Output
Power
Initiative
Dialogue
Task Allocation
Strategy
System Tasks
User Tasks
Shared Tasks
Software
Engineering
Issues
Program Design
Data Structures
Modules
Re-Usability
Off-the-Shelf
Options
Select
Analytical
Methods
Operations Research
Artificial Intelligence
Decision Analysis
Data Base Management
Statistical
Hybrid
Other
Maintenance/
Support Issues
Modifiability
Flexibility
Transportability
Hardware
Issues
Processing Speed
Input Devices
Output Devices
Initial
System
Concept
Defined?
NO
YES

Figure 1b.

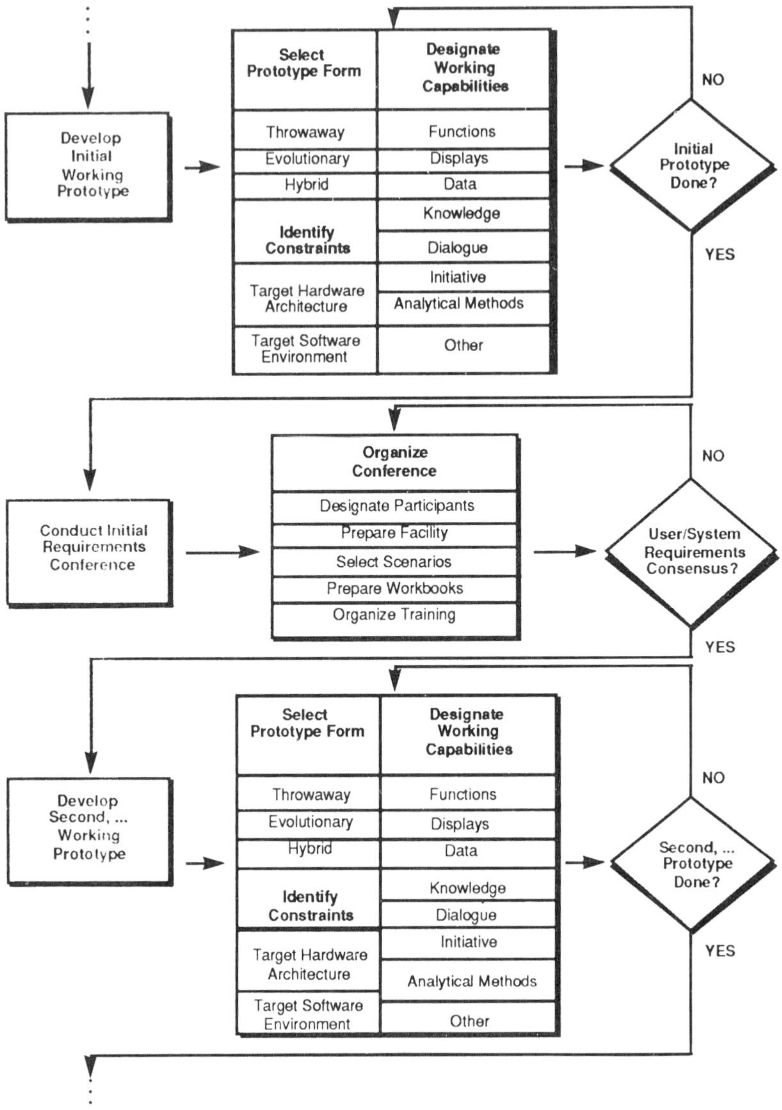
Develop Initial Working Prototype
Select Prototype Form
Designate Working Capabilities
Throwaway
Evolutionary
Hybrid
Identify Constraints
Target Hardware Architecture
Target Software Environment
Functions
Displays
Data
Knowledge
Dialogue
Initiative
Analytical Methods
Other
Initial Prototype Done?
NO
YES
Conduct Initial Requirements Conference
Organize Conference
Designate Participants
Prepare Facility
Select Scenarios
Prepare Workbooks
Organize Training
User/System Requirements Consensus?
NO
YES
Develop Second, ... Working Prototype
Select Prototype Form
Designate Working Capabilities
Throwaway
Evolutionary
Hybrid
Identify Constraints
Target Hardware Architecture
Target Software Environment
Functions
Displays
Data
Knowledge
Dialogue
Initiative
Analytical Methods
Other
Second, ... Prototype Done?
NO
YES

Figure 1c.

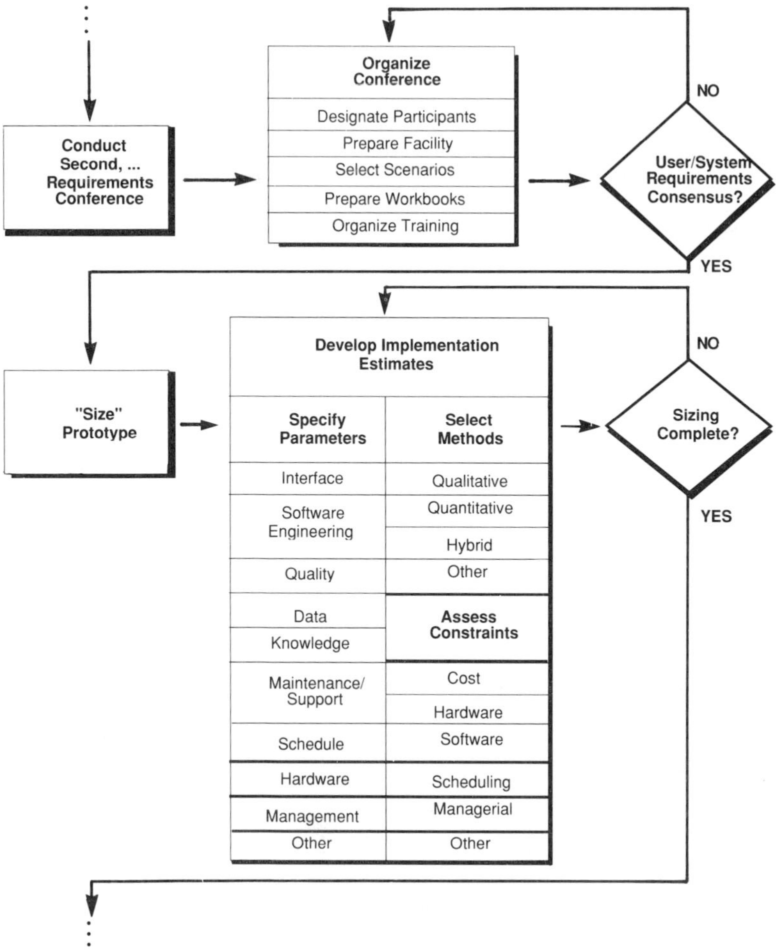
Conduct
Second, ...
Requirements
Conference
Organize
Conference
Designate Participants
Prepare Facility
Select Scenarios
Prepare Workbooks
Organize Training
User/System
Requirements
Consensus?
NO
YES
"Size"
Prototype
Develop Implementation
Estimates
Specify
Parameters
Select
Methods
Interface
Software
Engineering
Quality
Data
Knowledge
Maintenance/
Support
Schedule
Hardware
Management
Other
Qualitative
Quantitative
Hybrid
Other
Assess
Constraints
Cost
Hardware
Software
Scheduling
Managerial
Other
Sizing
Complete?
NO
YES

Figure 1d.

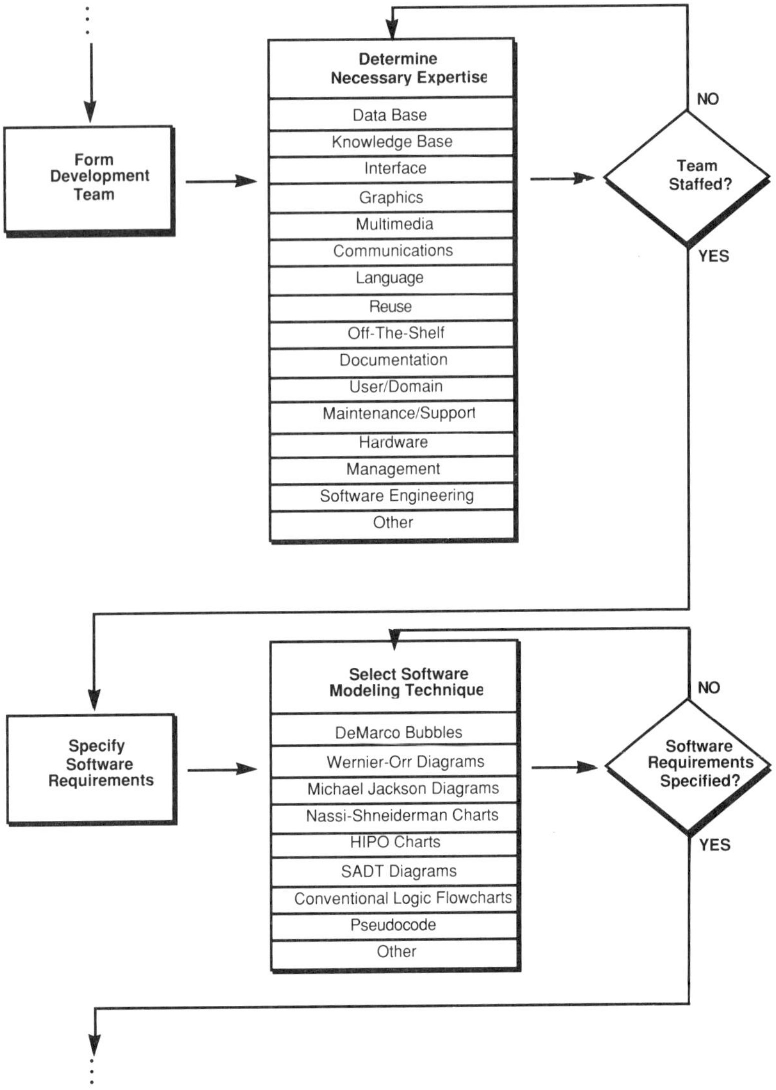
Form Development Team
Determine Necessary Expertise
Data Base
Knowledge Base
Interface
Graphics
Multimedia
Communications
Language
Reuse
Off-The-Shelf
Documentation
User/Domain
Maintenance/Support
Hardware
Management
Software Engineering
Other
Team Staffed?
NO
YES
Specify Software Requirements
Select Software Modeling Technique
DeMarco Bubbles
Wernier-Orr Diagrams
Michael Jackson Diagrams
Nassi-Shneiderman Charts
HIPO Charts
SADT Diagrams
Conventional Logic Flowcharts
Pseudocode
Other
Software Requirements Specified?
NO
YES

Figure 1e.

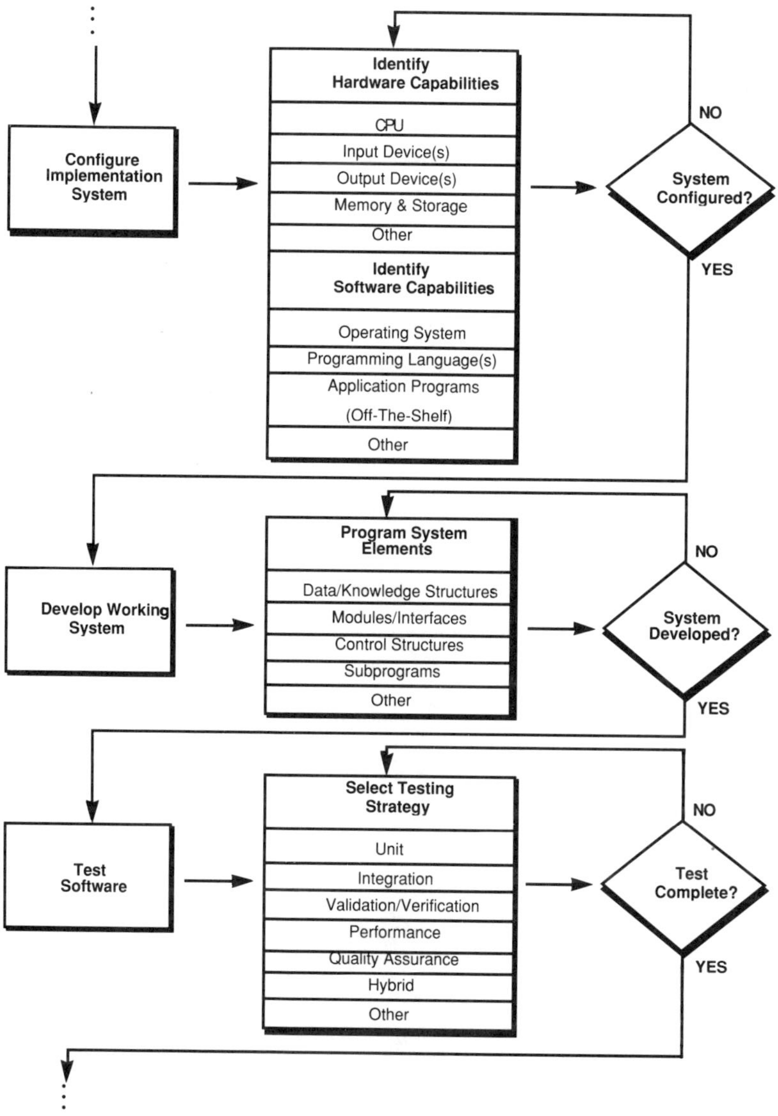
Configure Implementation System
Identify Hardware Capabilities
CPU
Input Device(s)
Output Device(s)
Memory & Storage
Other
Identify Software Capabilities
Operating System
Programming Language(s)
Application Programs (Off-The-Shelf)
Other
System Configured?
NO
YES
Develop Working System
Program System Elements
Data/Knowledge Structures
Modules/Interfaces
Control Structures
Subprograms
Other
System Developed?
NO
YES
Test Software
Select Testing Strategy
Unit
Integration
Validation/Verification
Performance
Quality Assurance
Hybrid
Other
Test Complete?
NO
YES

Figure 1f.

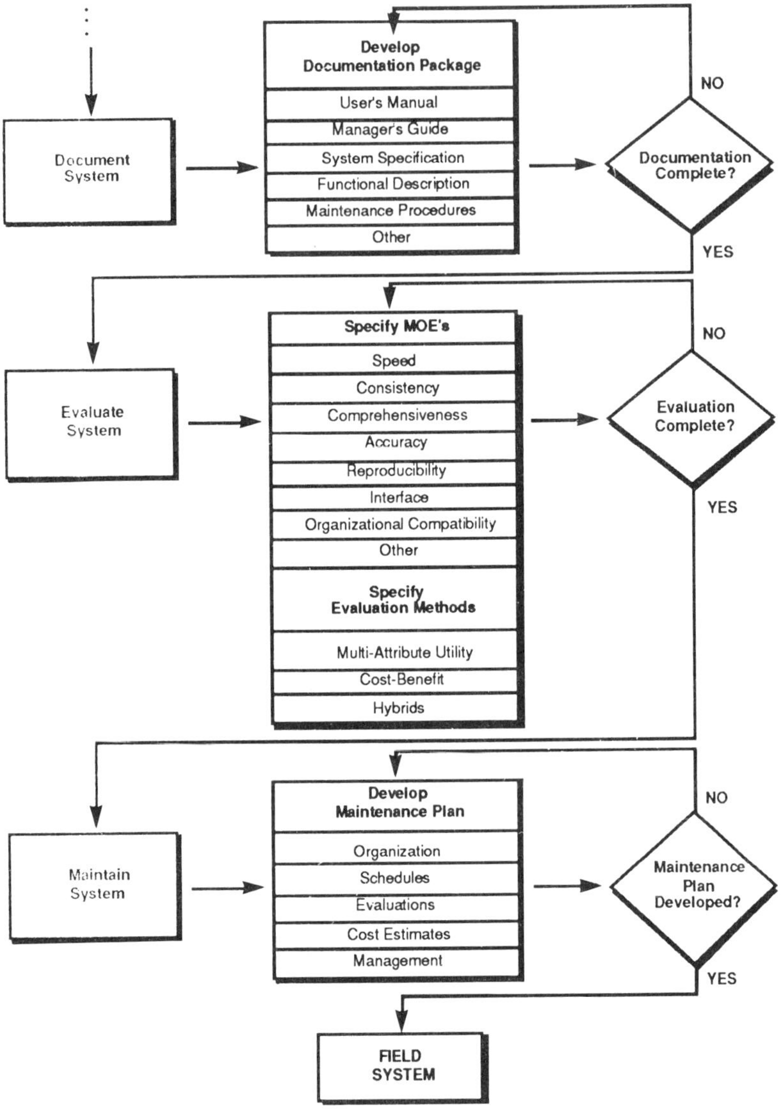
Document System
Develop Documentation Package
User's Manual
Manager's Guide
System Specification
Functional Description
Maintenance Procedures
Other
Documentation Complete?
NO
YES
Evaluate System
Specify MOE's
Speed
Consistency
Comprehensiveness
Accuracy
Reproducibility
Interface
Organizational Compatibility
Other
Specify Evaluation Methods
Multi-Attribute Utility
Cost-Benefit
Hybrids
Evaluation Complete?
NO
YES
Maintain System
Develop Maintenance Plan
Organization
Schedules
Evaluations
Cost Estimates
Management
Maintenance Plan Developed?
NO
YES
FIELD SYSTEM

Figure 1g.

Note that we are still light years away from the specification of software. The methods in this step are oriented to the elicitation of data, information and knowledge from real organic end users. The objective is to identify the tasks they will perform on the system, their particular capabilities and deficiencies, and some insight into the nature of the organization the system is intended to support.

Step four calls for the modeling of user requirements. As the figure suggests, there are a number of available modeling options ranging from simple narrative and flowcharting techniques to IPO and storyboarding methods. The selection of a method (or methods) will depend upon available expertise and—as always—the nature of the requirements to be modeled.

Step five invites creative thinking about what the "system concept" is all about. What kinds of input techniques are most appropriate? How should tasks be allocated? What analytical methods might best support system functions? What are the software engineering implications of alternative system concepts? What are the central hardware issues? What about maintenance? When all of these questions have been answered proceed to step six, which requires you to build your first (throwaway or evolutionary) prototype. By this time, the life cycle feels you should be ready to convert all of the requirements information you gathered into some kind of working model of how the system, when fully evolved, will behave. The first thing you must do is assess just how much you think you know about user requirements. Are you comfortable with your requirements data/knowledge base? Or do you feel that huge gaps still remain in your understanding of what the users really want and need? If you are pretty sure you've got it, then try an evolutionary prototype, but if you're not so sure, then try a cheap fast throwaway. When you build the first prototype, remember the real world. Just because an Apple Mac II can deliver 16 million colors on a single screen, there is no reason to make certain that every screen receives its just due. Just because animation is now easy and a lot of fun, only use it in the system concept if it satisfies some specific requirement(s) and it can be cost-effectively supported in the target hardware/software architecture. If you are one of the dwindling lucky ones who can specify the configuration after your prototype(s) have been approved, then use whatever capabilities that can be supported and make good requirements sense. But even when there is hardware/software selection flexibility, there always will be constraints that must be negotiated.

Step six is fun. Call in the users (that are not already members of the design team), sit them down in some plush high-back conference chairs and play the prototype. Record their comments, listen to their complaints and prepare for the next iteration. It is impossible to spend your time more cost-effectively than in requirements conferences with users. Make sure that conference facilities are first-class and that you have had a dry-run of the conferencing agenda. A large screen display system helps a lot, as do professional quality prototype workbooks.

You may have to go through the prototyping/conferencing cycle a few times before you get it right. The life cycle is flexible here. Continue for as long as necessary, provided you don't undermine the intent of the prototyping process. How do you know when you have prototyped and conferenced enough? When a consensus emerges, when subsequent returns on the prototyping/conferencing investments begin to fall disproportionately, and when your design team and users cease sounding creative, interested and/or coherent.

Step 10 (which might end up as step 11 or 12, depending on how well you do with your prototypes and requirements conferences) makes you take a good hard look at your prototype. If you have gone down the evolutionary path, then your sizing data should be pretty good, but if you have developed a bunch of throw-aways, you had better stop and assess the precise development implications of your sanctioned system concept. Remember, you should *never* develop a prototype that will be impossible to convert to a real system on a constrained architecture. The only greater sin is showing users a prototype with capabilities they can *never* have on the designated hardware system. Sizing is messy and essential. A whole host of functional and developmental parameters must be specified. Constraints should receive very special attention. There are quantitative and qualitative methods available to you; perhaps several should always be used to verify judgments about the prototype-to-system conversion process.

Step 11 suggests that you get rid of most of the members of the requirements/design team and hire some new firefighters. The long list of expertise should scare you. How many times in the past have you sought such multidisciplinary capabilities? When selecting the newer members of the set, consider the (potential) need for specialists in multimedia technology, knowledge base design and maintenance, off-the-shelf applications programs, and conceptual and algorithmic re-use. As always, your team had better have plenty of experience in good ol' fashioned software engineering. It had also better be very well managed.

Believe it or not, the proposed life cycle only now calls for the specification of software. If you have been following the blueprint up to this point, most managers probably would have fired you by now simply because your team has yet to crank out its first thousand lines of code. But if your management is enlightened, and you are still on the job, you will now have to think about how you are actually going to build the system.

There are many software specification methods around—DeMarco, Wernier-Orr, Michael Jackson, SADT, and Nassi-Schneiderman bubbles, charts and diagrams, among others. Note that the proposed life cycle takes some pressure off of these tools. Some life cyclers suggest using complex specification diagrams to communicate system functions and capabilities to users (Boddie, 1987). I don't think most users will understand the conventional flow chart or data flow diagram very well at all. But at this point in the process, you don't need to forge consensus about requirements via such diagrams; instead, they can be used by your development team to organize the implementation effort (which exploits their real strength).

The software specification step (still) does not call for programming. Instead, the life cycle expects you to identify and model what the system must do and then wait for the next critical step—configuration.

This next step makes you think about requisite hardware and software capabilities. It asks a simple question: If you had to build the specified system what kind of hardware and software capabilities would you need? Unfortunately, we are often told what the capabilities are even before we assemble the requirements team. Such is the fate of the modern systems architect—and the state of our sorry procurement system. When hardware and software is pre-specified, work toward optimization within the known constraints—it's all you can do (unless there is some extra money for appending the target system via input or output peripherals). In the never-never land of perfect MISE you will not even think about hardware or software until requirements have been manicured. But in the real world configuration constraints will haunt you throughout the MISE process. As always, optimization is the constraint watchword. Do what you can, when you can.

When pre-specification occurs, one of the things you can do is stay within the capabilities of the designated configuration. It makes sense to learn all you can about the configuration and convert the knowledge into complex filters through which all steps in the life cycle must pass. You should also characterize constraints as "fixed," "negotiable," or "soft," and then deal with them accordingly. You also can prioritize capabilities vis-à-vis fixed constraints, and develop a different ranking where there is more room to maneuver.

The next step—yes, finally!—suggests you develop the working system. Watch your programmers closely, consult regularly with users, monitor the budget and hope for the best.

Step 15 turns to testing. Does the thing work? Do the algorithms do what they are supposed to do? How well? How fast?

The next step often gets neglected in the "real world," usually because conventional life cycling leaves little money or energy left for documentation. But documentation is essential to system transfer. There are a variety of documents that together comprise the documentation "package." A good User's Manual is important if you want users to use your system. User's manuals can be in paper form or embedded as part of the system, or in both forms. One dividend from the prototyping strategy is the data you collect along the way. The system concept has certainly by this time been well described. Screen displays probably exist. These and other materials can be used to develop the User's Manual (and other kinds of documentation).

The next step shifts gears completely, steps back from the whole MISE effort, and organizes a series of tests, experiments and analyses designed to evaluate the system in toto. Does this kluge of hardware, software, displays and keyboards really support users? Does the thing "fit" in the organization? Is it a success? Most of the work we do in the area is targeted at individual system components; the evaluation step requires a macro assessment.

The final step requires a maintenance plan.

Is any of this hard and fixed? No. Like all life cycles it is a compass. It is explicitly multidisciplinary, iterative, and detailed. It is also flexible enough to accommodate a wide range of design challenges. It will change over time, hopefully evolving with the times. New methods will be added, while some old ones deleted. If we institutionalize the post mortem-to-memory process, we can feed good data into its continued refinement.

B. Situational Life Cycling

The proposed MISE life cycle can adapt to many design situations, organizations, and even management styles. It is intended to be flexible, iterative and explicit. This feature is important as we move through the 1990s because it is impossible to predict the precise challenges we will face. It is also impossible to predict where our methods, tools and techniques will be in five or 10 years. All life cycles should be living—and under constant scrutiny. The DOD life cycle (2167A) is reviewed only periodically, and while panels, task forces and working groups continue to assess its strengths and weaknesses, it cannot be modified without the written approval of most of the inhabitants of the free world. Modern life cycling should permit MIS engineers to "invent" new steps, methods and iterative procedures.

Figure 2 illustrates the MISE life cycle's situational flexibility. The shaded areas together describe the MISE problem. Iteration within the steps might well call for additional blocks to be shaded, reflecting how problems should redefine the MISE process. Such flexibility is crucial to our success. The proposed MISE life cycle suggests alternative methods when we fail to negotiate a particular step in the design or development process. As our mix of expertise or methods disintegrates, we need only turn to the model for ideas.

C. MISE Project Planning and Management

So how do we do all this? Our best management friend is the life cycle. It is an organizing device, a staffing agenda and a milestone tracker. The steps in the proposed MISE life cycle constitute a project planning and management strategy.

Explicit in the life cycle is the need for multidisciplinary talent. Staffing should revolve around MISE requirements. A lead MIS engineer should be designated and specific responsibilities should be identified.

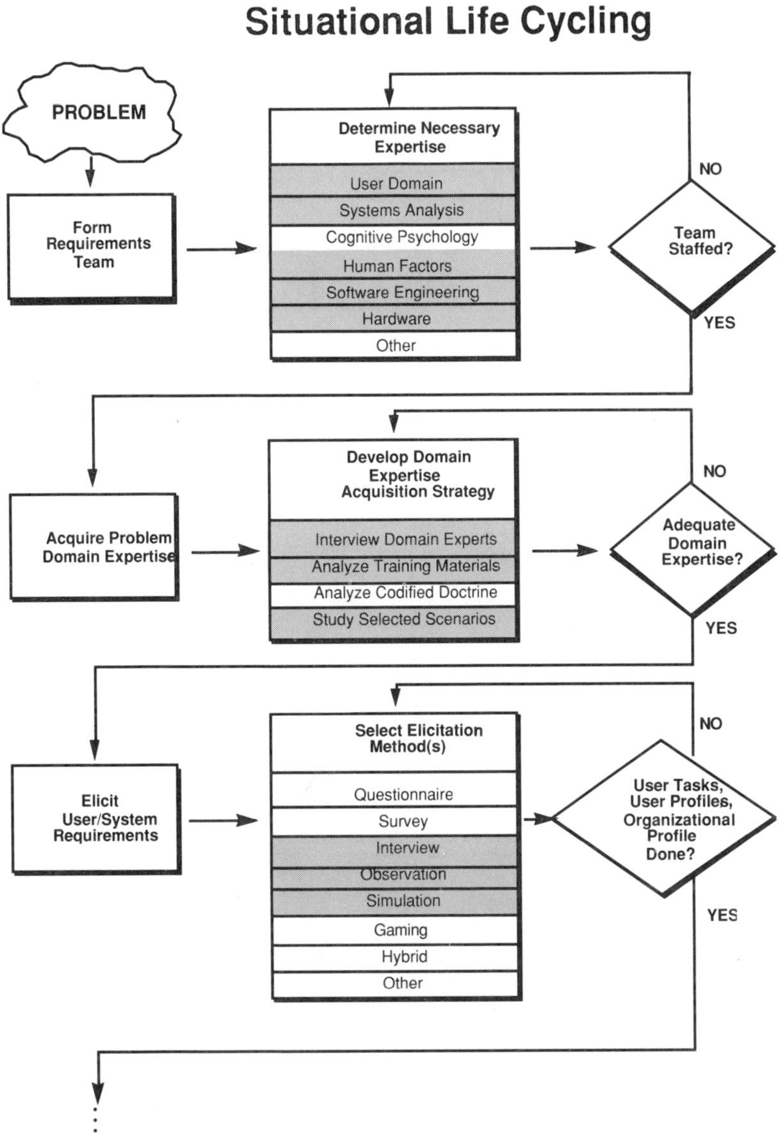
Situational Life Cycling
PROBLEM
Form Requirements Team
Determine Necessary Expertise
User Domain
Systems Analysis
Cognitive Psychology
Human Factors
Software Engineering
Hardware
Other
NO
Team Staffed?
YES
Acquire Problem Domain Expertise
Develop Domain Expertise Acquisition Strategy
Interview Domain Experts
Analyze Training Materials
Analyze Codified Doctrine
Study Selected Scenarios
NO
Adequate Domain Expertise?
YES
Elicit User/System Requirements
Select Elicitation Method(s)
Questionnaire
Survey
Interview
Observation
Simulation
Gaming
Hybrid
Other
NO
User Tasks, User Profiles, Organizational Profile Done?
YES

Figure 2a.

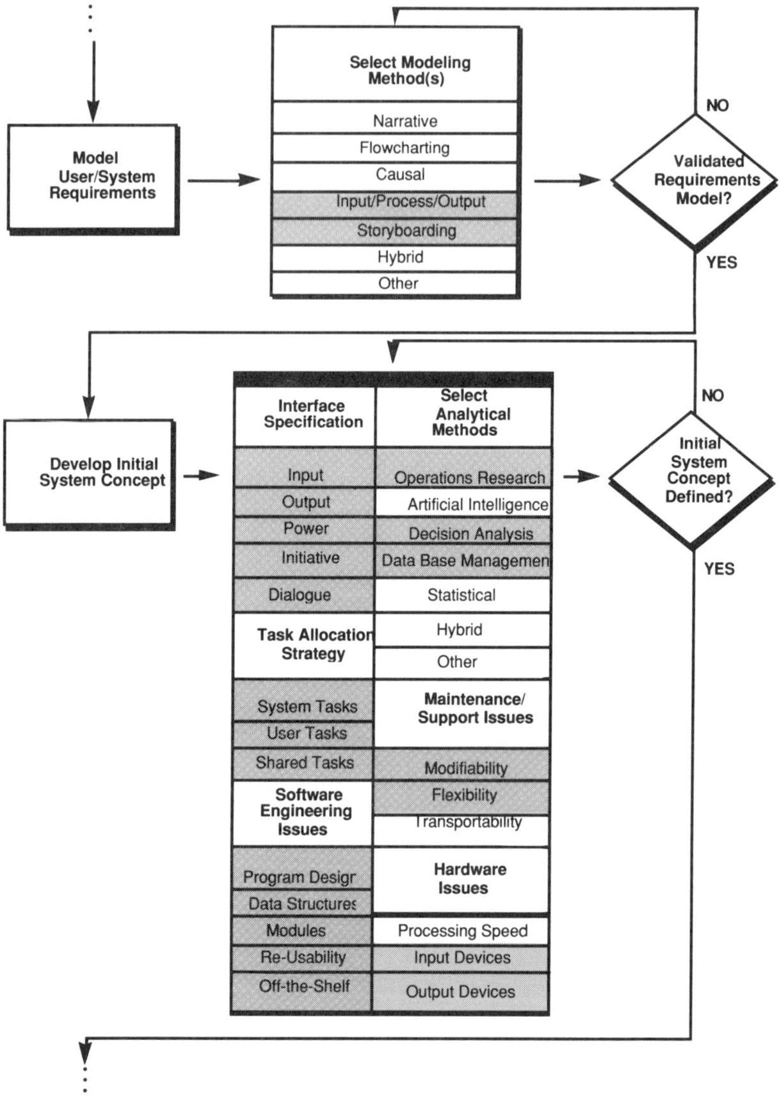

Model User/System Requirements
Select Modeling Method(s)
Narrative
Flowcharting
Causal
Input/Process/Output
Storyboarding
Hybrid
Other
Validated Requirements Model?
NO
YES
Develop Initial System Concept
Interface Specification
Input
Output
Power
Initiative
Dialogue
Task Allocation Strategy
System Tasks
User Tasks
Shared Tasks
Software Engineering Issues
Program Design
Data Structures
Modules
Re-Usability
Off-the-Shelf
Select Analytical Methods
Operations Research
Artificial Intelligence
Decision Analysis
Data Base Management
Statistical
Hybrid
Other
Maintenance/ Support Issues
Modifiability
Flexibility
Transportability
Hardware Issues
Processing Speed
Input Devices
Output Devices
Initial System Concept Defined?
NO
YES

Figure 2b.

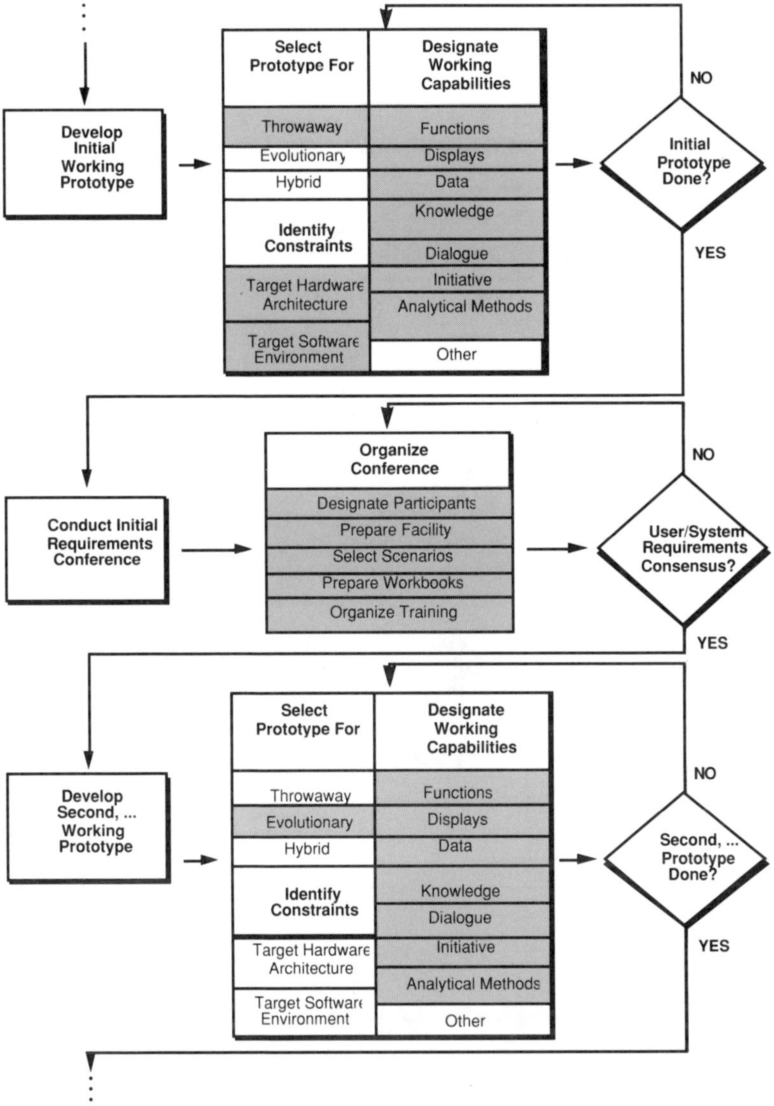
Develop Initial Working Prototype
Select Prototype For
Designate Working Capabilities
Throwaway
Evolutionary
Hybrid
Identify Constraints
Target Hardware Architecture
Target Software Environment
Functions
Displays
Data
Knowledge
Dialogue
Initiative
Analytical Methods
Other
Initial Prototype Done?
NO
YES
Conduct Initial Requirements Conference
Organize Conference
Designate Participants
Prepare Facility
Select Scenarios
Prepare Workbooks
Organize Training
User/System Requirements Consensus?
NO
YES
Develop Second, ... Working Prototype
Select Prototype For
Designate Working Capabilities
Throwaway
Evolutionary
Hybrid
Identify Constraints
Target Hardware Architecture
Target Software Environment
Functions
Displays
Data
Knowledge
Dialogue
Initiative
Analytical Methods
Other
Second, ... Prototype Done?
NO
YES

Figure 2c.

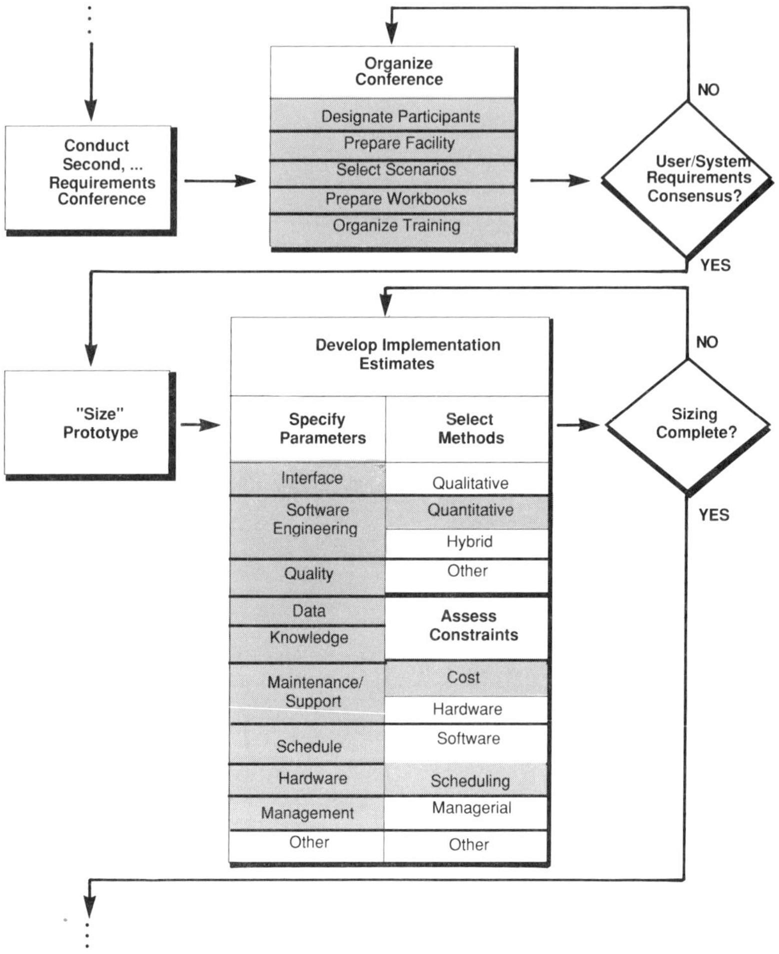
Conduct Second, ... Requirements Conference
Organize Conference
Designate Participants
Prepare Facility
Select Scenarios
Prepare Workbooks
Organize Training
User/System Requirements Consensus?
NO
YES
"Size" Prototype
Develop Implementation Estimates
Specify Parameters
Interface
Software Engineering
Quality
Data
Knowledge
Maintenance/ Support
Schedule
Hardware
Management
Other
Select Methods
Qualitative
Quantitative
Hybrid
Other
Assess Constraints
Cost
Hardware
Software
Scheduling
Managerial
Other
Sizing Complete?
NO
YES

Figure 2d.

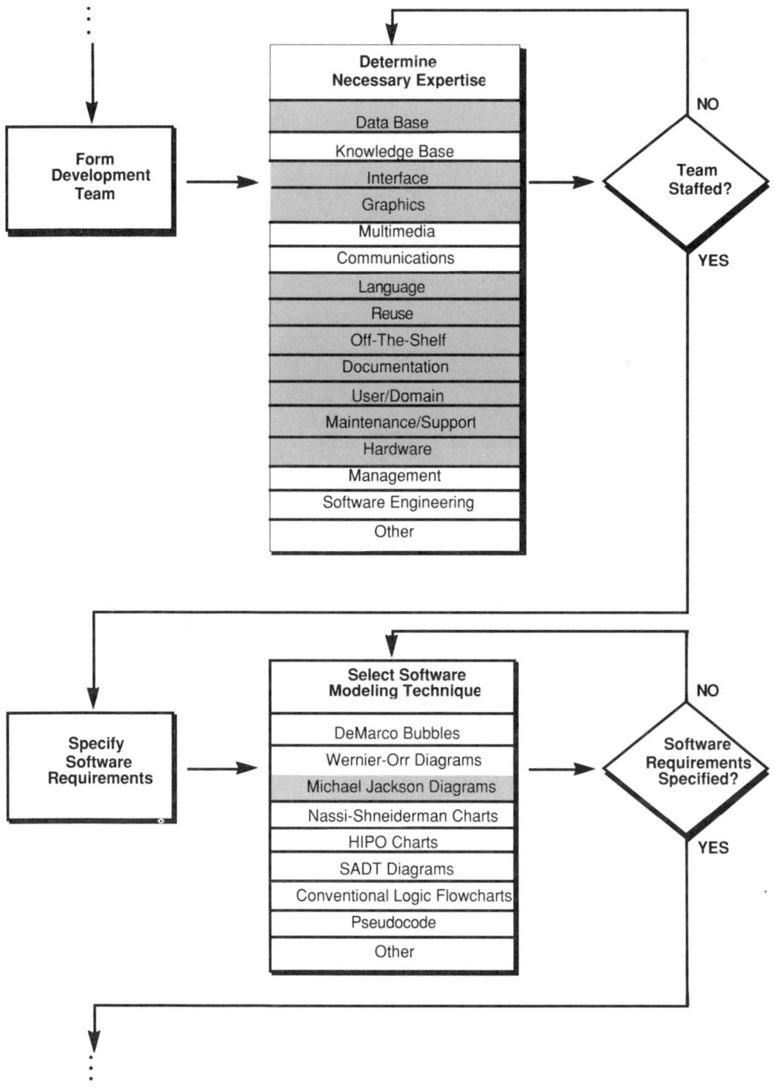

Form Development Team
Determine Necessary Expertise
Data Base
Knowledge Base
Interface
Graphics
Multimedia
Communications
Language
Reuse
Off-The-Shelf
Documentation
User/Domain
Maintenance/Support
Hardware
Management
Software Engineering
Other
Team Staffed?
NO
YES
Specify Software Requirements
Select Software Modeling Technique
DeMarco Bubbles
Wernier-Orr Diagrams
Michael Jackson Diagrams
Nassi-Shneiderman Charts
HIPO Charts
SADT Diagrams
Conventional Logic Flowcharts
Pseudocode
Other
Software Requirements Specified?
NO
YES

Figure 2e.

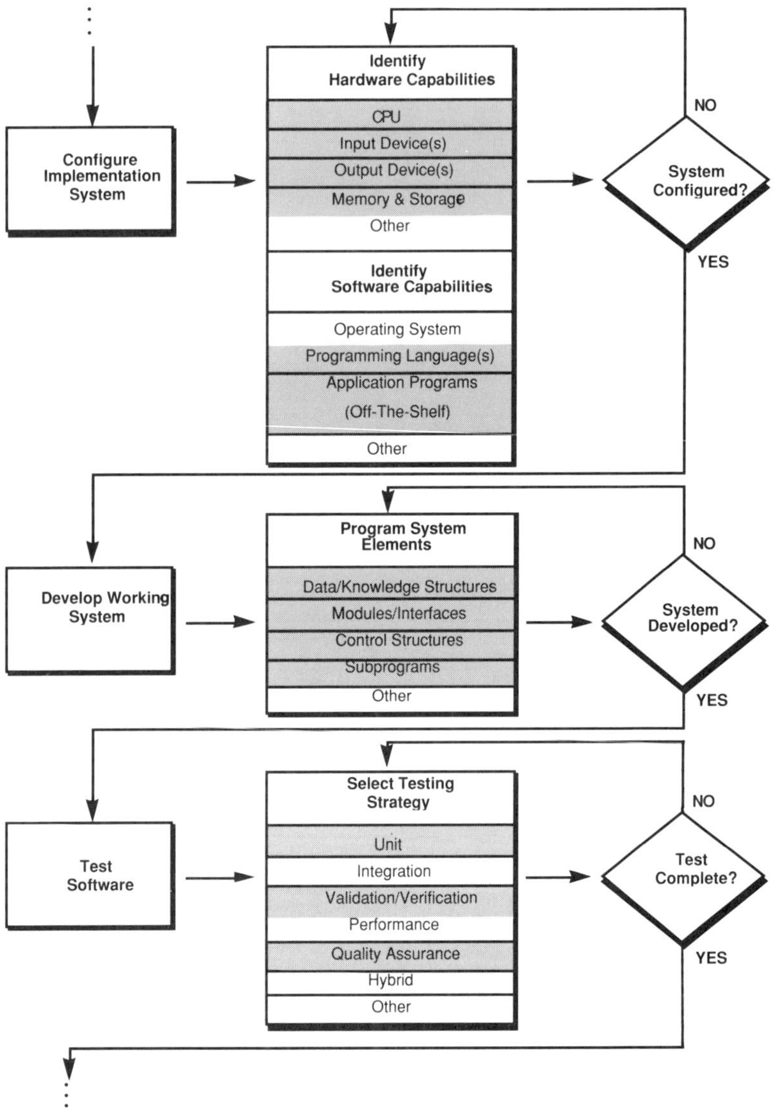
Configure Implementation System
Identify Hardware Capabilities
CPU
Input Device(s)
Output Device(s)
Memory & Storage
Other
Identify Software Capabilities
Operating System
Programming Language(s)
Application Programs (Off-The-Shelf)
Other
System Configured?
NO
YES
Develop Working System
Program System Elements
Data/Knowledge Structures
Modules/Interfaces
Control Structures
Subprograms
Other
System Developed?
NO
YES
Test Software
Select Testing Strategy
Unit
Integration
Validation/Verification
Performance
Quality Assurance
Hybrid
Other
Test Complete?
NO
YES

Figure 2f.

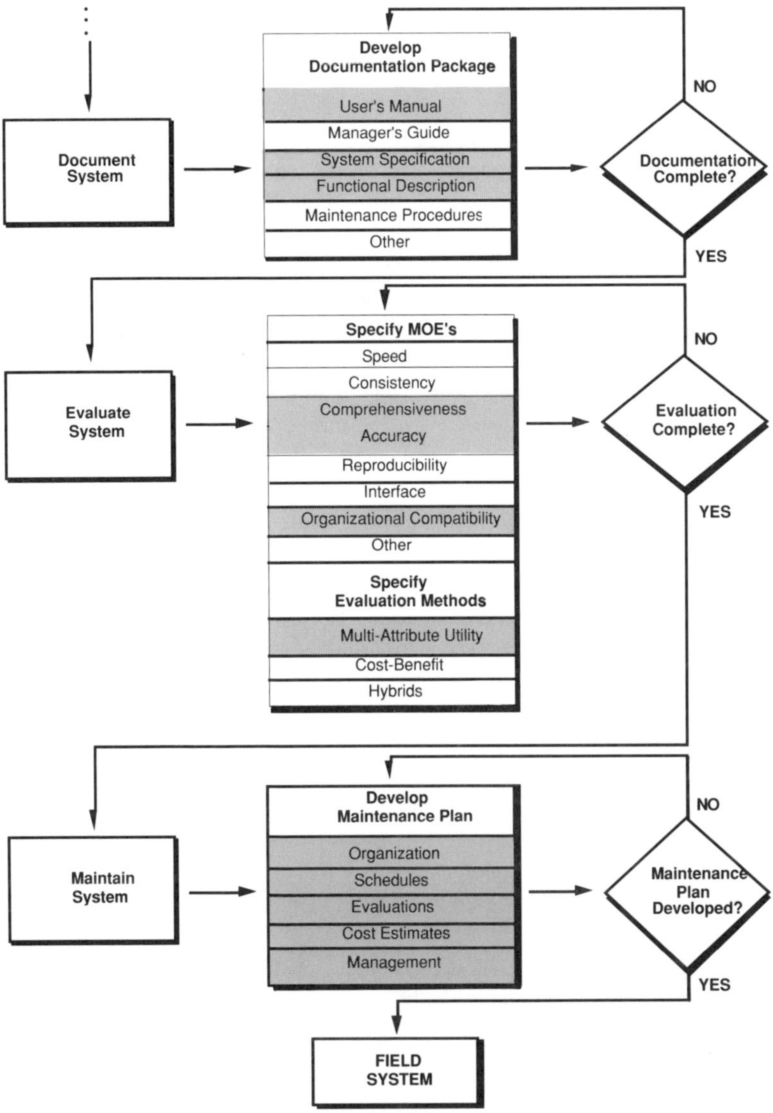
Document System
Develop Documentation Package
User's Manual
Manager's Guide
System Specification
Functional Description
Maintenance Procedures
Other
Documentation Complete?
NO
YES
Evaluate System
Specify MOE's
Speed
Consistency
Comprehensiveness
Accuracy
Reproducibility
Interface
Organizational Compatibility
Other
Specify Evaluation Methods
Multi-Attribute Utility
Cost-Benefit
Hybrids
Evaluation Complete?
NO
YES
Maintain System
Develop Maintenance Plan
Organization
Schedules
Evaluations
Cost Estimates
Management
Maintenance Plan Developed?
NO
YES
FIELD SYSTEM

Figure 2g.

The life cycle calls for at least 18 deliverables. Some of these will be in written form, some incarnated in software and some in multiple forms.

Scheduling is difficult because of the commitments to iteration and prototyping. It is impossible to predict how long it will take to elicit, model and validate user requirements; it is impossible to predict how many throwaway prototypes will be necessary or how quickly an evolutionary prototype will grow. It is impossible to estimate how complex the software will be, or how easy the whole system will be to configure. Most estimates about how long it will take (or how much it will cost) are at best educated guesses about what to expect. Unfortunately, client expectations are anchored in these estimates, estimates that frequently have to be "adjusted" time and again. The procurement process forces contractors to guess. Optimists hope for the best, while those that have been burned before proceed more cautiously (until the competition heats up).

The community provides some qualitative and quantitative estimation methods (Fairley, 1985), but most of them are imprecise and unverifiable. One is tempted to argue that the methods exist exclusively for the convenience of those who must churn out proposals—and those who must read them. At the same time, the real world calls for schedules, milestones and a management team organized around programmed progress. The life cycle permits the identification of task leaders (under the direction of the lead MIS engineer), tasks, deliverables and milestones. Estimates about precisely when each task will be completed are impossible to make. The steps that lead to "system sizing" might take months or years, depending on the scope of the project, the substantive nature of the computing challenge, and the talent of the design team. Always front-load the process with plenty of time, money and people. As suggested above, once a realistic system concept emerges, the conversion process is relatively straightforward. Problems have occurred so often in the past because we deprived the front-end of attention, "discovered" requirements well into development, and then overran all of our production estimates. The allocation of resources should be redressed toward the front-end, toward the identification, modeling and verification of user requirements.

Figure 3 suggests how the life cycle can be used to plan, staff and manage a complex MISE project. Remember, however, that since the life cycle itself is flexible, planning and management strategy and tactics must also be flexible.

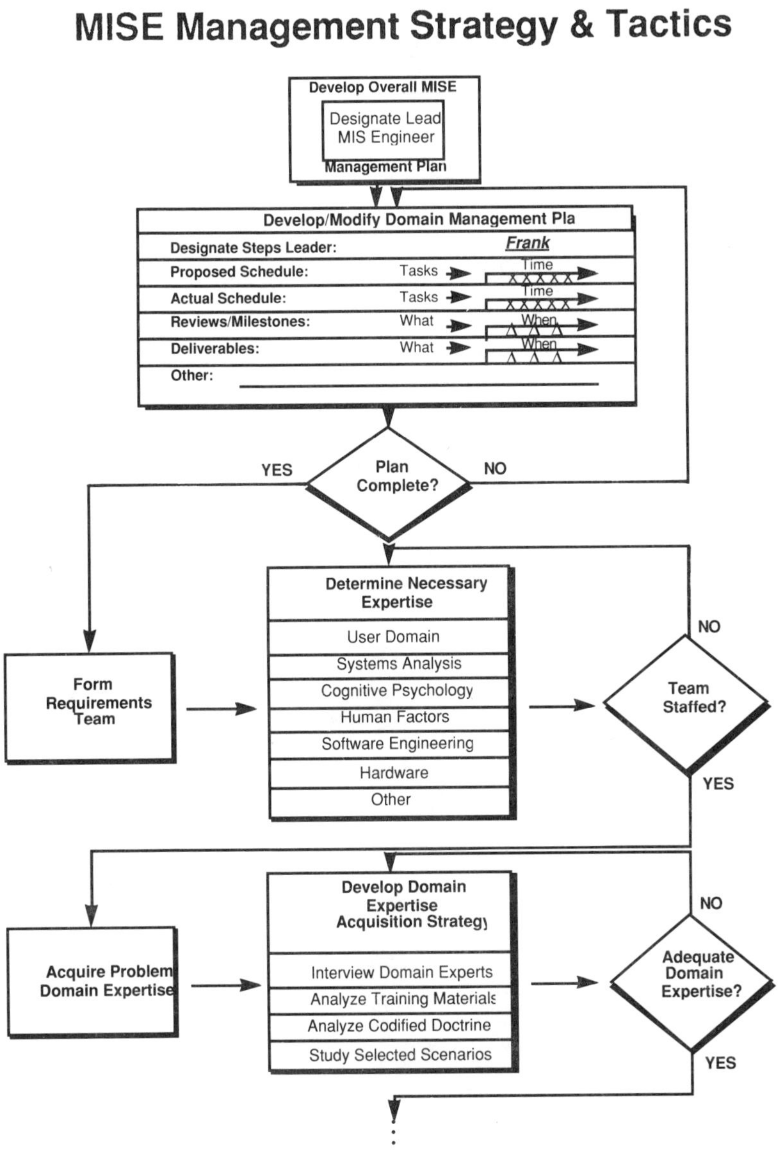
MISE Management Strategy & Tactics
Develop Overall MISE
Designate Lead MIS Engineer
Management Plan
Develop/Modify Domain Management Pla
Designate Steps Leader:
Frank
Proposed Schedule:
Tasks
Time
Actual Schedule:
Tasks
Time
Reviews/Milestones:
What
When
Deliverables:
What
When
Other:
YES
Plan Complete?
NO
Determine Necessary Expertise
User Domain
Systems Analysis
Cognitive Psychology
Human Factors
Software Engineering
Hardware
Other
Form Requirements Team
NO
Team Staffed?
YES
Develop Domain Expertise Acquisition Strateg)
Interview Domain Experts
Analyze Training Materials
Analyze Codified Doctrine
Study Selected Scenarios
Acquire Problem Domain Expertise
NO
Adequate Domain Expertise?
YES

Figure 3a.

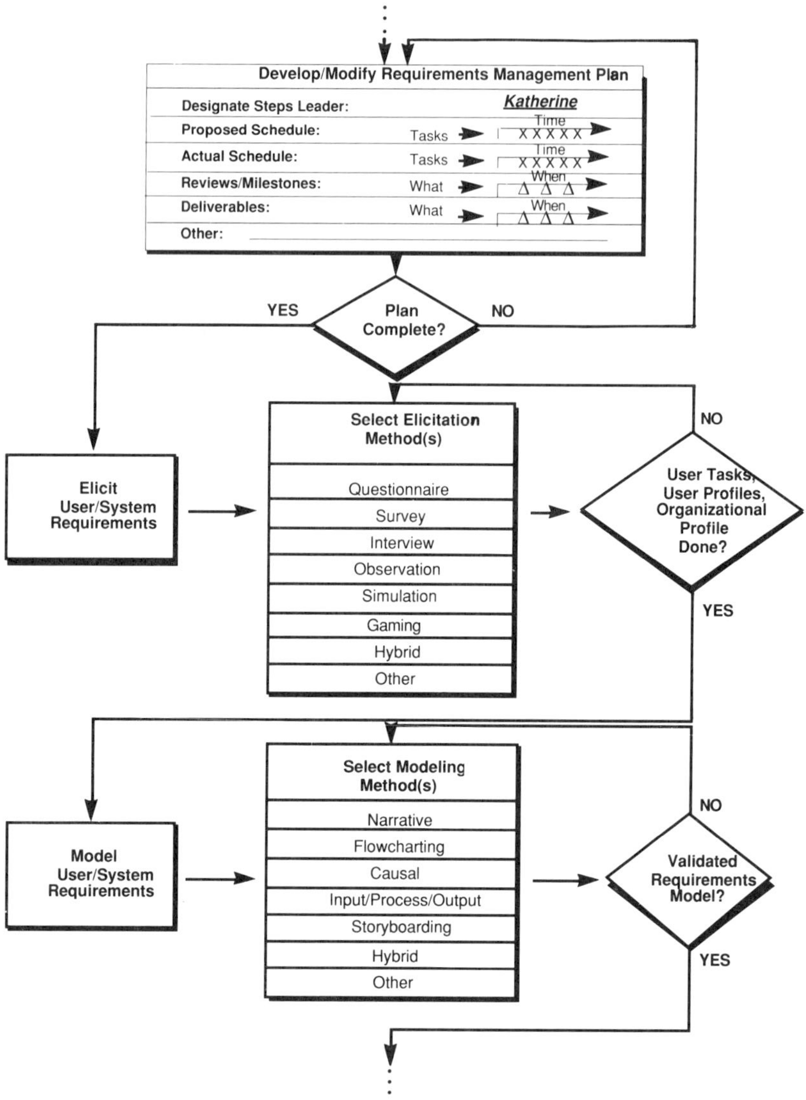
Develop/Modify Requirements Management Plan
Designate Steps Leader:
Katherine
Proposed Schedule:
Tasks
Time
X X X X X
Actual Schedule:
Tasks
Time
X X X X X
Reviews/Milestones:
What
When
Deliverables:
What
When
Other:
YES
Plan Complete?
NO
Elicit User/System Requirements
Select Elicitation Method(s)
Questionnaire
Survey
Interview
Observation
Simulation
Gaming
Hybrid
Other
NO
User Tasks, User Profiles, Organizational Profile Done?
YES
Model User/System Requirements
Select Modeling Method(s)
Narrative
Flowcharting
Causal
Input/Process/Output
Storyboarding
Hybrid
Other
NO
Validated Requirements Model?
YES

Figure 3b.

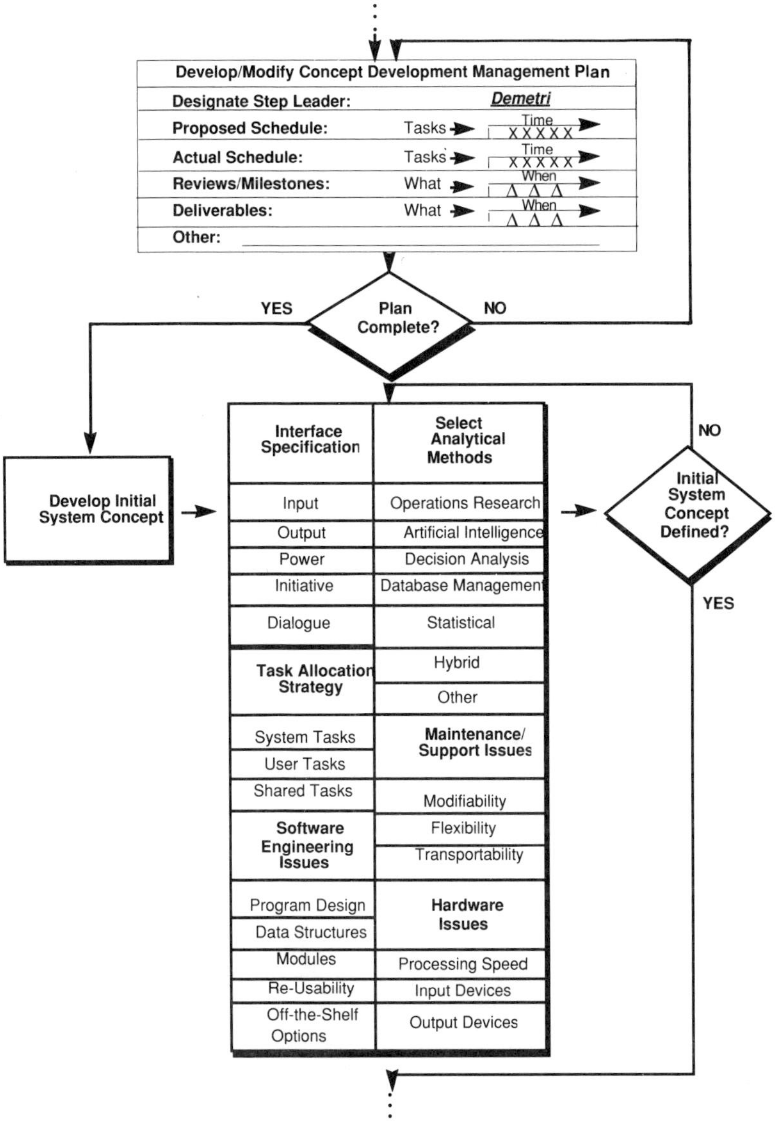
Develop/Modify Concept Development Management Plan
Designate Step Leader:
Demetri
Proposed Schedule:
Tasks
Time
X X X X X
Actual Schedule:
Tasks
Time
X X X X X
Reviews/Milestones:
What
When
Deliverables:
What
When
Other:
YES
Plan Complete?
NO
Develop Initial System Concept
Interface Specification
Input
Output
Power
Initiative
Dialogue
Task Allocation Strategy
System Tasks
User Tasks
Shared Tasks
Software Engineering Issues
Program Design
Data Structures
Modules
Re-Usability
Off-the-Shelf Options
Select Analytical Methods
Operations Research
Artificial Intelligence
Decision Analysis
Database Management
Statistical
Hybrid
Other
Maintenance/ Support Issues
Modifiability
Flexibility
Transportability
Hardware Issues
Processing Speed
Input Devices
Output Devices
NO
Initial System Concept Defined?
YES

Figure 3c.

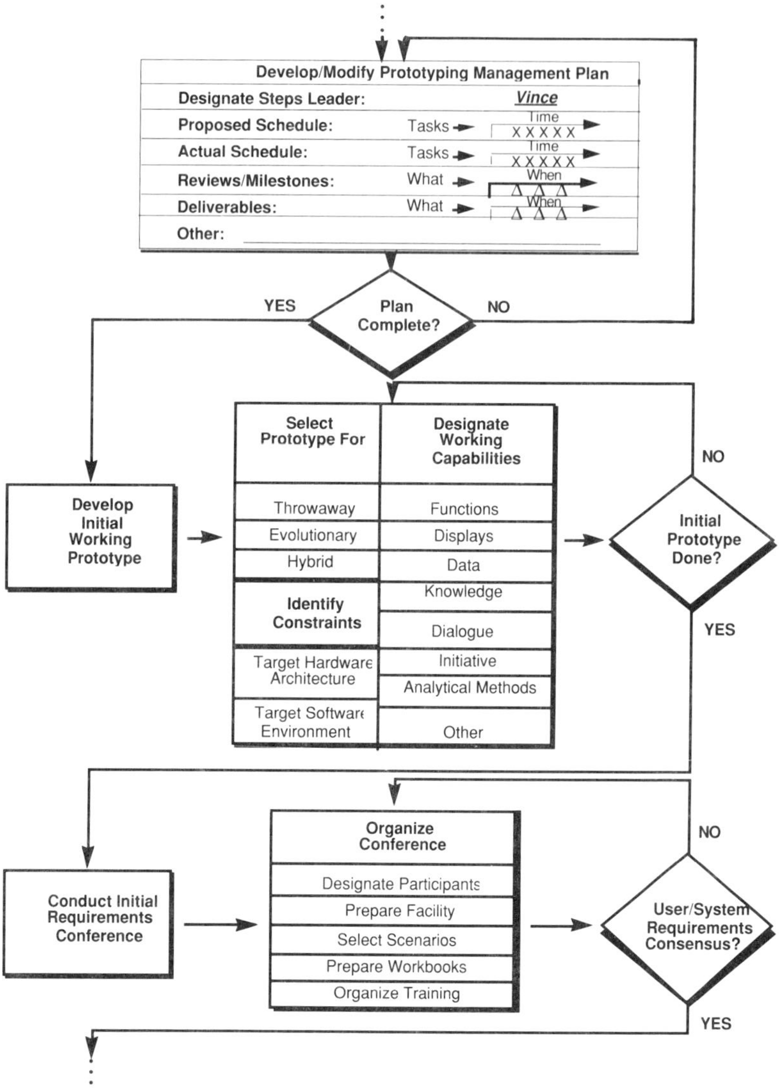

Develop/Modify Prototyping Management Plan
Designate Steps Leader:
Vince
Proposed Schedule:
Tasks
Time
X X X X X
Actual Schedule:
Tasks
Time
X X X X X
Reviews/Milestones:
What
When
Deliverables:
What
When
Other:
YES
Plan Complete?
NO
Develop Initial Working Prototype
Select Prototype For
Throwaway
Evolutionary
Hybrid
Identify Constraints
Target Hardware Architecture
Target Software Environment
Designate Working Capabilities
Functions
Displays
Data
Knowledge
Dialogue
Initiative
Analytical Methods
Other
NO
Initial Prototype Done?
YES
Conduct Initial Requirements Conference
Organize Conference
Designate Participants
Prepare Facility
Select Scenarios
Prepare Workbooks
Organize Training
NO
User/System Requirements Consensus?
YES

Figure 3d.

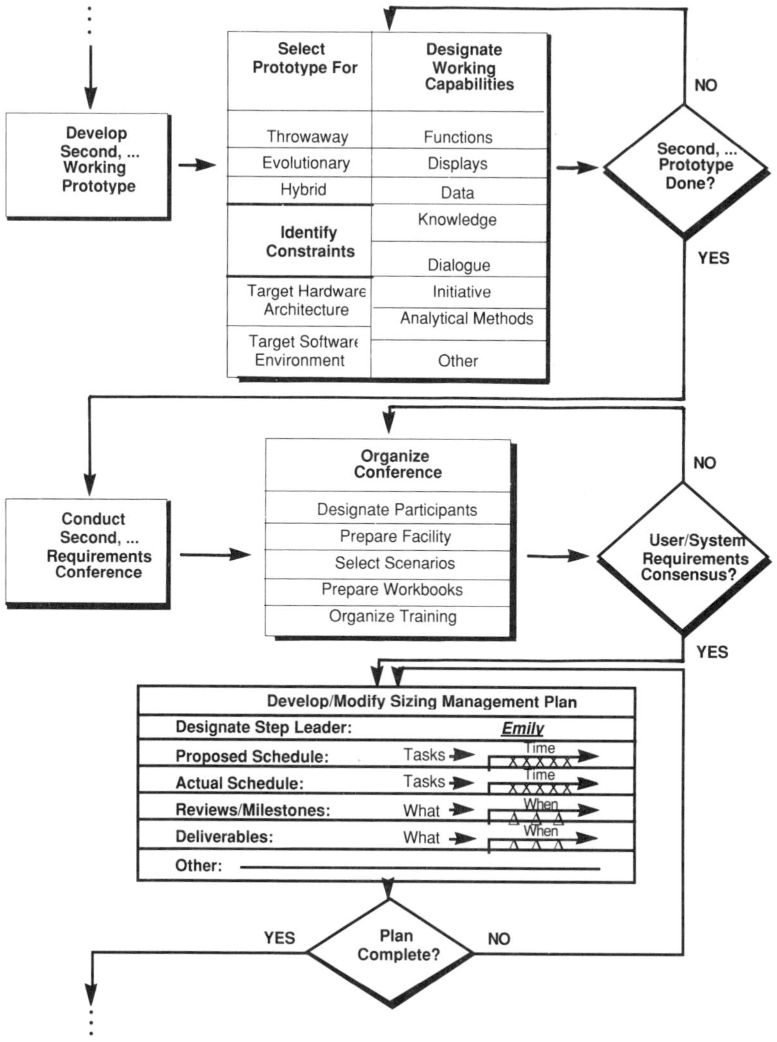
Develop Second, ... Working Prototype
Select Prototype For
Throwaway
Evolutionary
Hybrid
Identify Constraints
Target Hardware Architecture
Target Software Environment
Designate Working Capabilities
Functions
Displays
Data
Knowledge
Dialogue
Initiative
Analytical Methods
Other
Second, ... Prototype Done?
NO
YES
Conduct Second, ... Requirements Conference
Organize Conference
Designate Participants
Prepare Facility
Select Scenarios
Prepare Workbooks
Organize Training
User/System Requirements Consensus?
NO
YES
Develop/Modify Sizing Management Plan
Designate Step Leader:
Emily
Proposed Schedule:
Tasks
Time
Actual Schedule:
Tasks
Time
Reviews/Milestones:
What
When
Deliverables:
What
When
Other:
Plan Complete?
YES
NO

Figure 3e.

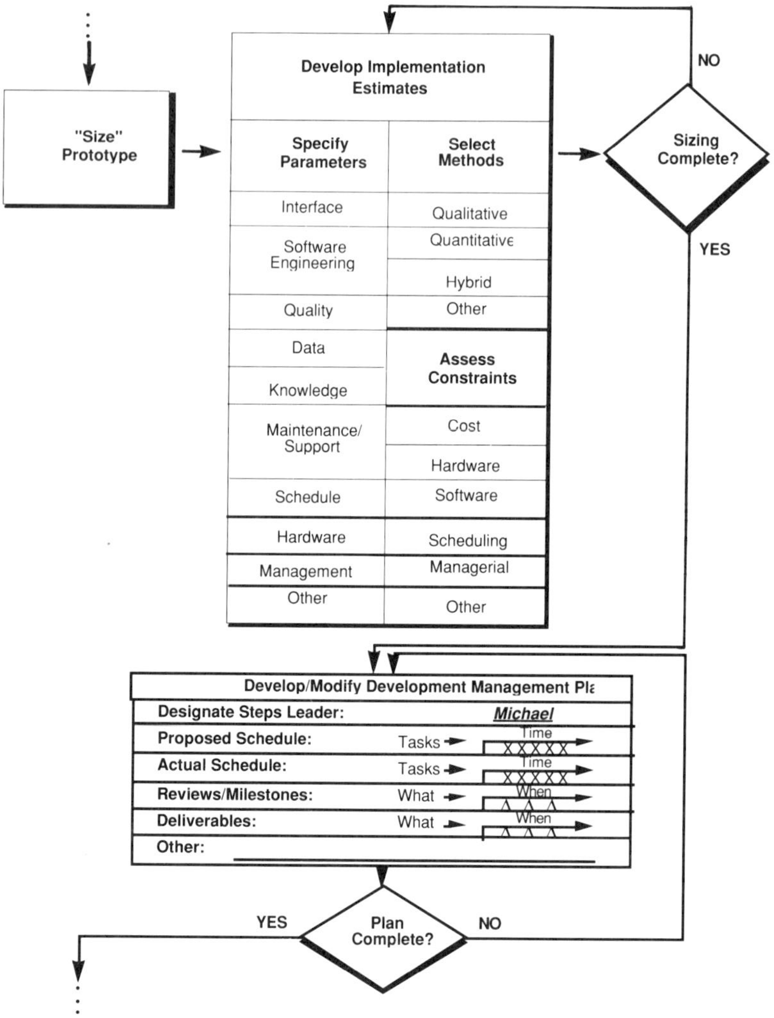
"Size" Prototype
Develop Implementation Estimates
Specify Parameters
Select Methods
Interface
Qualitative
Software Engineering
Quantitative
Hybrid
Quality
Other
Data
Assess Constraints
Knowledge
Maintenance/ Support
Cost
Hardware
Schedule
Software
Hardware
Scheduling
Management
Managerial
Other
Other
NO
Sizing Complete?
YES
Develop/Modify Development Management Pla
Designate Steps Leader:
Michael
Proposed Schedule:
Tasks
Time
Actual Schedule:
Tasks
Time
Reviews/Milestones:
What
When
Deliverables:
What
When
Other:
YES
Plan Complete?
NO

Figure 3f.

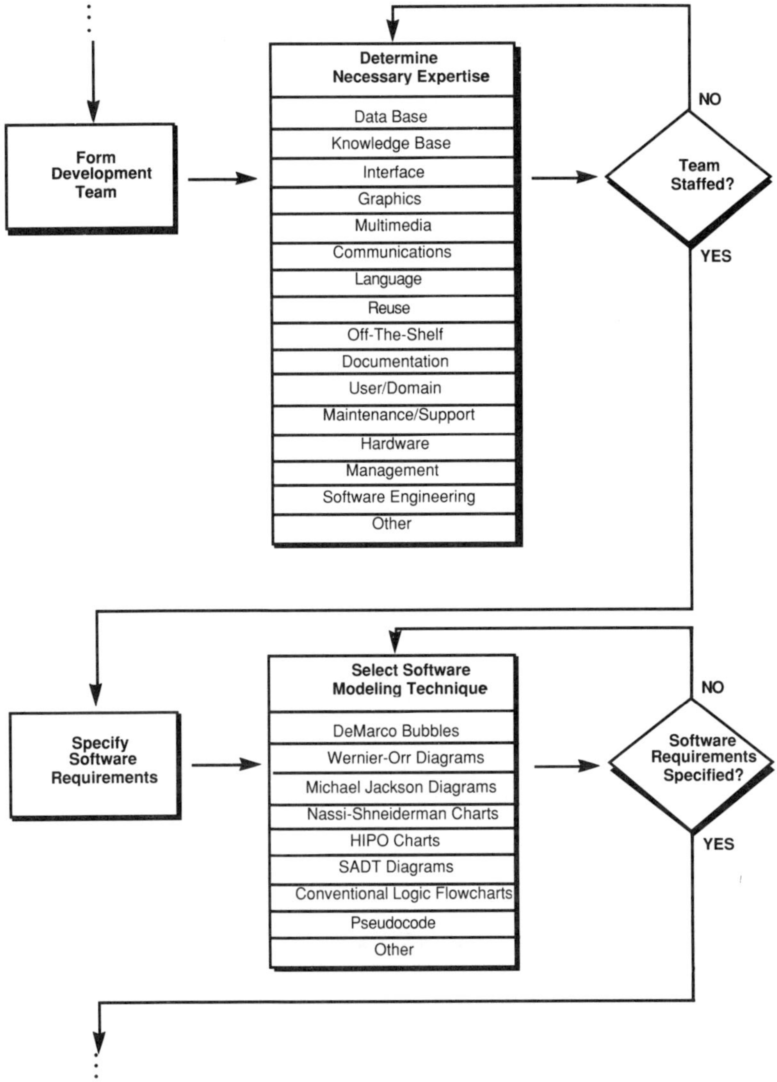
Form Development Team
Determine Necessary Expertise
Data Base
Knowledge Base
Interface
Graphics
Multimedia
Communications
Language
Reuse
Off-The-Shelf
Documentation
User/Domain
Maintenance/Support
Hardware
Management
Software Engineering
Other
Team Staffed?
NO
YES
Specify Software Requirements
Select Software Modeling Technique
DeMarco Bubbles
Wernier-Orr Diagrams
Michael Jackson Diagrams
Nassi-Shneiderman Charts
HIPO Charts
SADT Diagrams
Conventional Logic Flowcharts
Pseudocode
Other
Software Requirements Specified?
NO
YES

Figure 3g.

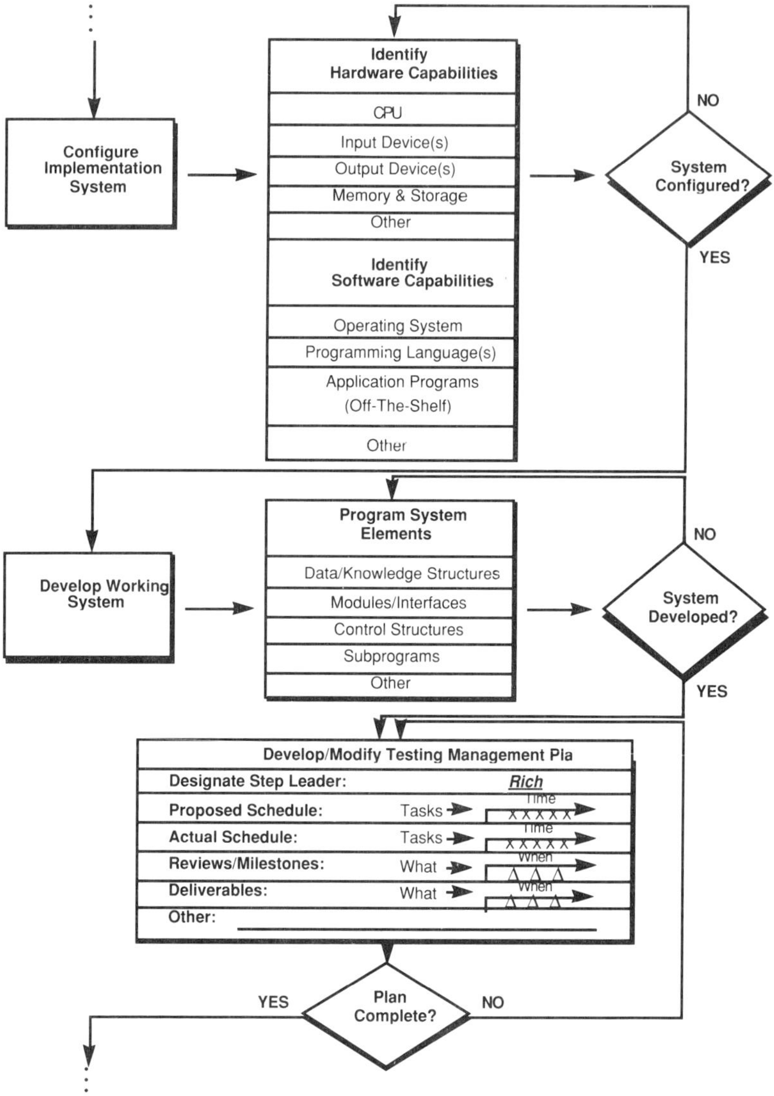
Configure Implementation System
Identify Hardware Capabilities
CPU
Input Device(s)
Output Device(s)
Memory & Storage
Other
Identify Software Capabilities
Operating System
Programming Language(s)
Application Programs (Off-The-Shelf)
Other
System Configured?
NO
YES
Develop Working System
Program System Elements
Data/Knowledge Structures
Modules/Interfaces
Control Structures
Subprograms
Other
System Developed?
NO
YES
Develop/Modify Testing Management Pla
Designate Step Leader:
Rich
Proposed Schedule:
Tasks
Time
Actual Schedule:
Tasks
Time
Reviews/Milestones:
What
When
Deliverables:
What
When
Other:
Plan Complete?
YES
NO

Figure 3h.

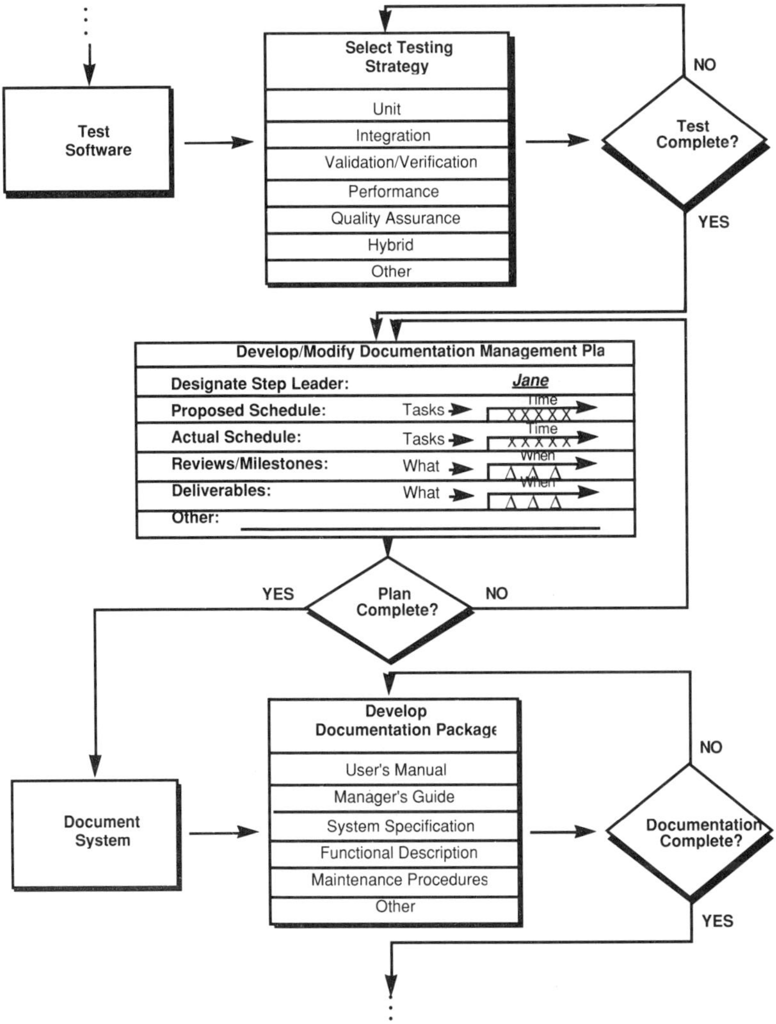

Test Software
Select Testing Strategy
Unit
Integration
Validation/Verification
Performance
Quality Assurance
Hybrid
Other
Test Complete?
NO
YES
Develop/Modify Documentation Management Pla
Designate Step Leader:
Jane
Proposed Schedule:
Tasks
Time
Actual Schedule:
Tasks
Time
Reviews/Milestones:
What
When
Deliverables:
What
When
Other:
Plan Complete?
YES
NO
Document System
Develop Documentation Package
User's Manual
Manager's Guide
System Specification
Functional Description
Maintenance Procedures
Other
Documentation Complete?
NO
YES

Figure 3i.

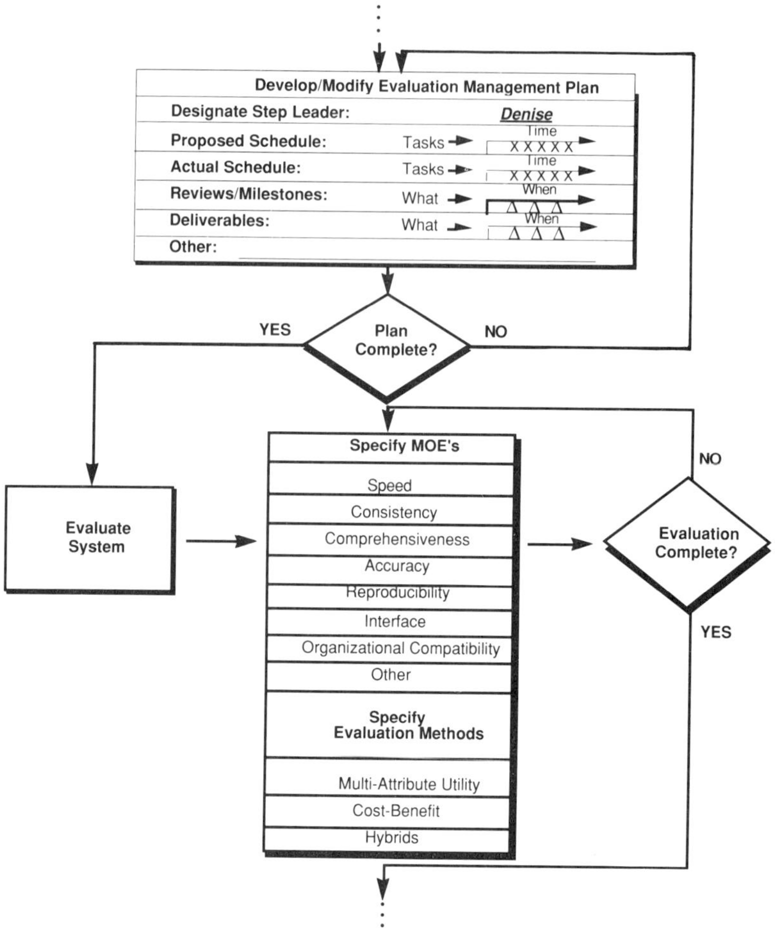
Develop/Modify Evaluation Management Plan
Designate Step Leader:
Denise
Proposed Schedule:
Tasks
Time
XXXXX
Actual Schedule:
Tasks
Time
XXXXX
Reviews/Milestones:
What
When
Deliverables:
What
When
Other:
YES
Plan Complete?
NO
Evaluate System
Specify MOE's
Speed
Consistency
Comprehensiveness
Accuracy
Reproducibility
Interface
Organizational Compatibility
Other
Specify Evaluation Methods
Multi-Attribute Utility
Cost-Benefit
Hybrids
NO
Evaluation Complete?
YES

Figure 3j.

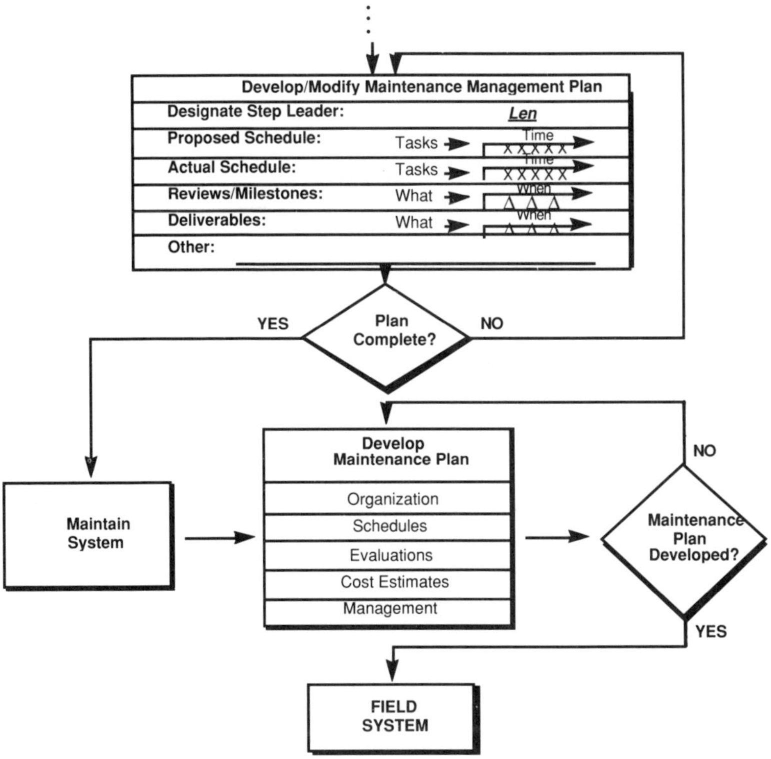
Develop/Modify Maintenance Management Plan
Designate Step Leader:
Len
Proposed Schedule:
Tasks
Time
Actual Schedule:
Tasks
Time
Reviews/Milestones:
What
When
Deliverables:
What
When
Other:
Plan Complete?
YES
NO
Maintain System
Develop Maintenance Plan
Organization
Schedules
Evaluations
Cost Estimates
Management
Maintenance Plan Developed?
NO
YES
FIELD SYSTEM

Figure 3k.

References

*This chapter draws upon S. J. Andriole's *Information System Design Principles for the 90s: Getting It Right!* Fairfax, VA: AFCEA International Press, 1990.

Boddie, J. (1987) *Crunch Mode: Building Effective Systems on a Tight Budget.* Englewood Cliffs, N.J.: Prentice-Hall.

Curtis, B., Krasner, H. and Iscoe, N. (1988) "A Field Study of the Software Design for Large Systems." *Communications of the ACM.* 31. 11. Pp. 1268-1287.

Davis, AM, Bersoff, EH and Comer, ER (1988) "A Strategy for Comparing Alternative Software Development Life Cycle Models." *IEEE Transactions on Software Engineering.* 14. 10. October. Pp. 1453-1461.

Department of Defense (DOD) (1987) *Defense System Software Development: Military Standard 2167A.* Washington, DC: US Department of Defense.

__________ (1986) *Defense System Software Quality Program:: Military Standard 2168.* Washington, DC: US Department of Defense.

Fairley, RE (1985) *Software Engineering Concepts.* New York, NY: McGraw-Hill.

Pressman, RS (1987) *Software Engineering: A Practitioner's Approach.* New York, NY: McGraw-Hill.

Royce, WW (1970) "Managing the Development of Large Software Systems: Concepts & Techniques." *TRW Software Series.* Redondo Beach, CA: TRW, Inc.

PART 2

C^2 Modeling and Simulation

War Gaming, Modeling and Simulation for Command and Control Training

Dr. Richard E. Hayes and Susan M. Horton

The Problem

The need for a finely tuned command and control (C^2) system is supported by history and a constantly evolving doctrine focused on reliable information processing and effective decision making. However, responsive command and control training has always been difficult to incorporate into a comprehensive training plan. In any military organization the commander and principal staff spend considerable resources assuring that subordinates are properly training, particularly in the technical tasks they must perform under the stresses of combat. "Blocking and tackling" skills such as proper use of management information and communications systems often consume the bulk of the time available. Very little training time is available for the commander and staff to develop their own war fighting skills: command and control.

Unfortunately, the most skilled professionals will experience great difficulty in modern combat unless their efforts are organized and managed effectively to fit into efficient multifaceted fighting organizations. Modern warfare is so fast paced, depends so much on intelligence from sources outside the fighting units, involves such a variety of hostile threats, includes so many different types of assets, and requires so much logistics support; that practice in battle management is essential. Many have argued command and control is the force multiplier needed to overcome the hostile environment and adversaries we face. We would contend that command and control is a force multiplier that ranges between 0.00

and 1.00; that is, effective command and control allows the unit to perform up to its capability, ineffective command and control reduces its performance.

Current Considerations

Time will always be a major constraint on command and control training. Current budgetary constraints and doctrine demand innovative training plans and require the increased use of technology to assist in training. Units can expect to have less actual field training than any time in the past decade. In addition, training budgets will continue to shrink. Political barriers exist in the exercise of joint staffs in some theaters of operation. Moreover, operations security considerations preclude the full application of intelligence resources during exercises. Indeed, the Defense Science Board Task Force on "Computer Applications to Training and Wargaming" concluded in May, 1988, that "computer based, simulated scenarios offer the only practical means to improve the training of joint operational commanders, their staffs, and the commanders and staffs who report to them."

The attraction of simulation and wargaming for command and control is not, however, confined to senior commands. DARPA has developed the SIMNET program from the bottom up—beginning with battle simulators of individual platforms and extending to teams, small unit tactics, and battle management. The *Army Times* (March 6, 1989) quotes the First Deputy Chief of the Soviet General Staff as saying "the skillful use of simulators, particularly those that use television screens to simulate the battlefield, significantly reduces the expenditure of ammunition, fuel, motor oil and other resources (required for training.)"

Requirements for Success

The use of simulation and wargaming for command and control training are hardly new to the U.S. military. The sand table games conducted at the Naval War College after World War I are credited with providing a rich understanding of the dynamics of combat in the Pacific during World War II. Accounts of extensive drilling during World War II are discussed in *Military Review* (November 1985, p.31) in a similar consideration of C^2 training, specifically: "We always used battle plays when we left our tank harbor in any formation other than a column or line...[the battalion commander] always used battle plays to control the battalion...the battle positions of a platoon, company, or battalion can be changed instantly by giving a battle play order of a very few words." The emergence of operations research as a tool to support military decision making during World War II legitimized the idea of expressing some aspects of combat in formal terms. Almost as soon as high speed computers became available, simulations were developed to evaluate alternative strategies, tactics, and force postures. They were particularly important where humans had least experience and greatest need to understand: nuclear warfare. However, much of the early experience with combat simulations was more valuable for what it taught about modeling than what it taught about combat.

War games, sometimes including low technology simulations to evaluate battle outcomes, but always relying heavily on humans for decision making, remained the dominant non-exercise method of command and control training well into the 1980s. Until very recently, simulations were simply not attractive as stimuli for command and control training. They failed to meet the six requirements needed to conduct effective training:

- Validity
- Credibility;
- Affordability;
- Flexibility;
- Evaluation; and
- Prompt Feedback.

The first four are tightly interconnected. Validity means simply the model correctly represents reality, at least in those aspects crucial to the application at hand. A model is, by definition, "a simplified representation of the entity it imitates or simulates." (Hughes, 1984) This leads Wilber Payne (1984) to note, "in one sense, all models are invalid." However, a model that does not represent those processes that impact on the battle management problem correctly lacks validity and will teach the wrong lessons. Unfortunately, most early models (and many today) have not been properly validated.

For effective training, simulations must also be believed valid by the commanders and staffs. If the model is not credible to its users, they will blame negative results on the model and ignore them. Equally important, if they are motivated to do well they will attempt to figure out how to do well on the model and not transfer their learning to combat performance.

Training also requires variety. Even when a military organization is given a primary mission, such as defending a particular part of the NATO front line, it must train under a variety of scenarios (time of year and weather, red force tactics, differential warning times, etc.). Moreover, the command must look beyond its first engagements at a myriad of possible situations. Equally important, all commands must train for contingencies. Early combat simulations that had adequate detail to train staffs were often inflexible, requiring long lead times and massive manpower to develop a single scenario.

Cost effective training also proved elusive. Early simulations that presented adequate detail to train staffs required huge support efforts; scenario writers, game players, analysts to examine output, just to name a few. In addition, computer systems were expensive and often not portable, which meant either major communications investments to move the information to the computers or the use of artificial sites and environments to conduct the training.

The last two requirements for success are also related. To be effective, training must include high quality evaluation and prompt feedback. Simulations, however, proved effective only in evaluating the mechanical aspects of warfare (loss ratios, FEBA movement, etc.) and not in providing rich feedback about command and control.

Commanders, whose enthusiastic support is essential to any training program, have proven reluctant to rely on simulation and modeling tools failing to satisfy these six criteria.

Activity and Trends

During the 1980s a variety of activities and trends have emerged that have improved the ability of combat models and simulators to act as drivers for command and control training. These include:

- Continued improvements in computer hardware and software; particularly greater speed, improved displays, better simulation languages, new data base management systems, and ruggedization. These have interacted to enhance affordability and flexibility.
- Acceptance of the ideas of evolutionary acquisition, rapid prototyping, and user involvement. These simplify the problems of developing valid models and improve credibility. They also force developers to confront cost issues and design flexible models to support military organizations.
- Commanders and key staff personnel with technical backgrounds have increasingly emerged. They are more able and willing to challenge the modeling community and guide its efforts. This is the maturing of our technical, computer literate society; not a criticism of earlier military professionals.
- The behavioral sciences have begun to make progress on the problems of command and control. As a result, tools are now available that will allow the use of simulation to evaluate command and control and provide timely feedback.

These are not merely abstract developments. They have resulted in a variety of real world programs that help to train military staffs today.

Naval command and control training, for example, involves simulations in some novel ways. The Naval War College not only runs its classic seminar war games with a host of supporting computer simulations to evaluate player moves, but also offers a system of battle cabins used to train battle group and fleet staffs. Second Fleet has run a series of exercises called Battle Force In-Port Training (BFIT) using a centralized simulation to drive sensors and communications systems on ships located along the East Coast. This allows staff training on scenarios of interest to the commander without putting to sea and allows precious exercise time to (a) run more smoothly and (b) focus on those things that can only be done at sea, such as flight operations and fleet maneuvers. The Naval Oceans Systems Command (NOSC) provides simulation facilities that can also be used for battle staff training as well as the evaluation of alternative scenarios.

The Warrior Preparation Center in Europe is another example of a major innovation that simultaneously filled a serious gap and acted as an important demonstration project for the community. It resulted from a recognition that there was no facility dedicated to training NATO commanders and their staffs. Command and control training was, therefore, confined to exercises, with all the cross pressures and lack of commonality implied by separate command approaches. Its emergence as a cost-effective approach to training senior commanders and their staffs was an important element in the Army's decision to implement the Battle Command Training Program to improve and standardize procedures and staff training at the Division and Corps levels. This program has progressed by incorporating the Army Command and Control Evaluation System (ACCES) and other tools that take advantage of the simulation environment to improve the evaluation and feedback to the command. They are also linking this effort to their lessons learned program so the benefits of the training can be shared.

Finally, the DARPA sponsored SIMNET program has developed into a major initiative with profound implications for the future of simulation as a tool for command and control training. This program originated from the very narrow problem of improving the cost effective training of ground platform crews, particularly tank crews. The original goal was to reduce the cost of driving tanks to train drivers, firing to train gunners, etc. As part of that process, however, the program adopted some assumptions with profound implications. They include:

- Selective fidelity: war fighters identify the elements to model in detail;
- Best commercial practices are applied, not military specifications;
- Rapid prototyping: technology expected to be thrown away at the end of five years;
- Acceptance by the engineers of the "60 percent solution:" rapid development to a "good enough" standard rather than long, costly efforts to get every detail correct;
- Users (warfighters) extensively involved; including continuous feedback and change; real evolutionary development;
- Systems designed for use by qualified warriors, not novices or trainees;
- No effort to introduce umpires or controllers; the battlefield and play result primarily from the acts of the players, secondarily from simulations of those actors not played by humans.

These principles have also been supported by the application of microcomputers and computer networks to minimize cost and enhance flexibility.

Over time, the SIMNET approach has shown itself capable of cost- effective training of ever larger combat forces. More effective field performance from platoons was followed by diversification to incorporate other combat functions (air defense, etc.) and different types of platforms (helicopters, ground support aircraft etc.) Recent developments include force on force engagements and the creation of long haul nets that permit units using simulators at different bases to participate in the same battle.

Problem Areas

While the recent trends and activities indicate simulation and wargaming are becoming effective tools for training in command and control, several problem areas remain. Commands planning to rely on models for training support need to be aware of those problems and the training and modeling communities need to be working together to resolve them.

Six particularly important problems need further attention:

- modeling "soft factors";
- quality of supporting data;
- better "red" players and simulations;
- understanding the uses and limits of wargames and simulations;
- evaluation and feedback; and
- simulation credibility.

"Soft factors" are those important components of combat for which there is incomplete or weak science, rendering the engineering problem of designing valid simulation modules difficult. Examples: include morale, leadership, and group decision making processes.

The recent Military Operations Research Society Workshop, SIMTECH-97, found it necessary to convene a special Working Group on soft factors, particularly command and control. They concluded the validity and credibility of military simulations depends on continued improvement in modeling soft factors and that a major program of research and development should be focused on these issues. (Results of the Workshop are being reviewed in a forthcoming article in *Phalanx*.)

Even in areas where strong engineering is available, there is considerable variability in data quality and parameters that underlie combat models. The Defense Science Board Task Force noted the models they reviewed tended to display Service bias: that is, Navy models had better data on naval warfare, Air Force on air warfare, etc. They have called for a shared repository, with each participant maintaining the data it knows best. The MORS Workshop made a similar recommendation.

Perhaps because it makes considerable demands on the heavily employed intelligence community, the problem of improving red play has not been addressed as clearly as some of the other problem areas. In many training simulations, the Red side is either played as a stereotype (lacking imagination or initiative) or as a perfect system, lacking human frailty. Some red capabilities, such as communications warfare, are routinely limited during training exercises. Effective training, however, requires both a worthy enemy and a realistic one. More effort is needed in this area. Any particular war game, simulation, and scenario (or set of scenarios) will have both appropriate uses and limits. If the goal of an exercise is to train three brigade staffs, the model and scenario must involve them all in realistic roles, not assign one to a reserve role. Hence, the training plan may require the insertion of a simulated unit to act as a reserve. Similarly, a model designed to simulate close combat is not a good tool for testing logistics plans. Models that try to do everything will either become so complex and burdensome to manage they are impossible to maintain, or grow so large we cannot interpret their results meaningfully. Modelers and trainers need better dialogue regarding the uses and limits of particular models and the warfare arenas for which they have been designed and validated.

Feedback and evaluation also remain problem areas. On the one hand, measurement of performance is essential to identifying strengths and weaknesses. Moreover, feedback, particularly feedback with impact, is an essential element of all learning. The success of the National Training Center results, in no small measure, on the opportunity for units to experience the consequences of their performance, both good and bad.

On the other hand, the military, like all professions, is leery of evaluation; particularly in a central area like command and control. The answer is two-fold (a) develop evaluation tools in the same way sound command and control software is built, with heavy involvement of military professionals; and (b) control, dissemination of evaluation information. New evaluation tools have emerged, particularly the Headquarters Effectiveness Assessment Tool (HEAT) and its Army variation, ACCES. These tools have been applied in a number of commands with some success, but they have not been brought fully into the military community.

As the Defense Science Board Task Force noted, simulations can be designed so they automatically collect information useful to support unit self-evaluation. (Time to carry out tasks, detection rates, resource consumption, timeliness of reports, etc.) There is no reason this information should be disseminated beyond the unit except in a form used to develop lessons learned on a non-attribution basis. Where performance evaluation is desired, the chain of command needs to accept its responsibility to protect data as it does in other areas such as alcoholism and crime rates. Training without effective evaluation and feedback will not improve performance.

The final problem area, simulation credibility, is by far the most difficult. Professional military officers, aware of the "fog of war," will continue to find it difficult to believe that a computer simulation correctly assesses combat processes. They can, however, take an active role in improving the quality of modeling efforts, to ensure the tools we use are the best we can devise. That effort and a healthy skepticism that insists on as much field time as possible to enhance what can be learned in war games and simulations will help training, not hinder it. Modelers, for their part, must accept that "elegant solutions" that confound experienced military judgment need to give way to "messy" models that reflect that experience.

The Future

Imperfect as they are, wargames and simulations are going to be used increasingly as drivers for future command and control training. The modeling community and the training community will need to work closely with the war fighters to ensure the lessons taught are the correct ones. There are promising and important programs well underway. If they are vigorously supported while being improved through constructive criticism and feedback, the quality of human performance in command and control will be greatly enhanced. This will mean lives saved and missions accomplished in future crises and combat situations.

References

Report of the Defense Science Board Task Force on "Computer Applications and Wargaming," May 1988, Office of the Under Secretary of Defense for Acquisition.

Adams Peter. "Training a Big Issue to Source Leaders Too," *Army Times*, March 6, 1989.

Hughes, Wayne P. (ed.). *Military Modeling*. Introduction by Hughes and Chapter 4. "Ground Battle Models" by William B. Payne. Military Operations Research Society, 1984.

Mountcastle, John W., LTC,USA. "Command and Control of Armor Units in Combat," *Military Review*. November 1985.

Report of the Military Operations Research Society Workshop on Simulation Technology 1987 (SIMTECH 97) summarized in *Phalanx*, forthcoming.

What-If Combat Simulation Support for Military Planning

Glenn Abrett, Stephen Deutsch,
Stephen Downes-Martin and Scott Swarts

Introduction

Several substantial computer driven wargaming and combat simulation systems have been developed over the past 10 years (for example, CASTFOREM, VIC, FORCEM, CORBAN, JESS, JANUS, etc.), some of which are used extensively in military training exercises. However, few are presently used for planning military operations in real-time situations. The What-If Combat Simulator, a combat simulation designed and developed by BBN Laboratories as part of the DARPA/Army ALBM project, is designed to provide a useful adjunct to the battlefield commander's operations planning effort. It does so by providing a two orders of magnitude faster than real time Heuristic Combat Evaluator (HCE) which explores courses of action and course of action fragments produced by a scenario generator or plan generator.

A critical component in the planning cycle of an operation is the G2/G3 manual wargaming of a proposed Course of Action to determine feasibility, strengths, and weaknesses of that proposed course of action. This manual wargaming, taught at the Command and General Staff College (CGSC), covers many hours of battle, and several branches of the game, in minutes. To do this the CGSC teaches a number of heuristics, (such as tables of advance rates, tables of comparative strength, probabilities of outcome given comparative strength, etc.), and relies on the tactical experience of the G2/G3 to carry out this wargaming process. This process can thus be characterized as a very much faster than real time combat simulation with what-if branching.

A combat simulation to provide support to the G2/G3 wargamers must have the following attributes; it must be orders of magnitude faster than real time, permit branch wargaming and scenario modifications (what-if) at arbitrary stages of the simulation, store the entire simulation game tree for retrieval, re-simulation and examination, provide the military experts complete control over the simulation at varying levels of interaction, and record any modifications to the simulation. In other words the simulation generates the space of possible tactical solutions and the military experts review that space for tactical solutions.

Ease of use and rapid execution are the first two requirements demanded by the ALBM project that HCE addresses. HCE generates an initial scenario from the Blue Course of Action and the proposed Red Courses of Action. In addition HCE provides powerful scenario editing facilities to enable the wargamers quickly to define and redefine the initial and intermediate situations, the course of action to be explored, and the most likely enemy course of action. To achieve rapid execution, HCE is a look-ahead discrete event combat simulation driven by contingent procedural scripts. The outcomes of military engagements are rapidly determined using heuristics developed under the direction of wargaming experts at the CGSC, Fort Leavenworth. The event driven character of the underlying simulation dictates that the computation be done only when required, thus leading to rapid execution.

The scope and complexity of military operations to be covered by a successful wargamer requires a significant military knowledge base. To meet this requirement a Contingent Hierarchical Script (CHS) Language and Editor was developed to provide specialist support for combat simulation. This Combat CHS subsystem has been developed as one of the significant support systems for HCE. It captures the procedural knowledge required to simulate complex military actions in declarative terms. Terminological knowledge, such as "a battalion-plus has five companies," is represented using flavors. Procedural knowledge, such as "an attack consists of an approach march, followed by a pre-battle and a battle march," is represented by CHS's goal/procedure language. CHS has a graphical editor to facilitate knowledge acquisition with military domain experts.

Wargaming, by its very nature, requires the exploration of many possible event outcomes. Each event outcome has an associated likelihood, attrition, engagement duration, and a statement of win/loss. ALBM's HCE provides an advance in wargaming technology by providing explicit tools to allow manual and automatic exploration of multiple outcomes. For each military engagement, several possible outcomes are determined, and unless told otherwise by the wargamers, the most likely outcome is the one explored. Alternatively, the system can be set up to automatically explore multiple outcomes above some wargamer set likelihood threshold in a depth first fashion with most likely outcome explored first. Most likely depth first is chosen so that the system reaches a final outcome as soon as possible. It is possible at any time to revisit any engagement and explore the other outcomes. Indeed, should the wargamers disagree with the computer determined outcomes, there is limited capability for them to redefine the outcome and continue wargame exploration. This ability to redefine outcomes and adjust the scenario are features under development to provide the operations planner with a truly powerful wargaming capability.

The core of the HCE technology is a three component audit trail, each one of which is coordinated with the others. First, a graphical representation of the game tree of wargame combat events is provided and saved as the wargame proceeds, the branches in the game tree corresponding to wargame branches. Second, for each wargame combat event a tactical overlay is presented and saved. Finally, a message log of the wargame actions, at wargamer controlled levels of detail, is also provided and saved. Selecting events or branches on the game tree allows the G2/G3 wargamer to scroll backward and forward through the wargame history, with the tactical overlay and the message log maintained in synchronization. At any point the wargamer may specify alternate branches to explore, or reset the system parameters.

The HCE is built in a modular fashion that places no requirements on the details of the tactics, doctrine, or wargaming algorithms. These may be modified or replaced at will by the military community. The HCE is a technology that allows the military tactician to explore the results of courses of action within a self-defined tactical world. The HCE is designed and built to be either integrated with a planning system, or to be used stand-alone as an independent wargaming tool, tactics trainer, or planning tool. In its stand-alone form it is especially useful as a tool to explore synchronization and coordination issues.

The essential HCE wargaming and simulation features defined here are explored in depth in this paper. The following section briefly discusses the military decision making process as described in ST 100-9 [CGSC 88], and the role of wargaming within that process. This is followed by the design philosophy of the HCE and description of a demonstration run through of a wargaming session, with a discussion of analysis techniques open to the wargamer.

The Role of Wargaming in Military Planning

The military planning process is summed up in figure 2-1 of ST 100-9, The Command Estimate [CGSC 88], see Figure 1. The analysis stage of the planning process is used in two subtly different manners. First analysis of proposed courses of action are carried out to determine changes to those courses of action, even leading to their rejection without the wargaming process. Secondly, courses of action which pass the initial analysis are wargamed to discover whether or not they are feasible, and if so, what are their strengths and weaknesses. The courses of action may then be refined by an iterative process of generation, analysis, modification, further analysis, etc.

The second purpose of the analysis stage—wargaming own courses of action—requires assumptions about the enemy courses of action. These are obtained from Intelligence staff, with most likely enemy courses of action being

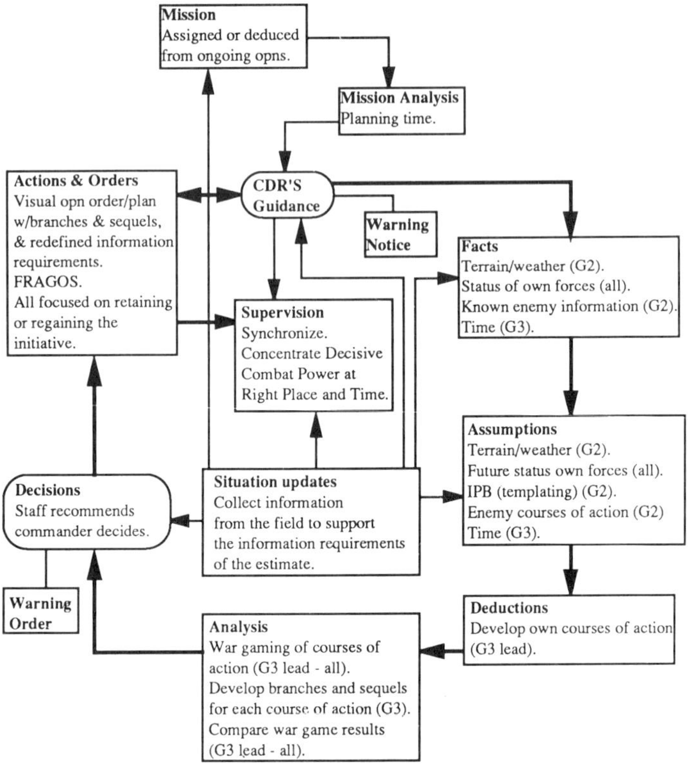

Figure 1. Manual Planning Process from ST 100-9 [CGSC 88].

considered first. Within each course of action pair (own and assumed enemy) branch wargaming is carried out to determine the robustness of ones own course of action. Branch wargaming refers to the consideration by the G2 and G3 manual wargamers of alternate outcomes to wargamed combat engagements and alternate decisions being made, especially by the enemy commander. Branch points are chosen as those where changes to the situation can lead to large changes in the wargame result, that is, branch points are high risk, high cost decision nodes and branch wargaming is a form of risk analysis or sensitivity analysis on the course of action.

It should be noted that the interactive process of developing a course of action, passing it to the analysis stage, and returning an assessment, is referred to as the wargaming process, while the major activity (the second purpose of analysis referred to above) of the analysis stage is a manual wargame. This manual wargame has to be carried out very much faster than real time—minutes of wargaming time covering hours of battle—and with multiple branches. Furthermore, the G2 and G3 do so with highly uncertain data about the combat situation and the enemy courses of action. In addition they must try to predict the behavior of men in battle by using large-scale aggregated data from history and combat trials. They integrate this by bringing to bear their own tactical experience and judgment.

These requirements of fast response times to what is a manual simulation using uncertain data have lead to the sensible decision to use tactical heuristics, or rules of thumb based on experience, rather than analytic and detailed simulations such as Lanchester equations [Taylor 80, 83]. This is no more than the realization that no calculation should be carried out to a greater accuracy than justified by the accuracy of the input data. Examples of the kind of heuristic taught by the Command and General Staff College can be found in ST 100-9 [CGSC 88], and in the open literature [Dupuy 85, 86, 89, Dunnigan 83, 89]. It is interesting to note the credibility accorded by the Soviets to numerical simulation applied to combat planning, using carefully designed nomograms and charts to make up for the lack of battlefield computers and calculators [Vayner 82, Donnelly 86, Kelly 87]. This raises the interesting question as to which set of tactical heuristics should be used when wargaming enemy courses of action; those believed by the wargamers to more closely represent reality, or those that the enemy will actually use to plan and execute their own courses of action.

Combat Simulation Support for Military Planning

Design Criteria

Simulation may be considered to be the attempt to predict the future of some given situation. It is this facet of simulation that drives requirements for using simulation as a support tool for planning. To be truly effective as a planning tool a combat simulation must satisfy at least two broad classes of criteria; those dealing with the nature of the soldier machine interaction and those dealing with the nature of combat event simulation. In HCE, the first class of criteria are collected in the concept of the Audit Trail and the second class are collected in the concept of the Combat Event. These criteria have a major effect on the design of the HCE, and so are discussed here:

- The Soldier Machine Interaction. The tool to support this is the Audit Trail. Criteria essential to the design of the HCE are:
    - —Place the soldier in the loop, avoid a black box design.
    - —Place the soldier in supervisory control of the wargame, allowing a spectrum of soldier interaction with the wargame.
    - —Permit the soldier to modify the combat situation being wargamed.

—Permit multiple branch wargaming.

- Combat Event Simulation. The tool to support this is the combat event. Criteria essential to the design of the HCE are:
  - —Match the level of simulation detail to the level of planning data available, do not oversimulate.
  - — Emulate the soldier G2/G3 wargaming process in the wargame.
  - —Provide a general simulation interface. Do not dictate the simulation method to be used, build the wargame so that it may interface to any set of combat simulation tools.
  - —Wargame faster than real time.

The Soldier Machine Interaction

Central to the requirements on the combat simulation design is the notion of the soldier in the loop. At the most basic level this is the statement that any decision maker in combat should at the very least be under the control of a soldier if so decided, that is, the machine decision maker must not be a black box.

There are two main reasons for aggressively inserting the soldier into the wargame simulation loop. The first is the recognition that unless one is willing to expend geologic time on software development, then software will not adequately simulate human behavior on the battlefield in all its complexity under the extreme stress of combat [Downes-Martin 89]. The battlefield is dominated by soldiers, despite the fact that we know infinitely more about simulating battlefield physics than we do about simulating the former. Therefore, a simulation to assist combat planning must take soldiers into account. The way to deal with this is to notice that machine and human performance overlap but are complementary. The overlap takes place at the soldier machine interface; the complementary nature of their respective performance lies in the fact that machines are very good at bookkeeping, and humans are very good at inspiration. The trick is to relieve the soldier of the burden of tracking the vast amounts of information which flow from a battle without allowing him to ignore or warp that information. Modern computer science paradigms such as artificial intelligence provide weak algorithms which permit the soldier to leverage his own expertise in a hybrid soldier-machine intellectual system, traditional computer science provides the strong algorithms used in well- understood domains.

The second reason is that warfighting is one of the oldest and best documented human activities, and the price of failure in warfare is the best motivation for doing as good a job at it as possible. Thus professional military expertise in combat and knowledge of the likely flow of battle should be utilized as much as possible by a combat simulation. This may be done in two ways; first permit the soldier to advise the simulation in those areas in which the simulation is deficient, and second to base the simulation algorithms on the rules actually used by the expert military professional when carrying out manual analysis of battle.

Supervisory Control

Computer systems are traditionally and rightfully held suspect when they are proposed as decision making entities in a high risk, high uncertainty environment. However, it is unreasonable to expect all decisions to be made by humans. The solution is to provide the human operator with monitoring facilities coupled with override functions. Under this paradigm the human may allow the machine to operate within its domain of expertise, but be able to interrupt and modify its behavior when it is observed to be deficient, or has requested help. This approach has the added advantage that it accommodates improvements in the machine decision making process by permitting the human operator to judge when those improvements have been obtained. Thus the human operator chooses to allow the machine more autonomy. In the combat simulation domain of planning support, the machine decisions to be under supervisory control are those of determining the actual outcomes of combat engagements, and which modifications to make to the initial and intermediate situations. Real combat is vastly more complex than any set of rules used by the machine, even those derived from CGSC experts. In any case, these rules or heuristics are used by experts to focus the mind and teach methodology: It is not claimed that they encapsulate all that is known about combat. Thus any machine based on such rules must be under the supervisory control of such an expert. Supervisory control of this type was first engineered by the authors in the DARPA funded Semi-Automated Forces component of the SIMNET program [Ceranowicz 88, 89], and is incorporated in the HCE.

A key aspect of supervisory control is the concept of providing control over how much autonomy the machine is granted, or rather providing control over how much interaction with the wargamer is required for the system to operate. Under different circumstances it is possible for different lengths of time to be available to the wargamers for analysis of Courses of Action. However, these available times are unlikely to be correlated with the difficulty of the course of action evaluation task at hand, or with the experience level of the available staff. Thus any combat simulation support tool for planning which makes use of supervisory control must also permit a spectrum of interaction levels between the man and the machine. This spectrum should span the two extremes of complete machine autonomy with no audit trail, to the human being asked at every decision point for a decision, along with an audit trail. Between the two extremes lies the preferred position of the machine making decisions under supervisory control and keeping an audit trail. The choice of precisely where in the spectrum to operate is left to the human operator.

The HCE user interface permits a wide range of autonomy to the machine, from fast blast wargaming where no audit trail is saved, all the way to every decision being referred to the human operator. The usual mode of operation is in the middle ground, where the machine is set up to carry out branch wargaming, with some likelihood threshold, to completion. The operator then reviews the wargame audit trail and either accepts the results or triggers further exploration by further wargaming from some point in the wargame history.

Situation Modification

A simulation run may be viewed as a tree of sequential moves from one situation state to a causally related situation state, finally ending up with a leaf node state. The leaf nodes of the tree are the possible final outcomes of the initial state, and the game tree branches represent different possible outcomes from each state. Supervisory control over the simulation run must not only permit control by the military expert of which branches are explored, but also must permit the military expert to insert possible outcomes not considered by the machine. The latter is nothing more than situation modification. Since this is allowed to take place within the simulation run, the military expert is also allowed to make minor modifications to the initial situations and Courses of Action (as the G2/G3 does in current planning).

The HCE permits minor modification of the situation at any point in the wargame. Changes in location and strength are currently supported to permit exploration of the effects of force correlation and timing. These changes are logged as user selected explorations in the audit trail, and the wargamer is encouraged to take account of such changes in the final analysis.

Branch Wargaming

The wargamers responsible for assessing the feasibility of a Course of Action are called on to wargame several branches. That is, to manually simulate different outcomes to significant combat events to test for plan robustness and to check for risks. Time constraints again restrict wargamers to a small number of immediately obvious branches. G2/G3 expertise in selecting wargame branches is provided by machine exploration of multiple branches, where such exploration is under the supervisory control. It is this feature that enables the planning staff to exploit its creativity and ingenuity in the wargaming process.

The combinatorics of a game can be defined by the depth of the game tree (the number of moves permitted the protagonists) and the branching factor (the number of alternate outcomes or results for each move). The combinatorics are potentially explosive, so pruning mechanisms are required. Either the machine support must be smart enough to recognize and report interesting events and branches, or the human operator must have control over the branching, because he is required to review the audit trail and must control the quantity of data he is faced with. The HCE provides simple but effective controls on the branching factor, which are under wargamer control:

- Permit only the most likely branch to be explored.
- Permit all branches whose likelihood is above a wargamer defined thresh-

old to be explored.

- Permit all branches to be explored.

These controls are generalizations of each other, and where branching is permitted, are carried out depth first and in order of likelihood. Any branch is gamed to completion before starting another and the most likely branches are gamed first. This ensures that in a time restricted environment the simulation at least provides some answers in order of likelihood. Breadth first search runs the risk of having time run out before any answers are obtained.

Combat Event Simulation

Simulation may be viewed as a function whose output is a situation which is the future of an input situation. An implication here is that the level of detail of the future situation should not be greater than that of the past situation from which it was deduced. This is similar to the elementary dictum in high school science classes against calculating a result to more decimal places than is justified by the accuracy of the data. Furthermore it should be recognized that for any but very well-understood processes, the simulation process itself has inherent errors which grow over time. In most circumstances military planning takes place in the context of highly uncertain data concerning the location, intentions, and capabilities of the enemy, and even uncertainty about one's own assets. Thus it is pointless to use highly accurate combat simulations of individual vehicles and weapon systems to predict the future of postulated situations whose definition is uncertain. The level of simulation detail must be matched to the level of plan detail. This is the basic justification for the use of simple and fast table lookups based on heuristics taught at the CGSC.

Emulates Manual G2/G3 Wargame

There are three reasons for the wargame to emulate the manual G2/G3 wargaming process. First, in a hybrid man machine system in which the human can (if so chosen) be in supervisory control of the machine, this control is easier to apply if the machine decisions are based on processes that share the same intellectual background as the human. The machine processes should not be opaque to the wargamer, and this applies equally to the underlying algorithms for calculating battlefield effects, such as attrition, and to the meta-level wargaming process itself. Second, the manual wargame is based on a large corpus of expert military knowledge and it makes sense to start with this when supplying automated aids to the process. And third, an automated process that shares the same intellectual background as its users can more easily be modified and expanded by those users.

The HCE makes full use at the process level and the tactical algorithm level of manual wargaming heuristics. However, it is designed in such a way that these tactical algorithms may be removed and replaced if so desired. The HCE technology is one that supports exploration of a wargame: it does not dictate the wargame algorithms.

General Simulation Interface

Any system approach to providing simulation support to military planning should not dictate the form of the simulation algorithms. These algorithms should be in a software layer that is transparent to the user and modifiable or replaceable by the user if so desired. It would be preferable if the interface between the simulation support for planning and the combat simulation algorithms used could be completely general purpose, that is, the simulation layer could be easily replaced by any chosen set of algorithms. While this is an ideal not easily provided within the scope of this work, the HCE system does maintain a separation between the technology that explores combat simulation outcomes in support of planning and the technology which implements the combat simulation details, making it easy to examine and alter the combat simulation algorithms.

Faster than Real Time

To be useful, prediction by simulation must be sufficiently faster than real time so that action can be taken to affect the future. In essence, simulation for planning support is being used in a manner similar to that provided by combat intelligence. It is trying to answer the question "What will the world do; when, where, how, and to what degree will it do it?" Like good combat intelligence, answers must be timely: Perfect knowledge is useless if it is too late. If one takes into account the requirement of branch simulation discussed above, then the requirement to run faster than real time is emphasized as each branch of the simulation itself takes up time. Parallel processing of the branches may reduce the simulation speed somewhat. Given the requirement placed on the planning process to produce plans in two hours covering thirty-six to seventy-two hours of future combat, and the recognition by the wargamers that they should have an assessment of any Course of Action ready in no more than about half an hour, then a simulation speed of approximately two orders of magnitude faster than real time is required.

This is achieved in the HCE system by a combination of look-ahead next event simulation, in which computations are only carried out when determined necessary by the branch being explored, and with the use of heuristics rather than analytic simulations. The HCE looks ahead and advances time to the next combat event. It then uses look-up tables from previously computed attrition models or heuristics to compute the likely outcomes of the engagement and the respective engagement durations; at which point the system advances time to the end of the engagement. Using these techniques allows wargaming large units hundreds of hours in minutes. The penalty is the use of heuristics rather than detailed analytic models, which we claim has been justified by the arguments above.

Heuristic Combat Evaluator Design

The design of HCE is organized around the fact that military units are striving to achieve goals. At any time a military unit is likely to have more than one goal and HCE accepts the fact that these goals may be in conflict (e.g. secure objective alpha and maintain unit strength.) HCE recognizes that military units are task organized, assigning missions to their subordinates in support of their own missions and, in turn, are responsible for completing missions in support of superiors. A military unit may have goals designed explicitly to prevent the success of the inferred missions of hostile units.

The Heuristic Combat Evaluator is a simulation of the attempts of military units to achieve their goals. These units have abstract plans consisting of subgoals and procedures. A language, Contingent Hierarchical Script (CHS) [Papazian 86, Saffi 88b] has been developed to represent these plans.

While owing some allegiance to old AI planning languages such as NOAH [Sacerdoti 77] or STRIPS [Fikes 71], CHS most closely resembles the newer procedural representation languages, typified by Georgeff and Lansky's Procedural Reasoning System (PRS) [Georgeff 86]. The portion of CHS used for representing plans to achieve goals is quite similar to PRS, but CHS goes considerably further in its explicit representation of lower level actions, its connection between goals and goal achievement plans and its carefully worked out constructions for representing parallel activities.

There are three building blocks out of which CHS constructs are composed; goals, procedures and nodes. Goals represent things that need to be accomplished. A goal is characterized by a set of achievement conditions that describe the state of the world under which the goal can be said to have been achieved. A goal defines a detailed script that describes how the goal is to be accomplished. It defines how the military unit plans to accomplish its mission.

Procedures are knowledge level descriptions of how actions in the simulated world are to be executed. They are used as pieces of a generic achievement plan and can be thought of as the methods by which goals are accomplished. CHS procedures capture in declarative form, knowledge that is usually hard coded and buried inside of computer programs.

Nodes are the building blocks out of which both goals and procedures are composed. A node is a named entity, unique to the goal or procedure to which it belongs. A goal or procedure is basically a graph of nodes. Within goals, the nodes define the structure and sequencing of the subgoals and procedures of the goal. Within procedures, nodes define programmatic structures such as flow of control, the invocation of procedures and functions and the definition of alternate worlds for what-if branches.

The most important node types are encapsulating nodes. Encapsulating nodes wrap around a goal, a procedure, or a primitive function and can be thought of as defining the environment within which a goal is activated, a procedure begun, or a primitive function called. An encapsulating node defines a mapping between the parameters of the object that it encloses and the local context of the graph to which it belongs. It can be thought of as an analog to the calling of a function inside another function. However encapsulating nodes are more powerful than function calls. As reified objects they can, themselves, be referenced in a variety of ways, or copied, modified, described and reasoned over.

The CHS Knowledge Acquisition Environment

Knowledge acquisition for a combat evaluator may be thought of as the process whereby the system developers acquire from subject matter experts, explicit knowledge of the doctrine and operating procedures that govern the actions of military units executing their missions and tasks on the battlefield.

The representation of both procedural and planning knowledge has historically been relegated to code in a computer language such as Lisp, or a collection of rule sets. This is unfortunate since most of the essential military knowledge about the world is knowledge of how to accomplish missions and tasks on the battlefield. It involves planning and procedural knowledge. Burying this knowledge in code or deproceduralizing it into disjoint rules makes it extremely awkward to acquire and virtually impossible to use as knowledge qua knowledge. One can execute the computer function or fire the rule set, but one cannot reason about what the computer function or rule set is doing in, or to, or with the world it is operating on.

The representation of military knowledge is addressed by the rich declarative semantics of CHS for encoding missions and tasks. To make the capture of this knowledge feasible, the CHS knowledge representation language is coupled to a knowledge acquisition environment specifically tailored to the CHS language. The knowledge acquisition environment uses a broad collection of BBN tools (for example, see [Sussman 88]) and relies on a basic knowledge acquisition strategy, known as knowledge editing, that was first used in the KREME system [Abrett 87].

The CHS knowledge acquisition environment is a graphically oriented, integrated editing tool that visually represents the syntax of the CHS language. The editor is used to graphically and interactively define military missions and tasks as CHS goals, goal achievement plans and procedures. It provides a joint workplace for system developers and military subject matter experts to develop, review, and extend the knowledge base for the Heuristic Combat Evaluator.

A CHS Example of a Tactical Procedure

A red division on attack presents a carefully orchestrated sequence of actions on the part of its first and second echelon maneuver regiments, possibly supplemented by a reserve unit. First echelon red maneuver regiments are supported by their regimental artillery groups to prepare and support their attack. In this example, we will explore the coordinated actions of one of the division's motor rifle regiments and its regimental artillery group as expressed in a CHS procedure.

A graphical editor is available for developing CHS goals and procedures. Figures 2 and 3 provide two views of the goal ATTACK-WITH-ARTY-TASK which is the CHS representation of the coordinated attack of a red motor rifle regiment and its regimental artillery group.

The graph in Figure 2 provides a view of the full depth of ATTACK-WITH-ARTY-TASK, although for the purposes of this discussion we will only be concerned with ATTACK-WITH-ARTY-TASK and its immediate children, the four nodes of the graph connected to it. The additional nodes provide the details of each of these children. This view of ATTACK-WITH-ARTY-TASK simply presents its four children, two of which are procedures (do arty move&fire orders, stop arty attack on target) presented in rectangular boxes, and two of which are subgoals (move to contact, attack) presented in hexagonal boxes. The user has requested from the system further information about ATTACK-WITH-ARTY-TASK, and has been provided with the three lines at the top of the graph. The first line simply repeats the name of the goal, while the second and third lines show the arguments to the goal and their type restrictions.

Figure 3 provides a second view of the goal ATTACK-WITH-ARTY-TASK. As explained in earlier sections, the graph may have and/or relationships among its nodes, as well as temporal precedence relationships. We see that the four children linked together by an "and" node indicating each of the children must succeed if the the goal is to succeed. If any child node fails, then any other active child nodes will be terminated and ATTACK-WITH-ARTY-TASK fails. The dashed lines tie the "and" node to its children. The solid lines provide the details of the temporal relationship among the nodes. They express exactly the relationship one would expect, issue the orders for move and fire, move-to contact, and then simultaneously order the fire lifted and continue the attack. Thus the carefully coordinated actions of the maneuver regiment and its artillery group is simply expressed using the CHS language as presented by its graphical editor.

The Audit Trail

The core of the HCE technology is the three component audit trails, each of which is coordinated in time with the others. A graphical representation of the game tree of wargame combat events is provided and saved as the wargame

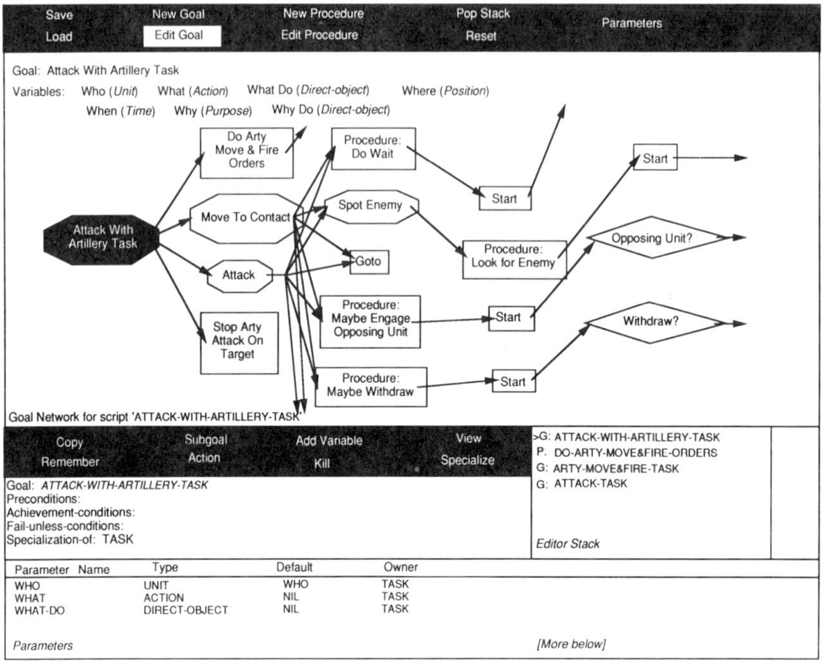

Figure 2. Main View of the Red Attack with Artillery Task.

proceeds; the branches in the game tree corresponding to wargame branches. A tactical overlay is presented and saved for each wargame combat event. Finally a message log is also provided and saved at wargamer controlled levels of detail. It shows what each unit is doing and the outcome and reasons behind each combat event. Selecting events or branches on the game tree allows the wargamer to scroll through the wargame history, with the tactical overlay and the message log also scrolling to the state of the game at that point. The wargamer can then interact with the tactical display and the game tree to extract information such as correlation of force, duration of event, who won, attrition, etc. At any point the wargamer may specify alternate branches to explore, or reset the system parameters.

There are two reasons why an audit trail is a vital part of the What-If Combat Simulator. The first has to do with the breakdown of labor between the machine and the man. The machine is tasked with rapid non-inspirational processing of large amounts of data in order to provide the human with a tactical space in which to exercise his inspiration. This large quantity of data and the lines of reasoning connecting them make up this tactical solution space and thus, by definition, must be audited. It is inspection of this space at different levels of detail by the human, which permits the human to apply their own expertise. The second

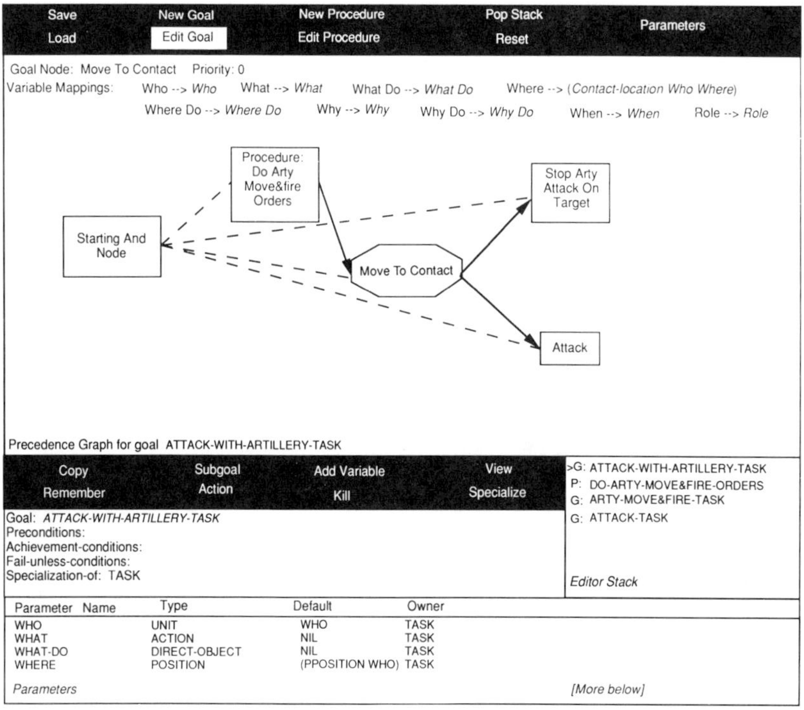

Figure 3. Precedence View of the Red Attack with Artillery Task.

reason has to do with the supervisory control activities of the human. Any changes to the machine decisions made by the human must be logged and audited for future analysis, either by the same human or by a replacement. The audit trail is used for three purposes; post simulation review for re-simulation from chosen decision points; providing justification for the final decision concerning the original Courses of Action; and logging and reviewing of operator instigated alterations to the simulation run.

The audit trail provides the user interface to the system. Figure 10 (see page 144) shows this interface, and provides an example of the audit trail in a specific state:

- The top right pane of Figure 10 is the game tree of the HCE system. It is in the form of a tree oriented left to right. Each node represents the state of the world when a combat event or engagement is taking place. Each arc out of the node represents the outcome of that engagement and carries with it the probability of that outcome. Within each node the cumulative probability, or expectation is indicated. Time flows from left to right. This game tree is used to control the wargamer exploration of the simulation history. When the simulation has run to completion, or been interrupted by the wargamer, then by using this tree the wargamer can set the simulation back to any previous state represented on the game tree. When this is done, the tactics display is reset to show the state of the world at that time, and the message log scrolls to show the actions taking place at that time. The G2/G3 may then explore alternate possible outcomes by triggering the wargame to simulate forward from that situation with the alternate outcome assumed.

- The top left pane of Figure 10 is the tactics overlay. This is an intelligence map of the situation as wargamed by HCE, able to show terrain, unit locations, control measures, graphics, etc.
- The full width pane across the middle is the message log of the system. It describes the actions of the HCE at levels of detail which can be set by the wargamer as the HCE runs. The major contents of this log are concerned with the actions carried out by units and the results of those actions.

Combat Events

The HCE supports the concept of defining any required battlefield event as an HCE combat event. In the basic system as built and delivered, combat events are chosen to be engagements, although it must be stressed that this is not a necessary restriction. Any other kind of event, such as receiving intelligence of a certain kind, or certain units arriving at certain locations, can be defined as combat events. The combat event, or engagement event, is classified according to the missions being undertaken by the engaging units. For example, two units advancing to contact and which engage each other will generate a meeting engagement event. An attacking unit meeting a defending unit will generate an assault event. Each combat event generates a probability distribution of outcomes, where each outcome is defined by attrition on both sides, the duration of the event, the likelihood of that outcome occurring and a text message stating who won or lost that engagement. Based on this information and the units missions the simulation will generate actions for both sides which depend on the results of the engagement.

The simulation system in HCE is a look-ahead, next event, goal driven simulation. Each unit has its missions and tasks defined at wargame initialization. Based on this information, the simulation system attempts to execute the mission using a contingent hierarchical script known as CHS [Papazian 86], where unit behavior is modified by terrain issues. The overall behavior of the system is derived from the fact that these units have conflicting missions (are enemies). The system generates movement for each unit and looks ahead for the conditions for triggering the next combat event: an engagement for example. The state of the world is then advanced by movement of units to the point in time when this next combat event is detected to occur and the engagement is triggered.

The algorithms used by the HCE to compute the outcomes of combat events are derived from heuristics taught at the CGSC,[CGSC 88] and at the Voroshilov Staff Academy [Hines 85, Kardashevskiy 78, Vayner 82, Donnelly 86, Kelly 87]. These heuristics give a probability distribution on the outcome of an engagement as a function of the correlation of forces of the event. The correlation of force in a combat event is computed by comparing the effective combat strengths of the units involved in the engagement. The effective combat strength of a unit is derived from a set of combat multipliers applied to the actual percentage strength of the unit. These multipliers include:

- The combat unit comparison value, for example if a BTR Bn has a baseline value of 1.0, then table 4-1 in ST 100-9 [CGSC 88] gives a comparison strength of 2 to a US M2 Bn.
- Mission dependent multipliers, for example, a unit in a prepared defensive position picks up a multiplier which grows as time available to prepare the position without being engaged.
- Deception dependent effects, for example, successfully employing surprise in a flank counterattack will grant a unit an additional force multiplier.

The final effective combat strength of a unit is thus the product of these factors. It should be stressed that the actual values of these multipliers is a matter for expert military opinion. The HCE is built in such a way that it is easy to modify these values, and to insert additional force multipliers.

From the correlation of forces, the HCE determines the outcome from a set of look up tables. It then determines from the likelihoods which outcomes should be explored, based on the wargamer settable controls on how to branch wargame. Given this final list of outcomes, the HCE moves the simulation forward and looks for the next combat event. The look up tables used to determine the probability distribution of outcomes may easily be replaced by the expert military tactician.

A minor complication needs to be dealt with. This occurs when a unit joins in a combat engagement during that engagement, for example, relieving forces or reinforcements or a flank counterattack of an attacking unit. In this case a simple exponential interpolation is carried out based on time to update the original combat event. The new unit is folded into a new combat event with the original units who have had their strengths updated. The interpolation is exponential rather than linear to ensure that the final outcome of an interrupted event with no new units joining the engagement is the same as though the engagement was not interrupted.

Multiple Worlds Implementation of Branch Wargaming

When the simulator encounters a combat event during the course of executing some unit procedure the outcomes function associated with that combat event is run. Depending on the state of some user definable system parameters, the system will generate outcome branches according to the rules described above.

The outcomes activity stores a snapshot of the entire state of the simulation just after the outcomes function was called and an outcome chosen. The outcomes activity also stores all the potential outcomes and their associated probabilities, marking each outcome as explored or unexplored. A visual representation of the outcomes activity—called a world node—is displayed on the game tree of the audit trail. The G2/G3 can examine a world node to determine which outcomes have been explored and which remain to be examined, or may halt the simulation, examine any world node and choose an alternative outcome. The simulation will then continue down that particular wargame branch.

The combat event outcomes function also computes the duration of the engagement for each possible outcome. This duration is used to schedule the completion of the outcomes activity. The simulation then immediately advances in time to the next most immediate event. This is either the start of a combat engagement, in which case the outcomes functions are run, or the end of a combat engagement, in which case the final attritions, locations and engagement results are calculated. It is important to note that the final results of an engagement are presented at completion, not before, since any engagement can be interrupted.

The Heuristic Combat Evaluator Architecture

The global architecture of the HCE follows from the principle of basing it as closely as possible on military procedures as well as domain knowledge. It is no accident that the principles laid down in METT-T (Mission, Enemy Situation, Troops-Friendly, Terrain and Meteorology, Time) so closely match the technique for analysis and wargaming laid down in CGSC ST 100-9 Chapter 5 [CGSC 88].

There exist two loops in the HCE (see Figure 4). There is an outer loop in which the HCE client (such as a planning system or a scenario generator) calls the wargame with evaluation and control data and a tactical situation. At this level the G2/G3 via the client SMI has control over the environment passed to the wargame. An inner loop then takes over in which the inference engine is controlled by the control data under the supervisory control of the G2/G3 via the SMI at the wargame (see Figure 5). The client may pass an environment plus control data to the wargame covering the spectrum from a full task force plan to postulated situations involving small numbers of small units (for example, a single blue battalion defending a bridge against a red regiment). Thus the wargame is capable of exploring the consequences of any postulated tactical situation, not just a complete plan.

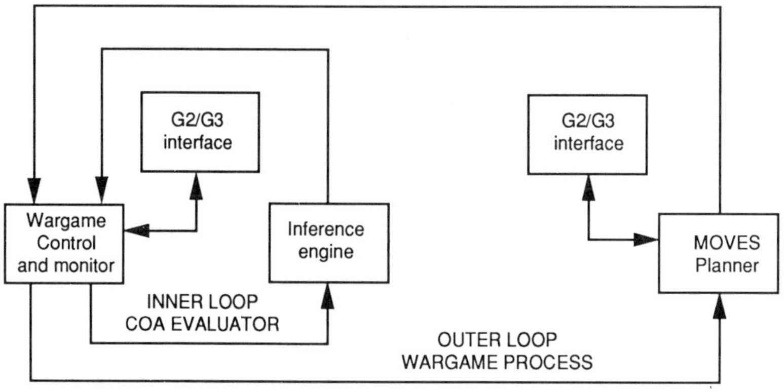

Figure 4. Wargame Process and Wargaming Loops.

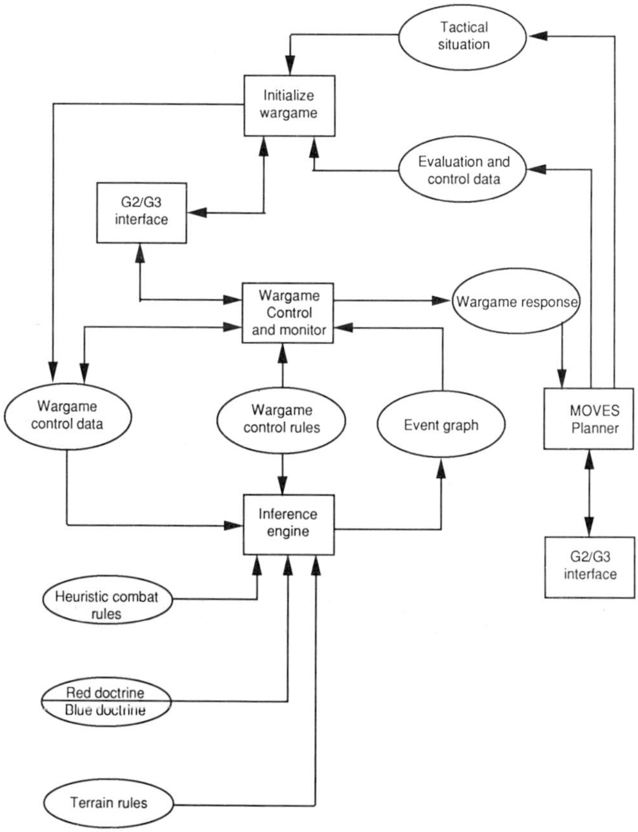

Figure 5. Wargame Architecture.

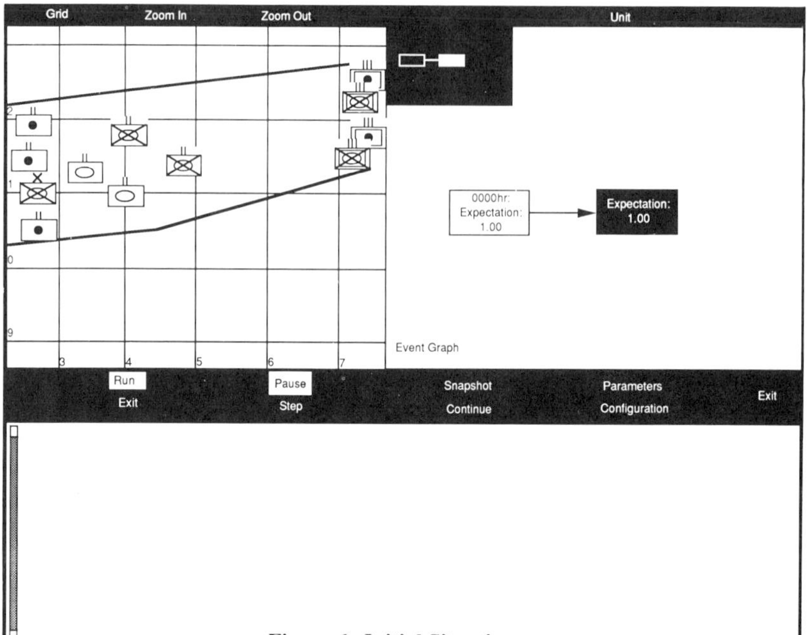

Figure 6. Initial Situation.

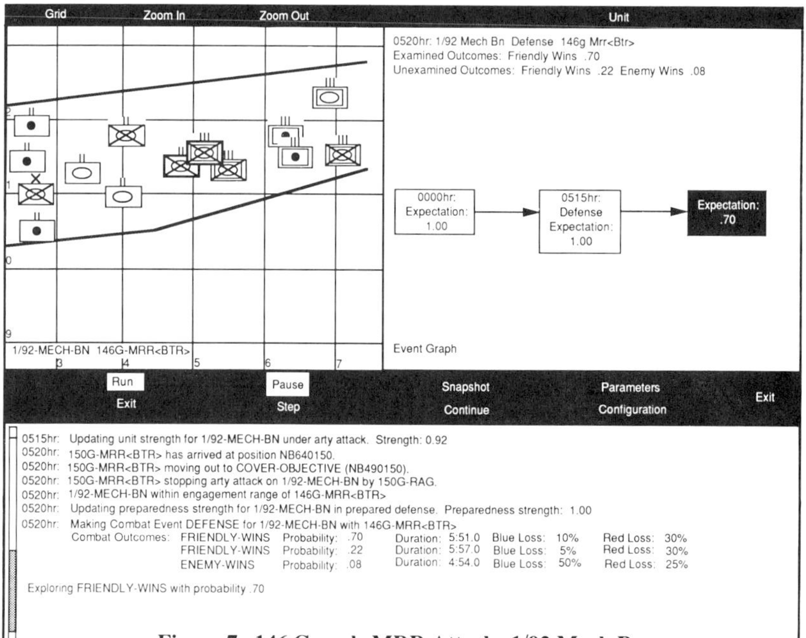

Figure 7. 146 Guards MRR Attacks 1/92 Mech Bn.

Branch Wargaming with the HCE

The Initial Situation

The initial situation is displayed in the tactics overlay of the interface (see top left of Figure 6). A blue task force is in defense with one covering force battalion with a delay mission and two battalions with defend missions in the main battle area. One battalion at the rear has a flank counterattack mission in support of the northern defending battalion. This flank counterattack will be launched if and only if both blue defending battalions are engaged. A red division is in the attack, two regiments with regimental artillery groups in direct support in the first echelon.

Note that no event has yet occurred, so the game tree (top right) and the message log (center) are virtually empty. The system is set up to explore most likely outcomes only.

The First Engagement

Figure 7 shows the state of play when the first engagement occurs. This is between the 146 Guards Motor Rifle Regiment and the 1/92 Mech Bn. The system determines the probability distribution of outcomes, and displays them in the message log with the statement that the system will now explore the most likely.

Final Situation

Now let us move ahead to the final situation as shown in Figure 8. We now have the state of the world when all units have either succeeded at or have failed at their missions. The HCE terminates at this point and freezes the units. The game tree can now be used to examine the history of the wargame, and to re-simulate any alternate outcomes the wargamer wishes to explore.

A Critical Combat Event

Examination of the fourth combat event to have taken place, (Figure 9) indicates that the 150 Guards MRR defeated the 1/10 Armor Bn by a narrow margin. An interesting question is what change in the final outcome would occur if the 1/10 Armor Bn had been successful in the defense?

Branch Wargaming

Figure 9 shows the result of branch wargaming from this engagement assuming that the 150 Guards MRR was actually defeated by the 1/10 Armor Bn.

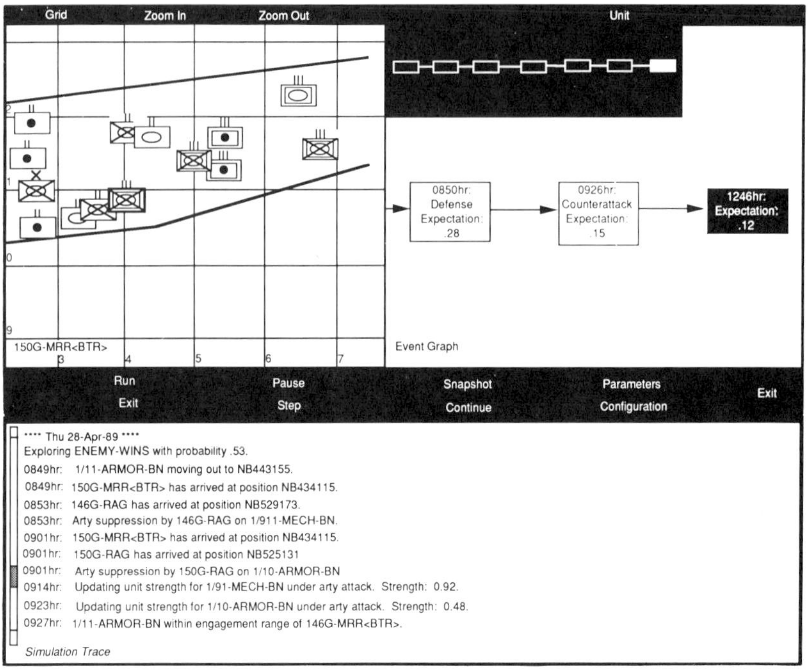

Figure 8. The Final Solution—all units have either succeeded or failed at all their missions.

Figure 11 shows the final outcome of this branch. Let us examine the differences between the final outcomes of the two branches by examining Figures 8 and 11. One immediate difference is that in the first case, when 1/10 Armor Bn is defeated, then the 150 Guards MRR advances to NB4110, while in the second case, when 1/10 Armor Bn defeats the attacking 150 Guards MRR, then the 150 Guards MRR is unable to advance, and withdraws to NB4811. Detailed examination of the data available from the interface reveals differences in times, durations of combat, and final strengths of the units.

This example, while not claiming to be tactically correct, gives an example of the analysis power made available to the wargamers.

Conclusions

The Heuristic Combat Evaluator provides a technology which attacks the

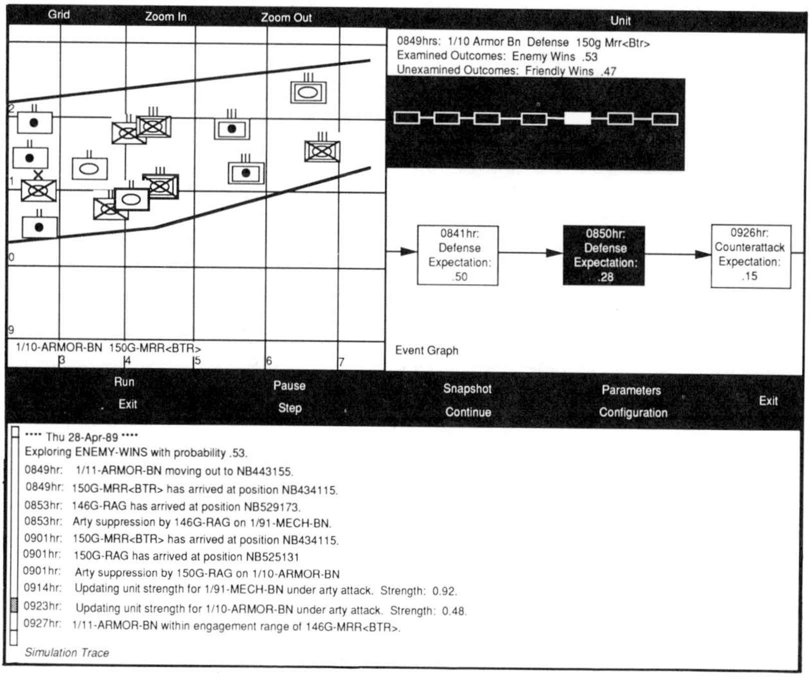

Figure 9. A Critical Event—nearly similar probabilities of mission success and mission failure

problem of freeing up the creative powers of the G2/G3 and the planner. It does this by providing in visual and text form the complete tactical solution space in the form of a fully audited branch wargame. The wargamer has complete control over exploration within that solution space, and can also trigger machine exploration outside that space. In effect, the HCE provides the catalyst for the G2/G3 to apply human ingenuity.

It does this by means of a three part audit trail. This consists of a game tree which maps out the state of the branching in the wargame process, and which is used to recall the other two parts of the audit trail, the tactics overlay and the message log. Using this system, the G2/G3 is able to examine in detail any event in the wargame, to explore any alternate outcome and to make modifications to the situation at any point. Alternatively, the wargamer may allow the wargame to run to completion without supervision, and then examine its audit trail at leisure.

This technology does not dictate the actual simulation algorithms (such as attrition) as it provides a clean interface to these algorithms. But, the HCE as currently implemented does use look up tables for combat event outcomes derived from U.S. and Soviet sources [CGSC 88, Hines 85, for example].

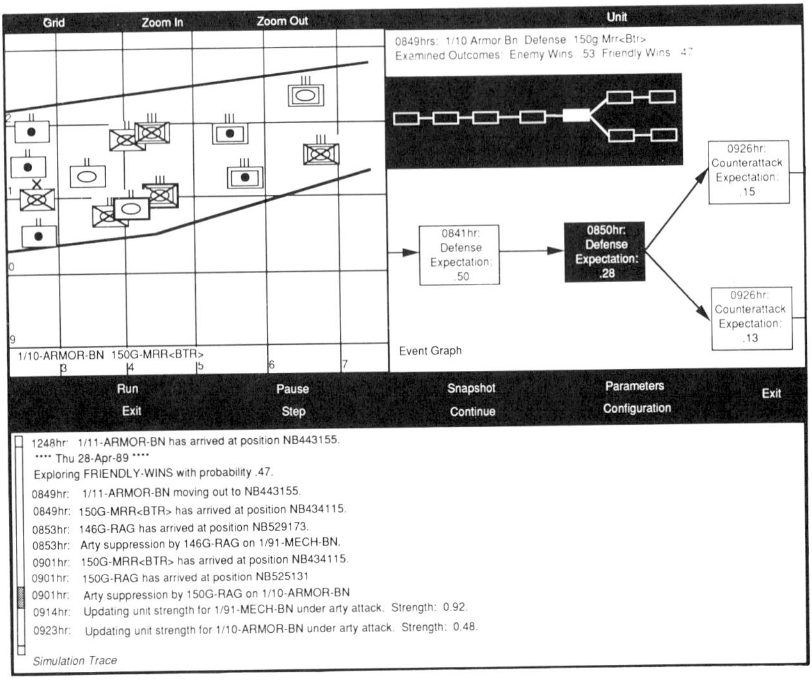

Figure 10. A Branch Point.

The HCE provides a technology applied to wargaming which is of use in a variety of domains. It can be used to examine synchronization and coordination, to teach tactics, and as a tool for planners. The technology is independent of actual tactics and algorithms implemented, which can be dictated by the military user community. The system has been demonstrated in both its stand-alone mode, where it is used as a planning tool by the wargamers [Saffi 88a], and as part of the ALBM system, where it is integrated with the maneuver planning component of ALBM known as MOVES.

Bibliography

[Abrett 87] Abrett, G., Burstein, M., "The Kreme Knowledge Editing Environment," *International Journal of Man-Machine Studies* 27:103-126, 1987.

[Ceranowicz 88] Ceranowicz, A., Downes-Martin, S., Saffi, M., "SIMNET Semi-Automated Forces Version 3.0: Implementation Issues," BBN Report 6940, 1988.

[Ceranowicz 89] Ceranowicz, A., Downes-Martin, S., Saffi, M., "SIMNET Semi-Automated Forces Version 3.0: A Functional Description," BBN Report 6939, 1989.

[CGSC 88] US Army Command and General Staff College., *The Command Estimate*, ST 100-9, 1988.

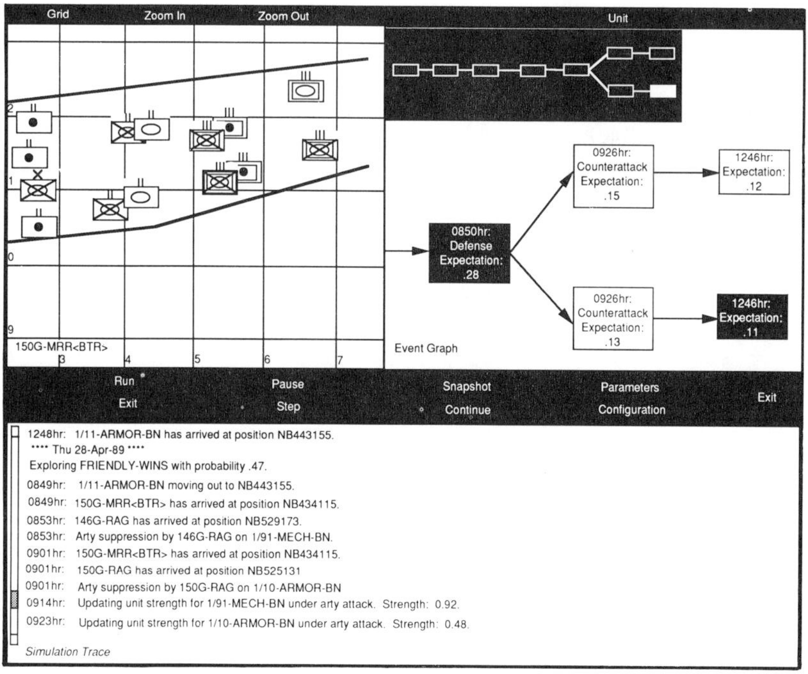

Figure 11. The Final Situation of a Second Warfame Branch.

[Donnelly 86] Donnelly, C. N., et al., *The Sustainability of the Soviet Army in Battle*, Soviet Studies Research Centre, 1986.

[Downes-Martin 89] Downes-Martin, S., Gilbert, A., Robbins, J., Payne, W., "Aggregation Issues for Command Modules in SIMNET," MORS Symposium MORIMOC II, February 1989.

[Dunnigan 83] Dunnigan, J. F., *How to Make War*, Quill, 1983.

[Dunnigan 89] Dunnigan, J. F., "Inserting the Human Factor into Combat Models," MORS Symposium MORIMOC II, February 1989.

[Dupuy 85] Dupuy, T. N., *Numbers, Predictions and War*, Hero Books, 1985.

[Dupuy 86] Dupuy, T. N., *Understanding War: History and Theory of Combat*, Paragon, 1986.

[Dupuy 89] Dupuy, T. N., "The Fundamental Information Base for Modeling Human Behavior in Combat," MORS Symposium MORIMOC II, February 1989.

[Fikes 71] Fikes, R., Nilsson, N., "Strips: A New Approach to The Application of Theorem Proving to Problem Solving," *Artificial Intelligence 2*: 189-208, 1971.

[Georgeff 86] Georgeff, M., Lansky, "A Procedural Knowledge," *Proceedings of the IEEE*, 74, 10, 1986.

[Hines 85] Hines, J. G., *Encirclement*, Soviet Studies Research Centre, 1985.

[Kardashevskiy 78] Kardashevskiy, Yu., Tvorcheski Planirovat' Ognevoye Porazheniye Tseley., *Voyenniy Vestnik*, 7, 64, 1978.

[Kelly 87] Kelly, M. E., Addendum to the Annexes to: *The Sustainability of the Soviet Army in Battle*. See [Donnelly 86],Soviet Studies Research Centre, 1987.

[Papazian 86] Papazian, B., V(INT)2 Phase 3 Developments: "Knowledge Representation and System Design," BBN Report 6373, 1986.

[Sacerdoti 77] Sacerdoti, E., *A Structure for Plans and Behavior*, American Elsevier, New York, 1977.

[Saffi 88a] Saffi, M., "The Wargamer Workstation: A User's Guide," BBN Report 7048, April 1988.

[Saffi 88b] Saffi, M., "The Contingent Hierarchical Script (CHS) Editor: A User's Guide," BBN Report 7049, July 1988.

[Sussman 88] Sussman, J., "The Grapher," BBN Report 6876, 1988.

[Taylor 80] Taylor, J. G., *Force-on-Force Attrition Modeling*, Military Applications Section, ORSA, 1980.

[Taylor 83] Taylor J. G., *Lanchester Models of Warfare*, Vols. I and II., Military Applications Section, ORSA, 1983.

[Vayner 82] Vayner, A. Ya., "Takticheskiye Raschety," *Voyenizdat*, Moscow 1982.

C^2 Simulation Soldier-in-the-Loop Requirements

Barbara A. Black

Introduction

The pace and lethality of the future battlefield will make increasing demands on elements of command and control (C^2) systems for lower echelon United States ground combat vehicles. Current voice and manual systems may not meet the challenges of this demanding new environment. New technologies such as microprocessors, reconfigurable software and digital burst communications provide the capability to rapidly and accurately distribute graphic and text information across the battlefield to enhance the effectiveness of new weapon systems. These technologies would provide a means for exploiting new doctrine and getting inside the enemy's decision loop. However, taking full advantage of these new capabilities is, in large measure, dependent upon individual soldier and commander performance, a consideration which often surfaces too late in the materiel acquisition process.

Until recently, few tools existed which allowed researchers and developers to adequately consider soldier-in-the-loop requirements for proposed new C^2 systems. The advent of networked, fully crewed, ground vehicle simulators and lower echelon command posts, such as those found at Fort Knox in the SIMNET-D facility, provides new opportunities to address soldier performance issues in sufficient time to influence the development of C^2 systems. This chapter will discuss the C^2 automation problem, how issues have been addressed to date using these simulators, and future plans for expanding their role in the C^2 arena.

Traditional C^2

Since World War II, systems supporting command and control for lower echelon maneuver units such as tank platoons, companies, and battalions have consisted mainly of paper maps and FM radios. The transmission of both text and graphic information must be accomplished in one of two ways, either through laborious verbal descriptions via FM radios or face to face discussions between commanders meeting at a common point on the battlefield to share grease penciled map overlays. Both these methods have inherent potential for errors in communication of the commander's intent. FM radio channels are often crowded and difficult to access. Lengthy codes are required to communicate information about geographic locations, for example, the nine digit grid system. In the case of the face to face meeting, if the tank company commander copies, in grease pencil, the overlays from the battalion map, then allows his platoon leaders to copy from his map and the individual platoon members copy from their platoon leader's map, it is easy to compound errors.

More recently, advanced technologies have been inserted in the command and control arena, for example, field artillery systems are now supported by TACFIRE, an automated targeting system. The problem has been that these new systems have been inserted in a piecemeal fashion and at different echelons. This has prevented inter-echelon and inter-vehicular integration and thereby, the full realization of the benefits of these technologies. For these reasons, traditional systems are still relied upon as the mainstay of C^2 execution.

Characteristics of the future battlefield such as closed hatch operations, resulting from either a nuclear, biological, or chemically (NBC) contaminated environment, will force reliance on the FM voice radio for communication of C^2 information. However, the threat's capability to employ electronic countermeasures may have a detrimental effect on speech intelligibility. This could increase the time required to send messages due to increased requirements for repetition. Countermeasures may also decrease the accuracy of the information received. These facets of the future battlefield make command and control difficult and critical.

Improving C^2

The key to overcoming many of these obstacles lies in the technology to automate command and control capabilities. Technology is currently available to allow each vehicle on the battlefield to know, not only its exact location, but also the location of other critical friendly vehicles. This technology is referred to as an inertial, mutual position navigation system. The technology also exists to provide digitized terrain base capabilities resident in each vehicle and to present them via a computer-based graphic map display complete with graphic operational overlays. Digitally stored text can provide preformatted menu-driven reports that could then be digitally burst to other vehicles and higher headquarters at the touch of a button. Microcomputer technology can now be made sufficiently robust, (i.e., can be militarized), to retain all this information in a hostile tactical environment.

There are pitfalls to the successful implementation of these technologies for automated C^2 systems. They lie in 1) failure to fully consider command, control and communication (C^3) system integration requirements for the particular vehicular application, 2) failure to fully consider the soldier operator, and 3) failure to adequately consider the training ramifications of C^3 interface design decisions (Black, 1989). The research necessary to address these issues and to describe the systems integration from the standpoint of soldier performance requirements must be conducted. Past excuses for not completing such research efforts include: 1) information concerning the functional requirements for each new technology not being available to the soldier performance researcher, 2) the necessary soldier-in-the-loop test bed not existing, 3) the funding requirements to support soldier-in-the-loop evaluations being considered excessive and 4) the notion that research takes too long to produce results.

The successful avoidance of these pitfalls is now possible with the advent of Simulation Network - Developmental (SIMNET-D).

SIMNET-D forces determination of system integration requirements because, in effect, the entire vehicle is built in simulation and must be integrated. Second, it provides the necessary, instrumented test bed capability to obtain soldier-in-the-loop evaluation data. And third, when compared to field testing, it is a cost-effective means of exploring alternative concepts and testing critical issues. To fully understand how SIMNET-D is the key ingredient to solving past C^2 research dilemmas, we must first look at the characteristics of the C^2 problem and then at the characteristics and capabilities of SIMNET-D.

The C^2 Automation Problem

Effective command and control for ground vehicles means the efficient, accurate, and timely movement of critical information around a battlefield containing large numbers of vehicles or nodes. Some inherent problems in automating such a system include: providing inaccurate or untimely information to key individuals, causing information overload, failing to ensure survivability of the information network, not providing redundancy of capability to account for destroyed vehicles or nodes, and providing information in a form that is unusable or inefficient.

In fact, automating C^2 systems for ground combat vehicles will result in revolutionary changes instead of the customary evolutionary changes. The Army must be prepared to exploit this situation. As Kroening (1986) states "Let us take full advantage of new technology and not unnecessarily constrain its application by simply automating the same manual processes that have been perpetuated since World War II." (p. 978). The Army has often had the tendency to become enamored of hardware technology to the possible exclusion of the leap ahead afforded in other areas. For example, Kroening (1986) makes the point that it is often considered "much easier to do business the same way, albeit perhaps faster with new hardware, than it is to do business in a different and perhaps more efficient way with new software."(p.975)

Automating command and control systems requires the implementation of computer-based systems which have significant soldier-machine-interface (SMI) requirements and which are geographically disparate. Heavy dependence upon user-friendly SMI for overall C^2 system effectiveness forces consideration of soldier aptitudes, abilities, and training. The fact that the systems are fielded in geographically separate locations, in hostile environments, forces reliance on indirect communication systems.

The SMI requirements for these computerized systems must take into consideration the ease of training and use of automated maps and attendant symbology, the preparation and transmission of text-based reports which were formerly voice based, and the selection of input devices which are militarized and usable by soldiers wearing cold weather or NBC gear. Performance of tasks by soldiers and commanders using proposed versions of automated C^2 systems must then be evaluated, not in isolation, but within the context of a battlefield environment. Demands must be placed on personnel in the evaluation which, as accurately and as safely possible, represent the demands of the future battlefield. In this environment, soldier performance and operator workload can quickly become the limiting factors in achieving full system effectiveness, rather than the performance of the hardware as has been commonly thought.

For the C^2 researcher, one of the most difficult problems to solve is the identification or development of useful, critical, representative measures of soldier performance for command and control tasks. These measures are necessary for differentiating between effective and ineffective technologies or soldier-machine-interface alternatives, and for determining training effectiveness. Judging C^2 performance has been a subjective task for many years. Attempting to develop an objective performance measure for a traditional C^2 task such as "carrying out the Commander's intent" is, to say the least, difficult. It is tantamount to asking these questions "what I said may not have communicated what I meant, but did you understand what I was trying to say?" and furthermore, "based on what you understood, did you execute the required tasks to bring about the outcome which I desired?"

This may appear to be an unreasonable example of the problem. However, when automation of the tactical and technological environment force communication of commander's intent into a few graphic symbols presented on a relatively small computer display in a ground combat vehicle, the software supporting the C^2 system must be extremely effective. The only way its effectiveness can be ensured is to devote considerable time and resources to the software development and evaluation process early in the acquisition cycle.

New state of the art capabilities in simulation like SIMNET-D provide the opportunity to complete the requisite software development and evaluation. In addition, this new simulation capability allows the researcher to record all, literally all, the events and communications on the battlefield, and to have real time perfect knowledge of those events. This may provide the data base necessary for establishing useful and critical measures of C^2 task performance. Only after those measures are determined can the relationships between stimuli and responses, between actions and outcomes, be scientifically and statistically validated. For too long, we have looked at variables which are easy to measure, such as rate of fire, number of hits, and loss exchange ratios, without thoroughly investigating the C^2 decisions which may have dramatically influenced those outcomes. SIMNET-D provides the opportunity to address the relationships between decisions and outcomes and to determine what information is critical as well as how best to present it.

SIMNET-D

Characteristics of SIMNET-D which support C^2 test requirements include the following. SIMNET-D contains fully crewed, soldier-in-the-loop weapon system modules or simulators, including the M1 tank, Bradley Fighting Vehicle and the Forward Area Air Defense System. SIMNET-D supports multi-vehicle, continuous combat operations in real time. The simulation network uses an Ethernet to link and update all simulator activities and weapon system effects. The image generators, in each simulator, provide real-time updates of the terrain features and weapon system effects within a 3.5 kilometer radius. A sound system in each simulator provides realistic battlefield acoustics such as weapons firing and vehicle sounds such as engine whine and track movement appropriate to speeds, terrain surface, steering and gear changes (Chung, Dickens, O'Toole and Chiang, 1988).

To the uninitiated, walking into a SIMNET facility is like walking into a giant video arcade (see Figure 1). Several players are required to operate each video game and all game booths (in this case vehicles) are linked in such a manner that when a player in one booth takes an action, like driving his tank, another player in a different booth can watch the result of that action on his video monitor (i.e., see a tank drive past him). Each vehicle simulator is fully interactive and can move, shoot, and communicate on the simulated battlefield. This capability becomes especially exciting when booths are divided into teams which develop and execute plans to fight each other, force on force. The number of

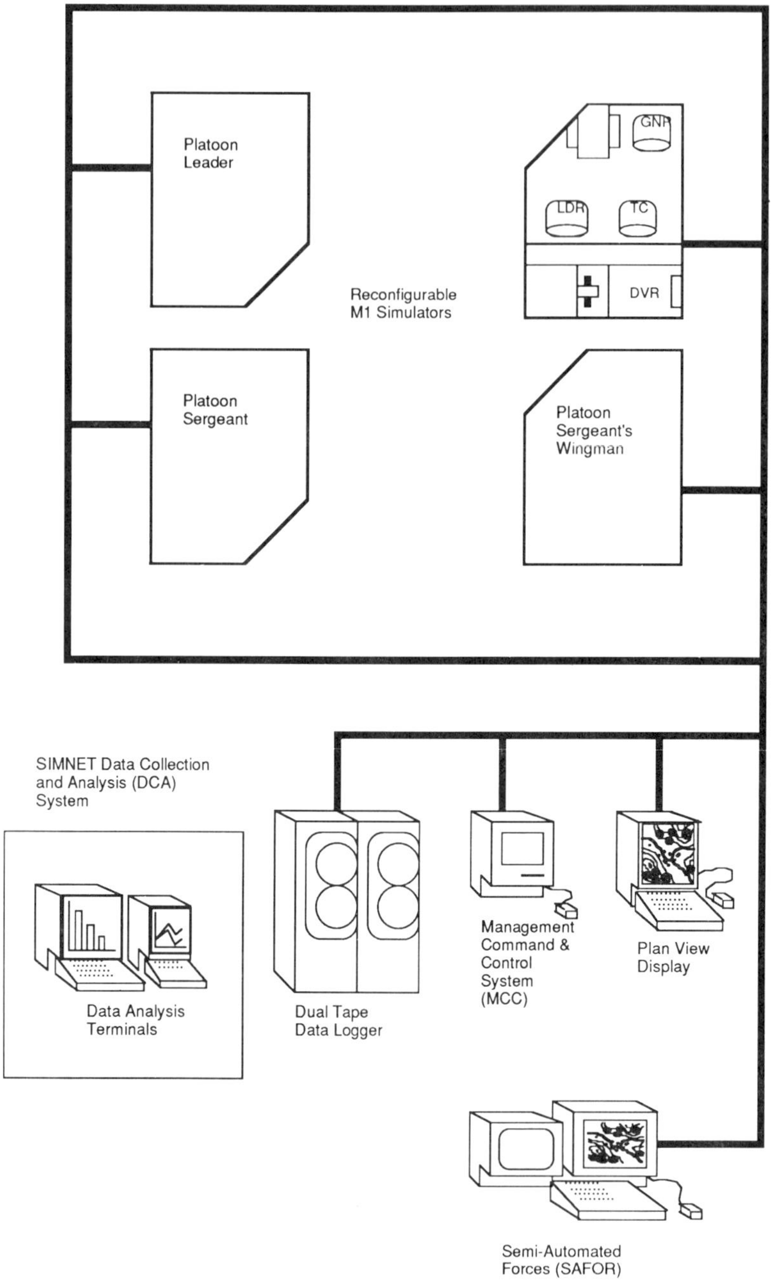
Platoon Leader
GNR
LDR
TC
DVR
Reconfigurable M1 Simulators
Platoon Sergeant
Platoon Sergeant's Wingman
SIMNET Data Collection and Analysis (DCA) System
Data Analysis Terminals
Dual Tape Data Logger
Management Command & Control System (MCC)
Plan View Display
Semi-Automated Forces (SAFOR)

Figure 1.

vehicles linked together (local area networked) determines the echelon at which the simulated battle is fought.

As the number of vehicles simulated increases, the simulation extends beyond the actual vehicle simulators to command posts and the tactical operations center (TOC) at the battalion level which provide for real time command, control and communications interplay. Using workstations in the TOC, battalion staff can provide logistics, artillery, intelligence and operations support. Residing on the Ethernet with the simulators and the TOC is a data capture capability which records all the data packets on the net. This capability, referred to as DataLogger, is compatible with a co-located data reduction and analysis system.

SIMNET-D applies the concept of selective fidelity to the development of vehicle simulators and TOC workstations. This concept is based on the assumption that many operational capabilities of the actual military systems can be "selectively" omitted because they do not, on a cost basis, play a significant role in the issue under investigation. For example, it may not be necessary to replicate the exact physical dimensions of an entire vehicle or system when the issue of interest involves only the commander's workstation.

Digital recording of both the information protocols sent between simulators over the network and analog recording of the voice communications provide the capability to instrument and measure all the activities occurring in the simulators. These activities include those which are psychomotor-based, such as target tracking, or cognitive-based, such as navigation errors or errors in FM voice reports. This provides the researcher with the requisite tools to evaluate command and control performance. Collectively, the features of SIMNET-D provide the researcher with the necessary test bed for addressing critical C^2 issues.

Process of Using SIMNET-D

The process of using SIMNET-D for addressing soldier performance issues is comprised of six basic steps: technical information gathering, simulation specification preparation, software development, data collection or experimentation, analyses, and interpretation of results.

Information must be obtained on the particular technology of interest, for example, information on position navigation (POSNAV) and heading reference systems. This information should detail the functional requirements for the technology in terms of its battlefield and vehicle specific capabilities. All too often there is considerable information available concerning the engineering capabilities of the technology and very little available on the operator interface requirements. In POSNAV for example, much of the documentation available concerned the accuracy of the system over time and distance, its physical space claim and weight. For this reason, development of the operator interface requirements is often the most difficult step.

The researcher must take information about the functional requirements of the technology and create a generic soldier machine interface (SMI) that allows the exercising of identified functions. This SMI should have as guidelines for its development, for example, the physical and mental ability range of the potential operators, the location and space claim in the actual vehicle work station, and the training requirements which might be imposed by a particular design configuration. The functional requirements and the SMI description are documented in a simulation specification.

After the simulation specifications are written, the software developer begins the iterative process of implementing and revising the SIMNET-D software. Revisions are based on continuous user reviews. As the final version of software emerges, the researcher must begin specifying the experimental design under which the capabilities of this new "simulated" technology for C^2 will be evaluated. In addition, training documents, media and methods must be developed to support the troops participating in the experiment. The experiment is conducted utilizing one or more alternative versions of the proposed technology tested against a baseline group of troops operating without the technology, e.g., with and without POSNAV. This experiment must consist of specially designed combat scenarios which force the utilization of the new technology under the most rigorous of simulated conditions. These conditions should tax the soldier operators in terms of workload, replicate the most deleterious of threat conditions and ensure testing of the systems under conditions which, from a safety standpoint, may not be possible to test in a field environment.

The analyses and the reporting of results must be technically sound and document the caveats or concessions made because the experiment was simulation-based and not a field test. Some examples of the conditions which could limit conclusions include unique features of the simulation that make it unrealistic, if such features exist. In SIMNET-D, such caveats include limits to target acquisition ranges resulting from limitations in the fidelity of the computer image generation system. Acquisition in the field can occur at ranges beyond 3000 meters, however, acquisition in SIMNET-D is extremely difficult beyond 2000 meters and averages about 1200 meters. In SIMNET-D, the terrain lacks sufficient fidelity to support use of the standard geographic military lap map, instead a map showing only terrain features that are represented in SIMNET-D is provided to each vehicle commander. The major question which must be answered in the affirmative, before SIMNET-D experiments are acceptable substitutes for or adjuncts to field trials, is whether vehicle operators and military commanders receive the necessary and sufficient stimuli or information to elicit decisions and responses in the simulation that would be elicited in the field environment.

Following this basic process, experiments can be conducted in SIMNET-D which provide information concerning: 1) the relative value of the technology to the performance of platoon, company, and battalion missions, 2) the effectiveness of particular SMI alternatives, 3) the tasks that are most difficult to train or to perform using the new technology, 4) the relative merits of future experiments using the technology and 5) the identification of additional critical issues.

Experiments in C² Using SIMNET-D

Several command and control experiments have been conducted by the Army Research Institute-Fort Knox in SIMNET-D. In describing each effort, a different portion of the SIMNET-D utilization process will be emphasized. However, it should be understood that each step of the process is necessary and has been or will be completed for each research effort.

POSNAV

Position Navigation (POSNAV) is a position location, heading reference system designed to provide information to the vehicle commander concerning where his vehicle is located and in what direction it is moving. This technology is integral to the implementation of an effective automated command and control capability in ground combat vehicles. Data were collected for this technology during the period January to June 1988.

The SIMNET-D capabilities were utilized to determine the value of POSNAV to tank commanders in the performance of navigational tasks and the training requirements which might accrue from the insertion of such technology into current vehicles. Two versions of soldier-machine-interface capability were evaluated (See Figure 2) using three groups of five platoons each. One group used a POSNAV display containing a digital map with contour lines, some cultural features and own vehicle icon, while the other POSNAV group used a display containing only the grid map with own vehicle icon. The third group completed the crew and platoon tasks without the aid of POSNAV in any form.

Results demonstrated significant improvement in tank crew and platoon performance of mission tasks when using the POSNAV system, either grid only or grid with digital map, as compared to control conditions where POSNAV was not available. For example, in command and control tasks, platoons without POSNAV took two and one half times longer to report their own location, on demand, to higher headquarters than platoons with POSNAV and they were much less accurate. Platoons with POSNAV were twice as successful at executing mission fragmentary orders and did so almost twice as fast. Platoons with POSNAV spent 37 percent less time and consumed 29 percent less fuel per mission segment completed than did their non-POSNAV counterparts (DuBois & Smith, in press). Differences between digital map and grid only were detected

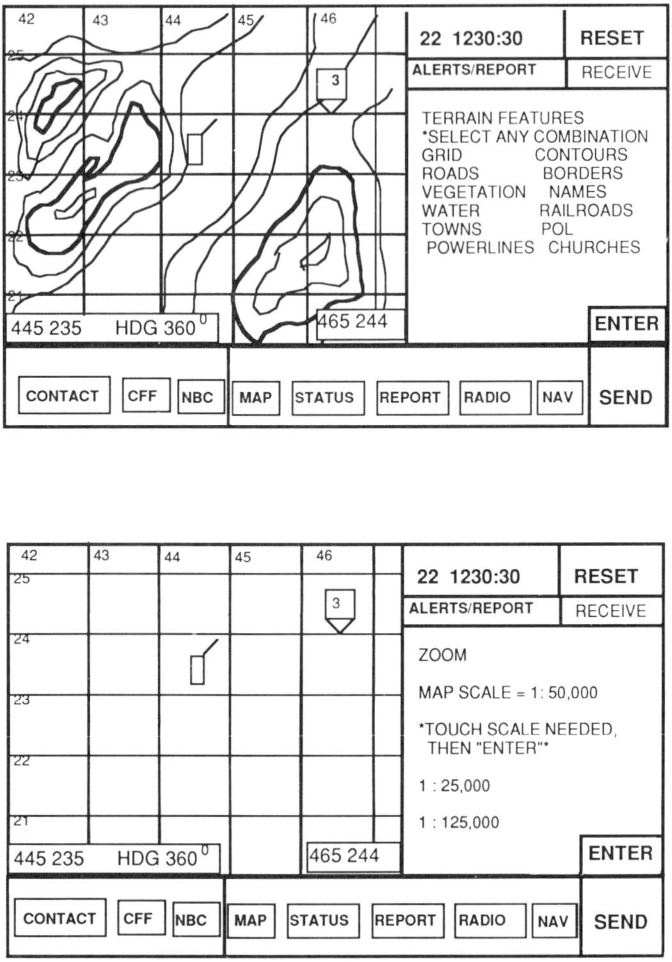

Figure 2.

for a limited number of C2 tasks. For each of those tasks the digital map was a significant asset.

These results indicate the level of specificity which can be achieved to measure small unit and crew performance in SIMNET-D. The results have proven extremely useful to the Army in making the case for a POSNAV system in ground combat vehicles. Caveats from the simulation included the substitution of SIMNET-D paper maps for military maps and that baseline M1 SIMNET-D simulators did contain azimuth indicators whereas actual M1 tanks do not. The relative cost figures for conducting this evaluation in SIMNET-D versus

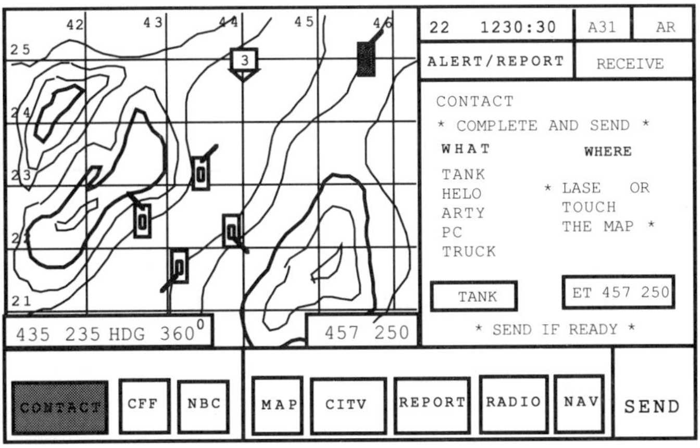

Figure 3.

attempting to obtain a similar quantity and quality of data from a field test have not been computed. However, a small scale innovative test conducted in SIMNET-D in 1987 reported a conservative 50 percent cost savings over a similar field-based effort (Schwab, 1987).

IVIS

The Intervehicular Information System (IVIS) provides a degree of automated command and control capability to the vehicle commander. It consists of a computer-based digital terrain map display, pre-formatted menu-driven reports, touch panel, and a radio interface unit for digital transmission of data between vehicles (See Figure 3). IVIS tasks are procedural in nature. Many tasks which tank commanders previously performed in a verbal mode using FM radio transmissions have been automated and reassigned to computer-based text and graphic modes using a touch interactive display panel.

ARI conducted an evaluation of a simulated IVIS system using four SIMNET-D M1 tank simulators. Six M1 platoons (24 tank crews) were trained and subsequently executed both crew and platoon missions using IVIS and six trained M1 platoons executed the same missions without IVIS. Data were collected on individual and collective performance during the time period October 1988 to February 1989.

One of the most important parts of the data collection effort for any command and control test involves determining what to measure and how to measure it. In the IVIS platoon level test, the measures identified for collection were divided into six categories. Specific data within each category were either collected automatically by instrumenting the simulated vehicles and by pre-programming the SIMNET-D data capture system or they were collected manually by observers and controllers. Categories of measures included: 1) Plan/Prepare for Tactical Operations, 2) Perform Platoon Fire Distribution, 3) Collect and Report Tactical Information, 4) Maintain Platoon Integrity, 5) Use Communication Nets Efficiently, and 6) Use Mission Resources Efficiently.

Some examples of automated measures included in this effort were: number of messages sent/received and time required to communicate changes in mission, that is, fragmentary orders, using IVIS; accuracy of initial call for artillery fire; fuel consumed; ammunition; distance travelled; and accuracy in maintaining platoon within axis. Manually collected measures included: number of messages sent/received and time required to communicate changes in mission without IVIS; time from receipt of fragmentary order until unit move out; and accuracy in maintaining platoon formations.

These exemplar measures indicate the level of detail required to begin to quantify command and control task performance. The researcher must carefully structure the scenarios to accommodate the performance measures while maintaining realism and the semblance of freeplay.

CITV

The Commander's Independent Thermal Viewer (CITV) was evaluated from April to August 1989 using M1 platoons in the SIMNET-D facility. Results from a CITV crew level test, using the Unit Conduct of Fire Trainer (Quinkert, in prep), identified several issues concerning soldier performance and training requirements for utilization of this device in platoon tactical missions. These included CITV's potential role in the following tasks: platoon fire distribution plan, consolidation, and reporting functions. SIMNET-D's capability to provide multiple vehicle implementation of new technologies, such as a CITV-equipped platoon, allows measurement of collective performance data which has not previously been possible.

Preparation of the simulation specifications for CITV (Quinkert, 1988) took approximately six months and involved the following steps. The information gathering step required obtaining engineering and functional descriptions from the tank program manager's office, from the Army's designated research and development agency, the Night Vision and Electro-Optics Laboratory, and from the potential users, namely, members of the Armor community. The second step was to develop alternative display switch arrangements by using a personal computer graphics program and working with subject matter experts (SME) from the Armor School. Alternative displays and functional capabilities were eliminated and refined in an iterative process as SMEs used mental rehearsal techniques to identify procedural or design flaws in the proposed display and control handle functions. The third, and final step, required integration of text description and graphic story boards into a single document to describe, for both the user community and the simulation software/hardware developer, simulation requirements for crew and platoon level implementation of CITV.

Drafts of the CITV simulation specifications were then coordinated with the simulation developer to ensure that requirements were feasible within the constraints of the simulation system from both an implementation and a cost standpoint. Requirements involved not only the software development but also the simulation hardware. Monitors and control handles for example, must be fit to the simulator much like the actual CITV hardware must meet the space and weight claims of the tank manufacturer. Development of the simulation specification document may be the most difficult step in the SIMNET-D utilization process because it requires a crosswalk between the operational world of the actual military equipment and the simulation world of computer-based illusion and selective fidelity. Selective fidelity, deciding what functions to make operational and what functions to ignore, is the key to success in the SIMNET-D simulation. The simulation specification writer must understand this and use it to an advantage from the standpoint of time, cost, and test validity trade-offs.

C^2 and the Future of SIMNET

Plans for the future of the SIMNET technology include long haul networking (LHN) of multiple simulator sites around the world to allow higher echelon warfighting. Simulators in different locations could fight on the same side or against each other. Simulators containing new technologies could fight fully crewed simulators containing current technologies. These LHNs provide the opportunity for contractor owned simulators at contractor sites to participate in warfighting demonstrations with government owned and operated simulators to test the efficacy of new technologies at minimal cost.

The ability to interlink different kinds of simulation devices through use of common SIMNET-D Ethernet protocols would provide the opportunity to conduct simulation-based warfighting exercises with other military services or to North Atlantic Treaty Organization allies who do not own SIMNET equipment. This is currently under consideration. The expansion into joint service and allied efforts will certainly challenge the capabilities of the SIMNET-D interface technology.

The Army plans to continue use of SIMNET-D capabilities to refine new materiel requirements as described in the POSNAV, IVIS and CITV efforts. However, efforts will be expanded to higher echelon units, tactical operation centers, and possibly to supplement operational testing of actual vehicles or hardware. It may also be possible to insert real military equipment into the SIMNET Ethernet to support embedded training of C^3 tasks.

SIMNET-D's hardware and software reconfigurablity, expandable computer-based network system, its simulated battlefield environment with task loaded workstations, relative low cost, and analytic capabilities make it an ideal medium for addressing soldier performance issues for automated command and control.

References

Black, B. A. (1989) "Using Simulation to Determine Future Training Requirements." Paper presented at the University of Central Florida Conference on Interactive Networked Simulation for Training. Orlando, FL.

Kroening (1986) "Army Command and Control Information Systems Requirements Definition." *IEEE Transactions on Systems, Man, and Cybernetics*, Vol. SMC-16, No. 6, November/December.

Chung, J. W., Dickens, A. R., O'Toole, B. P., and Chiang, C. J. (1988). "SIMNET M1 Abrams Main Battle Tank Simulation: Software Description and Documentation (Revision 1)." BBN Report 6323. Cambridge, MA: Bolt, Beranek, and Newman Systems and Technologies Corporation.

DuBois, R. S., and Smith, P. (1989). "Simulation-Based Evaluation of a Position Navigation System for Armor: Soldier Performance, Training, and Functional Requirements." ARI Technical Report 834.

DuBois, R. S. and Smith, P. (in prep). "Simulation-Based Assessment of Automated Command and Control Capabilities for Armor Crews and Platoons: The Intervehicular Information System (IVIS)." ARI Technical Report.

Quinkert, K. A. (in prep). "Crew Performance Assessment Associated with the Simulation of the Commander's Independent Thermal Viewer (CITV)." ARI Technical Report.

Quinkert, K. A. (1988) "Design and Functional Specifications for the Commander's Independent Thermal Viewer (CITV)." ARI Research Report 88-17.

Schwab, J. (1987) "Innovative Test of Simulation Networking—Developmental (SIMNET-D)." TRADOC TRMS No. 87-0000778. Fort Knox, KY: U.S. Army Armor and Engineer Board.

Human Performance in Military Command and Control

Michael A. Fineberg

Introduction

I believe that it is possible, looking into the future, to conclude that, almost certainly, a technologically advanced combatant in future warfare will be able to see all elements of an opposing force in real time, and will have at his disposal firepower means for reaching out to strike throughout the depth of the opponent's war-waging apparatus from his theater forces all the way back to his strategic reserves. Some naval officers have found it reasonable to say, vis-à-vis naval warfare, that it will be difficult if not impossible to steam around the seas with forces centered on a large-decked carrier, with protective rings of specialized air and submarine defense ships around that carrier. Some air officers have found it possible to say that we will have to find alternatives to the operation of air forces from large fixed airfields, whereon aircraft are processed for high sortie generation rates on something like an assembly line basis. I can assert for land warfare that the day will soon be gone when massed formations of armored vehicles will be able to swarm over the surface of the earth, trailing behind elaborate logistic tails. Instead we are going to have to move toward something like "distributed force," meaning that in order to provide protection we will have to disperse more broadly and thoroughly than ever before, and thus confuse the enemy as to which elements of the target array before them are particularly significant as threat. Our tactical dispositions will have to confront our foe with a large complex of target elements, each of which is potentially able to deliver punishing firepower, and each element of which could be capable of developing the intelligence requisite to the accurate delivery of that firepower.

(LTG Paul F. Gorman, 1981)

The fundamental premise of this chapter is that, even with optimal hardware/

software reliability and adequate quantity, quality, and timeliness of information, the achievement of command and control (C^2) effectiveness to the level implied by General Gorman remains problematic due to the inherent inconsistency of human behavior. The difficulty in predicting behavior within and between individuals is exacerbated by the hostile, highly stressful environment in which command and control tasks must be accomplished. As the demand for faster and more accurate C^2 has increased, attempts to meet it have most often been characterized by the development of hardware/ software solutions. In many cases these solutions have been appropriate and cost-effective when applied to such problems as the storage and retrieval of large volumes of data, the performance of myriad calculations, and the comparison/contrast of multiple sources of combat information. However, since no hardware/software (HW/SW) component performs perfectly, the effectiveness of the output of any C^2 system to which it belongs is reduced not only by this inherent imperfection, but also as a function of the required C^2 tasks and the human cognitive process on which they depend. Furthermore, since the applications of HW/SW mentioned above do not specifically require an analysis of operator tasks, any model developed to emulate these processes will be relatively simplistic with regard to the real world of C^2, and the contributions to model validity associated with consideration of the human component are not realized.

In addition to the perfunctory consideration of the operator in the types of applications mentioned above, there is a separate class of C^2 functions and tasks that is inherently unsuited to automation without an in-depth consideration of human cognitive parameters. This class is characterized by the higher order cognitive functions such as course of action analysis, targeting option selection, force allocation/deployment/employment and contingency response. To be sure, consideration of the operator has not gone totally unheeded, and some great strides have been made in the development of user-compatible interfaces and user/system integration software. What is still lacking is a deeper appreciation of the interaction between the internal capacity of the human operator and the external demands placed on him by the task at hand and by the stakes of the game. The concept of human component failure as opposed to human error is rarely employed to describe situations in which the capabilities of the operator are simply overcome by the situational demand. The need to place blame often overwhelms the valid requirement to establish cause and effect. Therefore, in order to be truly effective, any approach to the enhancement of C^2 system effectiveness must consider not only the threat driven modifications to the existing hardware/software, but also the fundamental capabilities and limitations of human operators and their interactions with situational stress. Furthermore, if a modeling approach is to be representative of the C^2 process it must be amenable to the analysis of human cognitive performance as well as machine performance.

In order to investigate these assertions, this chapter will first review the concept of C^2 and its increasing importance as we move into the 21st century. Following this review, some representative models of human cognition will be presented with their proposed application to the study of behavior in a C^2 environment. Finally, the idea of an overall model of cognitive performance and its integration into C^2 models will be discussed. In these discussions of human cognition three premises will be presented and are previewed here. First of all, it is proposed that the basic cognitive performance potential of the individual is constantly changing over time as a function of his internal emotional and psychophysiological state. Second, that this potential to solve problems is further mediated by psychosocial variables and one's own perception or appraisal of the situation. Third, that these internal variables interact with the real world demands of the situation at hand, such that actual performance is a joint function of capacity and demand. If these premises can be supported, then the implications are clear: As long as capacity exceeds demand, the human will perform within his prescribed limits. However, when situational demand exceeds the capacity of the operator to perform his assigned duties, the probability of system failure increases, and in theory, accelerates as the operator perceives his inability to meet job demands and realizes the consequences of that failure.

An Analysis of Human Performance Issues in Military Command and Control

This section provides an overview of the military threat, a brief history of C^2 from a human performance perspective, and a derivation of several human performance deficiencies that must be considered in the modeling of C^2 processes.

The command and control of maneuver units is an area of long-standing concern within the military services. Predictions of future battle scenarios from current Army AirLand Battle doctrine and from the Navy 21 study have revitalized, with a new sense of urgency, the military's commitment to develop more extensive and effective C^2 capabilities. Defined succinctly as the "exercise of authority and direction by a properly designated commander over assigned forces in the accomplishment of his mission" (JCS Publication 1), C^2 involves the coordination of information and people to facilitate correct, timely decision making and implementation. The objectives of any C^2 systems can then be restated as follows:

- To provide commanders with information on which to assess the capabilities and intentions of enemy forces,
- To correlate and manage information flow in order to generate viable options for action,
- To support effective option selection and operational battle planning, and
- To communicate the required actions to subordinate commanders.

The Threat to U.S. Supremacy in C^2

Warfare in the 21st century will be unlike any battle fought previously. The winner could be decided before the battle is joined since the combatant with accurate and timely knowledge of the other's position will have so great an advantage as to obviate the need to fight the battle. This paradoxical state of affairs will be brought about by the proliferation of highly capable information collection/dissemination systems, stealth technology, and low-cost, long-range, precision-guided weapons. The confluence of these three technological currents in the 21st century will make today's land, sea, and air forces immediately vulnerable to detection, recognition, and targeting by a sophisticated aggressor. The result of this threat is the expansion of the battle space in all three dimensions because the only defense against targeting will be dispersion of the force. Our vulnerability to accurate localization anywhere in the world and to the instantaneous lethal response of our enemy also blurs the distinction between our current concepts of strategic and tactical employment of forces.

Our response to such a threat against our land forces has manifested itself in the Army's AirLand Battle doctrine (TRADOC Pamphlet 525), the essence of which is an aggressive, dynamic, flexible use of forces with reliance on the competence and initiative of soldiers and the C^2 ability of their leaders. There are four guidelines which must be adhered to in order to succeed on the AirLand Battlefield: initiative, synchronization, agility, and depth (TRADOC Pamphlet 525-1).

Initiative implies an offensive spirit in the conduct of all operations. The underlying purpose of every encounter with the enemy is to seize or to retain independence of action. To do this we must *make decisions and act more quickly* than the enemy to disorganize his forces and to keep him off balance. (FM100-5)

Effective command and control is the channel through which this initiative must be applied to assure the successful conduct of the AirLand Battle. In order to make these decisions more effectively, a commander requires valid information at some optimal level of detail at the correct time.

Synchronization requires unity of effort throughout the force implying that every action of every element must flow from shared understanding of and commitment to the higher commander's concept. This means that all elements of combat power engaging the enemy must be committed toward achieving a shared and integrated set of objectives. In order to succeed on a battlefield in which synchronization is tantamount to survival, our military requires a C^2 system that provides *decentralized commanders/planners with instantaneous "shared perceptions" of the battle situation.* A unity of effort can only be supported in an organization that is knowledgeable of and committed to the objectives of the force as a whole.

Agility, according to FM100-5, "requires flexible organizations and quick-minded, flexible leaders who can act faster than the enemy," and *the key to agility is rapid decision making and execution.* Our military leaders must have timely, accurate intelligence, and must interpret and convert it into usable, understandable operations orders which can be executed by maneuver and firepower before the enemy has had time to react to the same situation. Because we must fight outnumbered, there will be more emphasis on maneuver on the battlefield. Warfighting will be increasingly decentralized and independent, yet based upon a common plan. Linear battle lines are giving way to mobile, fluid confrontations almost analogous to many-on-many dogfights. The terms "360 degree battle" and "pockets of conflict" are often used to describe the predicted conflict situation.

Depth on the battlefield refers to time, distance, and resources. The principle of depth implies delaying, disrupting, or defeating the enemy before he can engage and overwhelm friendly forces with his superior numbers. *It requires that the C^2 system predict/project in time.* In order to accomplish this, the C^2 system would be required to assimilate large amounts of friendly and aggressor combat information including numbers, armament, position, terrain, weather, etc. It also suggests that future C^2 systems may be guided by sophisticated expert systems which would peruse and integrate these data in regard to a set of decision rules.

The same precepts and principles described above for AirLand Battle theater level doctrine apply to the general concept of future strategic warfare. The key to this evolution is the changing world political environment and its implications for the military and the battlefield environment. The major implications for the military in general and for C^2 in particular are that:

- Our forces must be prepared to fight *anywhere.*
- Our forces must *win* the land battle. The only purpose of future military operations must be to win.
- *Parity* of combat power can only be maintained through quality of weapons, reliance on C^2 technology, tactics, leadership, and national will.
- Our forces must *avoid high combat losses* through maneuver, deception, and attacking the enemy's will to continue.
- We must *win the initial battles* because they have the potential to be so devastating that political settlements may be sought early.

The overall thrust of these statements is that we will emphasize agility, deception, maneuver, and precision firepower, in order to face the enemy with a succession of dangerous and unexpected situations more rapidly than he can react to them. In order to do this we must prevail in C^2 effectiveness. This requires an integration of man and machine capabilities to see, understand, decide, and act on the battlefield in a time cycle sufficiently faster than the enemy's so that he is limited to reacting and deprived of the initiative. Effective C^2 is the "high road" to achieving this advantage. (Fineberg, 1984)

In the battle scenarios of the early 2000s, conflict volume expands while time compresses. This results in more engagement volume to control within much shorter time windows. The extremely short time between detection and destruction obviates the segregation of C^2 by echelon, as has been the case. Under the new rules, all conflict may be viewed as "tactical" and therefore calls for the presence of command at the nexus of information collection. Therefore, new commanders in the battle for information must be trained in all aspects of C^2 and, more importantly, trained to view these aspects in a systems context. The computer-based, systems-oriented modeling to be discussed in this chapter is designed to support and train highly skilled and aggressive "information warriors" who will be on the front lines of the first and possibly only engagement in major conflicts of the future. The combatant who loses the battle for information will not survive as a viable force.

A Brief History of C² from a Human Performance Perspective

The one factor in the command and control of warfare that has remained constant throughout military history is the commander's need for accurate and timely location information. The combatant who knew where his enemy was, while remaining hidden from him, had a great advantage even if outgunned. Changes as a function of modernization have resulted primarily in an increase in the speed of information gathering, interpretation, evaluation, and dissemination, rather than in any qualitative, conceptual modification of the C^2 process. In the dawn of warfare the effectiveness of location information was purely a function of human performance capability. A scout was sent out in the general direction of the aggressor's homeland and instructed to seek out the enemy and report back. Basically, this was a patrol tactic and primarily defensive in nature. For example, a fleet might send out small, relatively fast vessels to scout ahead of the main body. Their job was to detect, recognize, and classify any ship that they came in contact with and then report back to the fleet commander as quickly as possible In the days of galleys and sailing ships, the time between detection of the enemy and bringing fire to bear on him could have been days.

This changed with the advent of various, primitive forms of remote communication such as signal fires, flares, flags, or heliographs. Since these all had to be line-of-sight communication, the time saved was the time it would have taken the scout to rejoin the main force. These few hours could have spelled the difference between victory and defeat. The scouts of opposing forces often met each other in their search, and soon became the first combatants in the battle for information supremacy.

The next major advance in reduction of information latency was the introduction of the balloons and airplanes to land and sea warfare. The airplane provided an incalculable advantage to its user since it could detect the enemy and report back in minutes, as opposed to the hours or days it used to take. The airplane was also able to cover a much larger area in much less time simply because of its vantage point.

Even with airborne information collection, the tempo of battle up until the time of World War II was slow enough that the commander, aided by his staff, could collect and process incoming information, develop options, evaluate them, select the optimal solution, and flesh out his plans and orders in real time. This situation began to change rapidly with increased maneuvering speed and improved collection capabilities. The increases in speed of maneuver reduced the time available to make decisions while the increasing volume of information multiplied the amount of work to be accomplished in that time.

In 1957 the "high ground" of information collection was opened with the launching of Sputnik. Since then the use of space-based information collection systems has become common-place. Overhead observation started out with somewhat primitive collection media and with limited coverage due to the small number of satellites in orbit and their narrow field of view. Since the early days of overhead information collection, great advances have been made in both the diversification of collection devices and the precision/accuracy of those devices. Tomorrow's information-collection technology will allow both sides to conduct all-weather, global, real-time surveillance of each other's forces. Coupled with global positioning systems, this information will enable real-time localization and targeting of assets anywhere in the world. The element of surprise will be removed from military endeavors, and with it the time to achieve a position of advantage with respect to the adversary's force. Therefore the first skirmish of tomorrow's battle may be fought between overhead systems in order to deny an adversary the information advantage. Information supremacy may become the most valid predictor of battle outcome, and the commander trained to achieve the most accurate and timely cognitive representation of the battle sphere will emerge the victor.

Human Performance Deficiencies in C^2 Derived From Future Combat Requirements

The approach to C^2 modeling taken in this chapter is based on relating human performance considerations to bona fide, C^2 requirements derived from the projected combat environment which will demand more of its combatants than ever before. As stated earlier, our force must take the *initiative* in all engagements in order to retain independence of action, it must have the *agility* in organization and leadership to act faster than the enemy, it must act in *synchronization* to maintain a unity of purpose while part of a "distributed force," it must understand *the depth* of the battle in space and time, and must be capable of *decentralized* operation in order to take advantage of fleeting opportunities (Fineberg 1984). The reasoning process employed to construct the linkages implied above is depicted in Figure 1.

The approach to constructing any set of C^2 human performance modeling

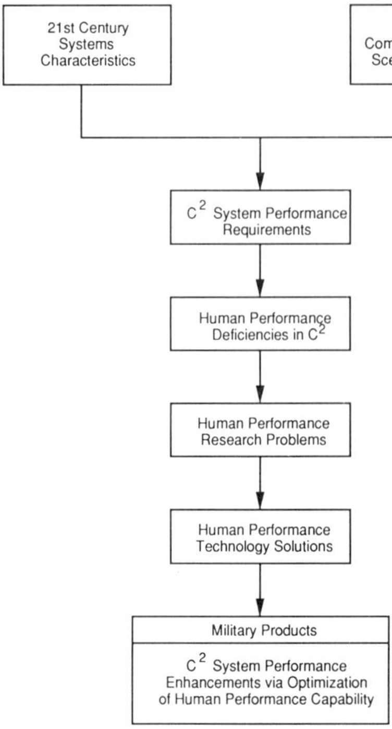

Figure 1. Human Performance Issues Development Process.

issues requires the integration of three major categories of information. The first, and by far the most important, category contains the deficiencies which the models are designed to resolve. The deficiencies must be derived from the combat requirements which are not filled by current capabilities. The C^2 requirements in turn must be deduced from the battlefield environment in which the system must function.

The second piece of information considered was the category of solution approaches. In the case of tactical C^2, these approaches are mostly technological and were derived from a review and extrapolation of the current state of the art in the field of information management. In the next section we will consider integration of the elements and illustrate this process with discussions of candidate model/constructs which may help to explain human performance deficiencies in the C^2 environment. A succinct presentation of the major human performance deficiencies as they relate to future combat requirements is shown in Table 1.

Table 1. Human Performance Deficiencies in C^2 Operation Derived from Combat Requirements.

The projected combat requirements for initiative, agility, depth, synchronization, and decentralization result in several deficiencies in the performance of the human component of future C^2 systems. Resolution of these deficiencies requires upgrades in user capability relative to the following areas:

1. Future battle managers, faced with the complexity and time constraints of the 1990s combat environments, will not be able to assign targets with the quickness and precision necessary to seize and retain independence of action. (Refers to the requirement for *Initiative.*)

2. Because of the numerous "pockets of conflict" on the integrated battlefield, the C^2 staff will be unable to analyze the battle situation quickly or accurately enough to generate effective options or plans. (Refers to the requirement for *Agility.*)

3. Given the compressed time frame and extended geography of the AirLand Battle, the C^2 staff will be unable to generate, select, and integrate battle plans with sufficient speed to optimize the deployment of limited forces. (Refers to the requirement for *Depth.*)

4. Because of the need for distributed combat power on the integrated battlefield, the commander will have difficulty communicating his plans and objectives quickly or accurately enough to maintain a unity of effort throughout the force. (Refers to the requirement for *Synchronization.*)

5. Given an enemy attack in depth, the commander will be unable to allocate his dispersed resources with sufficient accuracy and timeliness to put them into advantageous tactical positions. (Refers to the requirement for *Decentralization.*)

Approaches to the Enhancement Human Performance in the C² Environment

Human Performance Enhancement Technologies.

It is always much easier to levy requirements than it is to offer approaches to their resolution. However, in this case, a converging pair of trends—technological and doctrinal—does establish the framework for characterizing the information management systems of the 21st century. Accelerating advances in the development of compact, powerful, and fast computer systems suggest that every weapons platform or vehicle may have on-board processing capacity exceeding that which was available in all but the largest laboratories a few years ago. Reliable communications gear, using spread-spectrum and frequency-hopping techniques to avoid interception and interference, will allow these multiple processors to be netted together for real-time exchange of data.

While the optimistic technological projections of powerful on-board processors linked by advanced communications would seem to ensure that the necessary knowledge will be available for successful control and utilization of mobility and firepower, this "knowledge resource" must be carefully managed if it is to provide an advantage rather than a disadvantage. As the processing and communication capacity increases, choke-points in information flow and distribution will develop. Individuals will become overloaded beyond their capacity to process and communicate and will fall back on rough rules of thumb to select information which confirms their a priori view of the situation (Janus and Mann 1977).

Figure 2 illustrates the various sources of information that would constantly be coming to a tactical commander in such a "knowledge rich" situation. Essentially, the commander would be receiving information on a variety of dif-

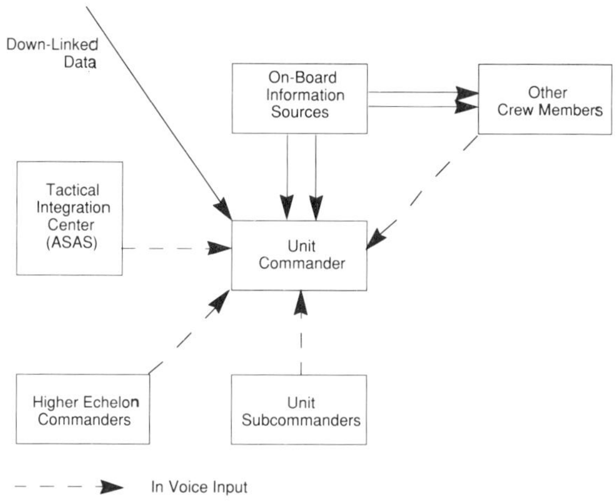

Figure 2. Unit Commander's Information Sources without "Intelligent" C² System.

ferent outside channels, any one of which could potentially require his full attention at a given time. In addition, the commander would need to monitor various displays of friendly force disposition and at the same time, be receiving constant updates on the overall battlefield situation from such capabilities as an All Source Analysis System (ASAS), perhaps downlinked through a Very Intelligent Surveillance and Target Acquisition (VISTA) system. Finally, the commander will be receiving commands or answering queries of higher echelon commanders, while directing and communicating with his subcommanders.

The technologies proposed in this section are the result of the painstaking literature review and in-depth deliberations carried out by a select group of C^2 experts in the summer of 1984. Known as the San Francisco Workshop, this effort brought together several of the nation's experts in C^2 requirements definition and C^2 technology development to share their perspectives and information and produce a comprehensive program of human performance enhancement. Their deliberations as reported by Fineberg (1984) have resulted in the selection of the technologies presented in Table 2.

The next natural step to formulating solution approaches is to relate the deficiencies described earlier to the technologies above in a manner which optimizes their interface. The deficiencies and technologies are arrayed in their respective rank orders along the two axes of a matrix, as shown in Figure 3. This enables the inspection of each of the 20 possible intersections (cells) that result from crossing four technologies with five deficiencies. Upon inspection of each cell experts were asked: "How valuable is this technology in resolving this deficiency?" For example, it was concluded that distributed information technology would be extremely valuable to the improvement of decentralized battle management. Therefore, an "H" was placed in the cell representing the interaction of these two elements. In a similar manner, each of the 20 cells was analyzed and marked.

Toward a Theoretical Model of Human Cognition in C^2

The foregoing discussions of human performance deficiencies and solution technologies was analytical in nature. In other words, the rules of logic were applied to the problem of C^2 in the 21st century and deficiencies deduced. In a similar manner, today's solution capabilities were extrapolated to predict the potential for enhancement of C^2 systems in the same time frame. This approach is purely descriptive and its results only as valuable as the underlying assumptions. In order to provide a more valid approach to the prediction and solution of C^2 human performance deficiencies, it is necessary to go beyond description to the level of explanation and that requires the development of theory. A theory of cognitive performance in C^2 has been sought for several years beginning with the SIMTOS studies (Halpin 1973?) and culminating in the ongoing research at

Table 2. Human Performance Technologies for Enhancement of C^2 Operations Ordered by "Effectiveness," "Maturity," and Feasibility.

Technology	Definition
1. Distributed information aids	Distributed information processing/displays, which support simultaneous decision making, e.g., electronic mail with confirmation/feedback loops
2. Expert systems and decision support algorithms	Decision protocols based on the rules used by human experts; procedure aids, artificial memory, information access aids.
3. Soldier machine interface optimization	Human factors engineering analysis,design, and assessment of the point of interaction between soldiers and computer.
4. Cognitive modeling	Analysis of expert problem-solving behavior and reconstitution of these elements into mathematical symbols and decision support algorithms.
5. Intelligent CAI in modular distributed applications	Computer-aided instruction conducted byan expert "tutor" system which can simulate the inputs of "virtual" team members in order to teach C^2 team operations to geographically dispersed trainees via netted simulation.

Human Performance Enhancement Technologies Ordered Based on: • Maturity • Feasibility • Effectiveness / Human Performance Deficiencies in DC^2 Operations — There is a Need for Improved:	Distributed Information Aids	Expert Systems and Decision Support Algorithms	Soldier Machine Interface Optimization	Cognitive Modeling and Simulations	Intelligent CAI in Modular Distributed Applications
Decentralized Battle Management and and Target Assignment	H	H	H	H	H
Situation Analysis and Distributed Decision Support	H	H	H	H	H
Information Management in Support of Option Generation and Battle Planning	H	H	H	M	M
Elaboration and Communication of Mission Objectives	H	N/A	H	N/A	M
Combat Resource Allocation and Reallocation	H	H	M	M	N/A

Legend:

H = High Value
M = Mid Value
L = Low Value
N/A = Not Applicable

Figure 3. Integration of Prioritized Deficiencies with Rank Ordered Solution Technologies.

the EDDIC facility at Fort Leavenworth. The purpose of this section is to describe several of the possible contributions to such a theory and then to integrate them within the framework of the appraisal process (Lazarus, Averill and Opton, 1974). In order to develop this theory we must fully understand the C^2 process, we must identify hypothetical constructs that explain C^2 functions, and finally we must integrate these constructs using an explanatory framework.

C^2 Function and Task Analysis Techniques

Prior to the development of any model, it is necessary to perform detailed function and task analyses on the process to be modeled. Fortunately, this difficult task has been initiated on many occasions. [Reason, Stewart, Tride, 1982; Fineberg, Conroy, Ross, 1984; Archer, Fineberg, Carter 1984]. An overview of these findings is presented as an introduction to theory integration.

The first level of analysis must be directed toward the overall job to be accomplished. The generic hardware, software, people, and procedures system that has evolved to support the C^2 mission is shown in Figure 4, in which the decision node or tactical operations center (TOC) is depicted as a situation anal-

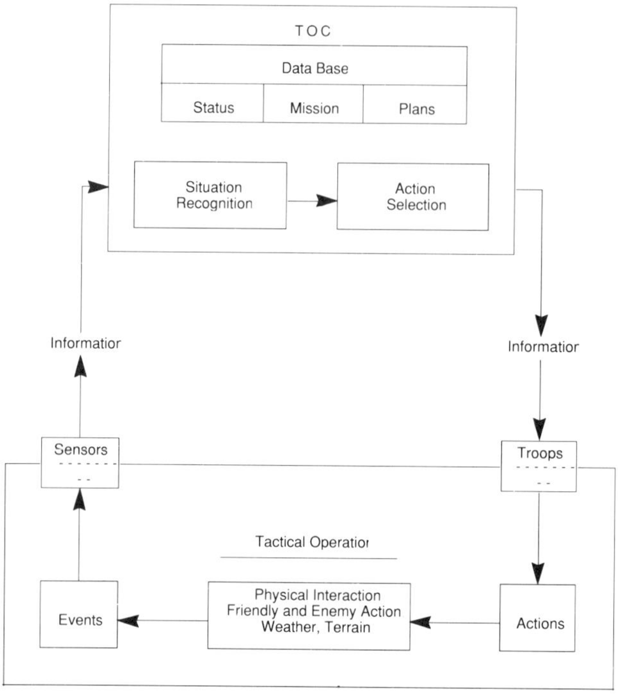

Figure 4. A Schematic of the C² System.

ysis, option generation, action selection process supported by a regularly refreshed data base. Today these actions are still performed "manually" with some automation support. The goal is to automate the process in the task areas most amenable to machine support and thereby to unburden the human decision maker to do what he does best—decide among a limited set of feasible options.

In order to develop meaningful models of the enhancement of human performance in tactical C², current procedures and techniques as indicated in Figure 4 must be understood as clearly as possible. The first step is to conduct an analysis of the various cognitive and behavioral tasks that must be performed to arrive at a clear perception of a tactical situation, generate, select, and integrate solution options, communicate those options, and allocate forces to implement them. Along with this, the essential elements of information needed to support tactical analysis must be determined, and some indication of the relative usefulness of different kinds of information must be provided.

The results of one approach to C² function and task analysis are shown in Figures 5, 6, 7. (Fineberg, Conroy, Ross, 1984) Figure 5 represents the highest level of analysis in which the fundamental command and control functions are

Command Functions / Support Elements	Perceive	Plan	Decide	Direct	Inform
Operations	• Aggregate FRENSIT Data • Evaluate Intelligence	• Keep FRENSIT • Evaluate Options • Plan OPSEC • Prioritize	• Query • Counsel • Select Options • Select Targets	• State Objectives • Allocate Resources • Prepare Orders • Coordinate and Communicate	• Interpret the Battle for Superiors • Inform Lower Echelons • Request Resources
Intelligence	• Observe • Query • Process/Correlate • Hypothesize • Present Results	• Keep ENSIT • Assess Intent • Plan Collection • Plan EW Activities	• Prioritize • Prioritize Processing	• Task Sensors • Support EW • Task Processing • Identify and Locate Targets	• Report Targets and Locations • Characterize Threats • Report on Resources
Communications	• Provide Sensor and Report Links	• Plan Links, Nets, and Relays • Develop Com. Options	• Establish Precedence Protocols • Select Circuits and Frequencies	• Manage Links and Networks • Execute Options	• Report on Status and Problems
Plans	• Provide Situation Models	• Develop Plans and Options • Assess Risks/ Rewards	• Prioritize Options • Prioritize Planning Resources Use	• Allocate Planning • Manage Plans Process • Evaluate Options	• Report Status of Plans and Options • Report Capabilities and Problems
Logistics	• Assess Logistics Status • Display Pertinent Data	• Plan Supplies • Plan Rendezvous • Plan Transport • Plan Maintenance	• Identify Resources in Accord with Battle Plan	• Task Depots • Task Transport • Task Maintenance Elements	• Report Status on Supplies, Transport, and Maintenance
Administration	• Provide Personnel and Administration Data	• Develop Admin- istration Plan	• Prioritize Support	• Manage Adminis- tration Functions	• Report Administra- tion and Personnel Status and Problems

Figure 5. C^2 System Function Review

	Perceive		Plan		Decide	Direct	Inform
Functions	Monitor Environment	Understand Situation	Generate Alternatives	Predict Consequences	Choose Among Options	State Objectives Allocate Resources	Report/Interpret
Tasks	Observe Query Receive Display Data	Structure and Review Data Hypothesize Situations Display Perceptions	Query Assess Formulate Describe Coordinate	Postulate Model Game Assess	Query Deliberate Counsel Select Option Coordinate Up and Down	Organize Prepare Orders Coordinate Transmit Confirm Receipt	Assess Prepare Reports Transmit Confirm
Measures	Accuracy of Data Age of Data Completeness	Accuracy of Perception Timeliness Correctness of Hypotheses	Completeness Timeliness Number of Options	Correctness Completeness Timeliness	% Life of Plan Selected Correctness Timeliness	Accuracy Timeliness Number of Queries	Accuracy Timeliness Number of Queries
Unmet Needs	Sensors --24 hrs --Al Wx --Cueable Automatic Interfaces Reliable Info. Transfer Syst.	Auto Commo Interfaces Correlation/ Aggregation Algorithms Purge Algor. Graphic Portrayal Hypothesis Tester	Plan-Gener. Algorithms Graphic Portrayal Elect. Mail Networking Protocols	Simulations Models Relational Data Base Graphics	Decision Supt Algorithms Graphics Elect. Mail Distributed DBMS	Elect. Mail Interactive Graphics Commo Relays	Elect. Mail Interactive Graphics Commo Relays Networking Protocols
	← Improved Personnel Selection, Training, and Motivation Techniques →						

Figure 6. C^2 Job Analysis for Operation Element

	Process	Man Unaided by Machine	Machine Without Man	Man/Machine Interaction Leverage
Input and Output	Receive, Transmit	Communicates mood of sender	Increased effective channel capacity; hard copy	Little*
	Verify	Slow; error prone	Fast; virtually error-free	Little*
	Input/Output Tag	Slow; error prone	Fast; virtually error-free	Little*
Pre- and Post Decision	Sort	Can sort on content; can generate sorting keys; slow; error prone	Can sort only on keys	Significant
	Associate	Can generate algorithms, but slow and error prone at associated operations such as File/Post/Plot/Retrieve	Limited to predetermined algorithms, but very much faster and error-free at associated operations	Significant
	Aggregate/Organize	Can generate algorithms and formats, but slow and error prone at associated operations: File/Post/Plot/Retrieve/Calculate	Limited to predetermined algorithms and formats, but very much faster and error-free at associated operations	Significant
Decision	Interpret/Validate	Only man can flesh out incomplete patterns and generate new hypotheses and tests for them	Can only extend human memory (associated operations) and facilitate hypothesis testing (calculation)	Tremendous
	Evaluate/Coordinate	Only man can interpret in context and generate hypotheses and insights	Can only extend human memory and facilitate coordination based on a priori rule	Significant
	Project/Extrapolate	Only man can define projection and extrapolation parameters; involves hypothesis generation	Can only extend memory and facilitates calculations	Tremendous
	Generate Alternatives	Only man can generate new hypotheses	Can only extend memory permitting consideration of larger data base; retrieve a priori alternatives for evaluation; facilitate rendering in hard copy	Tremendous
	Evaluate Alternatives and Decide	Man excels at structuring problems to be solved and establishing boundary conditions; man must make final decision	Can only permit much more rapid calculation with more sophisticated models; can apply decision criteria more rapidly; less error	Tremendous

* Machine can perform these processes without aid of man.

Figure 7. Comparison of Man and Machine Processing Capabilities

opposed to the commander's support staff elements whose duty is to implement these functions. The major point here is that each staff element has duties which both complement and overlap the duties of others. The apparent redundancy in functional responsibility is designed in so that if one element is unable to carry out its duties, they may be shared by those that remain. This inherent equipotentiality is a factor often taken for granted in military systems or other information management systems whose outputs are vital. In Figure 6 the same C^2 functions are presented only for the operations element of the command staff in order to allow for further analysis of those functions as an example of the depth that is necessary to support valid modeling. In addition to subfunctions and tasks, this figure identifies candidate measures used to quantify input and output data and presents some illustrations of C^2 needs currently not being addressed by current technology.

The next step in the analysis process would be to perform a function allocation study. A description of this process concerning the tactical command staff is presented below. As a reminder, the mission of the tactical staff is to plan, execute, monitor, and control battlefield activities through to the input, processing, and output of information.

Automation of C^2 systems impacts upon human reliability/performance. It is important that those tasks allocated to automation are those that a machine will do better, that is, more accurately/quickly. The human operator must believe that the automation introduced is capable of the tasks allocated to it. Further, the human operator must understand what the "black box" is doing so that he maintains a sense of the "whole" system. This is especially important when faults occur and diagnoses and remediation is needed. If these precautions are not taken, system reliability could be adversely affected, not improved, with automation.

Therefore, we now turn to the question of allocating these functions to people or automation which is illustrated in Figure 7. Listed in the first column are the information processes required for the input, pre-decision, and decision functions. The processes required for the post-decision and output functions have not been repeated since they are the same as for pre-decision and inputs, but in essentially inverse order. The second column lists the dominant characteristics of unaided man in carrying out each process, while the third column lists the dominant characteristics of the automated system (machine) by itself. The fourth column indicates the potential gain associated with combining the capabilities of man and machine,that is, of providing computer support to the process of decision making. (Lussier, 1986)

The figure shows that automation of the input and output offers improvement except for the loss of the abstract tone of voice information associated with verbal communication.

The next three processes, which are invoked for pre- and post-decision processing, are distinctly amenable to man/machine integration. While only man can provide the basis for sorting, the association algorithms, and the relationships needed for aggregating and organizing, he is very slow and error prone in the actual conduct of these processes.

Investigation of the last five decision processes shows that not only are man's creative talents required, but that they must be applied on line as the information is being processed. On the other hand, the computer provides the ideal medium to support the decision maker. Not only can it retrieve any data item in memory, but it can display it in whatever manner the decision maker desires. Further, the machine can operate on the data according to instructions and perform calculations without error, and, finally, it can accept new items created by the decision maker as he develops and tests alternative options. At still higher levels of artificial intelligence, the machine can store and retrieve both enemy and friendly tactics and status to further assist the decision maker. (Lussier, 1986)

Once these mission/function analyses and function allocation activities have been completed, the building blocks of a C^2 model become apparent. In other words, the modeler now knows the steps within the process and the inputs, outputs, and quantities necessary to their completion. Although the preceding discussion has demonstrated that automation can help reduce time lines, provide decision aids, and manage uncertainty of the data base, it has, however, also demonstrated that achieving these goals may not be quite as simple as we would like. Any system or organization composed of men and machines really has four distinct variables or dimensions that interact in ways which may be difficult to predict. These variables are:

- The human capabilities and limits
- The technological capabilities and limits
- The individual task structure
- The procedures for accomplishing the tasks required by the mission.

Any effort to improve system function involves a whole series of compromises and trade-offs between the capabilities and limitations of the system components. Change any one of these variables—for example, technology—and the compromises previously worked out may no longer be valid. In general, a change in any one requires changes in all the others in order to exploit the potential improvement to the utmost. Therefore, the C^2 system designer still needs the instructions for the assembly of the components. The integration plan or theory that will predict how and why these dimensions interact and will predict the outcome of their interaction. The next section is devoted to a discussion of theory development with regard to human performance in C^2. Several theories and hypothetical constructs will be reviewed, and an overall explanatory mechanism which links them all together and relates them to the results of the C^2 functional analyses will be proposed. As this mechanism is developed, it will become apparent that the concept we have called "human performance" heretofore must be reassessed relative to the tasks to be performed and the physiological/emotional "state" of the human operator. What is really at issue is not only the question of performance quality and timeliness, but also the reliability of that performance. Can we count on the operator to do the "right stuff" from time to time, task to task, and person to person?

The Elements of a Human Performance Theory

Human performance reliability is defined as the probability that some set of human actions will be performed successfully in a specified time or opportunity. Therefore, the human, with his "failure rates" and "repair times," may be seen as the most complex component in an even more complex C^2 system. In the early years of human reliability analysis (HRA), there was a dearth of human performance data with which to develop a model for human reliability assessment. This shortcoming led to development of the AIR Data Store (Munger et al., 1962). However, these task performance data were of limited applicability because they lacked a systematic approach to the classification of human performance by various tasks.

Berliner (1964) approached this problem by developing a classification scheme which allowed for tasks to be decomposed into elements from a behavioral perspective. This provided an effective schema for relating human performance data to various types of cognitive and psychomotor tasks. (See Table 3, from Woodson, 1981.)

Rasmussen et al. (1981), in response to the challenge posed by Three Mile Island, developed a model of human decision making which followed Berliner's classification scheme. Rasmussen began with the perceptual process split into detection, observation, and identification, and ended with a motor process he called execution. The model was a compilation of ideas from the information and cognitive sciences that represented the mechanisms an operator uses in moving from the detection of an off-normal event to acting on it. An example of an overlearned or "reflex" mechanism is indicated on the figure as a release of a preset response that connected the execution of the response directly with the detection of a need to respond.

A recent refinement of Rasmussen's taxonomy (see Figure 8, Embrey, 1986), divided Rasmussen's concept of rule-based behavior (which is implied by the emphasis on procedures) to include both knowledge-based and skill-based behaviors. Knowledge-based decisions occurred when no specific rules or procedures are available to assist the operator and he must rely on heuristics born of experience; skill-based decisions include pattern recognition, and actions which are manual, well trained, well-known, and practiced frequently. Quantification of these decision making processes was begun in order to model these cognitive mediators of human performance. The interaction of man and machine processes was now viewed as a complex information management system. A time reliability correlation (TRC) concept was adopted to model the time needed to pass and process information in order to respond to the question of how thinking time affects human reliability. Swain and Guttman (1983) in developing the Technique for Human Error Prediction (THERP), adopted a TRC called the "nominal diagnosis model" to be applied to the initial diagnosis needed to choose the correct procedure at the beginning of an off-normal event. These techniques may be applicable to a military C^2 situation.

Table 3. Berliner Classification of Tasks (from Woodson, 1981)

Processes	Activities	Specific Behaviors	
Perceptual	Searching for and receiving information	detect observe receive survey	inspect read scan
	Identifying objects, actions, and events	discriminate locate	identify
Mediational	Information processing	calculate compute interpolate tabulate	categorize encode itemize translate
	Problem solving and decision-making	analyze choose compute predict	calculate compare estimate plan
Communications		advise communicate indicate instruct transmit	answer direct inform request
Motor	Simple, discrete tasks	activate connect hold lower press raise	close disconnect join move set
	Complex, continuous tasks	align synchronize transport	regulate track

Methodological Uncertainties in the Investigation of Human Reliability in C2

Current human reliability theory is founded on the idea "that the seemingly immense variety of human errors observed may reflect complexity of the environment, rather than complexity in the psychological mechanisms involved." (Rasmussen,et al. 1981.) Therefore, popular models relying on task analysis such as Swain's THERP and models using task network simulation such as Siegel/Wolf and MICRO SAINT may not adequately account for all cognitive mechanisms involved in facilitating and mediating human performance. In order to progress toward a general theory of human cognitive performance, several of these mechanisms/issues are discussed below.

The first issue concerns the logic and mechanisms of the physical and conceptual apparatus that mediates cognitive behavior. Much of our memory seems to be associated with concepts, ideas, and beliefs that are abstract, imprecise, and based on uncertainty. Solutions to problems, then, are often based upon what we believe will happen rather than what experience, training, and logic may dictate.

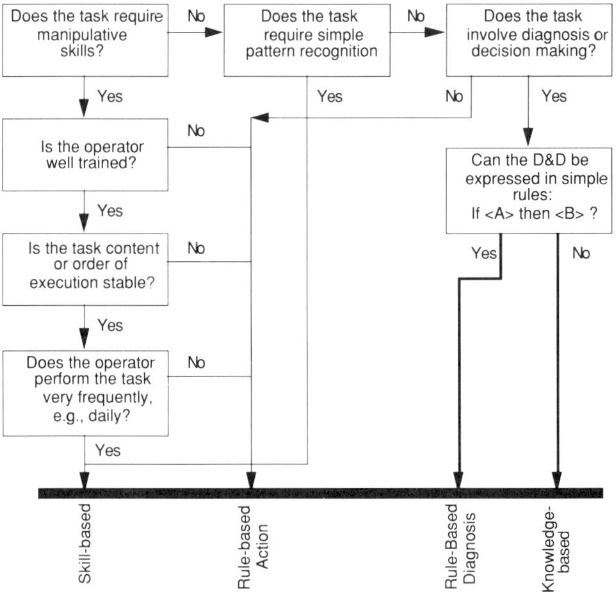

Figure 8. Embrey's Taxonomic System (Embrey, 1986).

Further, the human brain can hold inconsistent beliefs at the same time, often doggedly (Hoffer, 1951, and Janis, 1982). This tenacity of beliefs in the face of fact, and its influence on behavior can lead to decisions with disastrous results (e.g., the Vincennes disaster). This is especially true in complex systems when the demand of the situation may overstress our capability to respond rationally.

We must also be concerned with the linkages that mediate our relation to the world. Human reliability in automated battle management and C^2 systems will depend on the degree to which the products of our perceptual processes reflect "ground truth." Even so, people see, to a great extent, what they anticipate they will see (Neisser, 1976). They report what is consistent with their world view, often irrespective of the facts (Janis, 1982). Therefore, much of our cognitive processing and decision making may be based on long-held schemas and thus can go on without our consciousness of the activity. This not only means that people typically do not know how they do things, but also that they are uncertain as to why they do them.

In regard to the mechanisms underlying choice behavior, we find that human errors fall into two major categories (Norman, 1983). Knowledge-based and rule-based mistakes are failings or inadequacies in planning, decision making, and diagnosis, whereas "slips" are failings in the implementation and monitoring of actions. These behavior types seem to involve radically different mechanisms as noted by Rasmussen et al. (1981), Embrey (1986), and Reason (1983). Depending upon the mechanisms involved, radically different methods of enhancement—training, selection, automation, or some combination—may be required. In addition, making the right choice among options is often clouded by the emotional aspects of decision making. People make decisions routinely and unconsciously under the pressures of the moment. "Hot cognition" (Janis and Mann, 1977), not cold calculation, is the basis of most human decision making, especially in the operation of complex C^2 systems under wartime conditions. If the heat of combat has such a great influence, then design of automated systems depending on high levels of human reliability must account for this uncertainty. Currently, stress is modeled in THERP by increasing the error probability for a task if operator stress is assumed. In network simulation models such as Siegel/Wolf and SAINT, certain degrees of stress reduce the task accomplishment probability. Although these subjective estimates of the impact of stress on human reliability have not been measured empirically nor validated, their formulation is a testimony to the importance of stress as a decision mediator.

Furthermore, soldiers thinking and acting together do not necessarily do so harmoniously. Their group behavior may produce errors at higher rates than individuals exhibit and the group's behavior itself may even become irrational, ignoring and suppressing evidence against already held beliefs (Janis, 1982).

As we see from the foregoing discussion, there are many uncertainties associated with predicting the reliability of a person's internal problem-solving mechanisms as that individual operates in the real world. His responses may be based solely upon internal prejudgment at one extreme or driven totally by objective, external stimuli at the other extreme. In addition, each soldier brings his peculiar "wiring" or cognitive schema including knowledge, rules, and skills to the decision process which drives his choice behavior and the subsequent execution of tasks resulting from his decisions.

In summary, then, a number of factors contribute to an environment that fosters frequent and severe failures in C^2 systems performance generally, and human reliability in particular. These include:

- the integration of different individuals' internal problem solving and decision-making architectures with constantly changing situational variables created by the stress of combat;
- the notorious unreliability of complex C^2 systems; and
- the pressures of group decision making.

However, proper modeling of both intrinsic and extrinsic factors (Reason, 1983) matched to meet the demands of the task can assist in the early application of appropriate personnel selection and subsequent training and decision support.

As indicated by the foregoing analysis, the assumption that problems are solved logically and subsequent decisions made on the basis of accurate input is not always true. Often, an individual's belief in the face of other conflicting information, albeit valid, can lead to errors in the decision making process. Although this phenomenon is commonly exhibited across various tasks, environments, and situations, the reasons for beliefs which prevail in the face of facts are not well understood as the phenomenon varies both between people and within any one person across time (Streufert, 1978).

A Theory of Human Performance in C^2

In order to facilitate the development of a comprehensive theory or model of C^2 human reliability, several constructs or hypotheses have been identified which, when taken together, could provide a model of human performance in C^2 suitable for initiating further research and analysis. These components came from the areas of experimental social psychology, physiological psychology, and cognitive psychology, and are presented very briefly to provide the building blocks of a human performance model.

The first component concerns the concept of situational stress and the inherent hypothesis that stress is cumulative and additive. In other words, the general level of stress in the environment takes some toll on the individual's capacity and that his demand is exacerbated by the identification and recognition of additional, situation-specific stressors. For example, consider the Vincennes incident. The commander, already in a stressful environment produced by the recent disaster aboard the U.S.S. Stark, is programmed to protect his ship at all costs. He is under attack by several small craft and in the midst of this situation his sensors identify an unknown aircraft taking off from hostile territory on an intercept course. It is proposed that this last stressor was additive to, if not multiplicative with, the existing level of anxiety in the combat information center of the Vincennes.

However, it is the case that people behave appropriately under stressful conditions all the time. This observation then raises the question of internal capacity, that is, the emotional resources within and around the individual which can be brought to bear in times of stress to initiate and sustain adaptive decision making. These resources have been divided into group factors and individual factors by Lazarus (1974 and 1984) and Drum, Gal, Levin and Fineberg (1986). The individual factors include personal well-being, training, and role in the situation, while the group factors include leadership, commitment, and cohesion. These can be seen as antidotes to the deleterious effects of stress (Drum et al., 1986).

The relationship between the "venom" and the antidote can be viewed in terms of a capability versus demand model of behavior which provides the third component of the theory. The capability versus demand model as described by Santilli (1980) discusses this relationship in terms of a pilot oper-

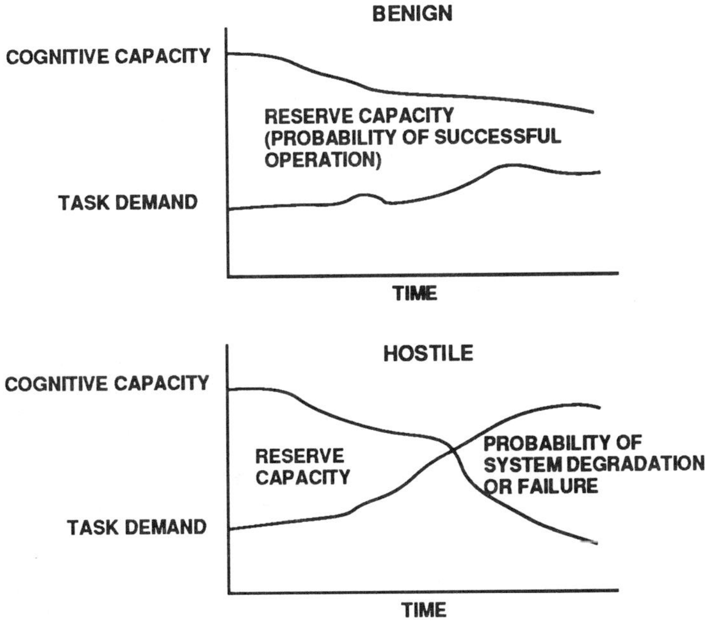

Figure 9. Relationship of Task Demand to Cognitive Capacity for a Benign and Hostile Situation.

ating a high-performance fighter aircraft; however, operators in the Abrams Tank or tactical decision makers in a combat information center (CIC) may face similar issues. The basic formulation attempts to explain the integration of intrinsic and extrinsic factors with regard to human performance reliability (Reason, 1983). Santillie's hypothesis states that as long as task demands stay below the cognitive capability threshold of the decision maker, the system performs within tolerance limits. As the demand approaches the threshold or as the operator's threshold is lowered due to fatigue, boredom, or toxic insult, the operator's ability to respond to task requirements is degraded. This relationship is shown in Figure 9.

Potential for C^2 system failures in this model rises sharply as the cross-over point for these two curves is approached. The area between the curves prior to the cross-over point can be defined as the probability of successful performance. The area between the curves after cross-over may be viewed as the probability of system failure.

To complicate the issue, the individual decision maker relies upon many different logic styles that imply different specific activities. The ratio of demand to capacity as it interacts with any one of these internal decision styles may lead to increased time and decreased accuracy. Indeed, the interdependence of these styles with the level of stress may provide different appraisals of the same situation from individual to individual.

An example of an existing cognitive schema that could play the role of

Table 4. Categories of Information Processing Style.

Unidimensional (Uses Only a Single Aspect of Information Available)	Multidimensional (Uses Two or More Aspects of Information Available)	
• Low--Uses discrete categorical judgments	Differentiator does not view multiple aspects of information as interactive	Integrator views multiple aspects of information as synergistic
• Normal--Uses continuous judgments	• General--Recognizes two or more but picks one-dimensional outcome to direct responses • Closed--Hierarchical--Uses dimensions whose priorities are predetermined and are not open to modification • Excessive--Uses large number of dimensions generating inordinate numbers of decision alternatives	• Low--Recognizes multiple dimensions, but uses two or more simultaneously to guide responses • High--Greatly integrative capability • Closed--Hierarchical--Integrates only predetermined priorities of dimensions which are not open to modification • Non-Closing--Uses multiple dimension. Integrates over several criteria. Lacks ability to close for selection of decision alternative

"appraiser" in the developing model is Cognitive Complexity Theory. The underlying construct of Cognitive Complexity (Streufert, 1978) asserts that one's ability to make effective, timely decisions is some interactive function of the environmental information load of stress and the organization of one's information processing apparatus. It may be that one's level of cognitive complexity can be a predictor of his ability to perform certain tactical decision tasks. Furthermore, this indicator may be useful for both initial selection and for measuring the effectiveness of training. Streufert's categories of decision making are presented in Table 4.

There are other explanations of cognitive appraisal (Lazarus, 1974, 1984; Drum et al. 1986); however, they all have one thing in common, and that is the proposition that people respond to their perception of any given situation, not to the actual/veridical situation per se. Furthermore, that appraisal is a function of learning, experience, emotional states, and inherent decision making mechanism and style. In other words, one person's challenge may be another's anxiety attack.

The final component of the theory is drawn from classic physiological psychology, and it is the contention that the act of decision making produces certain autonomic responses and automatic physical, physiological, emotional, cognitive, and social behaviors (Drum et al., 1986). Furthermore, that these changes in body status and in behavior are perceived by the individual either consciously or unconsciously and are interpreted as signs of the adaptiveness of the decision or action that preceded them. For example, if an individual makes an adaptive decision but is so anxious from the effort that the subsequent responses can be perceived as fearful, this information can be fed back into the decision process and cause the individual to review his decision. Humans are the only animals who can predict the outcome of their decisions albeit inaccurately.

The next step in the theory/model development process is to construct a framework within which to integrate these concepts that allows for their interaction and predicts decision outcome. Such an integrative model is shown in Figure 10. This model of human performance reliability in C^2 is based on the work of Lazarus (1974), who first proposed the appraisal concept, and on the work of Drum et al. (1986), who refined it for use in predicting the psychological response of troops in tactical nuclear conflict. As hypothesized in the figure, the general and specific stressors may interact with rumors and misinformation to impose a demand on the C^2 decision maker. The medium of this demand is viewed as stress which is an aggregate of the general stress inherent in the C^2 situation and tasks and the specific stress imposed by the immediate threat under assessment. Inputs also include the resources or capability of the individual. These capabilities are a result of group and individual support factors as described earlier. The stressors yield the intervening variable of demand while the resources yield capability. To this point the situation is rather straightforward. In fact, it can be viewed in a medical analogy. Consider the demand to be analogous to the presence of viruses or bacteria in the environ-

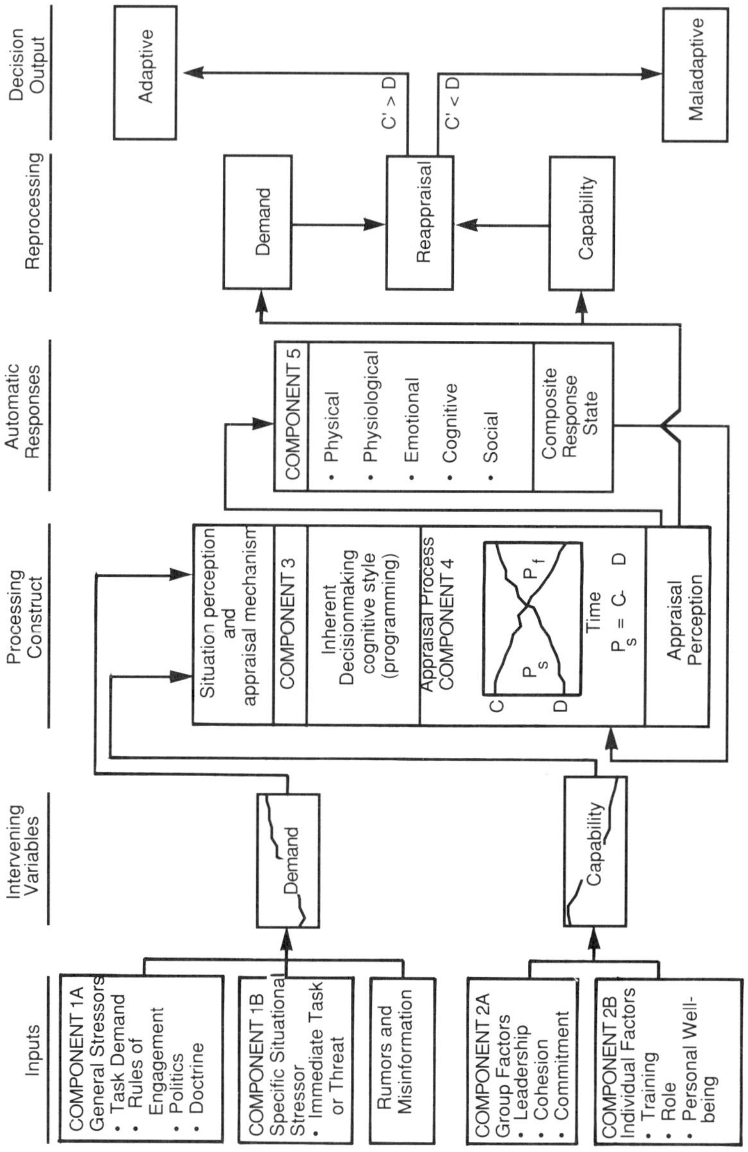

Figure 10. A Model of C^2 Human Performance in Decisionmaking

ment and the capability to be like the body's "resistance" or immune system. Most of the time the body's defenses are strong enough to fight off the daily attacks by environmental bacteria; however, there are times when a new strain of bacteria comes along or when the body's defenses are overtaxed. In these instances opportunistic infection may take place. In a like manner, there are times when the stress of the decision making task may outstrip the cognitive and emotional resources of the individual, thus lowering the constantly changing probability of C^2 system success.

In the model, the intervening variables of demand and capability are compared during the appraisal process. It is here that the notional curves of stress and resources are drawn over time and compared using the internal decision style specific to that individual. In some cases this comparison is a very simple one-dimensional process, while in other individuals many different dimensions of the decision inputs may be studied. Single is not necessarily better or worse than multidimensional processing. The success of the decision often depends more on the relative speed of closure of the information-gathering effort, than on how many decision variables are "studied." On the one hand are the individuals who close prematurely, who in effect say "don't confuse me with more facts, I already made my decision." On the other hand are those individuals who continually collect more and more data on multiple variables but fail to make a decision within the window of opportunity,that is, they are "overtaken by events" (Streufert 1978). In any case, it is proposed here that the processing construct is composed of a decision making apparatus and decision style that is "hardwired" and pre-programmed for each individual and an appraisal process based on the comparison of perceived demand to one's assessment of his own capability.

If capability is believed to exceed demand, the output will result in benign responses signalling a state of well-being. Even though the aftermath of the decision may leave some stress indicators, the intensity of these responses is perceived as decreasing. If capability is perceived to be less than demand, the autonomic/automatic responses related to stress and now anxiety/fear rise exponentially. No matter whether these indicators rise or fall, their effect is felt at various levels of consciousness through its impact on the appraisal process via a feedback loop. As indicated earlier, only man knows when he is getting into trouble and can predict the outcome of his actions. If he thinks he is doing the right thing, the unconscious predictions of future well-being will be input to the appraisal process, thereby reducing the stress level under which the appraisal is taking place. On the other hand, if he sees that the outcome of his decision will lead to dire consequences, the automatic responses are those of fear and anxiety, and when inputs to the appraisal process raise the background level of stress, the probability of further adaptive decision making is lowered. In other words, there are at least two iterations of situation appraisal, the first comparison of demand to capability, then a second comparison made in the context of an adjusted ambient stress level which is a function of one's own automatic responses to the decision making process and one's prediction about the outcome of the situation. Once these iterations are complete and reappraisal is accomplished, the decision is made. If, as a result of this process, capability does in fact exceed demand, there is a high probability that an adaptive or correct decision has been made. If demand exceeds capability, the probability of a successful, adaptive decision may be reduced.

One must be reminded that all of the foregoing discussion is purely theoretical. But that is exactly what has been intended. It was the goal of this chapter at the outset to assess the relevance of human performance issues to C^2 modeling. From this discussion, it is obvious that C^2 system effectiveness may be seriously compromised by impacting adversely on several different areas of the decision making cycle. Conversely, it has also been shown that there are many approaches to optimizing the contribution of the human component in the C^2 system. Improvements in human decision making performance are attainable through better training and C^2 team building, through decision aiding and personnel selection procedures based on cognitive style. It is also possible to improve decision making effectiveness by short circuiting the negative feedback from automatic responses through desensitization/relaxation exercises employing guided imagery. In fact, it has been shown (Fineberg, 1989) that guided imagery is effective in establishing a specific mental set in experimental subjects. In any case, the days of representing human performance as a zero or a one in C^2 system performance prediction models must come to an end. The concept of human performance in C^2, no matter what its ultimate explanation, can no longer be viewed as a discrete variable either 100 percent on or off. It must be recognized as a multidimensional component of C^2 system performance with continuous measurement on each of its many aspects. This will undoubtedly add to the difficulty of C^2 system modeling and to the complexity of the resulting model. However, to ignore the inherent richness and variability of human performance in C^2 is to delude ourselves into overestimating the effectiveness of our present C^2 hardware/software systems and to deny ourselves an extremely cost-effective medicine for C^2 system performance enhancement.

References

Archer, M.A., Fineberg, M.L., Carter, C.F., 1984, *Identification and Analysis of Battalion Command Group Behavior*, SAIC Final Report 1984

Army Field Manual FM100-5 Operations

Berliner, C. et al., "Behaviors, Measures and Instruments for Performance Evaluation in Simulated Environments," paper presented at a Symposium and Workshop on Quantification of Human Performance, Albuquerque, NM, 1964.

Drum, B.H., Gal, R., Levin, S., and Fineberg, M.L., *Psychological Response Study: Annual Summary Report #1*, SAIC Report No. DNA 001-86-6-0295, McLean, VA, October 1987.

Embrey, D.E., "SHERPA: A Systematic Human Effort Reduction and Prediction Approach," Proceedings of the International Topical Meeting on Advances in Human Factors in Nuclear Power Systems, American Nuclear Society, April 21-22, 1986.

Fineberg, Canoy, Ross, *Training Manager's Guide for C² Training*, SAIC Report MDA 903-81-C-0254, 1984.

Fineberg, M.L., *An Integrated Plan for Human Performance Enhancement in C² Operations*, SAIC Report No. 84/1825, November 1984

Fineberg, M.L., Levin, S., Gal, R., *Psychological Response Study Annual Summary Report #2*, SAIC Report No. DNA-001-86-C-0295, in preparation.

Hoffer, E., *The True Believer*, Harper & Row Publishers, New York, 1951.

Janis, I.L., *Victims of Group Think*, Houghton Mifflin, Boston, 1982.

Janis, I.L., and Mann, L., *Decision Making*, Free Press, New York, 1977.

Lazarus, R.S., Averill, J.R. and Opton, E.M. (1974), "The Psychology of Coping: Issues of Research and Assessment;" in G.V. Coelho, J.E. Adams and D.A. Haneburg (eds.) *Coping and Adaptation*; New York, Basic Books.

Lazarus, R.S., Fohlman, S. (1984) *Stress Appraisal and Coping*; New York, Springer Publishing Co.

Lussier, J.W., 1986 Guidelines for Automation: *A How To Manual for Units Receiving Automated C² Systems,* ARI Research Product 86-24, ARI, Fort Leavenworth, KS, 1986

Munger, S.J., Smith, R.W., and Payne, D., *An Index of Electronic Equipment Operability*: Data Store, AIR-C43-1/62-RP(1), American Institute for Research, Pittsburgh, PA, January, 1962.

Norman, D.A., "Position Paper on Human Error," NATO Conferenace on Human Error, Bellagio, Italy, September, 1983.

Rasmussen, J., Carnino, A., Griffon, M., Mancini, G. and Gagnolet, P., *Classification System for Reporting Events Involving Human Malfunctions*, RIS-M-2240, RISO National Laboratory, Denmark, March, 1981.

Reaser, J.M., Stewart, S. and Tiede, R.V., *A Methodology for Identification of Command and Control Group Behaviors*, USARI, Alexandria, Virginia, 1982.

Reason, J.T., "Absent-mindedness and Cognitive Control", in *Everyday Memory, Actions and Absent-Mindedness*, Academic Press, 1983, pp. 113-132.

Santilli, S.R., "Critical Interfaces between Environment and Organism" in *Class A Mishaps: A Retrospective Analysis*. USAFSAM-TR-80-3, Brooks AFB, TX, 1980.

Streufert, S., *Human Component in Decision Making, Managerial Control and Organization Democracy*, H. V. Winston & Sorr, 1978

Tradoc Pamphlet 525-1, *Airland Battle Doctrine*

Woodson, W.E., *Human Factors Design Handbook*, McGraw-Hill, 1981.

Performance Assessment Framework

Eva L. Baker, Harold F. O'Neil, Jr., and Robert Linn

We view a definition of performance assessment as consisting of two components: performance, which is the application of knowledge and skills to appropriate stimulus problems; and assessment, which is the valid measurement combined with judgment of quality and decision.

The Performance Assessment Framework (Baker & O'Neil, 1989) was developed with assistance from experts in measurement, assessment, and information science. The complete Framework is provided in Appendix 1. Our claim is that this framework may be used with minor modifications to investigate performance assessment issues in Command, control, communications and intelligence (C^3I). There were three related purposes for the design and development of a framework in the performance assessment area. These purposes are listed below to explain the motivation behind the design of the Performance Assessment Framework.

To facilitate access to existing databases archiving research and development studies related to performance assessment.

This use led us to use information science conventions in the organization of the framework. It also required the inclusion of descriptors from existing information systems (e.g., NASA) to provide access to the relevant databases. The decision to use existing terminology from existing information systems resulted in broad searches with high retrieval (i.e., many documents or hits) but low precision (i.e., few relevant documents). The best strategy for testing the framework's utility to access literature may be to use free text searches (of titles and abstracts) based on specific performance measures. In fact, this approach was used to generate a Performance Assessment Literature Review (O'Neil, Kazlauskas and Baker, 1989). While these terms are usually much less precise than those we would like, their omission would have rendered our framework less than useful for accessing extant literature. Thus, as a result of this decision, we also included structures and terms that relate to more general aspects of assessment.

To define with key words the critical attributes of the performance assessment domain for use in the design of a specific DOD-useful database in the area.

This purpose of potential use to our customer drove us to organize the framework and to select terms that emphasize the specific attributes of performance assessment that distinguish them from other kinds of measurement. The result of this decision was the inclusion of terms that are less likely to access extant literature in a computerized free text search of information systems.

To use the structure and terms of the Performance Assessment Framework as a conceptual framework to judge the following: a) extant R&D studies of performance assessment in general, b) the R&D plans for new performance assessment measures, and c) R&D studies of a specific performance assessment measure.

Specific performance assessment tests normally get judged by ad hoc criteria developed by special committees or by the reviewer locally. Our framework could be used as a standard reference checklist to aid designers and reviewers in their judgment of the quality of extant or new performance assessment approaches. Therefore, we have attempted to include elements of importance to the National Academy of Sciences Committee on the Performance of Military Personnel (Green & Wing, 1988).

Organization of the Framework

The Framework is organized into six top levels:

1.0 Performance Assessment Targets—the focus of the assessment (e.g., individual job performance).

2.0 Performance Assessment Domains—the substance of what is being assessed (e.g., common tasks).

3.0 Performance Assessment Approaches—the types of assessment and methods and instruments for data collection (e.g., types of tests).

4.0 Psychometrics—the technical issues pertinent to performance assessment (e.g., validity).

5.0 Interpretation and Reporting—how meaning is attached to the data and the formats and users for reports (e.g., norm-referenced).

6.0 Costs—cost considered as an element in performance assessment (e.g., research and development costs).

Within these categories are subdivisions, divided to usually no more than five additional levels. In general, the more specific elements, at levels four and five, are provided as illustrations rather than as an exhaustive list. These levels are meant, in part, to provide slots into which additional specific information could be added by either users, researchers, or developers. The format of the framework represents these levels doubly for ease of understanding. First, consistent with other DOD document organization, a decimal-based numbering system is used. Second, consistent with information science conventions, indentation is used to indicate additional levels.

Validation of Performance Assessment Framework

This section describes our efforts to assess the utility of the Performance Assessment Framework, developed for the area of performance assessment (Appendix 1). Three approaches were taken to accomplish the validation goal. First, the classificatory system was reviewed in terms of standards for considering taxonomies in general, as well as in the context of the specific purposes of our system. Secondly, the major analysis of the framework occurred in the context of a particular performance assessment technique, namely, the Skills Qualification Test. Third, subsets of the Framework and the Skills Qualification Test literature were reviewed independently by two experts (Drs. Baker and Linn) in assessment to identify gaps and to assure that the system would signal future research directions. The outcome of these three exercises was to generate the revised version of the Performance Assessment Framework and is documented in more detail by Baker and O'Neil (1989).

Internal Validity

Without question, the most relevant and comprehensive line of inquiry pertinent to our task of assessing the quality of the Performance Assessment Framework has been conducted by Fleishman and others over the last twenty years. In their 1984 book, *Taxonomies of Human Performance*, Fleishman and Quaintance describe in great detail alternative uses and structures of systems designed to organize phenomena in psychology and other scientific areas. Early in their book, they present a "Hierarchy of Taxonomic Terms" (p. 22) that distinguish among the theoretical science of classifications, their product (such as classificatory systems), processes (such as identification and classification), and the objects of classification, called units. Fleishman and Quaintance focus on the critical issues in developing a system, particularly the structure and relationships of constituent objects, with regard to critical issues such as boundaries, level of specificity, and size of categories. The intellectual task is to keep in mind what attributes of similarity and difference should be optimized in a classificatory system (p. 23). An example might be the terms house, mouse, grouse, and mice. Under a system where life science objects were paramount, mouse, mice, and grouse would be grouped together. Under linguistically oriented systems, house and grouse, as words that serve both subject and verb functions, would be grouped together. Finally, if we were counting letters, the three classifications would consist of grouse/house-mouse/mice. Classification depends upon the attributes of interest.

Thus, the uses of classificatory systems influence their structure. The justification for their design focuses on the benefits of economy and communication. For instance, ease of summarization and of information retrieval would be criteria consistent with these uses.

Validity is a comprehensive term that denotes "soundness," that is, a valid object is well-grounded in principles, logic, or evidence. Campbell and Stanley (1963) wrote about psychological experiments in terms of internal validity (their quality as judged by experts) and external validity (the quality of the results they generate). The concepts with classificatory systems are defined similarly. Internal validity relates to the organization, logic, and coherence of elements within the system. External validity refers to the extent to which the system is confirmed by application to broader phenomena, that is, whether it meets its objectives.

Special attributes of internal validity should also be considered. One is reliability and that means here that classificatory terms are used consistently as they appear in the system, and that items assigned to the system would be classified similarly on multiple trials. Fleishman and Quaintance (1984) also discuss the need for mutually exclusive categories on the same horizontal level, so that a unit could be placed in one and only one category. An additional issue involves the criterion of exhaustiveness, that is, whether all units can find a home in the categories. Finally, the authors refer to Cronbach (1971) and the notion of internal construct validation. Essentially, this process refers to having experts review and adjust the classificatory system to reflect the best understanding of the principles underlying the system.

One of the complexities that emerged in our analysis was the multipurpose nature of the system we were developing. We became aware of the need to juggle the conflicting demands of a system that was to be useful both for descriptive and research planning purposes and for information retrieval purposes. For the former purpose, an organization that moved from general to specific with equal numbers of subordinate categories would be desirable. For the information retrieval purpose, however, influenced as that process is by the nature of the architecture of existing databases, the desire is to get directly to the lowest level in the system that would access with the fewest trials the literature of interest. Compromises were made in attempt to meet both sets of needs.

We chose to review our framework in terms of standards for the development of systems for information retrieval, since we believed that this purpose was of principal interest to our sponsors. These standards have been described elsewhere (op. cit. 1984) and have been modified to our purposes in Table 1:

Table 1

Requirements of Systems for Information Retrieval

Large number of studies in the area (or anticipated growth)
Absence of systematic attempt to organize the area
Wide variety of tasks
Interest in the scientific, technical, and policy community
Terms should be those already in use in the research literature

Again, these criteria demonstrate that the information retrieval requirements conflict with the research direction-setting function of the framework (e.g., large number of studies in the area, common terms). This will be a persisting tension for any multipurpose scheme.

Internal validation of the Performance Assessment Framework occurred in two forms, judgmental and empirical. On the judgmental front, two experts in performance assessment (Drs. Baker and Linn) reviewed the framework, each on two separate occasions, with a three-month interval between reviews. The second review resulted in the proposed addition of related terms (6) that is, functional synonyms, the addition of new terms (12), and the identification of a section that was redundant and might be consolidated in another category (4 terms in the section). A third review resulted in the final addition of 15 relevant terms, 19 new terms, and one consolidation of one category. The framework in Appendix 1 reflects these changes.

Two different empirical exercises were conducted. First, the original Performance Assessment Framework (without the recommended modifications above) was transformed into an alphabetical list, including duplicated terms. The list found in Baker and O'Neil (1989) contains 380 words, 358 unduplicated, or a 94 percent rate of uniqueness. These results for uniqueness are relevant to the mutually exclusive criterion for internal validity. Inspection of the duplicates and their placement in the framework suggest that the level of redundancy is not excessive and that the particular terms are major constructs related to performance assessment and thus are appropriately repeated.

Given the results of data and review processes we put in place, we believed the findings were sufficiently positive for us to move toward the external validity issue.

External Validity Test

We used a literature review (O'Neil, Kazlauskas and Baker, 1989) to focus the selection of a specific performance technique to be used as a context to judge the utility of the framework. With respect to the selection of a specific technique, we suggested the following criteria: (a) select an individual performance assessment technique because the psychometrics are worked out; (b) select an institutionalized technique rather than a research technique because the "lessons learned" would be more useful; (c) select a technique that has an extensive literature so as to provide a good test for the "validity" of our framework. The only performance assessment technique that met these criteria was the Skill Qualification Test (SQT).

The literature on the Skills Qualification Test was secured in the manner specified in a prior report by O'Neil, Kazlauskas and Baker (1989). They used the term SQT to search against both NTIS (National Technical Information System) and ERIC (Educational Resource Information Center) databases using free text searching. The number of hits in the NTIS database was 41. The number of hits in the ERIC database was 7. Each record was printed in full. These citations were read by Dr. O'Neil and categorized as relevant based on the abstract. There were 40 relevant SQT documents in NTIS and seven relevant documents in ERIC. However, six of the seven documents in ERIC duplicate exactly documents in NTIS. Thus, the SQT population for our review from these databases consists of 41 relevant documents (40 from NTIS and 1 from ERIC). The abstracts for the relevant SQT documents can be found in O'Neil, Kazlauskas and Baker (1989). Hardcopy of the remaining documents was requested from NTIS (we had the one from ERIC). We required hardcopy to facilitate the review. ADI had in its library several of the documents. NTIS provided hardcopy of all required documents with three exceptions. Thus, 38 documents constituted the sample from the information systems. However, we noticed that there were no SQT regulations in the sample. Thus, we queried a colleague at the Institute for Defense Analyses (Dr. Dexter Fletcher) regarding such documentation. He provided four documents regarding SQT regulation from the Pentagon library.* In summary, our sample consisted of 42 documents (38 from Information Systems; 4 from the Pentagon library).

A major technical problem was to determine what data inherent in SQT documents could assist us in an empirical examination of the soundness of our Performance Assessment Framework. The use of the SQT was selected because of its maturity as an individual performance assessment technique and the relatively large documentary base it had accumulated in the area of military performance assessment. However, we immediately perceived that one limitation of the SQT with respect to the ADI Framework was the fact that the SQT is

---

* We wish to thank Dr. Fletcher for his assistance.

focused on a restricted domain and circumscribed by certain facts: it is criterion-referenced, focused on procedural skills, targeted for individuals, and limited to the Army. In contrast, the Performance Assessment Framework attempts to be inclusive, relating to many forms of testing that apply across a broad range of tasks. It also systematically addresses units, organizations, and the man-machine interface as opposed to only individuals. It obviously attends to Defense-wide issues. Thus, the utility of the SQT documents for assessing in a comprehensive way the utility of the Framework is clearly limited.

However, two uses for the SQT documents were conceived. One use is to provide a context to assess analytically the use of the framework for accessing existing research documents in computerized databases. A second use is to review the SQT research documents to determine the extent to which the framework offers new areas for systematic research on performance assessment instruments in general. In the second function, it must be reiterated that our role is not to make judgments about the scientific quality or practical utility of previous SQT efforts. Rather, we are interested in whether the framework serves the purpose cited by Fleishman and Quaintance (1984) of "exposing gaps in knowledge" (p. 6). Our approach would require that we review the SQT documents in the light of research areas identified by experts

One obvious approach to judging the usefulness of our framework to access existing literature is to obtain by searching on the name of a performance assessment technique (e.g., the SQT) all available documents in existing government databases and to search the databases again using terms from our framework to conduct the search. This approach, if properly done, would be iterative, requiring successive passes using subsets of terms, and would permit the direct computation of the percentages of the population of SQT literature accessed through our terms. Unfortunately, this approach is inappropriate for a number of scientific as well as practical reasons. First, given the organization of the databases for the National Technical Information Service (NTIS) and ERIC, the best solution to obtaining SQT documents is to do what we did already in O'Neil, Kazlauskas and Baker (1989), that is, search for them by name. Thus, the problem of finding all SQT documents is a spurious one. What would be of interest would be to obtain documents on performance assessment that were similar on the dimensions signified by our terms (for example, cost studies, research on individual assessment, computer-administered performance assessments). Such studies would be prohibited by limitations in time and available resources.

Rather, we chose to use inferential approaches to assessing the quality of our framework. We accomplished this part of our task by comparing terms in our framework to two sets of other terms: titles of SQT research, given by the author(s) of the documents, and key words assigned to the documents. We compared the terms in our framework with these other sources and in the process exposed issues that directed us to options for revision or future use of the Framework.

Our first task was to reread all obtained SQT documents and then sort them into as many mutually exclusive categories as possible. Forty-two documents were included in this phase of the study. An initial sort was made into four categories: Research and Evaluation; Development and Validation; Descriptions and Guidelines; and Administration and Regulations. A second sort was conducted and four documents were reassigned. A third sorting validated the distribution.

We believed that dividing the SQT documents into these categories would help determine if differences occurred among categories and would assist us in refining our framework and identifying research gaps.

The attributes that differentiate the categories are relatively clear. Category I, the Research and Evaluation category contains studies that investigate performance on the SQT as related to other student measures, report feasibility studies on SQT expansion or extension to new problems or contexts, or conduct impact studies on the SQT. In addition, some studies focus on other areas and use the SQT primarily as a dependent measure for other interventions. Most, but not all of the work in this category is empirical. Almost all of it is relatively short-term and very applied. Studies in Category II, Development and Validation studies, are divided between heavily empirical studies and descriptions of procedures that should be used in the design of SQT or similar instruments. Validation procedures are discussed, and psychometric issues such as the relationship of the SQT and ASVAB (Armed Services Vocational Aptitude Battery) and the establishment of job performance standards are discussed. Category III, Descriptions and Guidelines, tends to have a relatively limited empirical base, and its documents focus on how to implement the SQT or descriptions of ideal considerations in the practical development and implementation of the SQT. Category IV simply consists of regulations and changes thereof that specified procedures for the development and implementation of the SQT policies.

The distribution of documents to category is presented in Table 2.

Table 2 Distribution of SQT Documents by Category of Report

Category	Number of Documents
Research and Evaluation	21
Development and Validation	6
Description and Guidelines	11
Administration and Regulation	4

Our second task was to begin to assess the nature of the SQT document database as a means to improve our own framework. To do so, we created a file of the content words contained in the titles of the reports; this file simulated a population for free text search using our terms. To insure a cost-effective methodology, we excluded the abstract and other information that a complete computerized free text search would accomplish. Our technique of focusing only on titles allowed us to use capabilities of our word processing software and thus did not require the purchase or development of specialized software. (As will be noted later, our approach is conservative and will underestimate the degree of relationship between our framework and the literature.)

To construct the file, we identified all semantic units of meaning, and applied exclusion rules to function words (e.g., articles, prepositions). Content words then were listed. Our only major difficulty concerned boundaries among individual words and terms that have specialized meaning in SQT or Army contexts. For example, "training" and "evaluation" can be cited separately, but "training evaluation" has special meaning in the defense environment. We relied on the expertise a staff member with more than ten years of DOD management experience (Dr. O'Neil) to review our division of terms. This content word list formed the database to which we compared our framework terms. We must reassert that our terms were developed for a wider set of content than the SQT.

Table 3 summarizes the information in this list by document category.

Table 3. Frequencies of Title Content Words by Document Category.

Category	Total Words	Duplications	% Unique
Research & Evaluation	79	24	72
Development & Validation	27	16	41
Description & Guidelines	53	14	74
Administration & Regulation	16	8	50
Total for All Documents	175	62	66

Table 3 suggests that a relatively low rate of redundancy occurred in the title content word lists excepting the Development and Validation category. Furthermore, 42 percent of the redundancy resulted from the use of "SQT" in 20 document titles. An additional 40 percent can be accounted for by the use of three more terms: "effectiveness," "training," and "development."

The next logical task was to compare the terms in the Performance Assessment Framework and SQT title content word lists. The purpose of this comparison was to determine the match between the ways authors categorized their documents and the analyses we conducted to develop the framework. Defined match meant exact match between semantic units. A restricted list of tolerated differences was permitted: 1) in form (procedural/procedure, ASVAB); and 2) in number-singular or plural (procedure/procedures).

Our finding of 30 exact matches, excluding duplicates, shows that our Performance Assessment Framework overlaps with 26 percent of the unique content words in SQT document titles. (25 percent if the terms "Skills Qualification Test" or "SQT" are eliminated from both lists.) In order to determine whether this overlap is sufficient for search purposes, we then compared the documents with the exact match list to ascertain how many titles would have been discovered by searches (excluding the Skills Qualification Test and SQT terms) using the terms of the Performance Assessment Framework. (Recognize that the requirement for retrieval in a search does not call for a complete match for all content words but only that the searched term appears at all in the title.) On the basis of our analysis, we determined that we would have retrieved 34 documents, or 81 percent of the universe obtained through the SQT title search, using the overlapping terms in our Performance Assessment Framework. Four of the documents were not in computerized databases; if they are deleted from the computation, our percentage drops slightly to 79 percent.

We believe that this rate of projected retrieval provides a conceptual validation for the information retrieval utility of our framework in part because we believe this percentage is a serious underestimate of projected performance. For example, consider the difference between titles and abstracts as the objects of free text search. The longest title consisted of eight content words. On the other hand, we might expect between five and 10 times that number of content words in an average abstract. Nonetheless, we discovered that only minor modifications in our framework would dramatically increase our already high projected retrieval rates. For example, adding the term "system" raises the percentage to 84 percent; adding "soldier" raises the percentage to 89 percent, and adding "correlation" results in 92 percent projected retrieval or a total of 35 of 38 documents. However, in a real search, this additional retrieval would be at the expense of precision.

A second analysis was done using the SQT documentary database. Its purpose was to explore choices of additional terms for inclusion in the Performance Assessment Framework and to judge empirically our list on one of Fleishman and Quaintance's criteria (1984)—the use of a common vocabulary by users. For this analysis, a list of key words provided by the SQT document authors was created. (This list is presented in Baker & O'Neil, 1989.) Key word provision by the SQT document author is a requirement for inclusion in the database we searched, but anomalous as it is, only 24 of the 38 retrieved documents came with a key word list. A total of 183 key words were provided, 81 of which were unique, resulting in a 44 percent uniqueness factor. An exact term match was conducted, using the same procedure described above for the title content word lists, except that this time key words and Performance Assessment Framework terms were matched. The number of exact matches was 10.

In addition, we computed the exact matches among all three lists. Such terms would be useful for titles of future documents or descriptors for constrained searches for literature in the performance assessment area.

Based on analyses of content words in titles and our framework, we have projected a very positive rate of retrieval for the system, thus in part validating our selection of terms. Clearly, such work would need to be replicated with literature from other sectors (e.g., teams, units, Navy) to be confirmed. Furthermore, looking at the key words provided by authors provided us with a list of words that might be recommended for consideration, as appropriate, if a goal is retrieval of literature in computerized searches. Furthermore, all our findings must be considered gross underestimates since we were restricted by time and budget to titles and did not use the text of abstracts in our study. We could project that virtually all documents would be retrieved using our terms and a normal free text search of title, abstract, and funding agency text. However, the precision would be very low (see O'Neil, Kazlauskas and Baker, 1989, for a discussion of this issue).

In this section, we did not deal with well-known issues in the information science literature such as the influence of the level of domain knowledge on the efficiency of database searchers, the problems of ambiguity and technical knowledge of abstract writers, the need for domain knowledgeable librarians to assist in searches, and so on. One non-obvious issue is the extent to which authors seek to characterize their work in ways that they perceive to be prestigious or to meet supervisors' expectations, a practice that results, for example, in titles containing the term "Research" even though the document in our eyes does not qualify as actual research.

Another purpose of this activity was to review the utility of the Performance Assessment Framework for research and development planning. This aspect of our work is very speculative, since the overall impact of a classificatory system on research and development planning is likely to be marginal at best when compared to real requirements-driven R&D effort. Nonetheless, in the review of the Performance Assessment Framework in the context of the SQT document database, we considered its planning utility.

Three experts (Drs. Baker, Linn, and O'Neil) reviewed the documents and the SQT materials, made notes, and discussed their opinions. The review involved focusing on the research and validation documents and the entire framework. In contrast to previous sections, this aspect of our work was holistic and speculative. The issue was whether direction could be provided by the framework. For example, one finding might have been that certain categories were very well-represented by the literature. Our inference then would be that the direction setting function was not relevant for these categories.

The overall belief was that the framework could be very useful in setting directions for R&D in this area. For example, one source for this opinion for the distribution by R&D category of documents in Table 2. What was surprising was the relatively small amount of validation work or experimental studies present in the development of the measures themselves. In fact, many of the studies that were conducted in the development and validation category were not focused on the SQT but rather used the SQT as a criterion for assessing predictions by the ASVAB, for instance. Thus the existence of our framework, particularly section 4.0 Psychometrics, 5.0 Interpretation and Reporting, and 6.0 Costs, would be sparsely filled sections, based on the SQT documents alone. In addition, although the largest classification of studies was in the research and evaluation area, these studies often evaluated something else as well (e.g., the Soldier's Manual), or conducted non-databased feasibility studies rather than addressing the dimensions included by experts in the Performance Assessment Framework. For example, written component, reading skill, and readability were variables of interest in some of the research. Clearly, attributes related to section 3.0 of the Framework, that is, the section on Performance Assessment Approaches, could identify other serious issues for attention for research, issues that are driven either by existing policy (e.g., personnel reports) or that would explore the use of new forms of performance records. These statements are meant to be heavily qualified and are based on reviews conducted without knowing the policy context guiding the development of the documents we reviewed.

In conclusion, we believe that our Performance Assessment Framework in the command, control, communications and intelligence area could be productively used in the following manner: a) to facilitate access to existing databases archiving research and development studies relating C^3I to performance assessment by concatenating a specific system with specific terminology from our framework b) design a C^3I and performance assessment database by using the Framework terminology with a detailed requirements analysis, and c) use the Performance Assessment Framework as a conceptual framework to judge the following:

- extant R&D studies of C^3I and performance assessment,
- R&D plans for new C^3I performance assessment measures, and
- specific R&D studies of extant C^3I performance assessment measures.

References

Baker, E. L., and O'Neil, H. F. Jr. (1989). *Performance Assessment Framework*. Sherman Oaks, CA: Advance Design Information.

Campbell, D. T., and Stanley, J. C. (1963). Experimental and quasi-experimental designs. In N. L. Gage (Ed.), *Handbook of Research on Teaching*. New York: Rand McNally.

Cronbach, L. J. (1971). "Test Validation," in R. L. Thorndike (ed.), *Educational Measurement*. Washington, DC: American Council on Education.

Fleishman, E. A., and Quaintance, M. K. (1984). *Taxonomies of Human Performance: The Description of Human Tasks*. Orlando, FL: Academic Press.

Green, B. F., Jr., and Wing, H. (1988). *Analysis of Job Performance Measurement Data: Report of a Workshop*. Washington, D.C.: National Academy Press.

O'Neil, H. F., Jr., Kazlauskas, E. J., and Baker, E. L. (1989). *Performance Assessment Literature Review*. Sherman Oaks, CA: Advance Design Information, Inc.

Appendix 1: Performance Assessment Framework

1.0 PERFORMANCE ASSESSMENT TARGETS

(R) Performance Evaluation

1.1 Individual Job Performance
- (R) Employee Evaluation
- 1.1.1 Employee Efficiency
- 1.1.2 Employee Productivity
  - (R) Employee Proficiency
- 1.1.3 Can Do (Used to limit the target of the Joint Services Job Performance Project to what military personnel can do under supervision.)
- 1.1.4 Will Do (Used to identify performance on a day-to-day basis—more apt to be captured by ratings than by performance tests.)

1.2 Team Performance
- (R) Unit Performance
- (R) Group Performance
- 1.2.1 Readiness
- 1.2.2 Sustainability
- 1.2.3 Effectiveness
  - 1.2.3.1 Mission Accomplishment

1.3 Human-System Performance
- (R) Human System Evaluation
- (R) Man-Machine Performance
- (R) Man-Machine Evaluation
- (R) System Performance

1.4 Equipment Performance
- (R) Equipment Evaluation
- 1.4.1 Weapons Systems
  - (R) Weapons Systems Evaluation

1.5 Military Training
- 1.5.1 Army Training
  - 1.5.1.1. Soldier Training
- 1.5.2 Marine Corps Training
- 1.5.3 Air Force Training
- 1.5.4 Navy Training
- 1.5.5 Joint Training
- 1.5.6 Rank
  - (R) Paygrade
  - 1.5.6.1 Enlisted
  - 1.5.6.2 Non-commissioned officers
  - 1.5.6.3 Officers
  - 1.5.6.4 Civilians

1.6 Technical Training
- (R) Training Evaluation
- 1.6.1 Teacher Effectiveness
- 1.6.2 Health Personnel Performance
  - 1.6.2.1 Medical Personnel Performance
- 1.6.3. Operator Performance
  - 1.6.3.1 Pilot Performance
    - 1.6.3.1.1 F-14 Pilot Performance
  - 1.6.3.2 Astronaut Performance
- 1.6.4 Maintainer Performance

2.0 PERFORMANCE ASSESSMENT DOMAINS

2.1 Common Tasks
- 2.1.1 Common Tasks—Individual
  - 2.1.1.1 Safety and Survival
  - 2.1.1.2 Identification of Friend and Foe
  - 2.1.1.3 Communications

- 2.1.1.4 Physical Fitness
- 2.1.1.5 First Aid
- 2.1.1.6 Nuclear, Biological and Chemical Warfare

2.1.2 Common Tasks—Team

- (R) Unit
- (R) Group
- 2.1.2.1 Orientation
- 2.1.2.2 Organization
- 2.1.2.3 Adaptation
  - (R) Flexibility
- 2.1.2.4 Motivation

2.2 Performance Attributes

- (R) Task Analysis
- (R) Job Analysis
- 2.2.1 Task Knowledge
  - (R) Job Knowledge
  - 2.2.1.1 Procedural
  - 2.2.1.2 Declarative
  - 2.2.1.3 Strategic
    - 2.2.1.3.1 Mental Models
- 2.2.2 Processes
  - 2.2.2.1 Procedures
  - 2.2.2.2 Skills
  - 2.2.2.3 Motor Performance
- 2.2.3 Outcomes
  - 2.2.3.1 Acquisition
  - 2.2.3.2 Retention
  - 2.2.3.3 Transfer
    - (R) Generalization
  - 2.2.3.4 Adaptation
    - 2.2.3.4.1 Metacognition
- 2.2.4 Soft Skills
  - 2.2.4.1 Leadership
  - 2.2.4.2 Management
  - 2.2.4.3 C^3 (Command, Control, Communications)
  - 2.2.4.4 C^3I (Command, Control, Communications, and Intelligence)
  - 2.2.4.5 Discipline
  - 2.2.4.6 Attitudes
  - 2.2.4.7 Military Bearing
- 2.2.5 Job Functions
  - 2.2.5.1 Maintenance

2.2.5.1.1 Troubleshooting
2.2.5.1.1.1 Electronic Trouble-shooting
2.2.5.2 Operations
2.2.5.3 Planning
2.2.6 Critical Tasks
(R) Critical Incidents

3.0 PERFORMANCE ASSESSMENT APPROACHES

3.1 Types of Tests
(R) Types of Measures
3.1.1 Criterion Referenced
3.1.2 Norm-Referenced
3.1.3 Psychomotor Tests
3.1.4 Achievement Tests
3.1.4.1 Computer-administered
3.1.4.1.1 Adaptive
(R) Tailored
3.1.4.2 Paper-and-Pencil Tests
3.1.5 Competency Tests
3.1.6 Work Sample Tests
3.1.7 Skills Test
3.1.7.1 Skills Qualification Test
3.1.8 Occupational Test
3.1.9 Situational test
3.1.9.1 STX
3.1.10 Hands-on Test
3.1.10.1 ARTEP
3.1.10.2 Benchmark Measure
3.1.11 Walk-through Test
(R) Walk-through Performance Test
3.1.12 Simulation Test
3.1.12.1 Simulators
3.1.12.1.1 Flight Simulator
3.1.12.1.2 Gunnery Trainer
3.1.12.2 Computer Simulation
3.1.12.2.1 Radar Trainer
3.1.12.2.2 SIMNET
3.1.13 Administrative Records
3.1.13.1 Attrition
3.1.13.2 Reenlistment
3.1.13.3 Personnel Reports
3.1.13.3.1 Officer Efficiency Reports

3.1.13.3.2 Disciplinary Actions
3.1.13.3.3 Rank
3.1.13.4 Competitions

3.2. Performance Records
3.2.1 Videotape
3.2.2 Audiotape
3.2.3 Paper
3.2.4 Electronic Records
(R) Computer

3.3 Performance Assessment Instruments
3.3.1 Self Report
3.3.2 Observational Instruments
3.3.2.1 Behaviorally Anchored Scales
3.3.2.2 Check-lists
3.3.2.3 Structured Observations
3.3.2.4 Field Notes

3.4 Scoring
3.4.1 Score Development
3.4.1.1 Expert Scoring
3.4.1.2 Match-to-Sample Scoring
3.4.1.3 Competency Scoring

3.5 Rating
(R) Judgment
3.5.1 Product Rating
3.5.2 Performance Rating
3.5.3 Peer Rating
3.5.4 Self-Rating
3.5.5 Supervisor Rating
3.5.5.1 Commander Judgment
3.5.6 Test Administrator
3.5.6.1 Number
3.5.6.2 Selection
3.5.6.3 Training
3.5.6.3.1 Cost
3.5.6.3.1 Duration
3.5.7 Rater
(R) Observer
(R) Judge
3.5.7.1 Selection

3.5.7.2 Qualification
3.5.7.3 Training
3.5.7.3.1 Cost
3.5.7.3 2 Duration
3.5.7.4 Reliability
(R) Agreement
3.5.7.4.1 Interrater Reliability
3.5.7.4.2 Drift

3.6 Test Administration
3.6.1 Testing Conditions
3.6.1.1 Timing
3.6.1.2 Duration
3.6.1.3 Frequency
3.6.1.4 Job Performance Aids
(R) Aided Response
(R) Help
(R) Prompts
3.6.1.5 Feedback
3.6.1.5.1 Group
3.6.1.5.1.1 After Action Review
3.6.1.5.1.2 Debrief
3.6.1.5.2 Individual
3.6.2 Security

3.7 Indicators

4.0 PSYCHOMETRICS
(R) Measurement
(R) Test Theory
(R) Test Development

4.1 Validity
4.1.1 Fidelity
4.1.2 Criterion
(R) Criterion-related
4.1.2.1 Predictive
4.1.2.2 Concurrent
4.1.3 Content
(R) Content-Related
(R) Domain
4.1.3.1 Performance Sample
4.1.3.1.1 Representativeness
4.1.3.1.2 Random Sampling
4.1.3.1.3 Judgmental Sampling
4.1.3.1.4 Comprehensiveness

- 4.1.4 Construct
  - (R) Construct-Related
- 4.1.5 Validity Generalization
  - 4.1.5.1 Artifacts
  - 4.1.5.2 Transportability
  - 4.1.5.3 Situational specificity
- 4.1.6 Cross Validation
- 4.1.7 Synthetic
- 4.1.8 Differential Production
  - (R) Placement

4.2 Reliability

- (R) Generalizability
- (R) Dependability
- 4.2.1 Alternate Forms
  - (R) Parallel Forms
  - (R) Equivalent Forms
- 4.2.2 Internal Consistency
  - 4.2.2.1 Split-half
  - 4.2.2.2 Kuder-Richardson
  - 4.2.2.3 Coefficient Alpha
- 4.2.3 Stability
- 4.2.4 Inter-Rater
- 4.2.5 Errors of Measurement
  - 4.2.5.1 Standard Error of Measurement
  - 4.2.5.2 Variance Components
- 4.2.6 True Scores
  - (R) Universe Scores
- 4.2.7 Generalizability Coefficients
- 4.2.8 Attenuation
  - 4.2.8.1 Correction for Attenuation

4.3 Bias

- (R) Fairness
- 4.3.1 Item Bias
  - 4.3.1.1 Differential Item Functioning
- 4.3.2 Test Bias
  - 4.3.2.1 Differential Validity
  - 4.3.2.2 Differential Prediction
  - 4.3.2.3 Model Definition
- 4.3.3 Culture
- 4.3.4 Gender
- 4.3.5 Racial/Ethnic Group

- 4.4 Scaling
    - (R) Scoring
    - 4.4.1 Standardization
    - 4.4.2 Scaling Weights
    - 4.4.3 Judgmental
    - 4.4.4 Competency Scoring
- 4.5 Equating
    - 4.5.1 Linear
    - 4.5.2 Equipercentile
    - 4.5.3 Item Response Theory
    - 4.5.4 Equivalent Groups
- 4.6 Item Analysis
    - 4.6.1 Difficulty
    - 4.6.2 Discrimination
    - 4.6.3 Item Response Theory
- 4.7 Error
    - 4.7.1 Classification Error
- 4.8 Data Analysis
    - 4.8.1 Aggregation
        - (R) Units of Analysis
        - 4.8.1.1 Multi Level
        - 4.8.1.2 Subgroups
        - 4.8.1.3 Individuals
    - 4.8.2 Range Restriction
        - (R) Selection Effects
        - 4.8.2.1 Corrections for Range Restriction
            - (R) Adjustments for Range Restriction
    - 4.8.3 Statistics
        - 4.8.3.1 Descriptive
        - 4.8.3.2 Inferential
            - 4.8.3.2.1 Univariate
            - 4.8.3.2.2 Multivariate
            - 4.8.3.2.3 Regression
                - 4.8.3.2.3.1 Linear
                - 4.8.3.2.3.2 Multiple
                - 4.8.3.2.3.3 Logistic
                - 4.8.3.2.3.4 Nonlinear
            - 4.8.3.2.4 Correlation
                - 4.8.3.2.4.1 Linear
                - 4.8.3.2.4.2 Multiple
                - 4.8.3.2.4.3 Logistic
                - 4.8.3.2.4.4 Nonlinear

5.0 INTERPRETATION AND REPORTING

- 5.1 Norm-Referenced
  - 5.1.1 Comparisons
    - (R) Control
  - 5.1.2 Longitudinal
  - 5.1.3 Projections
- 5.2 Criterion-Referenced
  - (R) Competency Scores
- 5.3 Standards
  - (R) Criteria
  - (R) Quality Standards
  - 5.2.1 Standard Setting
    - (R) Passing Score
    - (R) Cut Score
    - (R) Minimum Qualification
    - 5.3.1.1 Expert Standard Setting
    - 5.3.1.2 Empirical Standard Setting
      - 5.3.1.2.1 Hierarchical Standard Setting
- 5.4 Report Formats
  - 5.4.1 Document Design
    - 5.4.1.1 Computerized Reports
      - 5.4.1.1.1 Hypertext Reports
  - 5.4.2 Data Reporting Formats
    - (R) Symbolic Representation
    - 5.4.2.1 Graphs
      - 5.4.2.1.1 3-D Graphs
      - 5.4.2.1.2 Computer Graphics
        - 5.4.2.1.2.1 Animated Computer Graphics
      - 5.4.2.1.3 Profiles
    - 5.4.2.2 Tables
    - 5.4.2.3 Pictures
    - 5.4.2.4 Summaries
      - 5.4.2.4.1 Executive Summaries

5.4.2.5 Briefs
5.4.2.5.1 Decision Briefs
5.4.2.5.2 Information Briefs

5.5 Users

5.5.1 Policy
5.5.1.1 Accession
5.5.1.2 Training
5.5.1.2.1 Basic
5.5.2 Program
5.5.2.1 Training
5.5.2.2 Test Development
5.5.2.3 Systems Acquisition
5.5.3 Office of the Secretary of Defense
5.5.3.1 Training Performance Data Center
5.5.4 Army
5.5.5 Navy
5.5.6 Air Force
5.5.7 Marine Corps
5.5.8 Allies
5.5.8.1 NATO
5.5.9 Congress
5.5.9.1 General Accounting Office
5.5.9.2 Office of Technology Assessment

6.0 COSTS

6.1 Research and Development

6.2 Initial Investment
(R) Acquisition

6.3 Operating
(R) Operating and Support

6.4 Salvage

6.5 Cost Formulae
- 6.5.1 Cost Effectiveness
- 6.5.2 Cost Benefit
- 6.5.3 Efficiency
- 6.5.4 Productivity

6.6 Test Development Costs
- 6.6.1 Specification
  - (R) Design
- 6.6.2 Item Development
- 6.6.3 Validation
- 6.6.4 Operation
- 6.6.5 Scoring
- 6.6.6 Change Costs
  - (R) Update

New Technologies and Tools for Operational C^3I Simulation and Modeling

Col. Alfred J. Mallette, USA and LTC. Guy L. Copeland,

Simulation and modeling technology rapidly has been advancing in military systems research, development, test and evaluation (RDT&E) applications although it is used mainly in the scientific research and engineering arenas. Unfortunately, less progress has been made in advancing simulation and modeling benefits to the services' user representatives who must determine how well a C^3I system design ultimately will fulfill its operational requirements and contribute to the overall force mission success. Serious deficiencies still exist in using automation to analyze concepts and evaluate emerging capabilities of complex new C^3I systems.

It often is difficult to perform complete operational evaluations on engineering development models without expensive commitments of large numbers of C^3I test equipment, platforms, affiliated weapons systems, personnel and other support resources. This dilemma appears in a number of major C^3I programs today. The Army is grappling with the question of how best to structure its area communications systems, data distribution systems and combat net radio systems as it changes C^2I architectures and signal equipment inventories over the next 10 to 15 years. Likewise, the Joint Tactical Information Distribution System (JTIDS) and its NATO counterpart, the Multifunctional Information Distribution System (MIDS), are programs that seek answers to many emerging complex network employment questions for all U.S. services and participating allies. Alternative analysis, evaluation and planning methodologies must be looked to for solutions, and simulation tools now have become readily obvious requirements.

Better Capabilities for Critical Needs

The Department of Defense and the individual services are intensifying their examination of simulation and modeling potentials and how they can be developed and better managed in support of C^3I system procurements. The Army Model Improvement Program (AMIP) is moving forward with plans for progressive development of interoperable battlefield functional area (BFA) and tactical environment models that can achieve more effective combined arms combat modeling.

The February 1987 National Security Industrial Association (NSIA) "Simulation and Modeling Study," reported to the Assistant Secretary of Defense for C^3I an assessment of how simulation and modeling might better be employed in DOD to support the resource allocation processes for major C^3I system acquisitions. While the report acknowledged good use of simulation in some areas, it pointed out that simulation has not been used effectively to determine how much a particular C^3I system makes a difference in the expected combat outcome. The NSIA report also suggested certain new technology capabilities that could improve simulation effectiveness in this deficient applications area, recommending wider use of computer graphics, digitized maps, artificial intelligence, object oriented programming, distributed simulation techniques, and rapid prototyping capabilities.

There simply remains no viable alternative. Without improved simulation support, concept planning for major C^3I system operations will continue to be too complex. The full scope of the most critical risks will be increasingly difficult to comprehend, hence the validity of solution assessments will suffer proportionately.

Given the enormously growing complexity of the electronic battlefield, innovative computer modeling techniques will be needed to evaluate performance impacts of the sophisticated Battlefield Automated Systems (BAS) on overall tactical force mission effectiveness in various candidate C^3I architectures of the planned war fighting doctrine. In many cases, simulation based analyses already offer the only hope for adequately defining the large networked system functions, the required mixes/modes of communications resources, the optimized battle management processes, the complex force control and system interoperability requirements in joint/combined operations, and the comprehensive operational test plans to validate the systems at critical milestones in the acquisition cycle. The Strategic Defense Initiative (SDI) may be the ultimate example of simulation need; however, theater/tactical C^3I may be becoming equally dependent on simulation.

To complicate the issue further, many-on-many engagement scenarios are essential drivers of C^3I system effectiveness evaluations. Engagement play on that scale rarely can be accomplished in live C^3I test operations but must be integrated with C^3I process evaluations almost exclusively via computer modeling.

The risk of not being able to do this well today is the risk of perhaps someday finding C³I programs doing operational testing (OT) in combat, as disastrous as on-the-job training (OJT) in combat.

Door to New Capabilities

A new era is dawning for the development and application of versatile, cost-effective simulation and modeling capabilities for the user representatives, who have been the traditionally tool-poor members of the defense RDT&E community. The C³I battlefield architects, the interoperability analysts, the operational testers, the system trainers and the tactical deployment planners are long overdue in acquiring simulation support tools that are easy to understand, easy to set up and operate and easy to adapt to new task applications, without requiring those users to have engineering or computer science degrees or to undergo extensive formal simulation system training. Advanced microprocessor hardware and simulation software technologies recently have arrived to support such tool developments.

Opportunities are available for major leaps forward in the users' ability to perform the large-scale, complex operational modeling functions that still remain well beyond the limited power of personal computers. Relatively low cost multiprocessor hardware equipment has become available with versatile operating systems. Application developers now have far more cost-effective computing power available than when they previously had only the larger minicomputers and mainframe systems for moderate and larger sized simulation jobs.

Today's new multi-microprocessor systems, such as those built around the Motorola 68020 microprocessor and the VME parallel processing architectures, allow distributed simulation designs that can cut run times to a fraction of the period required under single processor system executions. Distributed processor operating systems make programming and operation of these configurations relatively straightforward and efficient. Network file software is available to build networked processing systems where any or all processors can share common files anywhere in the network as if they were local to each processor. Industry-standard local area networking (LAN) can be implemented with virtually all the advanced multi-micro systems on today's market.

AI shells afford the opportunity to factor human processes into the C³I system model designs to a greater degree than previously attempted. Object oriented programming techniques and hierarchical structuring of the network system models can allow users to develop their simulations for modeling down only to the level of subcomponent detail (or echelon) necessary to obtain useful results. This is more and more important as the scope of operational modeling objectives moves up to subtheater levels with the requirement for many-on-many engagement play.

Increased internal processor memory allows direct accessing of much larger information arrays, such as digitized terrain data bases. This enables terrain related functions to operate across greater tactical map regions at higher levels of terrain model resolution. The ability to process simulations rapidly over a quarter million square mile digitized map model, with nominal horizontal data resolution of 100 meters, is readily achievable.

Powerful computer graphics engines, running at processor clock speed exceeding 20 megahertz (Mz), rapidly execute three-dimensional graphics manipulations and display functions from very large scale integrated (VLSI) hardware rather than software. While three-dimensional displays are not necessarily essential requirements for C^3I operations modeling, these graphic drivers support other needed features such as display zoom/dezoom, panning and altering view planes for more effective monitoring of the graphic objects, their simulated environments and digital map backgrounds.

Graphic windows and view ports provide a range of interactive input/output interface functions at single and multiple work stations. Display resolutions of 1,024 × 1,024 directly addressable pixels allow viewing hundreds of C^3I simulation elements at object sizes that can be selected to balance good legibility against clutter conditions. Output ports are available for projecting displays via large screen projectors, which are declining in price while increasing in performance.

High resolution color graphics copiers also perform quality hard copy reproduction of graphic screen displays. Graphic outputs also may be recorded on standard videotape through special signal encoders, thus affording user analysts an additional degree of post simulation functionality.

Interactive graphics are key to effective simulation operability. Graphic interfaces allow users to have immediate results; whereas results from non-graphic interfaces always have left the results embedded in the computer and accessible only through reams of printer listings.

The hardware and support software technologies are postured to launch a new wave of simulation and modeling tools achievements across a broader spectrum of subscriber needs, and C^3I system analysts and decision makers should be prime beneficiaries.

The Network Assessment Model

Two years ago, Teledyne Brown Engineering of Huntsville, Alabama, began an initiative to capitalize on its extensive simulation experience in strategic systems development and to transport powerful, distributed, graphics oriented simulation techniques from that advanced research domain to the tactical systems engineering environment. The result was a basic simulation and modeling concept, generically titled command, control, communications and computers simulation (C^4SIM). It was recognized by the U.S. Army Signal Center at Fort

Gordon, Georgia, as an opportunity to meld its technology advancements into a unique Army user tool.

As a result, the Signal Center embarked on a high priority program to develop from this concept a system called the Network Assessment Model. The first phase of the program was successfully completed in eight months, and the initial system capability was delivered to the Signal Center in March 1987. A second phase is continuing under expanded Army requirements monitored by the C^4 Models Advisory Group headed by the Combined Arms Center at Fort Leavenworth, Kansas.

The Signal Center carefully has guided development of the Network Assessment Model as a C^3I simulation environment, a multifunctional system of modeling support tools. Its primary objective is to support definition and detailed assessment of the Army's information exchange requirements and the network structuring of all current and developmental communications systems to meet those requirements—a mammoth task now well beyond classic manual analysis methodologies. To do so, the program is developing in the system an array of capabilities for the user to build any desired modeling elements, orchestrate them into scenario simulations, interact with the simulation as it runs and get quick, easy to understand results on C^3I measures of performance (MOPs).

The Network Assessment Model has a functionally modularized design for flexible tailoring of simulations to various levels of desired operational and technical detail. However, the most important aspects of the concept are its wide C^3I applications versatility, extensive employment of interactive computer graphic interfaces, ease of operability by nontechnical users and low cost microprocessor based hardware.

To begin with, the Network Assessment Model (Figure 1) provides the analyst with a model maintenance subsystem to build up to 10 million individual models of various type simulation elements for later scenario simulations. These include: models of tactical operational facilities (OPFACs), including mobile units, at any echelon down to individual commanders in all battlefield functional areas (BFAs), and containing the personnel, equipment and functions unique to each OPFAC model; models of specific communications equipments and their network management processes; models of digitized terrain regions; and models of electronic warfare threat jammers.

The user interacts with the graphic interface to draw OPFAC models from the library, perform the force laydown over digitized map areas to theater size and connect all OPFACs with tactical information needlines by automatic interface with the Army Communications Requirements Database (CRDB). Communication equipment models are then allocated on-screen to the OPFACs, and networks are graphically engineered for each separate communication system's planned employment. Special network planning features support proper siting of terminals, transmission elements, antennas and relay units for radio frequency line-of-sight (LOS) over planned links (Figure 2).

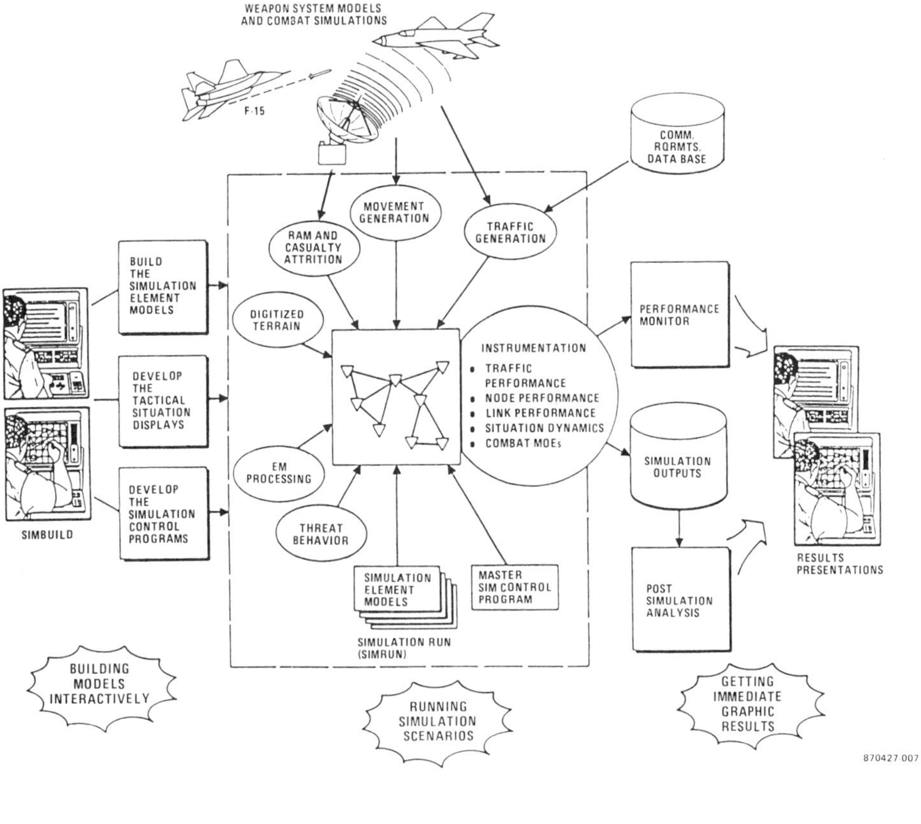

Figure 1. A high level functional representation of the Network Assessment Model.

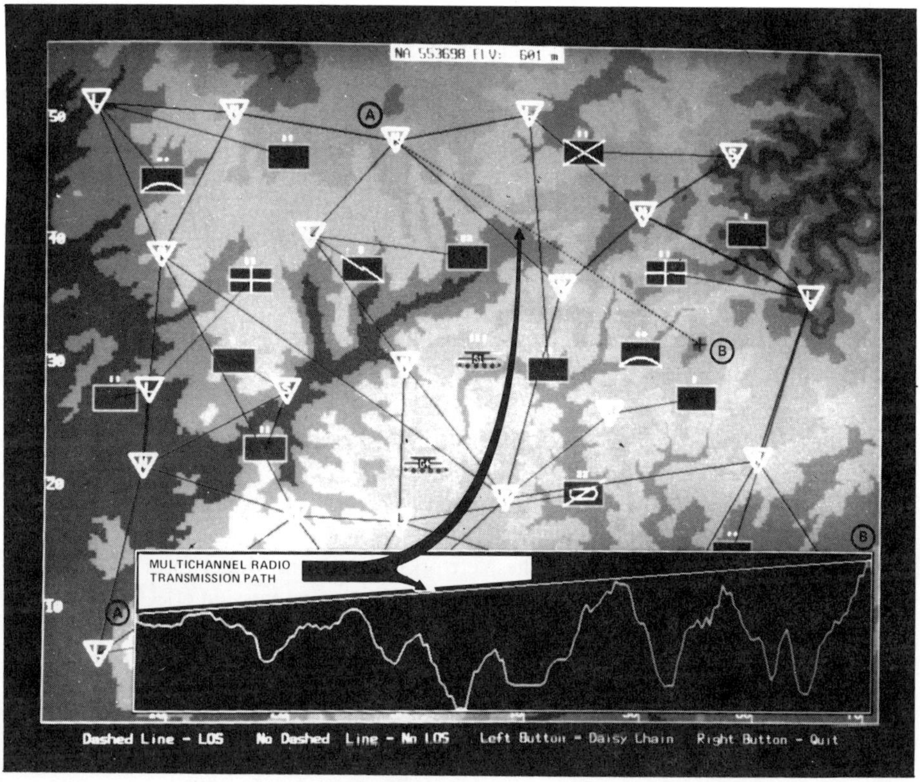

Figure 2. An on-screen Mobile Subscriber Equipment (MSE) network.

Mobile Subscriber Equipment (MSE) network modeling is implemented fully at this time. Mobile unit affiliations, multichannel link transmissions and detailed node processes for traffic switching under flood search routing are performed. Other models currently are being added for Tri-Service Tactical Communications (TRI-TAC), Packet Switching, JTIDS, Enhanced Position Location Reporting System (EPLRS), Single Channel Ground-Air Radio System (SINCGARS) and other emerging systems. The ultimate capability will be to define what message traffic is best served by specific system networks under varying network stress conditions in each BFA.

Unit movements and other scenario activities are graphically programmable and can be stored as complete scenario files for reruns, with easy reconfiguration for desired modeling parameter changes. Execution of large scenarios, exercising tens of thousands of message transmissions, is accomplished in minutes. End-to-end message transactions are simulated from their call originators through the terminal devices, through transmission, through switching and control elements, across terrain and electromagnetic environment models and against realistic simulated jammer interference at receivers.

Results of tactical activities are monitored on the system displays along with real time graphic indicators of communications node use and link performance. Instrumentation models log information on all facets of network operations for exhaustive post-simulation analyses supported by the model's versatile internal relational data base manager.

The Army's investment in the Network Assessment Model has returned a cost-effective tool system giving a first time capability to analyze, with a high degree of confidence and flexibility, the hybrid communications network structures within evolving combined arms C3I architectures for the 1990s.

Modeling JTIDS in Theater Air Defense

Early this year, the Office of the Assistant Secretary of Defense (OASD) for Command, Control, Communications and Intelligence (C3I) initiated a program with the Signal Center to adapt the Network Assessment Model for a special evaluation of JTIDS potential contribution to total tactical mission effectiveness in theater level, joint/combined air defense operations. The JTIDS Operational Performance Model (JOPM) is guided by a Supervisory Working Group (SWG) commissioned by OASDC3I, with membership from the Army, Navy, Air Force, Organization of the Joint Chiefs of Staff (OJCS), the Joint Tactical Command, Control and Communications Agency (JTC3A) and other DOD agencies.

When modeling most weapon systems, the performance of a C3I component often can be assumed; however, modeling operational effectiveness of a major C3I system also necessitates modeling its affiliated weapon systems to a high degree of fidelity, if valid C3I results are to be obtained across the potential intensity spectrum of combat actions. This JTIDS modeling application recognizes and implements that fundamental but complex requirement.

Supported by JTIDS networks modeled to the individual time slot levels, the system is being configured to model a many-on-many engagement of U.S. and NATO air defense/counter-air operations against an intensive hostile air attack and electronic warfare (EW) threat. In this large simulation, command, control and intelligence (C^2I) targeting traffic is generated dynamically from the air/ground combat actions and transmitted over the modeled voice and data networks between all participating sensor systems, C^2 (OPFACs), fighter aircraft, surface-to-air firing elements and other supporting units (Figure 3).

Operations are modeled over a digitized area covering a major part of Western Europe with the full theater air defense C^3I architecture of joint/combined forces represented. The computation of radar and communications intervisibility between hundreds of air and ground sensors, threat and friendly aircraft and communications nodes at all participating OPFACs requires processing at more than 30 million instructions per second. The model's expandable distributed processing power, which is configuring seven parallel microprocessors for this heavy demand within a multistation hardware suite costing less than $150,000, is unique.

The model's use of artificial intelligence (AI) components to represent the many human actions involved also is unique. More than 600 rule based modules govern not only the C^2 processes at controlling OPFACs (target assessment and weapons assignment) but also the smart pilot maneuvering and firing actions of friendly and threat aircraft, as well as the complex engagement kill scoring algorithms. Use of these AI rule sets provides the means to integrate dynamic but predictable human actions fully into C^3I evaluation, while the combat drivers and other functions can be iteratively changed (Figure 4).

The JTIDS/Air Defense model ultimately will provide results that effectively translate JTIDS communications performance to tangible and quantitative measures of improved air defense mission success, represented in finite and graphically portrayed assessments of how better to posture ourselves for the threat, engage and kill the threat, to keep from being killed by the threat, and to keep from shooting down our own forces and allies (Figure 5). These outputs are expected to be important and may even contribute to the forthcoming Defense Acquisition Board decision on JTIDS low rate initial production for two wings of Air Force F-15s.

With this capability established, the model is a firm foundation for other tools needed by operational JTIDS network managers and system deployment planners.

Building on This Foundation

The Network Assessment Model and its JTIDS/Air Defense modeling adaptation are helping to validate, and have focused strongly upon, the methods for

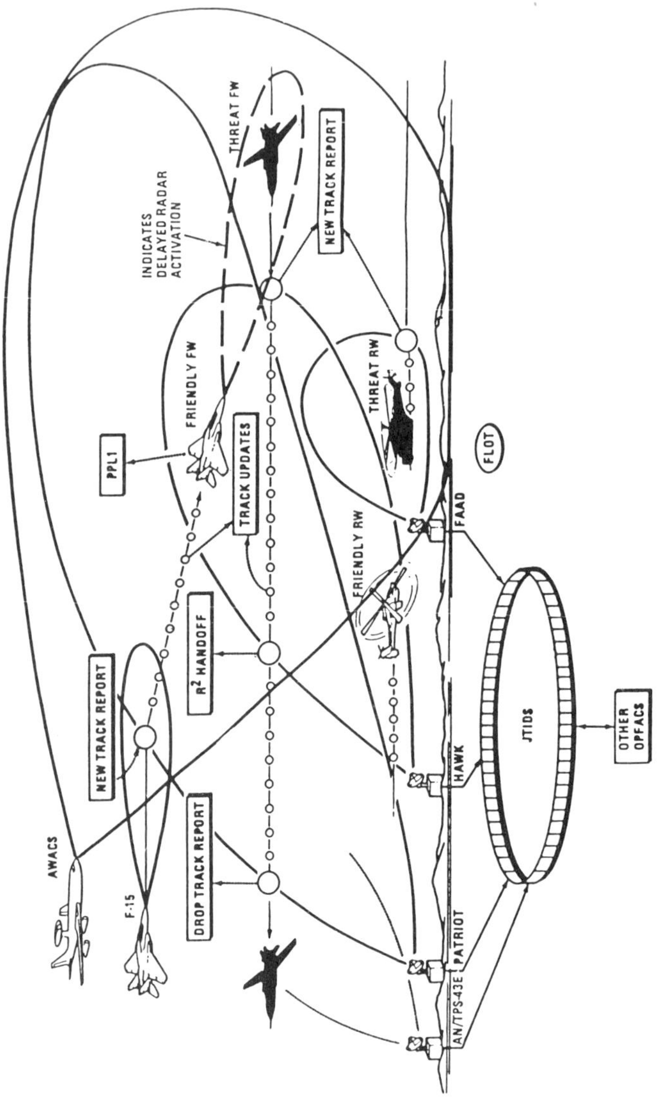

Figure 3. Dynamic targeting traffic generation in the JTIDS Operational Performance Model.

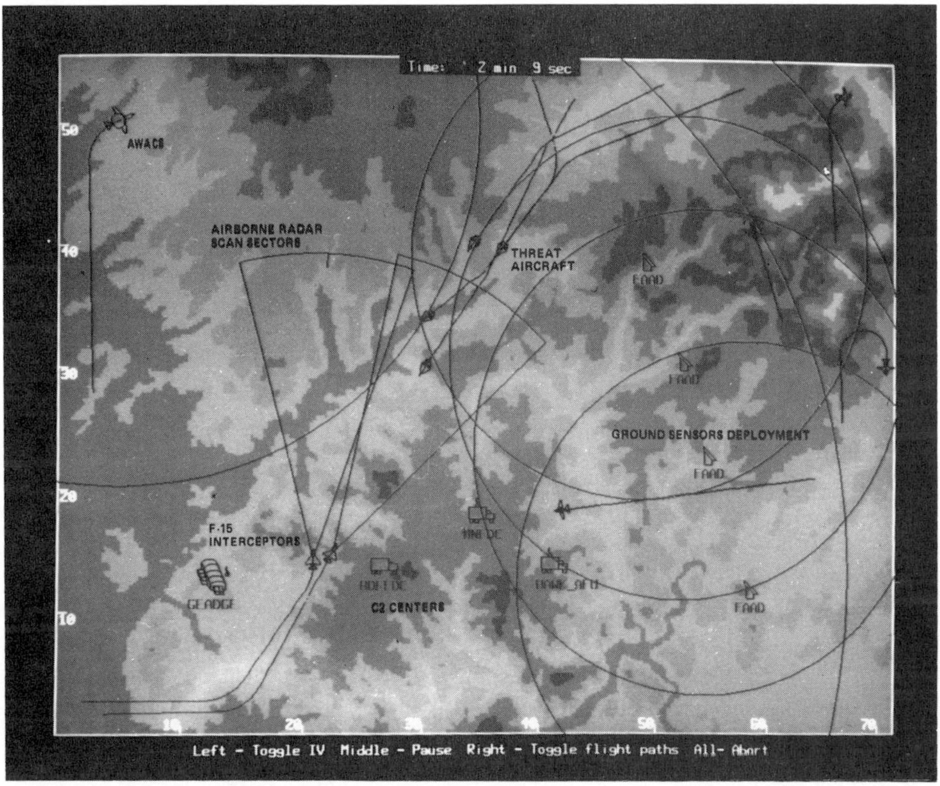

Figure 4. An on-screen representation of air combat simulation under artificial intelligence rules control.

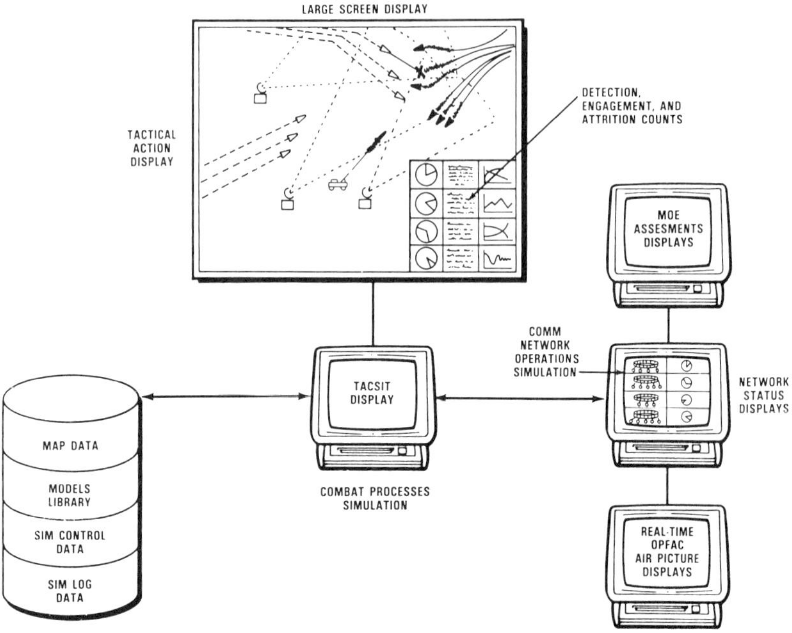

Figure 5. A typical distributed simulation configuration.

improving simulation and modeling effectiveness that were recommended by the NSIA study group in its "Simulation and Modeling Study" mentioned earlier. Additionally, the program is a firm start toward the capabilities and integrated modeling operability goals established by the Army Models Improvement Program.

The innovative simulation hardware/software technologies and programming techniques applied in these two modeling applications offer versatility for effective support of other jobs that must be performed by the user representatives who will eventually field the C3I systems, plan their employments and train their people to operate them.

Many other new capabilities are being swiftly introduced that will continue to enhance the technical posture to build and employ these useful, low cost, user oriented tools—tools with expansion potential for our growing needs, growing appreciation for simulation products and, inevitably, growing dependence on comprehensive computer modeling as an essential contributor to the C3I system acquisition processes.

This article originally appeared in the July 1987 issue of SIGNAL.

PART 3
Intelligent Systems Technology

Moving from Expert Systems to Intelligent Systems Technology in Command/Control Contexts

John W. Sutherland

Though some vendors are already at work trying to market what they refer to as "intelligent" systems, these are—even at their most sophisticated—usually just conventional expert systems to which have been appended some simple neural network type facilities. While such hybrid programs may offer certain marginal performance advantages over traditional knowledge-base constructs, the level of intelligence they exhibit may be considered trivial by other than their authors or promoters. More specifically, none can be relied upon to perform the function that serious system architects would allocate to true intelligent systems in command/control contexts: assist users in the rapid, rationalized resolution of problems that fall outside the bounds of authority of any knowledge-base facilities that might be available.

This perspective on the mission for intelligent systems is the only one that aligns directly with the classical definition of human intelligence as a capability for generating original responses to novel or unprecedented situations.

This paper will first suggest why Intelligent Systems in C^2 contexts ought to be pointed toward this particular challenge, and then offer some ideas as to the technical innovations that might eventually allow this challenge to be met in some practical way.

Key Characteristics of Command/Control Contexts

To ask first about the properties that distinguish organizations as proper command/control systems, at least the following would have to be mentioned:

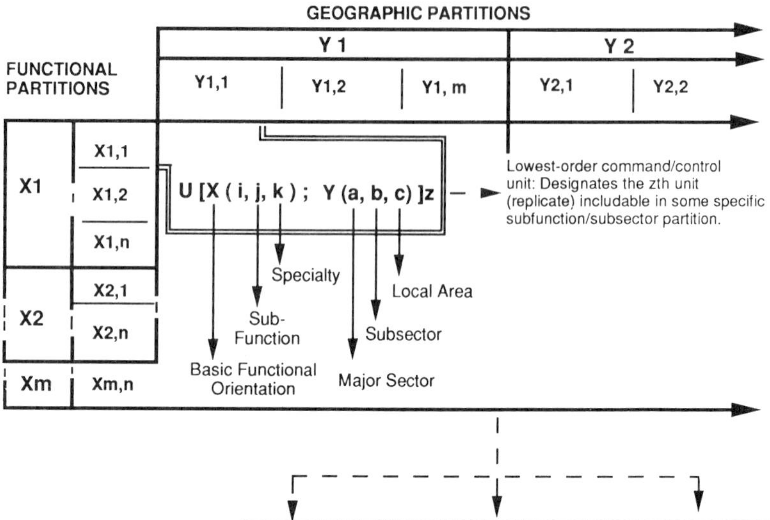

NOTATIONAL DESIGNATIONS:	MILITARY	COMMERCIAL	GOVERNMENTAL
C({X} x {Y})...commands all units in all sectors	Pentagon/ Joint Chiefs	Corporate Headquarters Level (CEO)	Secretaries of Cabinet-level Departments
C({X} :Ya) or C(Xi:{Y})... commands all units in a sector or all Xi-type units in all sectors	Theater Commander (over combined forces: Ground, Air and Naval)	The head of a major functional division (e.g., Vice President of marketing or production, etc.)	Directors of key Agencies, Bureaus or Services/Asst. Secretary level appointments
C({Xi,j}:Ya) or C({Xi} : Ya,b) ...commands all Xi,j-type units in a sector or all Xi-type units in a subsector	Force-specific commands (Corps or Fleet level in a theater)	Manager of a major product line/regional sales manager/plant manager	Regional Authorities/heads of sub-directorates
C({Xi,j} : Ya,b) or C({X} : Ya,b,c) ...commands all Xi,j-type units in a subsector or all functional units within some local area	Senior Field Commands (of an Infantry or Armor Division, an Air Wing or a Naval Battle Group)	Managers of retail outlets/department heads (e.g., director of advertising for a Product/market sector)	Heads of local Branch Offices (e.g., the agent-in-charge of FBI field office or Forest Service station)
C(Xi,j, {k} : Ya,b,c) ...commands two or more low-order specialized (replicate) units in a sub-function/subsector partition	Basic Line Level Commands (of a Light Infantry regiment, a fighter squadron or ASW assets of a battle group)	Local area sales managers; heads of departments in retail outlets or other first-line supervisory level personnel	Field and/or shift supervisors for a branch office or station, etc.

Figure 1. Expansive Hierarchical Matrix Structure for Command/Control Systems.

• The units will be both geographically dispersed and functionally specialized; that is, units may be distinguished either in terms of the particular location in which they are resident (or, more precisely, the sphere-of-interest they are assigned) and/or in terms of the specific types of operational activity(s) they perform. As for their basic configuration then, command/control organizations might best be modeled as an expansive hierarchical matrix, as per Figure 1.

• Each of the line (versus staff or ancillary support) units constituting a command/control organization will have "critical" missions to perform. The consequences of any default—through erroneous decision, untoward delay or ineffectiveness—may imperil not only individual units, but the organization itself.

• The organization must contend with one or more ingenious, aggressive adversary, with few if any effective constraints on the nature of the competitive initiatives they might author (i.e., the range of threats to which an organization is subject may expand to include those of distinctly predatory potential). This same condition may also apply to its constituent line units.

• Commonly, the key working challenge for the organization in aggregate and each of the units of which it is comprised is to be "ready" to deal with any threatening initiatives that might be mounted by their *peers* in adversary organizations (i.e., competitor units located at approximately the same level of the organizational hierarchy and operating within similar geographic and/or functional bounds).

• To tie dispersed units together, a wide-area communications network for near real-time interchanges given whatever capacity constraints and linkage protocols apply (these latter serving to specify which units may talk to which others along which paths).

Of the above properties, that which is most significant for system architecture is the convention whereby the "readiness" focus for each unit in a command/control system is on the range of prospective threats that might be authored by its adversary organization and/or the array of events that might arise in its geographic area.

The set of these threat/event possibilities would then become the key requirements against which management support system facilities allocated to any unit would be arrayed. And this would hold true not merely for traditional system assets (i.e., distributed data bases and analytical/quantitative decision aids), but also for knowledge-base constructs.

Knowledge-Base Constructs in Command/Control Contexts

As detailed in Figure 2, it's useful to distinguish between two basic types of knowledge-base constructs, d*ecision tables* and *decision trees*. The key featurc of the former is that they can allow for a simple "directed" or essentially automatic decision along the following lines:

$D = f([e^*] \leftrightarrow [r^*])$,

such that for any actual event (V) that maps well to one of the situations

		DECISION "TREE" PROGRAMS	
	DECISION TABLE $([e] <-> [r])$	**DIAGNOSTIC** $L[\{c\} -> e_i]$	**REMEDIAL** $L[r_i -> \{c\}]$
MAPPING MECHANICS:	**DISCRETE-BIPARTITE** (directed connections via simple associative inference) If $[X_1 \ldots X_n]$ then $P(Y_i) -> 1.0$	**FORWARD CHAINING** (data-driven search to effect convergence on a most probable problem-definition... search patterns are basically deterministic via user-supplied—a priori—node/arc valuations	**BACKWARD CHAINING** (hypothesis-driven retro-convergence operations following pre-determinable paths from conclusions to premises...simple deductive inference mechanics to trace syllogistic patterns)
PREDICATES: **(STARTING-STATE/INPUT CONDITIONS)**	**DETERMINACY** indicators (I) imply that the event (V) maps unambiguously to one or another of the situations that have been predefined in the set [e]: $I -> V = e_i \varepsilon [e]$	**INDICATOR AMBIGUITY** currently available indicators are sparse, such that there are two or more situations defined in [e] that have a significant probability of being the "true" event: $I -> V = P[e_1, e_2 \ldots]$	**RESPONSE AMBIGUITY** given the determination of a maximum-liklihood event definition, there are two or more response options that are posed as being the maximal-utility/optimal course of action to take: $P\hat{}(e_i) -> U[r_1, r_2 \ldots]$
CONCLUSIONS: **(DECISION PREMISES)**	**DIRECTED DECISION** $e_i \varepsilon [e] -> r_i \varepsilon [r]$	**MAXIMUM-LIKLIHOOD EVENT/PROBLEM-DEFINITION** $P\hat{}(e_i) \varepsilon [e] : Ey$	**EXPECTEDLY MAXIMAL-UTILITY TREATMENT OR COUNTER-ACTION OPTION** $U\hat{}(r_i) \varepsilon [r] : Rx$

Ey = the specific application DOMAIN (basic class of phenomena) to which a forward-chaining program is confined

Rx = a RANGE constraint (in terms of some singular family of response options) for a backward-chaining program

Figure 2. Classes of Knowledge-base Constructs.

defined in [e*], the operator (f) will point to one and only one response side correlate as some rj included within [r*].

Decision tables thus define a one-to-one proper relationship between a limited array of discretely-defined problem (event or threat) and solution (response) options. Decision table constructs can inject considerable efficiency into managerial decision making. For it is a facility that can allow for the more or less automatic (rote) handling of routine, iterative or otherwise highly predictable situations. But a decision table cannot *recognize* anything but the limited set of discrete events it was expressly programmed to comprehend. Nor can it deal with ambiguities.

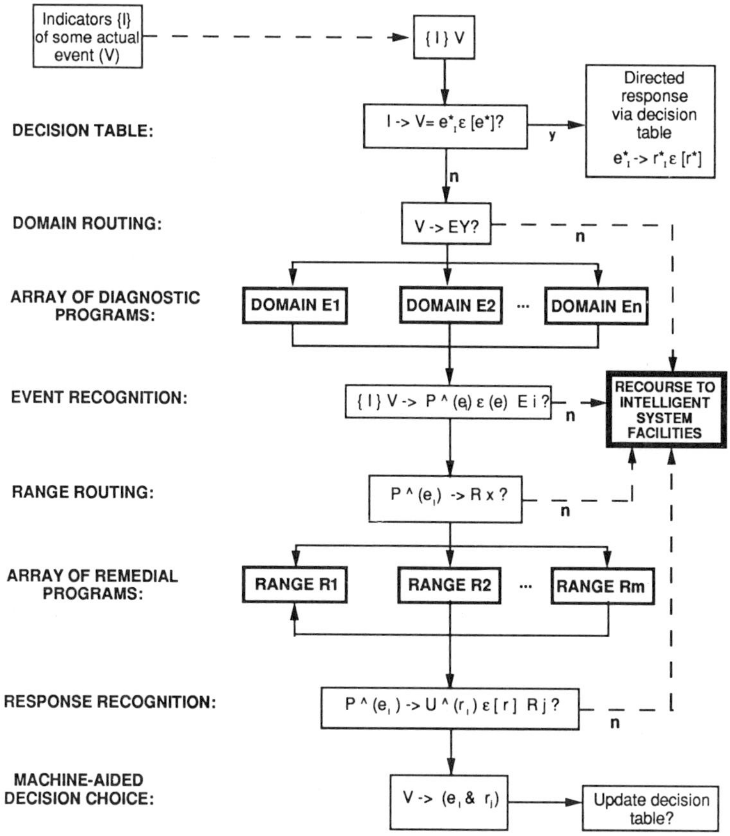

Figure 3. Key Features of a Knowledge-base Centered Management Support System.

Therefore, the likelihood of being confronted by some event/threat that does not map directly to any of the discrete possibilities a decision table has been programmed to recognize is then an occasion for development of an expert system qua decision tree construct. But while much has been written about the employment of expert systems based on decision trees, virtually all the mainstream literature looks at such systems in effective isolation (i.e., as autonomous and independent facilities). But with command/control contexts it is necessary to move beyond this focus and consider what a knowledge-base centered management support system might look like. For such would involve equipping any unit of a command/control organization with a multiplicity of knowledge-base expert system constructs, perhaps along the lines of the configuration shown in Figure 3. It is within the confines of this configuration that the practical reach of knowledge-base decision support facilities must be appreciated.

Given a knowledge-base centered management support system as shown in the exhibit, any decision exercise would be initiated by the appearance of some set of indicators (I) relating to some event/threat (V) that has occurred. As a first step, one might then test to see if the event maps to any of the events a decision table can recognize. For if it does, then the decision on response will have already been made.

Should the event/threat fail to align with the situations in the domain of the decision table, we would turn to another decision tree oriented expert system with which a knowledge-base centered management support system has been invested. Such systems have been so widely discussed of late that we need consider only their readiness-related implications for command/control contexts. Specifically, our interest is in the contribution they make to two basic analytical tasks over which any formal decision exercise can be defined:

- Mapping from a set of at least partially "sparse" indicators to a maximum likelihood (most probable) problem or event/threat definition;
- Mapping from the maximum likelihood event/threat definition to a maximal utility remedial or response option—a singularly best course of action.

In a knowledge-base centered management support system, if no match is found in a decision table, the next step is to pass it off—via the Domain Routing operation—to a forward chaining (diagnostic) program for treatment.

Diagnostic Expert Systems

As regards the first of the above tasks, a unit commander would call upon a *forward chaining* expert system program to help resolve event-side ambiguity along the following lines:

$$\text{Given: } I \longrightarrow P\,(ei \text{ or } ej \ldots) \quad \varepsilon\ [e] \subset Ex,$$
$$\text{Find: } P^{\wedge}\,(ei) \quad \varepsilon\ [e] \subset Ex.$$

The first expression is a case where available indicators (I) are sparse or imprecise, and there are two or more causal alternatives (e1 or e2, etc.) assumed to have some probability (via P) of ultimately proving to represent the true event/threat. As a critical qualification, all the alternatives recognizable by any forward chaining program will generally have to be intelligible as variations on some singular generic class of phenomena. Hence the term Ex may represent the specific applications *domain* over which the system has been defined.(i.e., some singular class of phenomena that serves to put qualitative constraints on the search space over which any forward chaining system may wander). As such, the members of the set [e] are the array of variations on that theme that the program can recognize by virtue of its diagnostic rules. For example, were we to take the domain (Ex) to represent the universe of respiratory maladies, then an instance of event-side ambiguity may arise were a patient to exhibit symptoms consistent with either a cold (e1) or an allergy attack (e2), etc. The objective is to converge on a singular maximum likelihood alternative—some P^(ei)—drawn from the set of possibilities [e].(i.e., decide on either a cold or an allergy). When employed interactively, diagnostic expert systems may direct the user as to what additional diagnostic data are needed to refine the level of certainty associated with the P^ alternative.

Remedial Expert Systems

Given convergence on a maximal likelihood event/threat definition as some P^(ei), a determination must then be made as to the *backward chaining* expert system to help formulate a response; this via the Range Routing operation shown as ei -> Ry (where Ry is a *range* constraint that bounds the response alternatives a backward chaining construct can comprehend). Assuming there were some backward chaining construct to which ei might be routed, then we might look to such a facility to help in resolving response-side ambiguity. This would arise whenever there is some uncertainty as to the best treatment/remedial option, that is:

$$\text{Given: } ei \longrightarrow U\,[r1, r2 \ldots rm] \subset Ry.$$
$$\text{Find: } U^{\wedge}(rj)\ \varepsilon\ [r]\,/ei$$

where r1, r2, etc., represent response alternatives—all definable as variations on some general theme, Ry— each of which is assumed to have some degree of positive utility (U) as a counter to the event/problem at hand (ei). The objective is to isolate some singular maximum likelihood alternative, U^(rj).

Suppose, for example, it had been determined that a patient had a specific bacterial infection, but there was doubt on the most favorable therapeutic tactic. If Ry were to stand for the universe of antibacterial therapies, a backward chaining program might help converge on an ostensibly best option as determined by expert or consensus opinion—in terms of a specific drug administered in a specific dosage, etc.

The Role for Intelligent Systems in Command/Control Contexts

If this is what diagnostic and remedial expert systems can do for decision authorities, equally obvious are the conditions under which they would fail to provide for readiness. As for diagnostic systems, forward chaining constructs cannot recognize any actual or prospective situation that is not definable within the domain over which the system was defined. Nor can they *reach* any event/threat definition (ei [e] Ex) that was not explicitly anticipated by the designers; that is, they can do no significant "learning" on their own during any diagnostic exercise (and this holds true irrespective of whether or not there is some sort of neural network facility in place). Essentially the same limitations proscribe the utility of backward- chaining expert system constructs. The only response options they can recognize are those imposed a priori by programmers; that is, they cannot be relied upon to endogenously generate alternatives.

Moreover, as a practical matter, expert system constructs do not usually operate entirely on their own. Rather, it is common for forward chaining or diagnostic programs to demand that the user input a hypothesis about the nature of the problem/event at hand, and for backward chaining programs to require that the user supply an initiating hypothesis setting out a priori judgement as to a best remedial option. Working expert systems vary widely in sensitivity to the quality of user-supplied hypotheses. In many cases the efficiency (and sometimes the basic effectiveness) of the programs will be so intimately tied to the quality of such inputs that the users of expert systems must themselves be experts. For example, when supplied with an ill-considered starting-state hypothesis by an inexpert user, many backward-chaining constructs may fail to arrive at any actionable conclusion.

Hence the several "no" paths in Figure 3 along which a knowledge-base centered management support system can be expected to go into default:

- when an event does not fall within the domain of any of the diagnostic expert systems;
- when an event fails to map to any of the situations a forward chaining construct was preprogrammed to recognize;
- when the event is one for which there is no appropriate standing remedial referent, either because it belongs to a class of phenomena for which there is no corresponding backward chaining construct (no Range-level correspondent) or because it represents an event/threat for which no situation specific remedial provisions were made.

In any of the above cases, the array of expert system constructs will fail to provide the wherewithal for either an automated (directed) or guided resolution to the situation at hand. Commanders would then be required to turn elsewhere for support.

But as shown in the figure, it is precisely the three default conditions noted just above that set us on a path toward a second recourse—intelligent system facilities intended to assist users in instances when expert system constructs are unable to cope. That is, intelligent systems would specifically be charged with helping command/control authorities do disciplined *exception handling*.(i.e., perform the diagnostic and/or remedial analytical functions in a reactive mode).

As such, these intelligent systems would have to entail an order of resolution power well beyond the traditional knowledge-base constructs from which they would take over.

Suggested Technical Properties of True Intelligent Systems

To echo earlier arguments, the key contribution of true intelligent systems in command/control contexts would be to take over the decision support role where traditional knowledge-base constructs have been defeated. And again, expert systems must be expected to face defeat whenever reality presents a situation that is partially unanticipated...alien, in some salient respect, from the events/threats that were expected. In such a case, the challenge for command/control authorities is to get "ready" to handle the situation on an after-the-fact versus preadaptive basis. And given the premium on responsiveness associated with the highly competitive arenas in which command/control organizations tend to operate, this may require that decision authorities perform the mapping from INDICATORS -> PROBLEM DEFINITION -> COURSE-OF-ACTION in real time. And from here we once again return to the key performance requirement against which the pragmatic contribution of intelligent systems facilities is to be assessed:

The extent to which they can expedite the formulation of a well-rationalized, at least partially original solution to an at least partially novel problem.

If this is what true intelligent systems would be expected to do for command/control, the focus now is on the sorts of technical innovations that might carry such systems out of the realm of academic speculation and into the arena of operationality. And of the lines of technical innovation that might be followed, two appear to be particularly critical:

- The development of facilities that allow systems to engage in ampliative inductive (versus simple associative) inference exercises.
- A shift in the focus of knowledge-base structures away from substantive (application related) content in favor of methodological (procedural) provisions.

And as a natural complement to these two technical requirements, there is this complementary, policy related initiative:

- The exchange of the traditionally passive orientation of machine-based decision aids for an "active" posture vis-à-vis users.

Facilities to Support Ampliative Inductive Inference

Despite some proponents' protestations to the contrary, no currently available expert system constructs have shown a capacity for anything other than simple associative inference. This the lowest form of cognitive capability in humans (being a product of classical or Pavlovian conditioning). What this means is that both diagnostic and remedial constructs are restricted to directed—i.e., deterministic or algorithmic— mappings between predicates and conclusions. This is true even of those decision tree constructs that are popularly promoted as being able to comprehend probabilistic versus merely deterministic mappings; this because the stochastic nodes a program might arrive at will have been assigned either direct or inherited static probability values. Systems equipped with only simple associative inference facilities, are capable of making any disciplined (versus casual) excursions beyond what they have been provided

But such excursions are exactly what are required following a defeat of expert system technology, because the emergence of a partially novel situation implies an event/threat that rests somewhere beyond the level of the diagnostic systems expertise. An adequate definition of the actual threat/event would have to include at least some properties that are not present in the program's existing knowledge base. And when it is necessary to go somewhat beyond that which is known a priori—to reach somewhere beyond the confines of the immediate facts of a matter or direct experience, etc—we have the occasion for recourse to that cognitive process known as *ampliative inductive inference*. Ampliative inductive inferences can be undertaken along three lines:

- *Longitudinal*: the generation of a set of prospective or future possibilities that are variations on themes we have direct historical experience of or which have definite empirical precedence.
- *Cross-Sectional*: the effect of arguing from the more particular to the more general, or the imputation of properties to a population as a whole on the basis of attributes or behaviors observed in some sample.
- *Analogic* using secondary evidence—observations made with respect to some other subject, information collected by other agencies or experiences undergone by others—to impute properties to some new situation.

What is immediately clear is that the majority of the statistical/probabilistic decision aids are expressions of ampliative inductive inference in its computational guise. For example, any of a wide variety of extrapolative/projective forecasting techniques can drive longitudinal inferences. Instrumental support for cross-sectional inference would include an extraordinarily rich collection of experiment related techniques, and most particularly those underlying ANOVA initiatives. Finally, instruments available to formalize analogic inference processes would include factor and cluster analysis initiatives, exploratory data analysis techniques, Bayesian learning protocols, fuzzy set operations, etc.

Therefore, the simplest and most direct route from expert system constructs to intelligent systems would be to supplement traditional rule-based structures with a *knowledge* of some set of statistical/probabilistic procedures, the net effect of which would be to develop machine-based systems that, given a hybrid instrument base, are capable of acceding to a higher level of effective intelligence than their predecessors.

There are two other areas of technical enhancement that might be distinctly more challenging requirements to be met in the drive toward true intelligent systems:

- As distinct from traditional expert systems, proper intelligent systems should have a capability for generating problem-definitional and/or remedial alternatives on their own. That is, true intelligent systems would be expected to generate (i.e., induct, deduct or adduct) some hypotheses for subsequent consideration by the man in the loop, whereas the opposite is true of expert systems (they are designed only to consider hypotheses introduced exogenously by users). And as for the specific technical advancement this implies, it suggests the requirements for some genuine progress in a subfield of artificial intelligence that has as yet been pursued only most tentatively, *discovery* system mechanics. For even in the very crude formulations and highly restricted applications that have been undertaken thus far, it is clear that enhancements along this line of development have a very good chance of leading to facilities that will allow computers to conduct meaningful endogenous realization exercises. Unfortunately, expert systems have so monopolized both technical attention and R&D resources that this potentially far more significant technology has been largely ignored.
- As a critical complement to the above, intelligent systems would also ideally supplement rule-driven knowledge bases with an *associative* (versus recursive) memory structure. At this point one might reasonably look for some ideas on how neural network theory concepts might enhance machine reasoning. For in an associative structure, properties that tend to occur together naturally will be stored together. Associative memory structures may then be comprised of intricate composites of properties drawn from different dimensions, such that the basic unit of access might be a proper "model" instead of isolated record-level entries or abstracted rules. Moreover, an associative memory structure might be configured (using vector-space network conventions) so that models or referents that have more in common are more proximate—both logically and physically. All these features may readily been seen as conducive to the generation of models or constructs that represent interpolative or extensive (ampliative) variations on existing referents.

As a final note, discovery system facilities and associative memory provisions are significant not merely for enabling machine-driven realization of new diagnostic or remedial alternatives, but also as features that could both expedite and help discipline interactive (man-machine) hypothesis formulation.

Intelligent Systems as Methodological "Experts"

Whereas traditional expert systems tend to have their knowledge fixed in a set of application-related or subject-specific rules, intelligent systems would serve as a source of expert advice on research and analysis methods in general—as purveyors of proper epistemological prescriptions, as it were. That is, we might think of the knowledge-base of intelligent systems consisting of *rules* that reflect prevailing methodological wisdom on the investigatory initiatives and procedures appropriate for solving different classes of problems under different contextual conditions and constraints. One aspect of this wisdom would, of course, be the knowledge of statistical/probabilistic techniques argued for earlier. Intelligent systems would then have to be equipped with a good working comprehension of the following:

- the procedural prescriptions associated with the various orthodox methodological protocols that the scientific community has devised, for example: the Positivist, Empiricist, Subjectivist and Action Research approaches.
- the reach and limits of the classes of decision aids associated with each of the protocols: Mathematical/optimization, statistical, Bayesian and judgemental, respectively.
- a basis for making a best choice—in terms of efficiency characteristics—among the various members of a given class of instruments for a particular task.

The urgency of equipping intelligent systems with a knowledge of such matters derives directly from what is becoming an everincreasing concern: the lack of technical significance among so many of those holding managerial positions in the commercial, governmental and even the military sectors and the equally troublesome tendency for today's technicians to be increasingly specialized in terms of both their instrumental and applications preferences.

But as a final aspect of the methodological focus of intelligent systems, equally if not even more critical is that they be equipped with some comprehension of unorthodox protocols, and most particularly those aimed at providing some sort of disciplined basis for conducting *heuristic* analysis/research exercises. Of such, perhaps the most appropriate is Action Research, this prescribing procedures for "learning while doing." Such an approach is well-suited to command/control contexts because of the frequency with which we must invoke some sort of action (e.g., "tweaking") in order to get the information on which to base diagnostic and hence remedial decision choices. Action Research initiatives are also the preferred basis for conducting the early stages of an inquiry into a significantly novel or unprecedented problem; this because of the tendency for problem-definitional and remedial functions to become partially codeterminate (e.g., a physician encountering a patient with a novel syndrome must be prepared to simultaneously treat whatever threatening symptoms are apparent and undertake obtrusive and perhaps even potentially perilous diagnostic tests in an effort to get at a true cause).

The Transition from Passive to "Active" System Orientations

Progress along the technical lines discussed above would suggest that intelligent systems could be designed to effect an orientation toward users that is substantially different from that which has been employed by any other family of machine-based decision aids. These latter, almost without exception, have been posed primarily in a simple *consultative* relationship, and so are distinctly "passive" in their implications for actual decision exercises. For example, when employed in a consultative capacity, an expert system is intelligible as a vehicle for testing the extent to which any user-generated hypothesis (diagnostic or remedial) aligns with the conclusion that some expert or collectivity of such would presumably have arrived at via the "rules" constituting the knowledge-base. But the user is under no compunction to heed the program's advice.

There are, however, at least two other orientations that might be adopted, *corroborative* and *compensatory*. While neither of these "active" options has been subjected to much conceptual or operational scrutiny, all are more appropriate than a simple consultative posture, given the role we have ascribed to intelligent systems in command/control contexts. The corroborative is the least exotic, aimed primarily at controlling the quality of ad hoc decisions by subordinate commanders faced with a novel situation. For under the corroborative approach, lower-order decision authorities would now be expected not only to be required to take recourse to some sort of intelligent system facility in formulating event/response alternatives, but would not be allowed to actually commit to any concrete course of action unless or until the intelligent system has validated his conclusions. In their corroborative guise then, intelligent systems would be designed as vehicles for both the conveyance and enforcement of normative decision discipline, this in the hope of being able to recognize and repair—before the fact—certain instances of prospective decision error.

When employed in their compensatory guise, intelligent systems would be allowed to substitute their judgement for that of a human decision-maker in either of two situations:

- When a human decision authority has failed to commit to some required course of action in a sufficiently timely way;
- When there is some reason to doubt the competence of the user with respect to some technical task.

The first case speaks to the prospect of machine-based facilities having to compensate for a lack of will or decisiveness in a human decision maker. In another variation, an intelligent system equipped with the technical features described might itself arrive at diagnostic and/or remedial conclusions that would be put into execution unless some superior man or machine intervenes.

The second of the two situations addresses an issue already raised: the widespread lack of mathematical, statistical and formal analytical skills among the general population of American managerial functionaries and the increasing parochialism of technical specialists. Hence, again, the urgency of equipping computer-based systems with a level of instrumental "intelligence" that may exceed that of the typical user. And to the extent that a machine-based system has a higher order of technical intelligence than a user, the relative balance of decision authority should be adjusted accordingly.

Thus, any intelligent system ought to be equipped with a facility for testing (in as unobtrusive a way as possible) whether the user is longer on charm than computational sensibilities. If so, the system might then itself undertake certain decisions—about what quantitative techniques to employ, about what level of precision to tolerate, etc.,that could otherwise be left to the man in the loop. And what this would imply, in terms of a basic technical provision, is that true intelligent systems be designed to be able to operate with some adequate degree of integrity even in the absence of starting-state hypotheses or prompts from a user.

There are two final points to make. First, true intelligent systems currently remain well beyond our immediate technical grasp. Secondly, if we continue to assume that they will eventually evolve somewhere along the technical trajectory that has been followed by those designing expert systems or arguing the omnipotence of neural network conventions, truly intelligent systems are also very likely to remain forever beyond our technical reach.

Automated Adversarial Planning Search Procedures with Provable Properties

Paul E. Lehner

Introduction

In artificial intelligence (AI) the problem of planning is generally defined as the determination of a set of actions, ordered over time, that will achieve a specified goal. An important category of planning problems are those where an agent must determine a plan in an environment where there exists a competitive agent or adversary simultaneously planning to achieve a competing goal. This type of planning is often referred to as adversarial planning.

Adversarial planning is ubiquitous. Many objectives (job promotions, contract awards, winning political elections, military victories, etc.) can be achieved only by denying opponents competitive objectives. If AI systems are to assist people in achieving their objectives, then these systems should be able to reason about an intelligent adversary that is intentionally trying to prevent achievement of an agent's objective.

From the perspective of automated planning, adversarial planning problems present some unique and interesting challenges relative to other classes of planning problems. Some of these are discussed below.

First, in nonadversarial domains the product of planning is generally the specification of a single sequence (or partial order) of actions. If an unanticipated event occurs during execution, replanning is performed as needed. In adversarial planning, it is often impossible to recover from unanticipated events, since such events often indicate that the adversary has found a course of action that the present disposition of agent resources cannot handle. Consequently, an adversarial planner must anticipate possible adversary actions and reactions, and specify appropriate contingency plans.

A second challenge is that of goal interactions. It is a characteristic of interesting planning problems that goals interact, that is, actions to achieve one goal will impact the feasibility and usefulness of actions to achieve another goal. Nonadversarial planning systems (e.g., Chapman, 1987; Wilkins, 1988) generally focus on the problem of harmful goal interactions, where actions to achieve one goal may make it more difficult to achieve other goals. Helpful interactions are accepted, if discovered, but the planners do not search for them. In adversarial planning, helpful goal interactions are essential to success. The most elementary tactics in adversarial planning domains involve some form of goal interactions (e.g., a fork in chess), where higher levels of planning skill generally reflect increasing sophistication in the identification and exploitation of subtle interactions.

Third, classical planning systems in AI (sec Chapman, 1988 for review) usually simulate a problem domain through action models. For each operator (i.e., type of actions) there is a well-defined set of statements or functions from which all consequences of an action can be deduced. Unfortunately action models are not very efficient in domains where (a) multiple agents act simultaneously and (b) multiple actions are needed to achieve some consequence. Since adversarial plans often require coordinated simultaneous actions to succeed, action models are not an appropriate approach to simulating adversarial planning domains. Consequently, in contrast to the classical planning approach, adversarial planning requires a domain simulation that is separate from the planning knowledge base.

A fourth problem is the tradeoff between soundness and completeness. A search procedure is sound if any solution it finds is guaranteed correct (relative to the definition of the search problem). A search procedure is complete if it is guaranteed to find a correct solution whenever one exists. Ideally a search procedure should be both sound and complete. Complete search procedures, however, are often too computationally complex to be of much practical use. Consequently, one generally seeks search procedures that are sound, but not complete. Unfortunately, in adversarial planning soundness and completeness are not separable properties. A noncomplete search procedure cannot be sound because it may have been incomplete in its search of an adversary's actions (reactions) to a proposed plan. In adversarial planning, a search procedure is sound only if it is both sound and complete.

This paper presents search procedures for adversarial planning that address these problems. The next section describes these procedures and the classes of adversarial planning problems for which they are provably sound and complete. Then we discuss how these procedures can be embedded in a practical adversarial planning system.

Search Procedures

The general class of problems which are defined as adversarial planning problems are described and formalized below. The generality of this formalism is also discussed.

An adversarial planning problem is composed of an agent and an adversary (competitive agent), where both are attempting to achieve competitive goals in a common environment. For the agent to effectively plan, he must at a minimum have the following elements:

• a model of the environment (domain simulation) that describes the agent's present perceived situation and the impact of any agent or adversary actions on the environment,

• a planning knowledge base that, for any given situation and agent (adversary) goal, can be "consulted" to identify possible subgoals and actions the agent (adversary) could pursue,

During planning, after each state change, the planner utilizes the planning knowledge base to enumerate possible agent and adversary actions. The planning knowledge bases, in turn, reference data available from the domain simulation to determine these possible actions. The objective of the planning agent is to identify a provably complete adversarial plan that, according to the environment model, leads to "success" no matter which of the perceived-as-possible adversary actions are encountered. Problems involving deception and uncertain outcomes can be handled in this framework through an appropriately defined planning knowledge base.

Formal Description

Assume that we have a problem domain with two agents: a planning agent (P) and an adversary or competitive agent (C). Both P and C can alter the problem state (s) and can perform multiple actions simultaneously.

For any situation, one or more goals can be posited for either P or C. A goal (g) is composed of a goal name and arguments (e.g., attack_bishop [e4], defend_border [France,Germany]).

We assume the existence of a planning knowledge base where, for any goal g, any subset of the following functions may be defined:

failed(g,s):	an expression that returns t if g has failed in state s; NIL otherwise (default = NIL)
succeeded(g,s):	an expression that returns t if g has succeeded in state s; NIL otherwise (default = NIL)
valid(g,s):	an expression that returns t if a goal g is valid for state s; NIL otherwise {default = ~[failed(g,s) v succeeded(g,s)]
feasible(g,s):	an expression that returns t if a goal is feasible

	for state s; NIL otherwise [default = valid(g,s)]
countergoal(g,s):	an expression that returns an adversary countergoal of g, for state s; a countergoal is defined by having directly opposite failed and succeeded functions (default = NIL)
subgoal(g,glst,s):	an expression that returns a subgoal of g, for state s, not in the list glst; will return NIL if valid(g,s) = NIL; will only return feasible subgoals (default = NIL)
action(g,s):	a function that specifies a state change procedure (default = NIL)

Each of the default values should be interpreted as a minimum constraint on the expression. That is, for instance, the feasibility test can be more constraining than a simple validity check, but cannot allow an invalid goal to be feasible.

The most direct way to define such a planning knowledge base is with a knowledge base of goal definitions, which specifies for each goal name a set of arguments and some of the above referenced functions. Figures 1 and 2 depict two simple goal definitions. For discussions here, we will assume the planning knowledge base is encoded as a set of goal definitions, although this assumption is not essential to the search procedures.

In terms of the discussion above, the problem state, s, refers to the domain simulation. In game playing, the problem state is usually defined in terms of a game board with perfect information on both sides. In more complex domains (e.g., military planning), a domain simulation may be used which includes explicit procedures for estimating both the agent's and adversary's perception of the environment. The planning knowledge base may differentiate between agent and adversary goal definitions to reflect the agent's understanding of the adver-

DEFEND_CORRIDOR (Corridor, Units, Until_Time)
Subgoals: Identify defensive positions.
Assign members of Units to each defensive position with subgoals of
DEFEND_LOCATION (Location, Units)
Failed if: Cannot find any locations to set up defense
Succeed if: Defensive positions are held Until_Time

Figure 1. Description of a domain specific goal definition.

ANDSIM (List_of_goals)

Subgoals:	Set equal to list of ANDSIMS of combinations of subgoals of each goal in List_of_goals
Failed if:	Any of the goals in List_of_goals has failed
Succeed if:	Any of the goals in List_of_goals has succeeded
Valid if:	All of the goals in List_of_goals are valid
Feasible if:	All of the goals in List_of_goals are feasible

Figure 2. Description of a generic goal definition.

sary's repertoire of strategies and tactics.

Within any state s, we say that g_2 is a subgoal of g_1 if for some list of goals, g1st,

$$g_2\text{=subgoal }(g_1, \text{g1st}, s)$$

Within any state s, we say that g_2 is a subgoal of g_1 if for some list of goals, glst,

$$g2 = \text{subgoal }(g1,glst,s).$$

The property of being a subgoal is transitive. That is, if g_3 is a subgoal of g_2 which is a subgoal of g_1, then g_3 is a subgoal of g_1.

To project the consequences of an action on the world state, it is necessary to specify the actions the agent and adversary are simultaneously pursuing. For instance, the impact of attack(cityA) will depend on whether or not the adversary's goal is to defend (cityA) or retreat from (cityA). Consequently, we define a state change function

$$\text{execute}(g,c,s)$$

that for any state s calls the domain simulation to execute the action functions defined in goals g and c. If g has an action function, then EXECUTABLE(g,s)=t; else it is NIL. The execute function can only be applied if either g or c is not NIL, and any non NIL goal is executable.

For the initial situation (s), the goal stacks for a goal g_1 is the set of all $\{g_n,...,g_1\}$ where

$$g_{i+1} = \text{subgoal}\ (g_i, g1st, s)$$

for some g1st, and

$$\text{EXECUTABLE}\ (g_n, s)=t.$$

During a lookahead, for a state s_i, let $gs=\{g_n,...,g_1\}$ and $cs=\{c_m,...,c_1\}$ be a pair of goal stacks generated from the pair or goals g_1 and c_1, and

s_{i+1} <— execute (first_of (gs),first_of (cs),s_i).

The *goal stacks for g_1 in s_{i+1}* is equal to the set of all goal stacks that can be generated from $g_1$in s_{i+1}, where g_1 is a member of the goal stack gs. It is important to note in this last definition that if g_2 is a subgoal of g_1 in state s_i and was part of the goal stack that resulted in state s_{i+1}, then g_2 is relevant to state s_{i+1}, *even if it is not a subgoal of g_1 in this new state.*

To illustrate why this is so, assume we have the simple maneuver planning problem depicted in Figure 3. Blue plans to break through Red's defensive lines

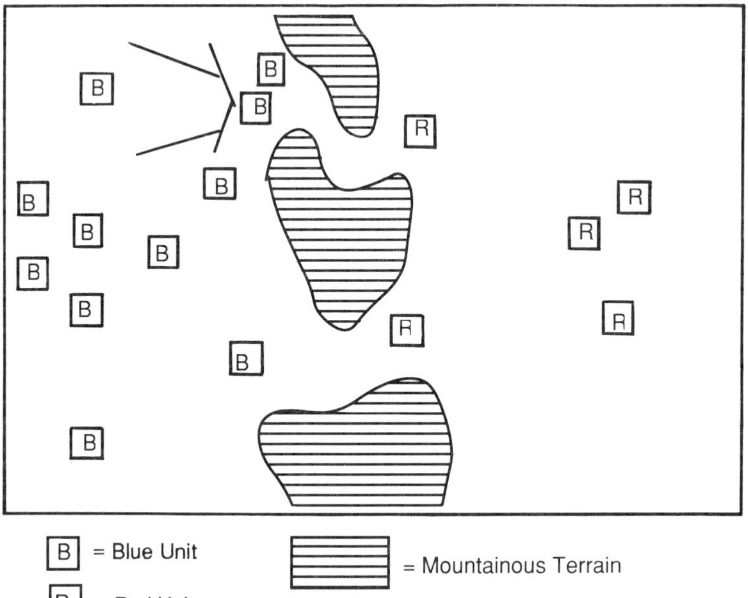

Figure 3. Initial Maneuver Planning Situation.

with And[Main_Attack(North), Defend(South)]. After a state change, however, the situation in Figure 4 occurs. Red has put virtually all forces to the North. Under this situation, the above Blue goal is not a reasonable attack scenario (an attack in the south should be included), and may not be generated as a subgoal. Nevertheless, since the purpose of the lookahead is to determine the extent to which this attack scenario can defeat possible enemy defensive schemes, this lookahead should continue until the attack scenario either succeeds or fails.

The *option space in s* is the set of all pairs of goal stacks from the possible goal stacks of g_1 and c_1 in s.

For an initial state (s_0) and a pair of goals g and c, the *search space* is defined as the set of all state sequences $\{s_0, s_1, \ldots, s_n\}$, where s_{i+1} was generated from a pair of goal stacks in the combined option space of s_i. That is, where

$$s_{i+1} = \text{execute}(gs, cs, s_i)$$

for gs and cs are members of the combined option space of s_i.

An adversarial planning problem is defined by an initial problem state, a planning knowledge base, and a pair of goals, g and c, where the goal definitions for g and c are such that, for any given state, s, one of the following holds:

failed(g,s)=succeeded(g,s)=failed(c,s)=succeeded(c,s)=NIL
failed (g,s)=T and failed(c,s)=T

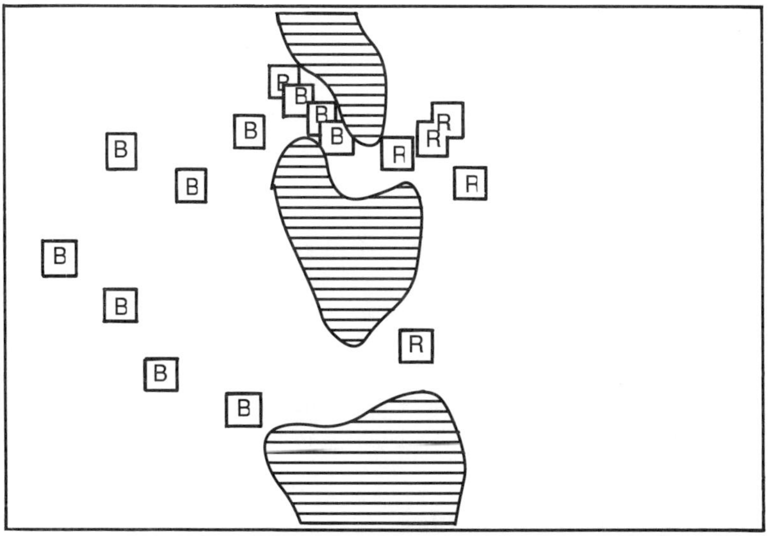

Figure 4. Situation after main attack initiated: A secondary attack in south would succeed.

succeeded(c,s)=T and failed(g,s)=T
succeeded(g,s)=T and failed(c,s)=T.

In other words, only one side can win, although both sides can fail.

The problem the adversarial planner must solve is to find a first action where, no matter what actions the adversary pursues, the planner has an appropriate response that will guarantee achievement of g. The set of preplanned responses (or response sequences) are referred to as contingency plans. It is assumed that the adversary's goal is similarly to achieve c.

Generality of the Formalism

This formalism for characterizing adversarial planning problems is quite general. It can be used to formulate planning problems in a variety of interesting domains, ranging from board games, such as chess and Go, to planning problems in military science, where multiple friendly and enemy units are moving simultaneously. It could, for instance, represent the knowledge encoded in the knowledge-based chess programs of Wilkins (1980, 1982); the Go programs in Reitman and Wilcox (1979) and Lehner (1983); as well as problems in ground maneuver planning (see example below). The formalism puts no specific constraints on the actual scheme used to represent knowledge relevant to planning. Instead it simply requires that the knowledge base (using whatever representation) be able to execute a subset of the functions noted above (i.e., generate subgoals, test feasibility, etc.).

Despite the generality of the formalism, we can specify contingency planning procedures that are theoretically correct for solving any adversarial planning problems consistent with this formalism. By theoretically correct, we mean that the adversarial planner is guaranteed to find a solution if one exists, or prove that one does not. These procedures are discussed in the next subsection.

Provably Complete Search Procedures

As noted previously, rarely in adversarial planning will there exist a single sequence of actions that will guarantee achievement of a goal no matter what an adversary does. Rather, an adversarial planner must generate contingency plans. Described below are several search procedures which will be collectively referred to as the CPx (Contingency Planner version x) procedures.

CP1: A Brute Force Algorithm

The recursive procedure, CP1, assumes the following inputs: gs and cs which are goal stacks in list form, and s which is the problem state. CP0 will return a 1 (agent wins), -1 (adversary wins) or 0 (neither side wins).

Recursive Procedure CP1 (gs, cs, s)

1. Check failure test for last goal in both gs and cs.
   Return 0 if both sides failed
   -1 if last goal in gs failed
   1 if last goal in cs failed.

2. Starting with gs (cs), use a last in-first out procedure to generate all possible goal stacks for side 1 (side 2). Call this set gstacks (cstacks). Mark each element in gstacks and cstacks "Undefeated."

   Let *stackpairs* be a list of all pairs of stacks in gstacks and cstacks.

3. If *stackpairs* is not empty, select and delete an element (gx,cx) from *stackpairs*.

   If either gx or cx is marked "Undefeated" then let
   r = CP1(gx,cx,execute(first(gx),first(cx),s),
   and
   if not r=1 (r=-1) then mark gx (cx) "Not Undefeated".

4. If *stackpairs* is not empty, go to step 3.

5. If there is a gx (cx) in gstacks (cstacks) that is labeled "Undefeated"
   Then Return 1 (-1).
   Otherwise Return 0.

CP1 is a simple procedure. Within each position, it generates all possible stacks for both sides. If there is an agent (adversary) stack that defeats all adversary (agent) stacks, it returns 1(-1). Otherwise it returns 0. For each pair of agent-adversary stacks, a recursive call to CP1 determines which stack, if either, wins.

Contingency plans are generated by recording each sequence of combined agent/adversary goal stacks. (For ease of exposition, this recording procedure was not included in the presentation of CP1.) As illustrated in Figure 5, these recorded plans can be summarized as a set of *Contingency Goal Trees* (CGTs).

It is straightforward to show that for the class of adversarial planning problems described, CP1 satisfies the following property.

Property 1: (general completeness)
Assume a domain simulation where state changes are deterministic. If the search space is finite, CP1 will find a solution for either side, if one exists, or it will determine that no solution exists. For an infinite search space, CP1 will only terminate with a correct result.

It is worth restating a point made in the introduction. Namely that in contrast to most other classes of automated reasoning problems, in adversarial planning,

GOAL STACKS FOR SINGLE SEARCH SEQUENCE

State	Agent Goal Stack	Adversary Goal Stack
s0	(g11, g1, G)	(c11, c1, C)
s1=execute(g11,c11,s0)	(g11, g1, G)	(c12, c1, C)
s2=execute(g11,c12,s1)	(g21, g2, G)	(c12, c1, C)
s3=execute(g21,c12,s2)	G succeeds	

CONTINGENCY GOAL TREE FOR SEARCH SEQUENCE

```
                G:C —— G succeeds
               /   \
          g1:c1     g2:c1
         /     \        \
  g11:c11   g11:c12   g21:c12
```

Figure 5. Representing goal stacks as contingency goal trees.

soundness and completeness properties are *not* separable. In nonadversarial planning, for instance, one can implement noncomplete planning procedures (not guaranteed to find a plan if one exists) that will only generate sound plans (solutions that are found are correct). This separation is not possible in adversarial planning. This is because any adversarial plan generated by an noncomplete planning procedure cannot be provably sound, since the planning procedure may have been incomplete in its exploration of possible adversary responses to the proposed plan. An adversarial planner is sound if and only if it is complete.

Completeness, therefore, seems to be a requirement of a sound and useful adversarial planning system. This is unfortunate since computationally achieving completeness is usually too expensive to be feasible. The inefficiency of complete algorithms is clearly reflected in CP1. Assume that each goal generates m subgoals, each stack is depth n and the search depth of states is arbitrarily cut off at d (at depth d+1 both sides automatically fail). Given these assumptions, CP1 will search between m^{nd} and m^{2nd} state changes. Furthermore, these two values represent the greatest lower and least upper bounds of any generally sound search procedure. Not very encouraging!

There are two ways to respond to this problem. First, one can simply give up on soundness, and employ heuristic techniques for trimming the search space. Although heuristic pruning at some level is essential, this approach is full of pitfalls. To illustrate one, consider the case of military battle planning. In military planning one can trim the search space by only considering adversary courses of action perceived as "likely." Unfortunately, military adversaries are also intelligent planning agents, and the most "likely" adversary courses of action are those that are the most likely to succeed. Consequently, reliable determination of an adversaries likely courses of action can only be performed after planning space has been explored, not before.

A second approach is to identify adversarial planning algorithms that satisfy restricted forms of completeness. This approach is explored in the next section.

Intentional Complete Search Procedures

Although generally complete, CP1 sometimes does more searching than appears necessary. To illustrate, consider the following CGT

```
        G:C
         |
        g1:c1
       /    \
  g11:c11  g11:c12
```

After the first state change (defined by g11:c11), the goal g1 may have failed, but CP1 never checks this, and therefore continues searching after g11:c11. As a result, CP1 may generate plans that have one or more intermediate failures. This behavior is necessary for completeness, since continued pursuit of a failed subgoal may still lead to an unintended success (e.g., lose the battle but win the war). In such cases, however, it is rarely the case that the failed subgoal was necessary to success. (In principal one can always replace it with an intentional sacrifice.) It therefore seems unlikely that an adversary would explore such options. Consequently, there is little risk in pruning the search whenever a failed subgoal is found. This is equivalent to the requirement that the search procedure not consider contingency plans (CGTs) with intermediate failures.

The following procedure is a modification of CP1 that does not generate plans with intermediate failures. The two new variables, pgs and pcs, are subsets of the goals in gs and cs that are "protected." That is, they cannot be popped from the goal stacks. Pgs and pcs are initially set to list(g) and list(c) respectively.

Recursive Procedure CP2 (gs,cs,s,cgt,pgs,pcs)

1a. Check failure test for any valid goal in pgs and pcs.

Return 0 if both sides failed
-1 if a goal in pgs failed
1 if a goal in pcs failed.

1b. Remove from pgs (pcs) any invalid subgoals and their subgoals.

2. Starting with gs (cs), use a last in-first out approach to generate all possible goal stacks for side 1 (side 2) that contain pgs (pcs). Call this set gstacks (cstacks). Mark each element in gstacks and cstacks "Undefeated."

Let *stackpairs* be a list of all pairs of stacks in gstacks and cstacks.

3. If *stackpairs* is not empty, select and delete an element (gx,cx) from *stackpairs*.

If either gx or cx is marked "Undefeated" then let
r = CP2(gx,cx,execute(first(gx),first(cx),s, gx, cx),
and
if not r=1 (r=-1) then mark gx (cx) "Not Undefeated."

4. If *stackpairs* is not empty, go to step 3.

5. If there is a gx (cx) in gstacks (cstacks) that is labeled "Undefeated"
Then Return 1 (-1).
Otherwise Return 0.

CP2 is identical to CP1 except that, for any problem state, the search is trimmed as soon as it is discovered that one of the goals that leads to that problem state has failed, but is still valid. This guarantees that none of the contingency plans generated will have intermediate failures. As stated in the following property, except for this restriction, CP2 is complete.

Property 2: (intentional complete)
Assume a domain simulation where state changes are deterministic and neither the planner nor the adversary generates plans with intermediate failures. If the search space is finite, CP2 will find a solution for either side, if one exists, or it will determine that no solution exists. For an infinite search space, CP2 will only terminate with a correct result.

The fact that CP2 is not generally complete implies that it cannot guarantee that the plans it returns will lead to a success. The plans it generates are not necessarily sound. However, although CP2 does not guarantee success, it does, in effect, guarantee that the planner will not be outsmarted. If CP2 finds a solution for the planner, then the only way the adversary (whose option space is as defined by the planning knowledge base) can succeed is to pursue a course of action that is expected to fail in its immediate objectives, yet somehow still work out in the long run. In effect, to delay the inevitable and hope for the best.

From the computational complexity perspective, CP2 helps considerably. If we assume that each stack contains k protected goals. Then the minimum and maximum number of states explored are $m^{n+(n-k)(d-1)}$ and $m^{2n+2(n-k)(d-1)}$. More importantly, one would normally expect that most of the goals in a goal stack would stay protected after an action is executed. Only the lowest level goals become invalid. Consequently, in any state after the initial state the number of state changes will typically be around, say m^3, rather than m^n, no matter what value n is. That is, the number of states searched will be relatively small and independent of the size of the goal stacks.

A Maneuver Planning Example

To further illustrate the operation of CP2, a brief maneuver planning example is provided. The maneuver planning example described below was actually executed by a predecessor procedure that is very similar to CP2. In this particular example, the results would be identical.

The domain is the simple computerized-wargame depicted in Figure 6. In this depiction, the lines indicate possible unit locations. The underlined units represent Blue units, while the other units are Red units. Units can move to adjacent locations. An x between two locations indicates difficult terrain to traverse. A * indicates impassable terrain. Each turn of the game represents a twelve-hour update. During each turn, a unit can move up to its allocated movement points. Each vertical or horizontal move costs one movement point. Each diagonal move costs 1.414 movement points. A move over an X doubles the cost of a move. Only one unit can be at any one location at a time. Blue and Red units cannot move through each other, but stop when they "collide." Opposing units must be on adjacent locations to attack. Factors, such as terrain, length of time the defending unit has been in position, relative strengths of defending and attacking units, etc., are all included in the calculation of an attack outcome.

On each turn, each player must specify a set of orders for each unit on his or her side. Once entered, these orders are executed by the computerized-game and the new situation is displayed.

In this example, Blue's top goal is to hold a line of defense along the rough terrain going north to south in the center of the board for 60 hours. Red's goal is to break through this defense line within 60 hours. A single knowledge base of goal definitions was defined for both sides. The knowledge base consisted of domain and generic goals. A domain goal has as its arguments characteristics of the state. Figure 1 describes the domain goal definition DEFEND1CORRIDOR which has as its arguments an identification of the corridor, the units available to defend the corridor, and how long the corridor must be defended. A generic goal definition has as its arguments other goals and are defined so that they could be used in multiple domains. Figure 2 describes the generic goal definition for ANDSIM, which has as its arguments a list of goals. Loosely described, ANDSIM defines a conjunctive goal, where all the goals in its arguments can be pursued independently and simultaneously. Figure 7 is a typical example of the type of goal CP2 would generate during search. Loosely interpreted, this goal corresponds to a plan of attack where corridors 1 and 2 are attacked, corridors 3 and 4 defended, and some units are kept in reserve to support whichever attack appears most likely to succeed. Success is obtained if one of the attacks succeed, and both defenses succeed.

```
—- —- —- —- —-*—- —- —- —- —- —- —- —-
   x x      x    * * * * *      x
—- —- —- —- —-*—- —-*7MD —-x—- —- —- —-
 x * * x        x   x          x * x
—-*—- —- —-4IDx—-x—- 6MD —-*—- —- —- —-
  x x           x   * x        x   x x
—- —- 3TDx—- —-*—-*—- —- —- —-x—- —- —-
              x * * * * * * *        x  x x
—-x—- —- 1ID —- —- —- 2AD 1MD —-   x—- —-
x x                                 x x x
—-x—- 1TD —- —- 1CR   1AD 4AD —- —- —- —-
   * x                              x x   x
—- —- —- 2ID —-*—-*—- 3AD 3MD 2MDx—- —- —-
  x * x         * * *
—- —- —- —-     *—*—- —- —- —- —-     x—- —-
   x  x         * * x *
—- —- —- —- —-*—-x—-x—-  4MD —- —- —- —-
    x           x  x x *              x x
—-x—- —2TD x—- —- —-    x—- —- —- —- —-
    *  x      * * * *     * * x          x  x * x
—- —- —- 3ID —x—- 5MD —- —- —- —-     x—- —-
    x            x x
—- —- —- —-*—-*—-*—- —- —- —- —- —- —-
```

—- = Possible Unit Locations

* = Terrain Impossible to Traverse   YYY = Blue Unit

x = Terrain Difficult to Traverse   YYY = Red Unit

Initial Blue Goal: Hold Defensive Line for 60 hours.

Initial Red Goal: Breakthrough Defense within 60 hours.

Figure 6. Initial position for maneuver planning example.

```
(ANDSIM
   ((ORSIM
       ((ATTACK1CORRIDOR (Corridor2 (1AD 2AD 3AD) 60))
        (ATTACK1CORRIDOR (Corridor1 (6MD 7MD 1MD) 60)))
    (ANDSIM
       ((DEFEND1CORRIDOR (Corridor4 (5MD) 60)
        (DEFEND1CORRIDOR (Corridor3 (4MD 3MD) 60)))
   (SUPPORTCORRDIORS (((Corridor1 Attack)
                                        (Corridor2 Attack))
                                       (2MD 4AD) 12)))
```

Figure 7. Sample goal from maneuver planning example.

In this demonstration, only the top goal had alternative subgoals. These alternative subgoals corresponded to different schemes of defense or attack. All other goals had a single subgoal, or no subgoals. CP2 initially paired a Blue shallow defense scheme against a Red massive attack in the central corridor. Figure 8 shows the situation after twelve hours. Playing out this option (with new goal stacks generated at each turn) resulted, after 48 hours, in the situation shown in Figure 9. Red breaks through. Since the defense fails, the planner backtracks on the goals it selected in each position until an alternative subgoal is found. This only occurs in the initial position, where an alternative scheme of defense is found. Figure 10 shows the position after 12 hours where an in-depth defense scheme is used. Note here that several units are held in reserve to support whichever corridor defense is most in need of support. During the next several turns, 1TD and 3TD move to support the central corridor defense. After 60 hours, the defense is still holding (although just barely). The resulting position is shown in Figure 11.

At this point, it is the massive main attack that failed. Consequently, the planner backtracks on the goals posited for Red to find alternatives. As shown in Figure 12, one option it tries is a dual avenue of attack. (The goal shown in Figure 6 is a subgoal of this attack scheme.) In the next several turns, it appears as though the central corridor attack is the one most likely to succeed. Consequently, reserve units from both sides are sent there. After 60 hours, the situation in Figure 13 resulted. Blue managed to defend the central corridor, but in the meantime, the defense of the northern corridor collapsed.

At this point in the demonstration, there are no other defensive schemes for Blue to try, so Blue cannot guarantee a successful defense. Note, however, that

```
—— —— —— —— ——*—— —— —— —— —— —— —— ——
   x x        x  *  *  *  *  *     x
 —— —— —— —— *—— ——  *—— ——   x—— —— —— ——
 x * * x          x      x          x * x
 ——*—— —— 3TD x 4ID x 7MD —— ——*—— —— —— ——
   x x        x         *   x   x   x x
 —— ——x—— —— *—— *—— —— —— ——  x—— —— ——
              x  *  *  *  *  *  *  *      x  x x
 ——x—— —— —— —— 1CR 2AD 1MD —— —— ——x—— ——
  x x                                 x x x
 ——x—— —— 2ID 1ID  1AD  4AD  6MD —— —— —— ——
     * x                               x x  x
 —— —— —— —— ——*—*3AD 3MD —— ——   x—— —— ——
    x * x        * * *
 —— —— —— —— 1TD*——*—— —— —— —— ——    x—— ——
      x  x          * * x *
 —— —— —— —— ——*——x——x2MD —— —— —— —— ——
       x          x   x x *            x x
 ——x—— —— —— ——x2TD 4MD ——x—— —— —— —— ——
        *  x     * * * *  * *      x       x  x * x
 —— —— —— —— 3IDx5MD —— —— —— —— ——x—— ——
    x              x x
 —— —— —— ——*——*——*—— —— —— —— —— —— ——
```

Figure 8. SHALLOW DEFENSE vs. SINGLE MASSIVE ATTACK after 12 hours.

at this point we do not know whether the dual avenue attack is a guaranteed success. This is because it was not tested against a shallow defense. After playing out this additional conflict, the dual-avenue attack does succeed, leading the planner to conclude that a dual-avenue attack is a success for Red.

A Note on Computer Game Playing

One area where problem solving in an adversarial environment has been intensively investigated is computer game playing. With relatively few exceptions, the work in this area has been in identifying efficient heuristic search algorithms for searching the game trees in two-person, perfect-information, games

```
--- --- --- --- ---  *--- --- --- --- --- --- --- ---
   x x      x     * * * * *  x
  --- --- --- --- ---*--- ---  *--- ---    x--- --- --- ---
  x * * x         x   x           x * x
  ---*--- --- --- 3TDx4IDx7MD --- --*--- --- --- ---
   x x            x       * x     x  x x
  --- --- ---  x--- ---*---*--- --- --- --      x--- --- ---
                x * * * * * * *                x  x x
  ---x--- --- --- ---   2ID 2AD 1MD --- --- ---x--- ---
  x x                                          x  x  x
  ---x--- --- --- 3AD 3MD 1AD --- --- --- --- --- ---
    * x                                        x x  x
  --- --- --- --- ---   *4AD*6MD --- --- ---  x--- --- ---
   x * x              * * *
  --- --- --- --- ---   *---*--- --- --- --- ---       x--- ---
    x   x             * * x *
  --- --- --- --- ---  *4MDx---    x--- --- --- --- --- ---
       x              x  x x *                        x x
  ---x--- --- --- 1TDx2TD 2MD ---x--- --- --- --- ---
      *  x        * * * *  * *         x      x      x * x
  --- --- --- ---  3IDx5MD --- --- --- --- ---       x--- ---
       x             x x
  --- --- --- --- *---*---*--- --- --- --- --- --- ---
```

Figure 9. After 48 hours:
SHALLOW DEFENSE fails, MASSIVE ATTACK succeeds

with alternating play of single moves (chess, checkers, othello, etc.) where virtually all knowledge embedded in the system is in a mathematical function for performing a static quantitative evaluation of a single position.

In the adversarial planning approach presented here, the existence of a predefined option space, such as the moves in a game, is not assumed. Rather, only options that can be generated through processing of the planning knowledge base are considered. It is, however, of interest to examine the behavior of adversarial planning algorithms when searching game trees.

CP1 and CP2 could be applied to alternating play games such as chess. The execute function would always be of the form

execute(Move(x),NIL,s).

```
___ ___ ___ ___ _*___ ___ ___ ___ ___ ___ ___ ___
  x x     x     * * * * *   x
___ ___ ___ ___  *___ __*___ ___x___ ___ ___ ___
x * * x          x   x          x * x
___*___ ___ ___ x4IDx7MD ___ *___ ___ ___ ___
  x x           x         * x   x  x x
___ ___ 3TDx___ *___*___ ___ ___ ___x___ ___ ___
               x * * * * * * *            x    x x
___x___ ___ ___  1CR 2AD 1MD ___ ___  x___ ___
x x                                 x x x
___x___ 1TD ___  1ID 1AD 4AD 6MD ___ ___ ___ ___
     * x                            x x  x
___ ___ ___ ___  *___*3AD 3MD ___   x___ ___ ___
  x * x          * * *
___ ___ ___ ___  *___*___ ___ ___ ___   x___ ___
     x   x       * * x *
___ ___ ___ ___  *___x___x2MD ___ ___ ___ ___ ___
     x           x   x x *                  x x
___x2ID ___ ___ x2TD  4MD ___x___ ___ ___ ___ ___
      *  x      * * * *     * * x      x  x * x
___ ___ ___ 3IDx5MD ___ ___ ___ ___ ___   x___ ___
     x           x  x
___ ___ ___ *___*___*___ ___ ___ ___ ___ ___ ___ ___
```

Figure 10. DEFENSE-IN-DEPTH vs. SINGLE MASSIVE ATTACK after 12 hours.

However, neither CP1 nor CP2 exploit the fact that simultaneous moves are not allowed or that each move reflects a single action. As a result, both CP1 and CP2 would spend considerable effort generating possible goal stacks for the side not on the move, even though none of these goal stacks could impact the game.

Modifying CP1 and CP2, so that neither would generate goals for the side not on the move is straightforward. The resulting search procedure would be similar to the procedure found in the Reitman and Wilcox (1979) Go program.

Practical Considerations

In the previous section, provably complete procedures for adversarial planning were presented. It was noted that an adversarial planning procedure should

```
--- --- --- ---   *--- --- --- --- --- --- --- ---
  x x      x    * * * * *       x
--- --- --- --- ---*--- --- *--- --- — x--- --- --- ---
x * * x        x    x          x * x
---*--- --- --- --x4IDx7MD --- ---  *--- --- --- ---
  x x          x        * x      x   x x
--- --- ---  x--- -*---  *--- --- --- --- — x--- --- ---
            x * * * * * * *                x  x x
---x--- --- -3TD 4AD --- --- --- --- --- ——   x--- ---
x x                                      x x x
---x--- --- -1TD 3AD 2AD --- --- --- --- --- ---
     * x                                 x x   x
--- --- --- --- --*1AD*6MD --- --- ---  —x--- --- ---
  x * x          * * *
--- --- --- --- ---*---*--- --- --- --- --- ———   x--- ---
   x   x         * * x *
--- --- --- --- ---*4MDx---x--- --- --- --- --- --- ——
     x           x      x x *                x x
---x2ID --- --- -x2TD 2MD ---x--- --- --- --- ---
      *  x      * * * *       * *x          x  x * x
--- --- --- ---3IDx5MD --- --- --- --- --- —   x--- ---
   x             x x
--- --- --- ---*---*---*--- --- --- --- --- --- ---
```

Figure 11. After 60 hours: SHALLOW DEFENSE succeeds, MASSIVE ATTACK fails.

be provably complete in order to guarantee that a solution, if found, is sound. Unfortunately, the search space of these algorithms can be quite large and for high-level goal pairs (e.g., a white win vs. a black win in chess) intractable even for algorithms that satisfy only intentional completeness. In this section, issues relevant to building a useful adversarial planning system are briefly introduced. As will be seen, many of these issues remain unresolved.

Level of Simulation

Clearly the CPx approach to adversarial planning is search intensive. It requires a domain simulator on which numerous possible futures are explored. An issue of obvious concern therefore is the extent to which a simulator needs to provide a detailed versus aggregate characterization of the domain. If a detailed

```
—- —- —- —- —-*—- —- —- —- —- —- —- —-
  x x          x    * * * * *      x
—- —- —- —- —-*—- 7MD*—- —-x—- —- —- —-
x * * x          x  x          x * x
—*—- —- —- —-x4IDx1MD —- —-*—- —- —- —-
  x x           x        * x      x  x x
—- —- 3TDx—-  *—- *6MD —- —- —x—- —- —-
               x * * * * * * *         x     x x
—x—- —- —- —- 1CR 2AD —- —- 2MD 4ADx—- —-
x x                                     x x x
—x—- 1TD —- —1ID 1AD —- —- —- —- —- —-
    * x                                 x x    x
—- —- —- —- —-*—*3AD —- —- —-     x—- —- —-
  x * x          * * *
—- —- —- —- —-*—-*—- —- —- —- —- —-    x—- —-
    x    x        * * x *
—- —- —- —- —-*—x—-x2MD —- —- —- —- —-
    x             x    x x *                 x x
—x2ID —- —-    x2TD 4MD —x—- —- —- —- —-
      *   x       * * * *       * * x        x   x * x
—- —- —- —- 3IDx5MD —- —- —- —- —-     x—- —-
      x             x x
—- —- —- —-*—-*—-*—- —- —- —- —- —- —-
```

Figure 12. DEFENSE-IN-DEPTH vs. DUAL ATTACK after 12 hours.

simulator is required, then search intensive procedures are infeasible. If a superficial (aggregate) simulator will do, then search intensive planning procedures become a viable option.

My own view on this issue is that the notion of a speed versus accuracy tradeoff in simulation systems is fundamentally misplaced. There is no inherent trade off—detailed simulation systems are usually *less* accurate. To illustrate why, consider the following simple scenario. Assume the following:

(a) that we are trying to predict the outcome of a battle between a specified set of Blue and Red forces,

(b) we have available a force ratio analyzer (set of equations) that is about 90 percent accurate in its Win/Loss predictions,

(c) we can decompose the analysis of a battle into a set of five component battles (each of which is also predicted with 90 percent accuracy),

(d) we know with certainty the relationship between the five component

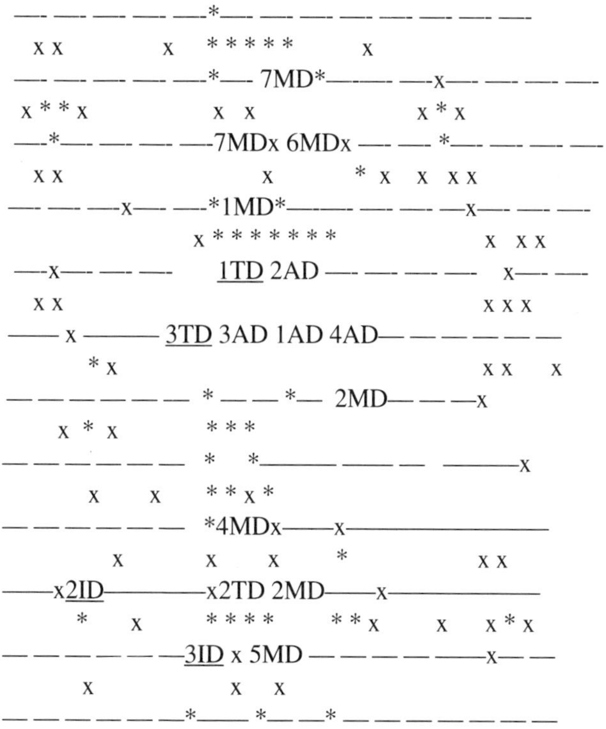

Figure 13. After 60 hours: DEFENSE-IN-DEPTH fails, DUAL ATTACK succeeds.

battles and the overall outcome, namely that Blue wins the overall battle if and only if Blue wins three or more of the component battle, and finally

(e) as a "technical" assumption, an equal prior probability on each possible pattern of wins and losses on each of the five battles (i.e., prob(wwwww) = prob(wwwwl) = prob (wwwlw) ...)

(a)-(e) above describes a situation where the more detailed simulator reflects a perfect model of the environment—the relationship between the components of the more detailed simulation and the overall outcome are known with perfect accuracy. Despite this, the overall accuracy of the more detailed, five component model is 84 percent—when compared to the aggregate model a six percent drop in accuracy! Adding more components or more levels to the hierarchy would further reduce accuracy.

In general it seems that the accuracy gains obtained from adding details to a simulator can be overwhelmed by the errors introduced by being unable to accurately predict the exact status of those details. The most accurate simulators may be the most superficial ones for which reasonably accurate predictions on the status of those input variables can be obtained.

Choosing Subproblems and Hierarchical Planning

A fundamental technique in problem solving is that of decomposing a large intractable problem into a set of smaller, more tractable subproblems. In adversarial planning, this approach is often instantiated by specifying a series of intermediate objectives to be achieved, where it is assumed that achievement of each intermediate objective will support attainment of the next intermediate objective until the final planning objective is reached. In chess, for instance, the opening sequence is generally oriented toward obtaining objectives such as control of the center squares and good piece development. Obtaining these objectives will support the transition to the middle game. Attainment of middle-game objectives, such as high mobility or control of an open file, can, in principle, be parlayed into a superior tactical exchange or end-game position.

The adversarial planning procedures presented above are certainly compatible with the use of intermediate objectives. It is not clear, however, the extent to which problems such as the blemish effect (Berliner, 1982) would be present. This remains an area of investigation.

A related approach, common to many nonadversarial planning systems, is hierarchical planning. Here, planning at one level of abstraction defines the sequence of intermediate objectives to be achieved at the next level of abstraction. Hierarchical planning appears appropriate in adversarial planning domains as well. Here, planning at higher levels can be achieved through definition of alternative simulation levels (if that is reasonable) or representative searching (Lehner, 1983) within a single level of abstraction (discussed below). In adversarial planning, however, planning at each level will result in a set of contingency plans, rather than a single plan. It is therefore likely that all but one of these contingency plans will be invalid after the first step is executed. Consequently, it is probably only useful to limit more detailed planning to the common first step of the higher-level plans.

Discovering Goal Interactions

Goal interactions are an extremely important element of adversarial planning. In adversarial planning, the most elementary tactics involve some form of goal interactions (e.g., a fork in chess), and higher levels of planning skill generally correspond to increasing sophistication in identifying and exploiting subtle interactions. Goal interactions involve circumstances where a step or sequence of steps supports achievement of multiple goals. When achievement of any one of multiple goals implies success, then it is a disjunctive interaction. When all the goals must be achieved, it is a conjunctive interaction. (A forking move in chess is a disjunctive interaction, while the need to find a defense against the fork is a conjunctive interaction.) Success in adversarial planning often means identifying and solving all conjunctive interactions, while finding disjunctive interactions that the adversary may have missed.

The CPx search procedures properly handle any conjunctive or disjunctive goals that are generated from the planning knowledge base. However, the fact that two goals interact is something that is often not derivable by examining an initial problem state. Rather it is something that must be discovered through search. For instance, a chess position may be such that a force capture of the opponent's (white) queen is feasible, leading to exploration of the adversarial planning problem of

Force_Capture(white_queen):Prevent[Force_Capture(white_queen)].

In the process of exploring one of the contingency paths, it may be discovered that a black move leaves black vulnerable to a quick mating combination. If threatening a mate is not a subgoal of the Prevent[Force_Capture(x)] goal and the possibility of a mate was not initially considered, then it was not represented in the initial planning problem. The discovery of a mating attack therefore represents a discovered interaction. It further suggests that the initial planning problem should be redefined as:

And{Force_Capture(white_queen),Prevent[Mate(white)]}:
Or{Prevent[Force_Capture(white_queen)],Mate(white)}

As this example suggests, a useful adversarial planning search procedure should not only solve the planning problem, but also identify goal interactions that help determine whether the planning problem, as defined, is even appropriate. Otherwise, the planner may find itself in the position of properly solving an inappropriate planning problem.

We have not identified a general, provably correct mechanism for identifying goal interactions. More than likely the problem is undecidable. One heuristic approach is to use *representative searching* (Lehner, 1983). This approach involves generating a single-thread lookahead (i.e., play out a scenario) and then looking around in the terminal position to determine if the playout resulted in any significant side effects. A side effect is defined as a change in the feasibility of pursuing a goal not represented in the original planning problem. Another approach (Reitman and Wilcox, 1979) is to record the objects and areas impacted by a search. If another search overlaps, then some goal interaction is involved and must be isolated. The CPx procedures can be easily expanded to incorporate these heuristics. Their, utility, however, remains unproven.

Incorporating Static Evaluation Functions

Most work in game playing involves the use of static evaluation functions to control search in an externally defined search space. In the present approach, knowledge of an externally defined search space is not presumed. Rather, the option space is defined by the planning knowledge base. However, this does not preclude the use of evaluation functions. Indeed, in many domains, feasibility, success and failure tests are defined in terms of evaluation functions. For instance, in ground maneuver planning, equations for calculating force ratios predominate. A high force ratio (>5:1) implies that an attack will likely succeed, and low force ratios (<3:1) suggest a likely failure. A force planner maneuvers to achieve a high force ratio at the location of an attack, maintains the attack as long as the force ratio is high, and discontinues the attack when the ratio is low. Lookaheads often focus on maneuvers to maintain or exceed a desired force ratio during an engagement.

Of interest here is to note that static evaluation functions can be used to implement an iterative deepening (Korf, 1985) version of CPx. In this approach, a global static evaluation procedure is used to prune search on the basis of such factors as overall probability of success (e.g., if a force ratio of 6:1 achieved, assume success), acceptable risk levels (estimated negative utility associated with if unexpectedly lose a battle), probability that adversary will pursue a possible goal (estimate likelihood of alternative adversary courses of action), etc. CPx is then repeatedly called with increasingly stringent requirements on the static evaluation criteria (e.g., accept decreasing levels of risk). Each call to CPx will increase monotonically the set of contingency plans explored, so there is no need to regenerate previously generated goal stacks. Eventually CPx will be unable to solve the problem, because the requirements for success have become too stringent (probability of success must exceed 99 percent, with virtually no risk allowed). At that point, the CGTs found in the previous search specifies an adversarial plan for which the worst possible outcome has been improved as much as possible (similar to a maxmin solution).

Using this approach, CPx will always generate a plan. For different situations, however, different plans will satisfy different criteria. If the agent is in a winning situation, it should generate a conservative plan that should guarantee an overall win. If the situation is desperate, however, then it should generate plans that are either unlikely to succeed, risky, or makes some strong assumptions about the adversary's likely actions.

Incorporating Beliefs About What an Adversary is Likely to do

In military science, it is common practice to engage in an intelligence analysis to predict an enemy's likely courses of action prior to planning friendly actions. While this approach does serve to severely constrain the search space, it is not without considerable risk. It predicts an adversary's actions on the basis of an evaluation of the adversary's present position, without lookahead. If the military planner finds a strategy (instantiated as a set of contingency plans) that defeat all of the Courses of Action (COAs) postulated for the adversary, one should not be too quick to accept that strategy. An equally valid conclusion would be that the adversary also engaged in a lookahead, realized that the COAs were defeatable, and therefore decided to generate and pursue an alternative COA. The very fact that a lookahead revealed that a likely COA is easily defeated, is a good indication that it is an unlikely COA.

This suggests that the analysis of adversary intentions should occur after planning, not before. At the very least, there should be considerable iteration between the two processes. (See Vane, et. al. 1989 for a discussion of how game theory relates to this problem.)

Before leaving this topic, we should note that the CPx approach does not assume a perfect information game. It assumes only that the simulation will allow execution of the procedures specified in the goal definition. If a goal definition (e.g., defend_corridor) references as one of its arguments a belief about the opponent (e.g., opponent believes attack imminent in corridor), then CPx, in effect, assumes that the inputs necessary to test the belief precondition are available. This permits the development of procedures for handling beliefs that are external to CPx, making it possible for CPx to handle problems involving information seeking and/or deception. It does not, however, allow second-order interactions between planning and reasoning about beliefs (e.g., "I can try to deceive him that I am attacking in the North. However, his planning should reveal that an attack in the North cannot possibly succeed, so he'll quickly figure out that the Northern threat is a deception.")

Summary

This paper has presented a general approach to adversarial planning that can be characterized as knowledge-controlled search. For the general class of planning problems consistent with a two-agent conflict, provably complete procedures were defined that will either find a winning set of contingency plans, determine that there is no way to guarantee a win, or continue processing.

As a final note, I would like to point out that the general CPx approach to planning is not necessarily limited to strictly adversarial planning problems. For instance, it is possible to think of a number of cooperative multiagent planning problems where (perhaps because of limited communication) the agents must reason about the possible goals and actions of other agents to find a plan or set of contingency plans that guarantees achievement of the agent's mutual goals. (Being disconnected in the middle of a telephone conversation is a good example of this type of problem.) The extent to which the CPx approach to adversarial planning can be extended to nonadversarial planning problems remains an open issue.

References

Berliner, H. "Backgammon Computer Program Beats World Champion," *Artificial Intelligence*, 14 (1982), 205-220.

Chapman, D. "Planning for Conjunctive Goals," *Artificial Intelligence* 32 (1987) 333-378.

Korf, R. "Depth-first Iterative-Deepening: An Optimal Admissable Tree Search," *Artificial Intelligence* 27 (1985) 97-109.

Lehner, P. "Strategic Planning in Go," in *Computer Game-Playing: Theory and Practice,* ed. M. Bramer, (1983) Ellis Horwood Limited: Chichester, England.

Reitman, W. and Wilcox, B. "Modeling Tactical Analysis and Problem Solving in Go," in *Proceedings of the Tenth Annual Conference on Modeling and Simulation* (1979) 2133-2144.

Vane, R., Laskey, K. and Lehner, P. "Game Playing and Game Trees in Knowledge-based Adversarial Planning," *Proceedings of the Sixth Annual Workshop on Command and Control Decision Aiding*, February 1989.

Wilkins, D. "Using Patterns and Plans in Chess," *Artificial Intelligence* 14 (1980) 165-203.

Wilkins, D. "Using Knowledge to Control Tree Searching," *Artificial Intelligence* 18 (1982) 1-51.

Wilkins, D., *Practical Planning*, (1988) Morgan Kaufmann.

Evaluation of Real-Time Expert Advisory Systems for C^2: A General Methodology and Case Study

Michael L. Donnell,
Richard H. Stottler,III and Christopher Barrett

Introduction and Overview

This article outlines an approach to a general methodology for the evaluation of Real-time expert advisory systems (RTEAS) for intelligent process control and, in particular, C^2. It addresses the myriad of complex issues encountered in attempting to evaluate the performance of a highly complex ensemble of men and machines, acting in time and space, under resource, doctrinal, and other constraints. Insertion of intelligent subsystems in such environments can have subtle and important behavioral and performance implications that, due to the inherent complexity found in these situations, demand a systematic approach to analysis. The general approach adopted relies heavily on computer simulation of tactical situations and C^2 system responses to them. The basic approach is an empirical one which progresses from a general, coarse grained, across the board focus to a specific, fined grained, narrow focus. It also moves from parametric studies of the unperturbed system to similar studies of well-defined perturbed system performance outcomes.

Motivation for Evaluation

What we will present is a methodology for objectively and quantitatively evaluating knowledge based systems performing real-time control of process. It is inherently and explicitly performance based and is therefore preoccupied with input/out structure. Our approach to evaluation does not require detailed examination of the knowledge base. Being performance oriented (which, after all, should be the primary focus of a system evaluator), it does not require familiarity with expert system technology. Not only is the approach we propose valuable once the system has been completed, but it should also prove useful during system development in order to uncover weaknesses requiring redesign and reimplementation. Furthermore, the methodology requires a separation of functions that is useful for questions of person-in-the-loop interaction and analysis.

A General Methodology

This article will document the evaluation of a RTEAS which supports the defensive aspect of the naval outer air battle. Various simulated threats were employed in order to provide a baseline measure of system performance against which modifications of the system and its embedded heuristics could be compared. Human interaction, either for the purposes of situation assessment or tactics, was not involved. In future studies, factors related to human interaction can be studied.

The simulation which provides the environment for the RTEAS evaluation consists of two parts, defense and threat. Each part has several components which must be instantiated for any particular evaluation situation. Three of these components have proven most salient in the development of an evaluation methodology for RTEASs. These three are context, doctrine, and decision interfaces.

Context refers to the methods of information gathering and resolution, the physical weapons platforms for offense or defense, the various weapons, the command and control hierarchies, and the communications, to name but a few elements. In the case study that will follow, contextual considerations include, among others, the experimenter's decisions to set a detection perimeter at 450 miles, to establish two 60 degree sectors each with four outer and two inner stations at 250 and 150 miles, respectively, to locate a continuous refueling station midway between outer and inner stations, and to allocate two types of missiles to each F-14.

Doctrine concerns the rules whereby an attack or defense is initially configured, how a threat is configured or responded to, and the principles for adjudicating between simultaneous—and possibly conflicting—goals, such as (a) eliminate all threats and (b) preserve defensive capability against the possibility of a future threat. In the present simulation the doctrine includes, among others, the high-level decisions to establish a vector logic screen and to vector up to a maximum of six aircraft independent of the number of threat aircraft. It is particularly important to distinguish between parameter-specific doctrinal rules and parameter-independent ones. The former include numerical constants in the situation evaluation or action parts of the rules (e.g., "If threat is detected at 60 degree bearing then ...") while the latter are generic.

Context and doctrine jointly determine most of the characteristics of the operational concept of the defense or threat simulation. The third factor, decision interfaces, refers to those elements of the operational concept in which response alternatives are generated/selected and evaluated and choices made among the alternatives. In the present system, the decision interfaces include the operator-element which decides whether or not to take the advice of the simulation and with what delay. Testing undertaken in the baseline effort did not evaluate effects of human interaction on system performance. For example, the system was set to accept all generated advice as quickly as evidence (known to be complete and error-free) was delivered.

More importantly, interpretive functions associated with simulation assessment were not included in this baseline analysis. The evidence delivered to the decision generator was complete, error free, synchronous and not time-lagged. It is important that the baseline study be performed under these conditions such that the behavior of the tactic alone can be isolated in performance MDEs. Further studies, indeed, different architectures, are required to sort out the details of the influences of interpretive functions on performance.

An important aspect of baseline analysis that was not performed in the case history presented here is the issue of decision time-lags. Systematic variation of delays in decision making is an essential factor in use of the baseline to assess human-induced system performance variation.

Initial Steps in the General Methodology

Our general methodology can be characterized as commencing with several general steps followed first by a phase of study involving assessment of unperturbed system performance and second by a phase of study involving assessment of perturbed system performance. The three initial steps in the general methodology are as follows:

- Characterize both the defense and the threat in terms of essential dimensions that constitute the context (e.g., what types of platforms with what weapons configurations), capturing the interrelationships among dimensions and the range of possible values. Thus, this can be viewed as a model in the sense of capturing class, entity, and relation structures.
- For the type of warfare under investigation (e.g., AEW, ASW, ASuW, AW) identify the major high-level doctrinal strategy elements such as "leaky"

screen (i.e., optimized for endurance) versus "non-leaky" screen (i.e., optimized for coverage), defensive screen, degree of expendability of defensive (or offensive) resources, and so forth, and identify the elements as heuristic or analytical statements in a computer program. Where particular elements are not readily identified, develop evidence generating procedures which, taking into account the specifics of the context, would permit objective determination of the elements by application of those procedures. For example, for defense, one element of a "leaky" screen doctrine could be identified as a decision rule stipulating that the number of defensive elements committed against a threat will increase with the size of the threat only up to some maximum, and that cut-off is less than commitment of total defensive capability.

- Identify as many dependent measures as possible which, on some logical basis, could reasonably characterize the performance of the defense and the threat.

Unperturbed Performance Evaluation

The primary aim of this first phase of RTEAS evaluation is to determine expert system performance in an unperturbed (no threat) environment relative to a number of key evaluating factors (dependent variables). The low level RTEAS performance data to be gathered might include, in a naval air C^2 domain, the following: For each aircraft, at each time increment, its location, fuel load, status, and weapons load. Basic data such as these would be used to calculate the key evaluation factors that would then be studied and modeled and used to drive the RTEAS perturbed evaluation phase. This unperturbed evaluation phase requires successively more in-depth focusing on aspects of performance identified as either weak or critical. In other words, finer grained sampling across the independent variables should be performed and new/alternative independent and dependent variables should be included as necessary to clarify RTEAS functioning in key areas. Dependent variables should be sorted into the generic categories of cost and benefit and the relationships of variables between and within these categories should be determined.

The explicit steps, then, in performing the unperturbed evaluation phase are three-fold:

- Run the RTEAS under conditions of no threat (that is, unperturbed by a threat) and no human involvement, for a period of time many times the length of the periodic cycle of platform resupply, on-station duty cycle, or other periods which have a cyclical nature and which are likely to influence the values of the measures identified in Step 3 above. Record the values of these measures at time intervals short enough to detect fine details of fluctuation.
- From the results of the above, characterize response surfaces in time and space, and generate an unperturbed steady state description of RTEAS performance. Minimum-strength points identified through this procedure indicate a local vulnerability in space or time to an attack. This vulnerability is determined

primarily by the contextual and doctrinal considerations associated with refueling and station keeping. If such contextual and doctrinal performance parameters are not explicitly known, then selecting various threat configurations—especially those predicting threat superiority (such as large numbers of threat platforms or weapons)—and engaging the defense in this time or space region can reveal some of them. The maximum-strength points identify a local time or space region at which the defense is at a relative maximum strength and least influenced by cyclical contextual factors. This is the region that should be used for mounting threats to determine the maximum capability of the RTEAS.

• The final step in our general methodology for studying the unperturbed state is to plot the findings on multidimensional response surfaces for human review, perform relevant statistical tests, and select for more refined sampling any inflection points or trends of interest.

Perturbed Performance Evaluation

The explicit steps in performing the perturbed performance evaluation phase are as follows:

• Performance Function Determination. The first step in the evaluation of a RTEAS under perturbed conditions is to determine the performance function or functions of interest. This function must be able to be calculated from measurable parameters of a simulation run. In simple cases it may be possible to develop one function of a number of measurable parameters which captures the desired performance. In most cases, multiple functions will have to be developed and analyzed. This may be because the interesting performance measures are unrelated, but it is more likely to occur because, in the domain, there is not an agreed on way to combine separate measures. The performance functions will become the dependent variables of the evaluation.

• Determination of the Threat Space of Interest. Identifying the threat space of interest involves two steps. First, the identification of the dimensions are important. These are parameters of the threat that should be varied. The second step is to identify the appropriate ranges for these dimensions.

The determination of the threat dimensions is extremely important. The threat dimensions become the independent variables which are varied to collect the dependent data (the performance functions). Care must be taken not to select too many independent variables because the number of simulation runs varies exponentially with the number of threat dimensions.

After the dimensions of interest have been chosen, the ranges over which the independent variables will vary must be chosen. The ranges should be chosen as small as possible without leaving out important areas of the threat space. The actual values delineating a certain range may depend upon the values of other variables. For example, Dimension 1's range may be from 3 to 30 and Dimension 2 might vary from 1 to 2 times the value of Dimension 1. This occurs in order to make the threat space as small as possible. It is important that all possible conditions under which the knowledge base will be expected to operate under are covered in order to reveal any errors or omissions present in the RTEAS's knowledge base as applied to certain areas of the threat space.

- Iterative Sampling. After the threat space of interest has been determined, sampling and analysis are required. It is recommended that these be performed interactively. The first iterations should sample large portions of the entire threat space in order to acquire a general understanding of the performance of the RTEAS (the knowledge base and simulation). More interesting areas of the threat space can then be examined in greater detail.

In real applications, there is never enough computing time to sample as much of the threat space as is desired. It is, therefore, important to limit the number of simulation runs whenever possible. Often it is possible to calculate rather simply the outcomes from certain simulation runs. This often occurs when very large or very small threats attack the defense, corresponding to overwhelming and underutilizing the defense, respectively.

There are two kinds of sampling, regular and random. In regular sampling, the samples are taken at regular intervals within the threat space. It is often very useful in two dimensions to lay out a grid of points from the threat space to be sampled. This grid is relatively easy to interpret graphically.

In random sampling, points are selected at random for sampling from the threat space, using a probability distribution. Often a uniform distribution of points across the threat space of current interest will prove most appropriate for later statistical analysis. Exponential distributions may sometimes prove appropriate because parts of the threat space will exist where it is assumed that the answers were known. This provides a check on those areas. Furthermore, these areas will only be lightly sampled.

It is important, in performing the random sampling, to have determined the independent variables that will be used in the analysis and to sample these directly from probability distributions. Parameters from individual parts of the threat can then be generated from these. If threat parameters, which are not used directly in analysis, are generated directly from probability distributions, then it is likely that, after the manipulations required to get the independent variables (sums, averages, differences), the independent variables will not vary over any range of significance (Central Value Theorem) and certainly not uniformly or according to a desired distribution.

- Stochastic Event Issues. Most discrete event simulations of C^2 situations will contain stochastic events. In order to perform an evaluation of the defense's performance against a single scenario, multiple simulation runs must be performed and an average taken. How many simulation runs should be performed? This number is likely to vary throughout the threat space. It will also depend on the required resolution of the evaluation. Variances, as well as means, should be considered when deciding on performance functions. It may be important to sample consistently regardless of the outcomes of stochastic events during a scenario.

- Analysis. There are two distinct types of analysis which we found useful in interpreting the large quantities of data produced by this evaluation methodology, graphical and statistical. Graphical techniques are important to acquire a quick understanding of the performance. A three-dimensional graphics package with dynamic rotation of points and animation is essential. Often the performance space will look very noisy so that averaging and other smoothing techniques can greatly increase the usefulness of the analysis. This noise is due to probablistic events in the simulation and the complexity of the performance surface caused by the fact that the knowledge base defines a surface much more complex than standard mathematical functions.

Various statistical tests are important to get a more quantitative understanding of the knowledge base. Statistical analysis also helps with the noise mentioned in the previous section. Averaging techniques are important in both reducing the number of dimensions to be understood at once and reducing the noise. Factor analysis can be especially helpful in reducing the number of dimensions which should be considered in the analysis.

A Case Study

The Airborne Early Warning/Real-Time Advisory System (AEW/RTAS) is an RTEAS concept demonstration of an employment of intelligent systems to assist the Navy's E-2C Combat Information Control Officer to aid in controlling the Outer Air Battle around a carrier group. The system provides advice on the movement of friendly Combat Air Patrol (CAP) aircraft within one half a second and considers such issues as which enemy tracks to intercept with what and how many CAP aircraft, maintaining a protective screen, and refueling.

For AEW/RTAS's development and testing there existed a simulation of the outer air battle scenario. The simulation revolved around the E-2C's orders. The actions were taken by friendly aircraft, based on orders given by the E-2C to the simulation. AEW/RTAS can provide advice on the screen of the simulation, and a person is in control of issuing the actual orders, or AEW/RTAS can issue the orders directly, without human involvement. This is the mode of operation which can be exploited in evaluating the RTEAS.

The simulation consists basically of friendly and enemy aircraft. The enemy aircraft are grouped together into tracks. In fact, the tracks are the basic simulation unit for the enemy. Tracks can consist of one or more aircraft. The type of aircraft may be Backfire bombers, Badger bombers, or pathfinders which are used to detect the location of the carrier group. Enemy tracks appear without warning at various distances from the carrier group. Enemy bombers which get to within 300 NM, release missiles directed at the group and then exit.

Defending the carrier group are F-14s which can be individually directed. Normally a screen of F-14s is maintained at 250 NM from the carrier group. The E-2C command aircraft, tankers, and communications aircraft are also present in the scenario. The F-14s are equipped with a number of air-to-air missiles of different types, which have different ranges and kill probabilities.

A typical scenario might consist of two simulated hours which correspond to two real-time hours on the host machine. By porting to a more powerful machine and making some adjustments to the simulation, a typical scenario could be run to completion in seconds. This gave us the potential of collecting a large volume of data.

The evaluation proceeded in two phases. In the unperturbed phase, no threats were presented to the defense in the scenario. The behavior of the aircraft and, in particular, the defense screen was studied as it evolved over time. From this study we gained a thorough understanding of the initial conditions during the appearance of the first track of a scenario.

The perturbed phase consisted of analyses of the behavior and performance of the defense against various threats. A threat consists of all the enemy tracks in a scenario. Threats varied in the number and type of aircraft in a track, the number of tracks, and their appearance time and location.

Summary of Study of the Unperturbed Performance State

In studying the unperturbed state as suggested in the three steps above, the basic approach attempts to provide an answer to the following question: "Within a given response time, measured from a specified point in time and space, how much firepower can be delivered to a specific point in space along a designated defense perimeter?" Firepower can be measured in expected kills, that is, the number of missiles released against threats, normalized by probability of kill. In the unperturbed state, with fueling and defense posturing proceeding systematically according to a fixed set of rules, this equates to halting all ongoing activities (e.g., refueling, flights from rear to forward stations or vice versa, flights from forward or rear positions to the carrier) and responding with maximum resources against a unitary threat. In essence, unperturbed performance is measured by examining the potential response of the heretofore unperturbed system in the face of a single perturbation. Responses to this hypothetical unitary threat are, therefore, not affected by past or potential future responses.

Besides the raw ability to muster strength, one should be interested in the periodic nature of this strength, as it might vary in space or time. Do any such variations exhibit regular, periodic properties and are any such variations statistically significant? Even if not of statistical significance, might an enemy be capable of exploiting them? And, finally, do the evaluation findings enable the ready identification of remedies to perceived performance deficits?

Procedures

What follows is a description of the specific procedures used by Donnell and Stottler (1989) for study of the unperturbed performance state of the AEW/RTAS.

The basic dependent measures suggested above were used to determine (calculate) a set of higher level unperturbed performance parameters. Donnell and Stottler were concerned with how responsive the system could be along all points on its defense perimeter. Their initial question was: "How much fire-power, measured in normalized missiles, can be vectored to each point along the 300 NM Weapons Release Line (WRL), at 5 to 10 degree increments, within varying response times (i.e., from 5 to 11 minutes, in 1 minute increments)?" No plane that could reach the specified point on the WRL with adequate fuel reserves remaining was excluded from consideration. In reality, many planes would almost certainly not be chosen for vectoring by the AEW/RTAS either due to their distance from the designated point or because of the resulting weakening of defenses in their current location due to their departure.

For a specific defense point, the calculation to be performed was as follows: Assuming that all planes flew at maximum speed to the defense point, determine those planes that arrive at that point within the designated time with some fuel remaining. Use this to determine the number of normalized missiles (# missiles PK) deliverable. The data for unperturbed performance were first used to determine the times and angles of maximum defensive response strength. They were then used to establish bounds for the study of perturbed performance parameters.

The above approach to studying unperturbed performance was designed to give insight into potential perturbed system performance, allowing Donnell and Stottler to optimize their study of the perturbed performance and to anticipate some of the more interesting results. It placed a very heavy emphasis on the study of the unperturbed performance parameters. The philosophy of this approach was that unperturbed performance (and the resulting defensive capability) can occur in only one way, as there are really no independent variables to adjust (other than the defensive assets and their rules for use, which are not free to vary during evaluation). Given this, one should be capable of complete and in-depth understanding of the performance of the unperturbed system.

In adopting this approach for the study of the unperturbed state, Donnell and Stottler were aware that once they began later to study the perturbed state, they would unfortunately find that every answer leads to two new questions, that they would constantly be asked to trade-off depth of study with breadth of study, and that far too many interesting variables would have to be held constant to study too few manipulated ones. In summary, their approach for the study of the unperturbed state was to study thoroughly a relatively bounded unperturbed performance domain and to use those results to guide their endeavors in the relatively unbounded perturbed performance domain where their studies could potentially go on indefinitely.

Outcomes

The initial study of the unperturbed performance state demonstrated the exceedingly systematic nature of system performance in the absence of threats. Ability to perform, measured in terms of the number of (normalized) missiles deliverable, was seen to vary systematically as a function of space and time. It was seen that a quantitatively significant modification to the rule base, involving the forward movement of replacement aircraft, failed to have a statistically significant effect on overall system performance, a demonstration of the robustness of the system as it was configured and implemented.

In statistical tests, the system performance differences found for angle, simulation time, and time to intercept were all shown to be of statistical significance using a variety of tests. The conclusion is that the system exhibits notably significant, systematic performance differences in space and time which an enemy might be capable of exploiting. This implies that an important next step would be to modify the system in an effort to smooth the performance space and render it not so readily exploitable.

Performance Function Determination

The primary mission of the defense is to protect the carrier group. This is accomplished by destroying enemy bombers before they reach their maximum weapons release distance from the group. Therefore, the number of bombers destroyed is of primary interest. Unfortunately, this number is highly dependent on the number of bombers in the threat and not, therefore, very suitable in evaluating the performance of the RTEAS (AEW/RTAS). Kill ratio, defined as the number of bombers destroyed divided by the total number of bombers in the scenario, became the main performance function of interest. A secondary interest of the defense is to utilize their resources efficiently. Therefore, total fuel consumed, total missiles fired, and other similar measures became secondary performance functions.

Threat Space Determination

The threat space of interest consisted of seven dimensions which fell naturally into four groups. The "number" group consisted of the number of tracks and the average number of bombers per track. The "time" group consisted of the appearance time of the first track and the average time between track appearances. The "angle" group consists of the average angle of all the tracks and the average angle between tracks. The last dimension, the percentage of Backfires in the attack was a group unto itself. There are many other possible combinations of the seven dimensions which could delineate the space.

The entire threat space that the AEW/RTAS might be subjected to had to be considered. Actual combat considerations would tend to limit a scenario to about six hours. A single track consisted of only one kind of bomber, but different tracks could be of different types. The percent of Backfires (versus Badgers) varied from 0 to 100 percent. The tracks could appear at bearings anywhere from 0 to 120 degrees, corresponding to the maximum spread of an expected enemy attack. The enemy track appeared at certain distances and altitudes based on bomber type. This information was derived from intelligence estimates of the capabilities of the enemy bombers and the capabilities of the detection equipment use by friendly forces.

Sampling and Analysis

The overall philosophy of the sampling and analysis was to start with simple subsets of the threat space and, as these became understood, to increase the complexity of the threat space studied. The first subset studied consisted of a single track. This subset was termed Concentrated Space, Concentrated Time (CSCT). The next level of complexity consisted of using multiple tracks. The two simplest scenario subsets were Concentrated Space, Distributed Time (CSDT), the tracks all appeared at the same bearing, but at different times, and Distributed Space, Concentrated Time (DSCT), the tracks all appeared at the same time, but at different bearings. Next, Distributed Space, Distributed Time (DSDT) subsets were studied. A random study was performed. After analysis revealed several problems, these were remedied, and another CSCT study was performed. Finally, a much more thorough random study was performed.

From an analysis of the probabilistic events for typical scenarios, it was decided that probabilistic runs should be performed five times and averaged. This was a compromise between desired accuracy and computer processing time requirements.

The first fact that became evident was how complex the performance space could be, even for simple threat subspaces. By shifting the track one degree, the kill ratio could jump considerably, because a fighter that was not within range in the previous scenario might now be. If a track's time of appearance was changed by one minute, one or two fighters which were refueling might now be back on station leading to a notable jump in kill ratio. The kill ratio decreases as the number of bombers per track increases, as the angle of attack approaches the edges of the defense screen, and as the time between tracks decreases. A clear drop in the maximum attained kill ratio was evident when the number of tracks was 10 or more. It was later discovered that the knowledge base was ignoring tracks appearing after the ninth one.

The problem of the knowledge base ignoring the tenth and later tracks was fixed. Two other problems in the knowledge base that occurred extremely rarely were uncovered and fixed. The discovery of these problems illustrates the kinds of catastrophic errors that this methodology could uncover relatively easily. It should be noted that all of these errors occurred in an area of the threat space in which the developers had not anticipated the knowledge base operating.

A set of runs was performed which was designed to verify the results of the unperturbed study, especially the surprising weakness of the screen down its center. In the case of the Backfire tracks this result was verified. Badgers take much longer to reach the weapons release line so that similar results for them were not expected.

We decided to study short time span attacks, where short meant that fighters did not have time to engage, return to the carrier for more missiles, and then engage again. This limited the total time during which tracks appeared. The unperturbed study had uncovered a period of oscillation for the strength of the screen. This period of time was used as the period during which the first track arrived. The earliest the first track could arrive was set to allow the screen plenty of time to "warm up." It was decided to vary the percentage of Backfires in an attack among five values, 0, 25, 50, 75, and 100 percent.

Five hundred different scenarios were selected randomly using the ranges developed in the previous section. There were 7,768 total tracks for an average of 15.5 tracks per scenario. Regressions were performed, using kill ratio as the dependent variable and using the threat dimensions as the independent variable. There were very low correlations, demonstrating that the surface is fairly complex. The results indicated the defense does better against spread out attacks and worse against Backfires.

Factor analysis was performed twice, once using the data from entire scenarios and once using the data from individual tracks. The former identified a factor which represented the expenditure of friendly resources during the defense's attack on incoming bombers and another factor which represented waiting for the incoming tracks to appear. The other factor analysis identified two factors as well. These represented bombers that were destroyed and bombers that escaped, respectively.

Nonlinear regressions were performed using kill ratio as the dependent variable and average angle of the attacking tracks as the independent variable. Similar linear regressions had a zero correlation. The fitted curve showed a drop-off in kill ratio at the edges that one would expect, given the results of the unperturbed study. When the nonlinear regression was performed on narrow attacks only (spread less than 15 degrees) the fitted curve had a relative minima at the middle of the screen.

The complexity of the performance surface created a large amount of apparent noise for the statistical tests. In order to reduce this noise, averaging was performed on the track data. Kill ratio was averaged for all tracks with the same arrival time. This was also done for number of bombers in a track and a tracks position in the order of appearance. Along with averages, standard deviations were calculated.

In the case of average kill ratio versus position, the kill ratio drops with later position in the order of appearance. Interestingly, the standard deviation of the kill ratio increases as the square of the position in the order. The later a track appears, the more difficult it is to predict the probability of destruction.

Summary and Conclusions for Case Study

The unperturbed study showed that the defense screen strength drops off significantly at its edges. More surprisingly, the screen is relatively weak in its center. Furthermore, the screen strength varies in a time-dependent and angle-dependent way. All screen strength differences noted here were statistically significant.

The perturbed study confirmed the screen strength drop off at the edges and relative weakness in its middle. The study also demonstrated the extreme complexity of the kill ratio performance function, with the exception that the defense is much more likely to destroy Badgers than Backfires. The study also showed that the defense performs better if the attack is spread out in time, involves more tracks, and spans a greater range of angles. The variance in kill ratio for an individual track seems to be parabolic with respect to the track's position in the order of appearance.

The methodology was also important in detecting errors which did not show up through normal testing. These included two errors that were caused by using a larger number of tracks than had been tested before and an error which caused erroneous interception calculations on very rare occasions.

Reference

Donnell, Michael L. and Stottler, Richard H., III, "Evaluation of the Airborne Early Warning/Real-Time Advisory System." Final Report 89-1, Donnell & Associates, Incorporated. Reston, Virginia.

Intelligent Systems for Command and Control

Robert G. Eisenhardt

Decision aids for command and control compensate for human deficiencies in handling multiple dimensional problems including incompleteness, extended temporal bases, and conflicting data. Such aids provide analysis and options to a decision maker in parallel with his reasoning and in parallel with other aids. At times the decision maker must arbitrate among sources of information. Typically these aids are computer-based deterministic algorithms harboring the logic of their creators and constrained to brittle shells bounding constituent options. This chapter suggests another approach to command and control assistance for decision makers that employs adaptive learning systems, working in consort with humans to jointly solve operational problems in real time. Such systems are effective through the formation of symbiotic relations with humans.

Important characteristics of systems are developed and a concept of intelligent synthetic systems (ISS) proposed with supporting definitions. It is shown that two important attributes of intelligent systems, biological or synthetic, are their indivisibility and ability to form social organizations, a necessary ingredient of cooperating man-machine endeavors. The concepts of unsupervised learning, adaptive, systems are beyond classical AI and computational constructs of neural nets.

System Architectures

The fundamental idea of any system is to achieve synergism through its constituent parts or hypo-systems.[1] If not carefully synthesized,[2] such systems "work" but lack the emergent properties supporting synergism. Frequently the impression of enhanced system performance results from the presence of man-in-the-loop where he has direct control over most sensor and effector systems. Historically, man has always been the pivotal point in any system, but this has been at the price of increasing workload as sensor and processing technology make available more data and information than humans can process. This is even more of a problem when the data are incomplete or wrong, or both. But there are ways systems can be made more effective in combination with humans as Figure 1 illustrates. To reduce human workload requires moving away from direct human involvement at low levels toward symbiotic concepts. This is easier said than done since this arrangement requires a system comparable to (1) or (2) as an integral part. If this state is achieved by adding a second human, we have only sidestepped the main issue and returned to a concept utilizing a potentially overloaded human-in-the-loop. A better solution implies *participation* by non-biological intelligent systems—the thrust of this chapter.

How can we improve system synthesis and the end product? To gain insight we need to dissect the concept of systems and expose vital parts. All systems of interest are active and, as such, should be viewed from three different perspectives. These are *function*, *process* and *form*. Each has its own different architecture.

• *Function* deals with the *why* of a system's existence and specifies goals it should (must) be capable of achieving.

• *Process* is that combination of intangible operations collectively implementing functional goal *achievement*.

• *Form* is the tangible substrate (hardware and software) *hosting* the processes.

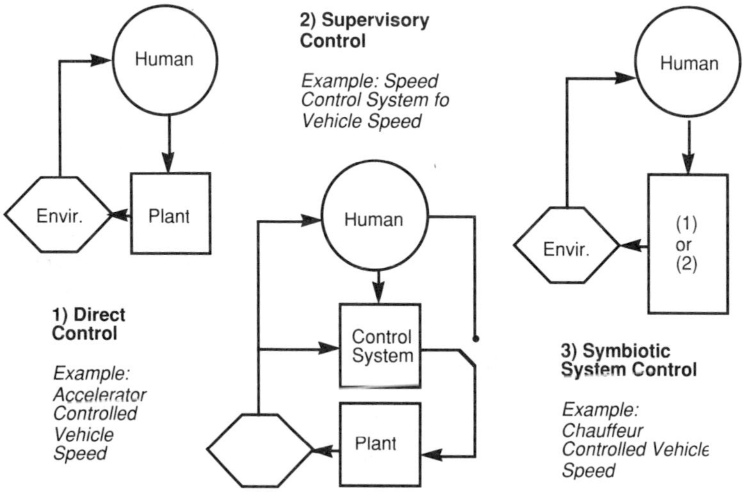

Figure 1. Typical Control Structures.

These perspectives are based on work by the late David Marr.[3] The novel feature is interposition of the third level (process) in the middle. Marr's thesis has been modified somewhat to account for differences between major systems and vision information processing—his forte. While an extensive analysis of the thesis is beyond this chapter, a few comments are in order.

One should not be tempted to force strong relationships among the perspectives as they are often only loosely related; for example, some phenomena may be explainable from only one or two of the perspectives and not others. Further, strong relationships erode the value of independent perspectives and force a linear relationship between function and form. In that case, process merely implements that relationship with no self value. It should be apparent that this is a paradigm now used in system design; that is, *form follows function*, for example, communication subsystem follows communication function and navigation subsystem follows navigation function. Opportunity for synergism from the process perspective is lost.

By not taking advantage of process synergism, system requirements can be established easily through successive decomposition of system functions. This permits *system design* to be accomplished simply by "integrating" the ensuing formalisms (i.e., subsystems). There is no system *design* per se; it simply emerges as a consequence of functional decomposition.

This paradigm leads directly to the "Performance/Effectiveness Syndrome," a (mistaken) belief that performance and effectiveness are linearly related. This fosters poor design decisions by assuming any increase in *subsystem performance* leads directly to *comparable increases in system effectiveness.* This is pertinent under the definition of intelligence developed by Dean Douthat[4] discussed later in this chapter.

Another property of active systems is that they interact with their environment in three stages; *sense*, *think* and *act*. These are implemented by what we have called the major hypo-systems. Their most interesting characteristics are processes which should be tangled and indivisible at both the process and formal levels as they are in biological systems.

• *Sense* is to detect and recognize measures of the environment. In many systems, particularly biological, sensed stimuli support more than one function and more than one process and can be local or remote.

• *Think* is the act of converting sense data to action (or non-action as the case may be) in a manner beneficial to the system. For non-intelligent systems "think" should not be construed in biological terms but simply as processes transforming sense data into action.

• *Act* can be paraphrased as a change in behavior, a direct modification of the environment, or environmental change indirectly through communicating with other systems. In all cases action is taken in support of some system goal. This phase-oriented view of hypo-systems is stressed because it gives perspective to system needs vis-à-vis its environment. Furthermore, it is a valid construct at all levels of detail. That is, within each of sense, think, and act, the same three may be recursively nested.

Three distinctly different system architectures have been introduced and reviewed: Collectively, they define a "hyper-system" architecture. This approach provides insight to systems from new perspectives and provides a more firm basis for system synthesis as opposed to functional decomposition. It serves as a framework for ISS development.

Starting with this framework we expand the concept of intelligent systems by defining key attributes distinguishing them from their more conventional counterparts. From that perspective, we briefly review the place of symbolic processing and neural networks as well as systems of collaborating intelligent entities.

Intelligent Systems

Definition of Intelligence

Intelligent systems, as distinguished from other systems, are defined by a concurrence of several major attributes. But first, *intelligence* must be defined.

People exhibit strong feelings about the definition of intelligence. For a majority there is a tendency to limit the definition to Homo sapiens or certain of their characteristics. In particular, many people relate it to cognitive processes in the conscious domain. In the absence of any other comparable term to use in its place, such definitions act as a barrier to understanding biological systems and their synthetic counterparts. Douthat proposes a definition of intelligence we will use in this chapter. It states that intelligence is "The difference between the effectiveness of the whole and that of the sum of its parts." While this definition bears little resemblance to concepts most people hold, it offers some significant advantages.

First, it is independent of any particular species or other division of the biological taxonomy. Second, it offers a metric thereby opening the possibility that intelligence can be measured for all systems, not just humans; namely, effectiveness in the system's environment. It may not be easy to measure in some cases, but at least a consistent metric exists. Furthermore, the definition allows measurement at the relative level as well as absolute. The relative measure may prove more valuable in a practical sense.

Intelligent System Attributes

Among the three system perspectives and hypo-systems are attributes shared by all living organisms. Others are unique to members higher in the taxonomy and, by this definition, some attributes are limited to the group defined as intelligent systems. The specific hypothesis: Nine attributes are necessary and sufficient to distinguish a class of systems we call intelligent from all others. Further postulate: Effectiveness of these intelligent systems will be orders of magnitude greater than *any* non-intelligent system designed to operate in the same environmental niche and satisfy the same functional goals.

With this definition, the group of intelligent biological systems covers a wide range of the biological taxa. The reason for this is to define a group of attributes permitting a variety of possible implementations such that success in synthesizing a low order system will not fall under the "fallacy of the successful first step."[5]

A major concern of AI efforts in the past has been the issue of scalability: to upgrade a system to higher levels of performance by extending principles demonstrated at lower levels of complexity. This concern has driven the definition to include a sufficiently large segment of systems as intelligent so that requirements for scalability are definable and the direction for system complexity growth is evident. Another reason for such a broad range is to permit developing systems composed of numbers of simple ISSs combined into a collaborative system as opposed to developing one giant ISS: a much more unlikely scenario. If the multiple ISS approach is to bear fruit, then the small ISS must be of the same source as their more advanced fellows, otherwise they cannot operate in a synergistic social organization.

In further support of this broad definition of intelligent systems, biological organisms of lower intellect are very effective in dealing with the more limited range of environmental situations their particular niche presents. No one would claim a cockroach is *highly* intelligent, but in its niche it has been an extremely effective system for over 300 million years. This is not to imply that extinct species were not intelligent; intelligence is no guarantee of survival at any level.

It might be argued that a very broad definition for intelligent systems invites research and development at levels so low they are of no practical military value. This is rebutted in two parts: First, while the equivalent of a synthetic ant, for example, may have no military value per se, it would represent the first real step toward more complex systems with military potential. Second, biological ants meet the requirement of having the ability to form social organizations whose collective intellect is greater than any one ant; so should synthetic ants. This attribute raises the level of effectiveness for lower order systems and lowers the required level of complexity of an individual ISS for military value. Further merits and limitations of single and collaborative systems are discussed later.

While the following attributes are considered necessary and sufficient to characterize all intelligent systems,[6] higher level intelligent systems possess special attributes necessary for their ecological niche. As noted above, such systems do not *alone* deserve the mantle of intelligence, but may be classed as special cases with unique capability.

The order of presentation was selected to build each attribute on the preceding attributes. Order is *not* essential however, only a convenience. But a note of caution about the attributes; there is not room in this chapter to discuss the full range of attributes and their significance. For example, metabolism is a characteristic of living systems, intelligent or not. In synthetic systems, a comparable source of "power" will obviously be required; and may need to be sought actively by the non-biological system. In large measure these are issues to be resolved in replication as they apply to meeting the critical attributes. This includes such items as the relative role of chemical, electrical, magnetic, etc. mechanisms in memory, motor behavior, et al.

Mobility

That mobility is one of the fundamental action functions of intelligent systems is put in perspective by Rodolpho R. Llinas:

> It is delightful to consider that development of a nervous system is in fact very much a property of actively moving organisms (i.e., non-sessile organisms), that is those that do not remain fixed to a particular place as is the case for plants. The best example of this principle is beautifully exemplified by the tunicate (a "sea squirt") with free swimming larvae. Herein lies one of the truly extraordinary lessons of biology. In ancestral tunicates as well as in some present forms, the sessile adult form (which is rooted by its pedicel to a stable object in the sea) carries out two main functions; it feeds by filtering seawater (a deeply bureaucratic type of function) and it reproduces by budding. Upon reproduction, larvae are free swimming and possess brain-like ganglion which can be informed about the environment by peripheral sensory inputs from a statocyst (organ for balance), a primitive eye, and a notochord (primitive spinal cord). These central nervous structures have the connections necessary to deal with the continuously changing environment, as this primitive tadpole-like larva swims through the water.
>
> Upon finding a suitable substrate, the larva proceeds to implant its head into the selected location and becomes sessile. In doing so, it absorbs much of its brain and returns to the rather primitive condition of the adult of the species (a process paralleled by some human academics upon obtaining university tenure). Under some circumstances, probably related to the ability that the larva has of utilizing its primitive digestive system, it may proceed to reproduce prior to becoming sessile. This step is believed to have generated a class of brainy organisms (our far ancestors) known as the chordate filter feeders. The lesson of this evolutionary stage is, of course, very clear. It is that brains are needed only in those multicellular beings that move actively. Indeed, it is only then that such forms must interact with the environment at speeds that are sufficiently fast to require a rapidly acting organ specialized for the forecasting of external changes with sufficient anticipation (prediction) to afford correct behavior.[7]

It remains to be seen if having learned, a non-biological system can still be effective absent mobility. This property exists in humans and may be a possibility in synthetic systems. Some have considered the option of satisfying the mobility need by causing the environment to "move" rather than the system. This approach does not appear to be a fruitful direction of investigation, but it does need to be considered.

Multiple Modality Sensing

If survival of mobile organisms depends on prediction, then before that process can emerge, a system must have the ability to sense multiple environmental signals for correlation, as sensing only a single signal requires temporal autocorrelation as the sole basis for prediction. Moreover, at least one modality must sense remotely to provide positive temporal sequencing. It is presumed the oldest biological local sense is tactile, and the oldest remote sense was chemical—now embodied in the olfactory sense.

Motivated by Innate and Acquired Drives

Here we take from A. Harry Klopf[8] who states that intelligent systems are drive motivated, not goal oriented; at least not at phylogenetically lower levels. Some drives are innate and commonly acknowledged such as eating and fighting. He extends this concept to the more general case saying that innate drives are represented by signals directly or indirectly attributable to measures of the environmental state to which the organism responds in predictable ways. Other drives are acquired through correlation with innate drives in a manner he shows to be identical in certain cases to classical conditioning in animals. The learning process is implemented by *drive reinforcement* or the change in input signal. While drive reinforcement is not the only learning method in living intelligent systems, it forms a basis for instrumental and higher forms of learning. In subsequent conversations, Klopf has said in this context that the high level hypo-system process architecture follows sense, *want*, think and act.

Capable of Learning While Operating

One of the best arguments for this attribute was given by Lashon Booker, David Goldberg and John Holland and is, in part, as follows:

> Consider the simply defined world of checkers. We can analyze many of its complexities and with some real effort we can design a system that plays a pretty decent game. However, even in this simple world novelty abounds. A good player will quickly learn to confuse the system by giving play some novel twists. The real world about us is much more complex. A system confronting this environment faces perpetual novelty—the

flow of visual information impinging upon a mammalian retina, for example, never twice generates the same firing pattern during the mammal's life span. How can a system act other than randomly in such environments?

It is small wonder, in the face of such complexity, that even the most carefully contrived systems err significantly and repeatedly. There are only two cures. An outside agency can intervene to provide a new design, or the system can revise its own design on the basis of its experience. For the systems of most interest here—cognitive systems or robotic systems in realistic environments, ecological systems, the immune system, economic systems, and so on—the first option is rarely feasible. Such systems are immersed in constantly changing environments wherein timely outside intervention is difficult or impossible. The only option then is learning or, using the more inclusive word, adaption.[9]

A system that must partition its time *between* learning and operating frequently cannot wait to learn before dealing with reality. In biological systems such a trait leads to early extinction of the species.

In most respects, the world is temporally ordered; that is, certain sequences of events are predictable. From this its inhabitants learn to act in predictable ways and these attributes are used by intelligent organisms, particularly predators. Hence it is not surprising the basic learning mechanism is stimulus/stimulus conditioning from which arises anticipation. In summary, learning for survival takes advantage and is attuned to equivalent regularities in the environment. Mobility permits an intelligent biological system to explore and learn from these regularities. To substitute a man-made environment, particularly one that does not contain temporal regularities, may prohibit natural learning mechanisms from working. (Reference *Mobility* above.)

Behavior Based on Internal Models

Arguments for internal model-based [10] operation by intelligent systems have gained considerable strength in the past five years. The concept is now widely, if not universally, accepted. First, based on observation and introspection it makes sense and is supported by empirical data,for example, Robert Eggleston's study in perception.[11] Second, it appears to be a necessary ingredient for non-associative as well as associative learning. It is arguable that sensory data alone are insufficient for decision making for survival. Such data are transient, localized, and subject to noise and other corruption either at the source or in the brain's processing systems. There is strong evidence model-based behavior exists as low as the insect level.

To provide a sound basis for decisions requires a warranted model of the environment of interest. Action is based on the perception of reality, not reality itself. Thus, in the sequential and simultaneous processes of sense, think, and act a major part of the cognitive process is building and/or warranting internal models. The act of warranting is inferred to be similar to that used by humans in warranting external models and, hence, sets certain requirements on internal models. These are that the models be capable of being parametrically "modified" in a manner permitting them to assume a configuration consistent with observed data. Furthermore, such models *stand in the way of action* rather than form part of the direct stimulus/response path as in the case for reflexive (innate) behavior.

Memory is Distributed and Layered

One practical limit to symbolic knowledge representation that AI researchers have encountered is brittleness of the structure when new data are added. If a system must be able to assimilate and relate a lifetime of information in a limited formal structure, then flexible methods of storing, relating, and accessing acquired information must be established. The move to neural networks from semantic nets was a step in the right direction.

Memory must be organized in such a way that search in the classical meaning is not required to gain access to memory or for making decisions about appropriate action in response to environmental changes. Distributed memories make this possible. Storage of representations must be extremely parsimonious and not constrained to expectations derived from conscious symbolic reasoning: it must be at the subsymbolic level. The concept of emergent properties implies layering where emergent constructs from one layer serve as input to other layers from which additional constructs emerge. This organization has been described as a "Tangled Hierarchy" by Douglas Hofstadter.[12]

This memory construct in intelligent systems provides rapid access to pertinent data for survival where search time through a large structured data base would, in all likelihood, be fatal. Empirical data support the claim that, from sense to action initiation in intuitive situations can involve less than five sequential steps. Such a construct also permits intelligent systems to deal with categories of things in a natural manner.

Self Organizing at Both Process and Formal Levels

The *raison d'etre* for self organization by intelligent systems is the need to continually allocate limited resources in the most efficient manner. This results in a highly complex system that is almost never in equilibrium. All intelligent biological systems exhibit this property, hence alternative solutions make little sense. The number of environmental situations is so large it is not physically possible to build an intelligent system pre-wired to handle all possible situations, and no single physically realizable *fixed* system could cope with more than a minuscule fraction of possible environmental conditions; thus denying its claim to being intelligent.

Communicate Symbolically and Form Social Organizations With Others of Their Species

Intelligent biological systems did not evolve, nor do they exist, alone. Part of their learning process consists of guidance at the hands of a tutor. In most animals this is a parent or parents. In higher systems it may include older siblings, and in humans it includes teaching via natural language. Obviously, such learning cannot exist without an ability to communicate among members of the species. And without learning, there can be no intelligent biological systems. It may be argued that all necessary learning can be accomplished through experience but no intelligent system operates in this fashion, particularly higher order systems. Darwin noted that if higher animals did not have parent tutelage they could not learn enough in their lifetime to survive. Without parents or surrogate parents human babies could not survive and the species would be extinct in one lifetime.

The requirement that intelligent systems be capable of forming higher order systems helps isolate that set of biological systems constituting the ensemble. For example, it excludes flies, moths, mosquitoes and the like from intelligent entities. It permits inclusion of ants, bees, and higher systems which also share all of the other special attributes. It is also an important attribute in extending the effectiveness of individual systems or organisms.

In addition to learning, communication is essential for communal activities, many of which are essential for survival of the individual. As Hofstadter said through the Anteater in the Ant Fugue:

> There is some degree of communication among the ants, just enough to keep them from wandering off completely at random. By this minimal communication they can remind each other that they are not alone but are cooperating with teammates. It takes a large number of ants, all reinforcing each other this way, to sustain any activity—such as trail blazing—for any length of time.[13]

At all levels of intelligence, communication among members of a species is symbolic where the symbolism is an *emergent property* of the system.[14] In higher level systems symbolism also facilitates internal communication, sometimes referred to as "speaking silently." There are two principal levels of communication; substantive communication to exchange information about internal models and, in higher level systems, meta communication to discuss communication issues. The ability to communicate should not be confused with an ability to communicate via natural language. That is, the ability to "think" is not required to communicate.

Possess a Central Nervous System (CNS)

This attribute sets another lower bound on living intelligent systems as well as contributing to their definition. This criterion is essential to intelligent systems as the most efficient structure for rapid processing of massive amounts of information essential for survival. A distributed system could not tolerate the associated transport delays nor the mass of connecting nerve fibers necessary to maintain a speed and accuracy of response for survival.

A property of systems with a CNS is that they are indivisible; the hypo-systems cannot be isolated, nor can their constituent parts. At best, if an intelligent system is partitioned only one intelligent entity can survive.

As intelligent as humans are, their ability to build complex physical and organizational structures, such as building the Brooklyn Bridge or operating the CVN Enterprise, depends on an ability for synergism among large groups of people with widely diversified skills. In these cases, it is not easy to evaluate, even subjectively, the relative intelligence of any one group member versus intelligence of the group as a whole. The task is easier when one examines a basketball team. Here the difference in intelligence (measured as the difference in *effectiveness*) between all players taken individually and the trained team is obvious and significant. This further illustrates that large numbers of intelligent elements are not necessary to make a significant difference in effectiveness when operating synergistically. Another example of a small effective "team" where skills are much more diverse than basketball players is a surgical team.

Possibly the most important reason for suggesting a concentrated effort in intelligent systems development concerns an operational issue—trust. Combat personnel at all levels depend on the ability of others for their well being: specifically, they depend on their support to not make mistakes. While humans are prone to errors of judgment and action, a natural consequence of being intelligent, this same intelligence frequently vindicates the human by permitting him/her to overcome such errors in the future—learning from their mistakes. As humans, we must and do anticipate this behavior, the building of trust among compatriots. Trust is not the anticipation of infallibility, but of an ability to adjust to new situations in an acceptable manner. Systems not capable of this attribute of overcoming errors will not develop trust in humans and will not be used. The system need not be complex. Many pilots do not trust their IFF and turn it off. The same often holds for electronic countermeasurers. The fact that these potentially life saving systems are not trusted carries strong implications for more complex systems, and particularly for those labeled as "intelligent."

In summary, the potential advantages of Intelligent Synthetic Systems over decision aids are:

• Continual learning under combat conditions while operating, that is, no requirement for human computer specialists to modify software or data and no requirement for alternating periods of learning and operating.

• A potential for specialization. One cannot expect the broad capability of humans in a synthetic system, but the possibility for intelligent specialization is a prime motivator for such systems.

- An ability to cooperate among themselves and with human superiors.
- A capability to communicate bidirectionally with humans.

Endnotes

[1] For a variety of reasons, some covered later in this chapter, the term "subsystem" is avoided as it has a special meaning not appropriate in the current context. Of all possible expressions, "hyposystems" seems to be the most expressive and simultaneously free of prior connotations.

[2] Simply connecting related parts (integration) does not guarantee the desired result.

[3] David Marr, *Vision*, (San Francisco: W.H. Freeman, 1982) 22-29.

[4] Deam Z. Douthat, "Paradigmatic Shifts: Moving Toward Next Generation Intelligent Systems," in *Defense Applications of Artificial Intelligence*, eds. Gerald W. Hopple and Steven J. Andriole (Lexington, MA: Lexington Books, 1988).

[5] Hubert L. Dreyfus and Stuart E. Dreyfus, *Mind Over Machine*, (New York: The Free Press, 1986) 7.

[6] However, they are not sufficient for all of the purely biological needs of *living* intelligent systems (e.g., metabolism).

[7] Rodolfo R. Llinas, "Mindness' as a Functional State of the Brain," in *Mindwaves: Thoughts on Intelligence, Identity and Consciousness*, eds. Colin Blakemore and Susan Greenfield (Basil Blackwell, Oxford Press, 1989).

[8] A. Harry Klopf, *A Neuronal Model of Classical Conditioning*, AFWAL-TR-87-1139 Information Processing Technology Branch, Systems Avionics Division, Avionics Laboratory, Wright-Patterson Air Force Base.

[9] Lashon B. Booker, David E. Goldberg, and John H. Holland, *Classifier Systems and Genetic Algorithms*, Technical Report Number 8, Cognitive Science and Machine Intelligence Laboratory, The University of Michigan, 1987.

[10] The term model is to be taken broadly. No universal term yet exists so for this chapter *model* must suffice to cover cognitive maps, representations, etc.

[11] Robert G. Eggleston, *Apparent Motion and Prior Correspondence Effects in Visual Perception*, Ph.D. Dissertation, Miami University, Ohio, 1983.

[12] Douglas R. Hofstadter, *Godel, Escher, Bach: An Eternal Golden Braid*, (Vintage Books, 1979) Chapter XX.

[13] Ibid, p 316.

[14] The entomologist may define the dance of a bee to be symbolic of the fact a source of food has been found but it need not be symbolic to the bees. However, among humans discussing these topics, the reference to symbolic communication should not be taken too literally.

A Natural Language Processing System for Intelligence Message Analysis

R. Bruce Loatman and Stephen D. Post

Intelligence analysts face a continuous flow of teletype message traffic. This information must be disseminated to the salient analysts; it must be recorded in their data bases, either as extracted data base records or in textual form; and it must be available for recall to support their analysis. However, even under the best state-of-the-art automated message handling systems, these processes can be inconsistent, labor intensive and error prone. This problem exists because the computer can search for keywords, perform word counts and so forth, but it cannot understand the content of the text as a human does. However, research for natural language processing within the domain of artificial intelligence (AI) is ongoing. A computer software system has been developed under a four year research effort toward the goal of understanding military message traffic for the purposes of dissemination, data base update and retrieval.

The I&W Problem

At the most general level, the indications and warning (I&W) problem is to correlate information from a variety of sources to recognize threatening or abnormal situations and to recognize enemy courses of action. For example, the disposition and activities of Soviet heavy bomber forces, organized as Air Armies under the title of Soviet Strategic Aviation (SSA), are central to I&W of imminent hostilities against North America. That warning problem is an important element of the Worldwide Warning Indicator Monitoring System. The force posture, dispersal, operations and Soviet doctrine underlying the mission of Soviet heavy bomber forces are integral to both the strategic and tactical warning problem. The development and execution of aerial refueling capabilities, and new, more sophisticated weapon systems have shortened the warning timelines and provided the Soviets with a more flexible, capable threat to North America. As a result, constant monitoring of SSA activities and facilities, particularly those of the Moscow Air Army Bear-H force, is required to provide strategic warning of potentially hostile operations. As improvements in collection capabilities continue, data volume will increase significantly and intelligence cycle timelines will decrease. The automation of the I&W process is critical to the provision of near real time intelligence to decision makers.

In the last decade, some headway has been made in the area of data bases to provide on-line information. This allows for the easy application of statistical and other algorithmic aids to the data. Intelligence data handling systems have provided means for passing the right data to the right people. However, these automated aids still are difficult to use and on occasion are more complicated and time consuming than the paper and pencil methods they replace.

During Soviet training, formal exercises and periods of increased tensions, the level of activity and the resulting Allied reporting can overwhelm current analytical capabilities. This creates a data backlog at the point of manual entry, which complicates tracking and analytical processes at the time when their outputs are most critical to the decision making process. Furthermore, the manual process often becomes error-prone as the pressure builds. If analysts must interrupt the process to update decision makers personally, which is quite likely, the process degenerates further. In summary, a system is required to alleviate this manual process to enable the analyst to do the analysis that cannot be provided through automation.

Much of the current work to enhance the usefulness of these systems, to make them more user friendly, is being performed under the broad heading of AI. A division of this technology is the area of natural language processing (NLP). The goal of most NLP work is to provide a natural language interface through which the user can perform data base queries. This article focuses on a different use of NLP—extracting data base updates from free text input in an accurate and timely manner. An I&W system is of little use if it does not contain the most recent information with all the nuances and uncertainties reflected in the source of the data. In a time critical situation in current I&W centers, analysts sometimes discard the computer aids and return to pencil, paper and direct reading of the intelligence messages. The reasons they give primarily are those of speed and accuracy.

"It takes too long for the data to be entered into the system, and I do not have the time in those situations to do it myself," says one U.S. Space Command analyst. "Even if the data were there, it takes too long to figure out the right keywords to retrieve them. And even if you do retrieve the data, you cannot tell how good the information is—you have to go back to the message itself to see if it came from a highly certain source, or if it is just a possible assessment." Analysts interviewed at Headquarters, Strategic Air Command and other I&W centers report the same problems.

These problems stem from the fact that the actual messages used to support I&W analysis may be either formatted or unformatted. For example, tactical report (TACREP), RAINFORM, Joint Interoperability of Tactical Communications and Control Systems and weather forecast messages are formatted (but may include unformatted comments); however, SPOT reports, E-Grams, imagery reports, intelligence summary reports and intelligence and AIRSEA Advisories usually are not. Indeed, even the highly structured message formats (such as missile report, Initial Photo Interpretation Report and TACREP) still include a field dedicated to analyst comments, or narrative, which is free text. This often is the field with the most interesting information that did not fit a formatted field, or the place where the analyst explains the amount of confidence associated with a given assessment. This unformatted information must be interpreted before any relevant data can be entered into the data base, and much of it may have to be ignored due to the design of the data base management systems (DBMS). The formatted information eventually will find its way into the data base, but even this often takes a couple of hours, perhaps days, for some information to be properly stored.

Admittedly, accessing the information still is difficult, and this problem needs addressing. But if the information is not in the data base, it does not matter how user friendly the data base query is—it will not be used. So, the more important problem lies in automatically entering data into the data base as rapidly as the messages are received at the intelligence center.

An NLP System for I&W Data Base Generation

The extraction of information from the content of free text emphasizes the ability to build structures that model real world events. Other aspects of NLP have considerably different emphases depending on the type of application. For example, NLP for DBMS query largely is concerned with converting sentences into DBMS query language for a functional domain clearly defined by the DBMS schema and data. The scope of quantifiers and alternative ways of referring to objects and attributes are key concerns. DBMS query systems are not made to handle verbs with high conceptual content, such as attack or communicate (as opposed to is and has) nor to model real world events. Similarly, NLP for operator interface is constrained to a small functional domain and largely is concerned with discourse and intentional level aspects of sentential input.

The area of NLP systems for data base update generation is relatively less developed than other NLP domains of DBMS query or operator interface.

However, useful systems in this field are beginning to emerge. One such system is the PRC Adaptive Knowledge-based Text Understanding System (PAKTUS). It has been applied to prototypes for message processing in several domains of relevance to I&W analysis and is being developed further under several government contracts to address specific message streams and data base requirements.

Several NLP methodologies have advanced to message understanding. Many of these focus on one aspect of the problem, such as conceptual analysis, syntax or knowledge about specific words. PAKTUS employs a hybrid approach, which is summarized in Figure 1. Processing begins with the arrival of an electronic stream of text, indicated at the top of Figure 1. The system first breaks down the stream of characters into individual words, sentences and messages. This is done in preprocessors. Next, in the lookup and morphology module, the words are compared to entries in a lexicon to retrieve information about their meaning and usage. Words that have no entry in the lexicon are analyzed morphologically. If this fails, the system deduces as much as it can from the syntactic and semantic context, later verifying its deductions through mixed initiative interaction with the user.

In the parser module the sentences are syntactically analyzed according to a grammar specification embodied in the system as an augmented transition network (ATN). An interpreter applies the ATN to the text. This process identifies the syntactic roles of the sentence, such as subject, main verb and direct object. Then the conceptual analysis module converts the syntactic structure into canonical, language-independent conceptual frames. This process is based on case grammar and a network of conceptual templates. The conceptual analysis module is invoked to test each candidate clause. If the syntactic parse cannot be mapped into a valid conceptual frame, it is rejected and the syntactic parser tries alternatives.

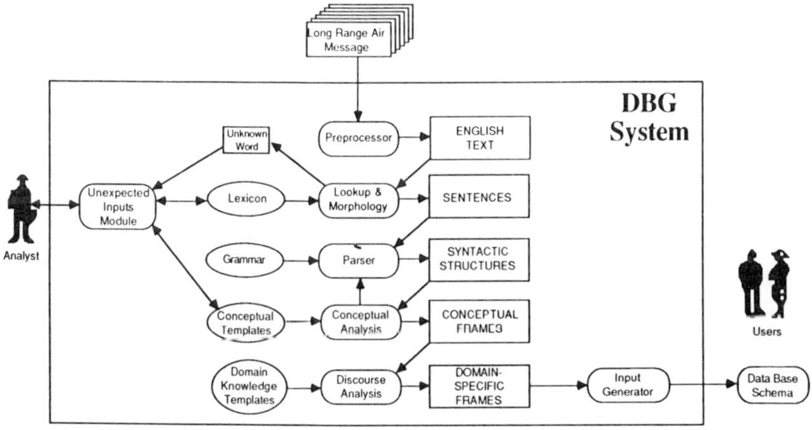

Figure 1. Overview of DBG Operation (Analyst-Attended Mode).

The conceptual frames are collected by a discourse analysis component using concept templates that spawn "demons" to fill in the blanks. The demons specify logical patterns that apply knowledge about the particular domain of the application system to integrate all the information, both explicit and implied, of the message into structures representing its meaning. These structures then may be reformatted into a data base update. Alternatively, they could be passed, perhaps in a slightly different but equally rich representation, to a knowledge based system (KBS), or they might be matched against I&W analysts' interest profiles, which are cast in terms of conceptual templates and routed accordingly. A simulated intelligence message as received by PAKTUS is shown below:

REPORT NO. MR1-1
REPORT DATE/TIME: 30-JAN-85 19:00

SEMAPHORE MESSAGE TRAFFIC FROM TARGET 1 TO THE TOLIMA DEFENSE COMMUNICATION CENTER WAS DETECTED ON TWO OCCASIONS BETWEEN THE HOURS OF 0900 AND 1300 ON 30-JAN-85. THE MESSAGES WERE IDENTIFIED AS SYSTEM TESTS.

The message consists of a simple header and two sentences that are the body of the message. Figures 2 through 4 are actual window images produced automatically from the results of the parse. Windows like these are intended for use by NLP system developers; the I&W analyst has no need to use them. The following points will help to interpret the graphs:

- C^ designates a concept; for example, C^ PERCEIVE is the concept embodied by the verb detect. Concepts are associated with all verbs and with those words that are derived closely from a verb, such as communications or testing.
- R^ designates a semantic role in a concept; for example, R^ INSTR is the instrument role of a concept.
- Words are represented by their root and their lexical category; for example, TARGET 1^ MILITARY-UNIT.
- The graphs depict major syntactic roles and conceptual frames; they are summaries that do not display much of the detail available. For example, the conceptual graphs show only the head and qualifiers (that is, relative clauses) of each noun phrase, although descriptors and determiners have been identified by the parser and are stored in the appropriate objects.

Figure 2 shows the syntactic and conceptual analyses of the first sentence. The conceptual analysis can be understood as follows. Perceive is the main concept of the sentence from the verb detect. Because the passive voice was used, the agent is shown as someone. The focus of the Perceive is an Express concept, from the noun, message. The agent and experiencer of the express (that is, message) are identified from within prepositional phrases. The beginning and

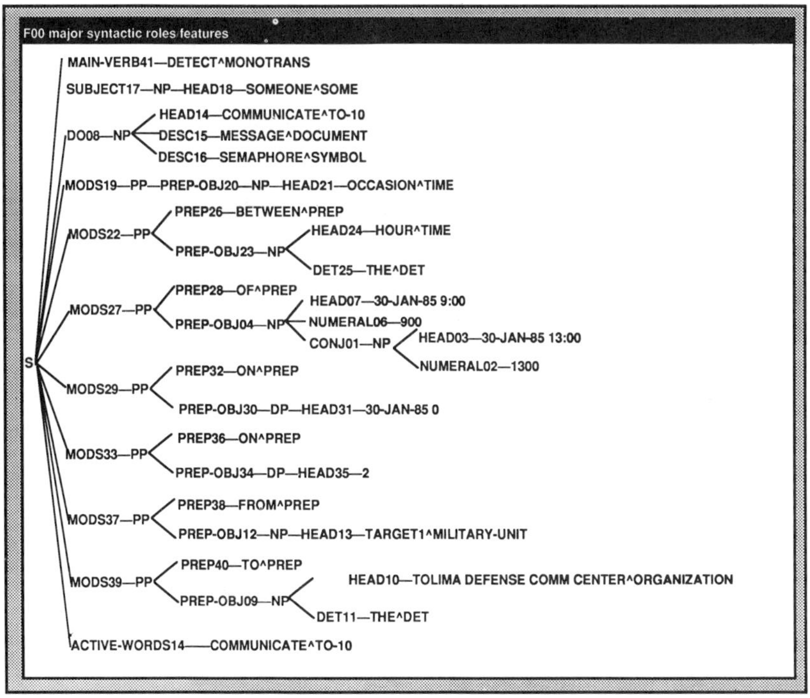

Figure 2a. Syntactic analysis of the first sentence.

ending of the Perceive are identified from within prepositional phrases including conjunctions. Figure 3 shows the syntactic and conceptual analysis of the second sentence.

Figure 4 shows a completed data structure that could be transmitted to an I&W system to generate a data base update. The Sender was located by finding an organization that is the agent of an Express. The Receiver was located by finding an organization that is the experiencer of an Express. The Channel was found as a descriptor in the noun phrase that represented the Express. The

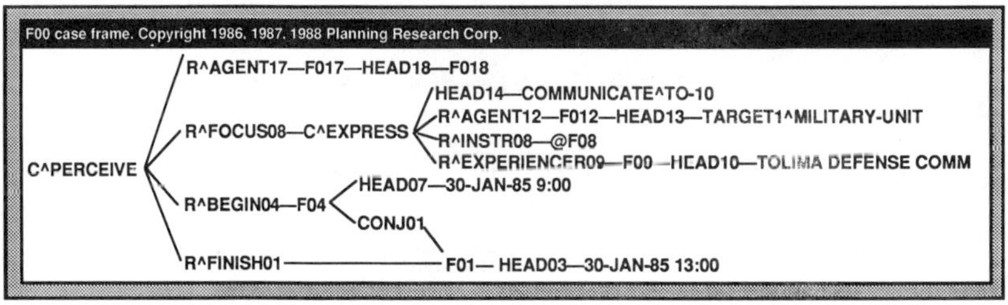

Figure 2b. Conceptual analysis of the first sentence.

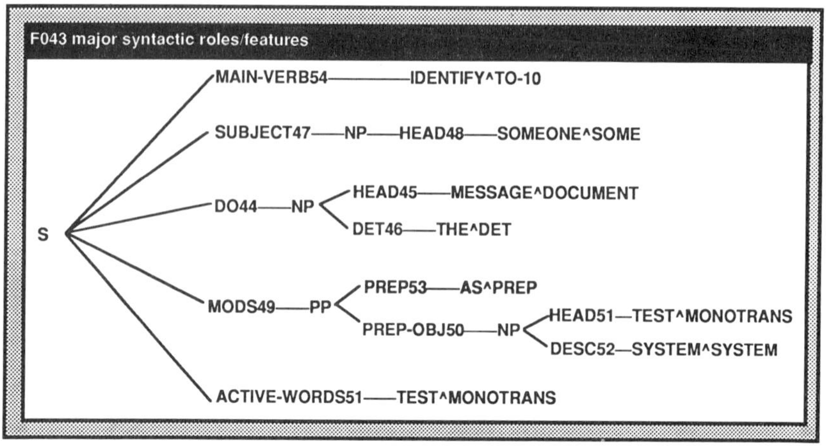

Figure 3a. Syntactic analysis of the second sentence.

Message-Count was identified in a prepositional phrase that modified the Perceive, as "on two occasions."

Applicability to I&W Analysis

PAKTUS has been used to develop a prototype NLP front end for a KBS for relocatable target analysis and also has been tested continually on text from a variety of sources, including the United Press International news wire, RAINFORM messages and examples from the computational linguistics literature. Moreover, it is being applied to actual message traffic under several contracts on a prototype basis. The lexicon includes about 2,200 root words with approximately 5,000 different senses. Many more inflected forms are recognized. About 380 core and subsidiary generic concepts exist. The grammar handles all the major English

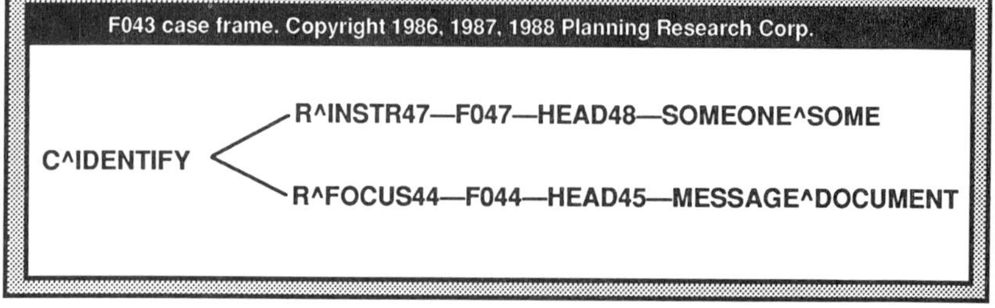

Figure 3b. Conceptual analysis of the second sentence.

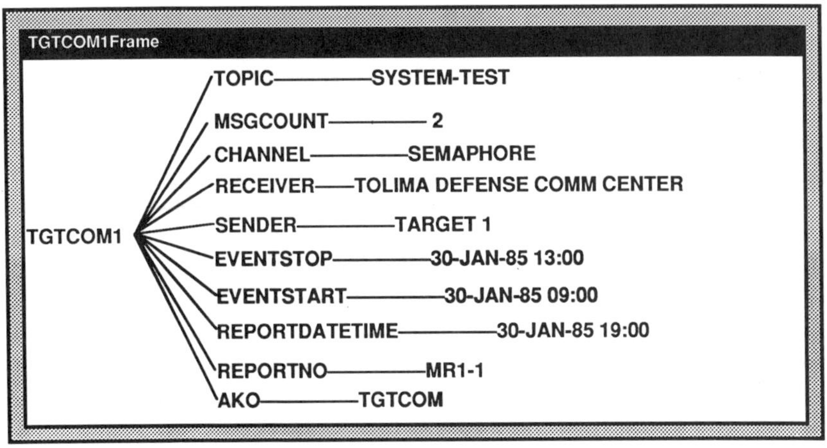

Figure 4. Data base update from message.

syntactic constructions, including such difficult phenomena as reduced coordinate conjunction and apposition, and many of the minor constructions.

PAKTUS has been exercised on several dozen messages of the type illustrated in the previous example, and it successfully extracts the information needed from all of them. In some cases, the syntactic or conceptual analysis is not quite correct or complete, but it is close enough for the discourse analysis component to extract the required information. In this respect, the modular approach achieves a robustness that would be difficult to attain with a monolithic approach. Additionally, some of the messages contain grammatically ill-formed phrases, such as "Duration 10 minutes." While such text has not been the focus of PAKTUS applications, preliminary modules for dealing with it have been incorporated and will be extended. In general, it has been found that grammatical input is processed in linear time (approximately one second per word on a Xerox 1108-111). Therefore, if a sentence is not parsed within that time limit, the parse is abandoned and a relaxed mode partial parse is attempted. This often is adequate.

PAKTUS incorporates an interactive graphic NLP system development environment and modules for semi-automatic extension. These are intended for NLP system developers, not for end-user analysts. The syntactic component contains basic English grammar that easily can be extended, modified and debugged in the interactive graphic environment. The lexicon subsystem can classify some previously unseen words (not just morphologically derived forms) based on the context in which they appear and can recognize new complementation types for individual words. Conceptual specification is assisted by context-sensitive help prompts and graphic browsers. Conceptual associations are loosely based on a psychologically motivated classification, and concepts are organized into a semantic network. PAKTUS is developed sufficiently so that English specifications of some concepts may be given in a few seconds, in lieu of complex patterns whose specification formerly required up to an hour of the developer's time.

PAKTUS also has an interactive lexicon-building module. It is designed to acquire new words as encountered and can learn words in a variety of ways and with varying confidence. The simplest method consists of recognizing an inflected form of a known root word. At the next level is regular morphological derivation, which PAKTUS can perform for 138 affixes. Such derived forms are not always reliable, however, so PAKTUS asks for verification of them. If a word cannot be morphologically broken down, either because its root is unknown or it is irregularly derived, the way it is handled is determined by a switch setting. PAKTUS either will ask to be taught the definition immediately (including any associated conceptual frame) and acquire it through mixed-initiative interaction, or it will guess as much as it can from the context in which the word is used and proceed with the parse by asking the NLP system developer to verify the guesses later.

The development environment includes visual debugging aids. These have become indispensable as the grammar, lexicon and concept net have grown quickly in efforts to process more sophisticated linguistic structures. One aid is a graphic trace of the parsing process. This trace mechanism blinks on each arc or state of the ATN graph as it is considered and prints the current word of input. The speed of blinking is controlled by the developer while the parse is in progress. The tracer also boxes seek arcs (recursive calls to the ATN) as they are considered. This mechanism has been extremely useful in debugging the grammar at a very detailed level. For more macroscopic debugging, another trace mechanism (that also operates while the parse proceeds) displays the conceptual frames that are constructed for each clause.

These tools recently have been used to adapt PAKTUS to Navy RAINFORM ship-to-shore messages quickly. These messages are quite different from the text on which PAKTUS was previously exercised. They tend to be ungrammatical and telegraphic in style. Constituents such as subjects, verbs, auxiliary words and punctuation often are omitted. Although PAKTUS previously was applied only to relatively grammatical text, 60 hours of effort were sufficient to adapt it to this new domain.

PAKTUS may be viewed from two perspectives. In one view it is seen as a generic environment for building NLP systems, incorporating modules for lexical acquisition, grammar building and conceptual template specification. The other perspective focuses on the grammar, lexicon, concept templates and parser already embedded within it and views it as an NLP system itself. The emphasis in developing PAKTUS has been on those components supporting the former view. Because of new terms, acronyms, jargon and domain-specific syntax characteristics, the grammar, lexicon and conceptual specifications never will be complete. Furthermore, the domain-specific knowledge needed to complete records for a specific application clearly is beyond that which can be included in a general NLP system. Consequently, what is needed is an adaptable, transportable NLP shell, and that is what PAKTUS is intended to be. The grammar and lexicon that form the common core of the English language, as well as the stock of generic conceptual templates, will continue to be extended but primarily as a side effect of the testing of extensions to the NLP system development environment. Some exceptional problem areas do require specific treatment, however, such as prepositional phrase attachment, compound nominals, temporal analysis and metaphorical usage.

Conclusion

Current natural language processing techniques such as those described in this article should make it possible to make information from free text available to a data base or expert system as soon as it is received by the intelligence center. In the near term, these techniques can be used to load current data bases from messages with all of the information the data base is capable of representing. In the long term, current data bases can be replaced by the richer representations used by these natural language processing systems themselves, allowing for access to all of the information and nuances currently contained in the original messages. These representations will allow for smarter data base searches and more accurate message (or information) distribution and will better serve the more advanced intelligence analysis aids being developed.

This article originally appeared in the September 1988 edition of SIGNAL.

PART 4
Decision Aiding and Support Systems Technology

Group Decision Support System Prototypes for Army Theater Planning and Counter-Terrorism Crisis Management

Stephen J. Andriole, Lee S. Ehrhart,
Peter H. Aiken and Walter W. Matyskiela

Introduction

The research areas of interest during the project included the need for interactive systems to support group problem solving, the need to design and develop systems that could be used in actual operational environments, and how large screen display technology can be exploited for enhanced human problem solving. The substantive areas of interest included tactical planning at echelons above Corps and counter-terrorism crisis management. Prototype group decision support systems were designed for each domain.

Requirements Analyses

We began the research with requirements analyses in both domains. The first analysis identified the tasks and sub tasks that tactical planners (at echelons above Corps) must perform to develop Concepts of Operations. A "requirements hierarchy" was developed that identified and described each task and sub task. We then developed a second hierarchy that focused exclusively upon user-computer interaction (UCI) requirements [3, 4]. Excerpts from both of these hierarchies appear in Appendix B.

We also developed two hierarchies for substantive and UCI counter-terrorism crisis management requirements [5, 6]. They too appear in Appendix B.

We then identified a set of group requirements as an overlay to the substantive and UCI requirements. Some of these include the (system's) capability to share data and knowledge, share and challenge perspectives, defuse biases, focus discussion, and present data, inferences, options, explanations and recommendations in ways compatible with the requirements, the characteristics of users, and the organization structures the systems might eventually support.

Our requirements analyses also developed some profiles of planners and crisis managers, which by and large characterized these users as inexperienced with advanced computing and likely to make intermittent use of interactive group decision support systems [11].

Substantive Requirements

The substantive requirements that we identified—as suggested in the hierarchies—were varied, reflecting the variety of tasks and sub tasks that decision-makers perform as they generate courses of action. Theater planners require information about terrain, adversary capabilities, their own combat capabilities, among other aspects of tactical planning, while counter-terrorism crisis managers require information about the nature of the terrorist threat, terrorist demands, their own counter-terrorism capabilities, and counter-terrorism policy. The substantive requirements hierarchies identify all such tasks and sub tasks, arranging them in way that permitted us to convert them into system (interface) characteristics.

UCI Requirements

The UCI requirements were decidedly visual, suggesting the need for interactive and animated graphic displays. This requirements finding was not surprising: planners and crisis managers are trained in the visual and graphic through the use of maps and the communication of complex concepts like risk, opportunity, and constraints.

Many different display requirements emerged. Some were anchored in the need for graphic equivalence [8] of complicated data sets. Some required animation, while others required graphics for direct manipulation of battlefield data and processes. The system concept that emerged reflected these requirements and our response to them.

Group Problem Solving Requirements

We identified a set of group problem solving requirements as well [3]. These requirements were screened with reference to the substantive and UCI requirements. The intention was to restrict group requirements to the domains, users and unique interface requirements therein; the system concepts that emerged reflect this screening.

Group decision support systems design and development has received some increased attention over the past few years [9, 10, 12]. Much of this literature is abstract and theoretical, while much less is applied. Our work attempts to exploit the basic for applied purposes.

Storyboard Prototypes

Storyboarding is a technique that converts requirements into a system concept, a model of what the system will do when it is programmed [2]. The working model of the system is a prototype designed to accelerate the requirements validation process. The purpose of the prototype is to enhance communication between systems analysts and users.

Master Menu Structures

The master menu structures were designed with inexperienced users in mind, with substantive and UCI requirements in mind, and with the intention of eventually deploying the system via a large screen display system. We thus gave up substantial portions of the displays—for stationary on-screen menu options—since the impact would be less keenly felt when projected on a larger screen.

Figure 1 presents the theater planning master menu, and Figure 2 the master menu for the counter-terrorism crisis management prototype. These menus suggest how substantive, UCI and group requirements were converted into a prototype designed to help validate requirements.

Prototype Capabilities

The storyboard prototypes were developed on an Apple Macintosh II. Color was utilized to communicate a variety of concepts, options, data, and information. The displays in Appendix A suggest how the storyboard can be operated, its interface capabilities, and what the displays look like to users. The menu structure permits users to execute commands by clicking on but a few icons that represent the elements of tactical planning and counter-terrorism crisis management and available options. Another way to describe the interface and operation of the prototype is to liken the elements to "objects" and system commands as "functions."

It is possible to convert narrative commands like: "show us the courses of action that G2 expects Red to consider" by clicking on two icons, "Show" and "Enemy COAs." It is possible to overlay weather onto the COAs by clicking

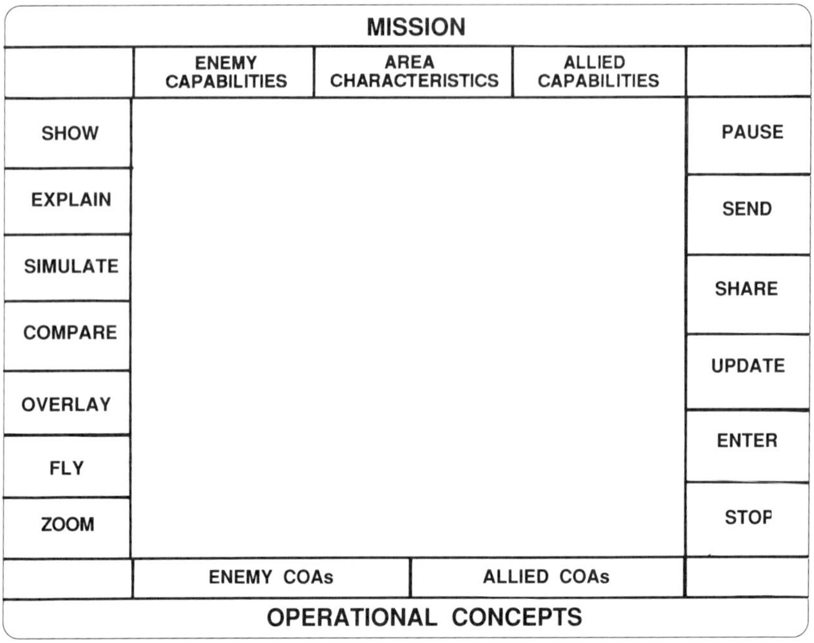

Figure 1. Theater planning master menu.

on "Overlay," "Area Characteristics," and "Weather." It is possible to "mix and match" commands to create great flexibility, and thereby permit users to access information in the system in non-sequential random ways. This flexibility is important to the overall system concept. Many decision support systems permit users to interact with the system's contents in rigid ways; we tried to design a system concept and interface that would permit users relatively unrestricted access to the system's data and knowledge bases.

Conclusions and Future Research

Key Findings

The storyboard prototypes demonstrate some new approaches to UCI design, group decision support, and the use of large screen display technology. The prototypes suggest that large screen-based decision support systems can be used to help solve analytical problems via option generation and inference-making. Previous large screen display-based systems have excelled in conventional data base management tasks but have failed to provide analyti-

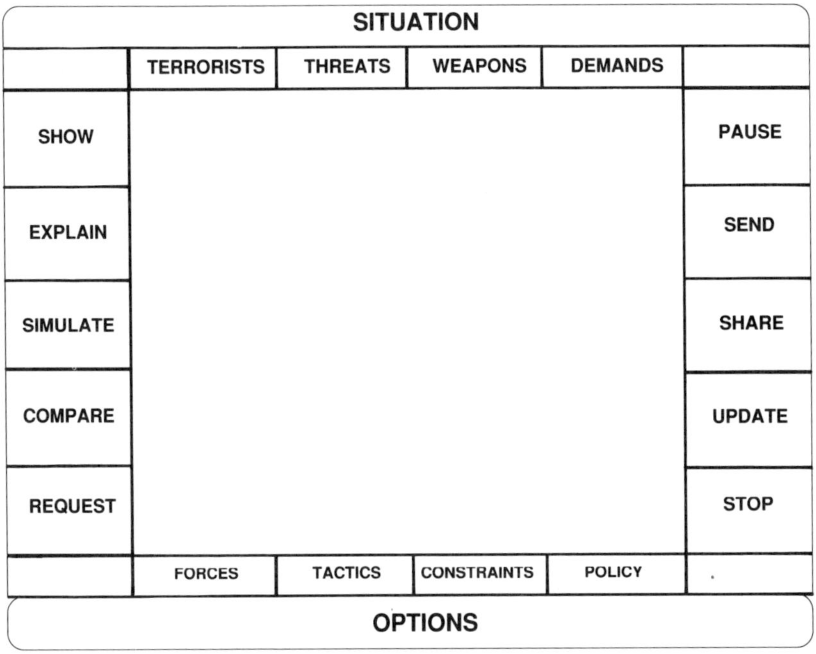

Figure 2. Master menu for the counter-terrorism crisis management prototype.

cal support to users. We believe it is possible to go far beyond conventional display capabilities.

We continue to evaluate the systems; plans are underway to conduct some experiments to determine the extent to which the prototypes satisfy requirements and will thereby enhance human performance.

Next Generation Group Decision Support

Many of the requirements identified during the requirements analysis phase of the project could not be satisfied with the project's hardware/software configuration. Here the reference is to visual requirements such as actual maps, photographs, video information and sound, among other "media." Next generation systems will incorporate multimedia technology directly into their designs. Planners will, for example, be able to drive down highways, peer over bridges, and assess key terrain via film footage of the area of operations. Counter-terrorism crisis managers will be able to see photographs of terrorists as well as film of the denied or threatened areas. Systems architects will be able to satisfy user requirements with a larger arsenal, providing users with presentation technologies that will narrow the communications bandwidth between users and system capabilities.

The following figures suggest how multimedia might impact the tactical planning and counter-terrorism crisis management processes [1]. They also suggest how multimedia can be accessed in a hypothetical system concept.

Acknowledgements

We would like to thank TRW, Inc. for supporting this research via its Independent Research and Development Program. This paper was presented at the 1989 IEEE International Conference on Systems, Man and Cybernetics.

References

[1] Aiken, P. (1988a) *An Evaluation of the Capabilities of a DSS Based on Hypermedia Technologies.* Fairfax, VA: George Mason University.

[2] Andriole, S.J. (1989) *Storyboard Prototyping for Systems Design: A New Approach to User Requirements Analysis.* Wellesley, MA: QED Information Sciences, Inc.

[3] Andriole, S.J. (1988) *User-Computer Interaction (UCI) Requirements for Group Problem-Solving.* Fairfax, VA: George Mason University.

[4] Andriole, S.J. (1987a) *Functional Requirements for Theater-Level Planning and Decision-Making.* Fairfax, VA: George Mason University.

[5] _________ (1987b) *Functional Requirements for Counter-Terrorism Crisis Management.* Fairfax, VA: George Mason University.

[6] _________ (1987c) *User-Computer Interaction (UCI; Display) Requirements for Theater-Level Tactical Planning.* Fairfax,VA: George Mason University.

[7] _________ (1987d) *User Computer Interaction (UCI; Display) Requirements for Counter-Terrorism Crisis Management.* Fairfax, VA: George Mason University.

[8] Andriole, S.J. (1986) "Graphic Equivalence, Graphic Explanations & Embedded Process Modeling for Enhanced User-System Interaction." *IEEE Transactions on Systems, Man and Cybernetics.* Volume SMC-16, Number 6, November-December, pp. 919-926.

[9] Bertcher, H.J. (1979) *Group Participation.* Newbury, CA: Sage Publications.

[10] DeSanctis, G. and Gallupe R.B. (1987) "A Foundation for the Study of Group Decision Support Systems." *Management Science,* Volume 33, Number 5, May, pp. 589-609.

[11] Ehrhart, L.S. (1988) *Storyboard Architectures for Group Problem-Solving.* Fairfax, VA: George Mason University.

[12] Gladstein, D.L. and Reilly, N.P. (1985) "Group Decision Making Under Threat: The TYCOON Game." *Academy of Management Journal.* Volume 28, Number 3, pp. 613-627.

[13] Hendrick, C. (1987) *Group Processes.* Newbury Park, CA: Sage Publications.

Appendix A

Storyboard Prototypes for Tactical Planning at the Theater Level and Counter-Terrorism Crisis Management

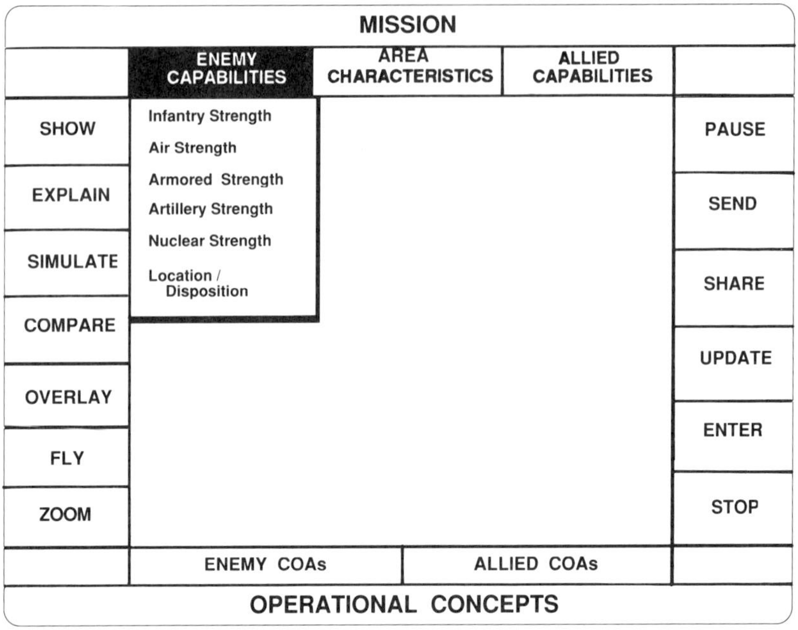
MISSION
ENEMY CAPABILITIES
AREA CHARACTERISTICS
ALLIED CAPABILITIES
Infantry Strength
Air Strength
Armored Strength
Artillery Strength
Nuclear Strength
Location / Disposition
SHOW
EXPLAIN
SIMULATE
COMPARE
OVERLAY
FLY
ZOOM
PAUSE
SEND
SHARE
UPDATE
ENTER
STOP
ENEMY COAs
ALLIED COAs
OPERATIONAL CONCEPTS

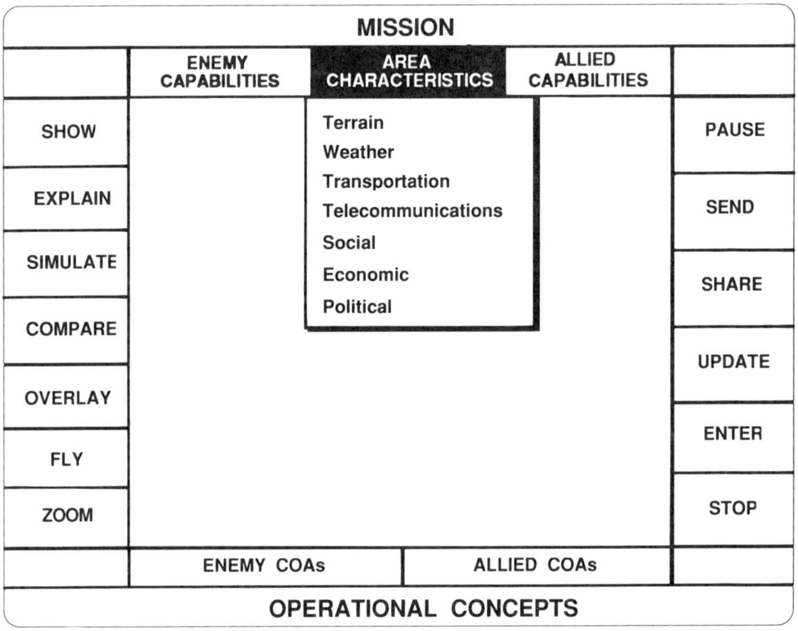
MISSION
ENEMY CAPABILITIES
AREA CHARACTERISTICS
ALLIED CAPABILITIES
Terrain
Weather
Transportation
Telecommunications
Social
Economic
Political
SHOW
EXPLAIN
SIMULATE
COMPARE
OVERLAY
FLY
ZOOM
PAUSE
SEND
SHARE
UPDATE
ENTER
STOP
ENEMY COAs
ALLIED COAs
OPERATIONAL CONCEPTS

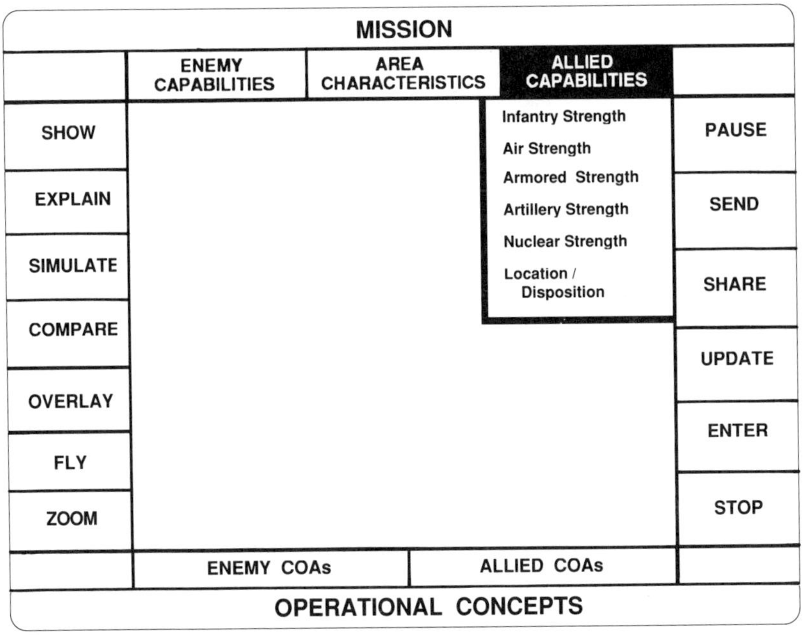
MISSION
ENEMY CAPABILITIES
AREA CHARACTERISTICS
ALLIED CAPABILITIES
Infantry Strength
Air Strength
Armored Strength
Artillery Strength
Nuclear Strength
Location / Disposition
SHOW
EXPLAIN
SIMULATE
COMPARE
OVERLAY
FLY
ZOOM
PAUSE
SEND
SHARE
UPDATE
ENTER
STOP
ENEMY COAs
ALLIED COAs
OPERATIONAL CONCEPTS

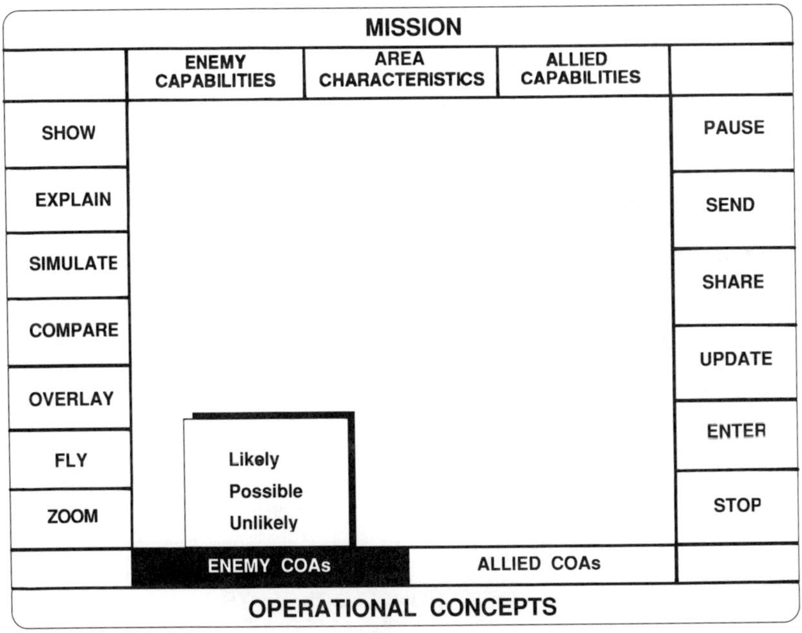
MISSION
ENEMY CAPABILITIES
AREA CHARACTERISTICS
ALLIED CAPABILITIES
SHOW
EXPLAIN
SIMULATE
COMPARE
OVERLAY
FLY
ZOOM
Likely
Possible
Unlikely
PAUSE
SEND
SHARE
UPDATE
ENTER
STOP
ENEMY COAs
ALLIED COAs
OPERATIONAL CONCEPTS

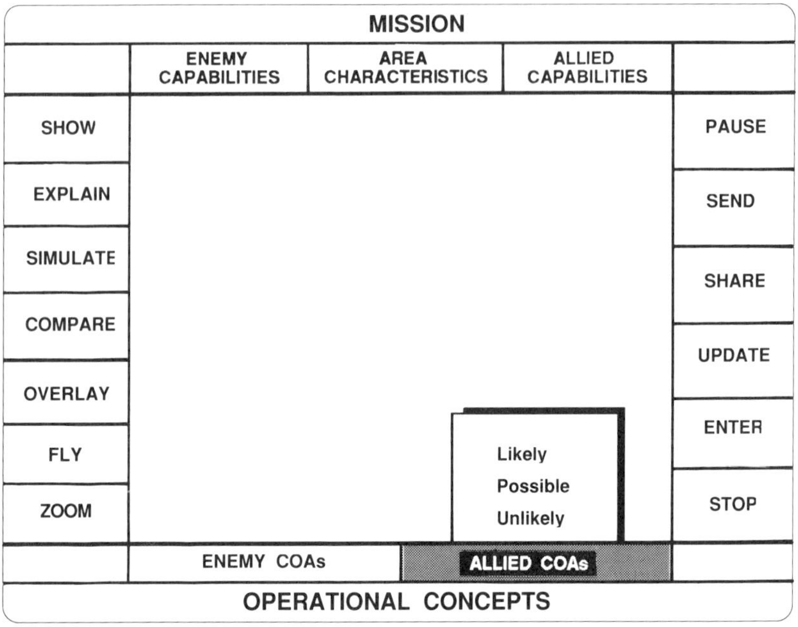
MISSION
ENEMY CAPABILITIES
AREA CHARACTERISTICS
ALLIED CAPABILITIES
SHOW
EXPLAIN
SIMULATE
COMPARE
OVERLAY
FLY
ZOOM
PAUSE
SEND
SHARE
UPDATE
ENTER
STOP
Likely
Possible
Unlikely
ENEMY COAs
ALLIED COAs
OPERATIONAL CONCEPTS

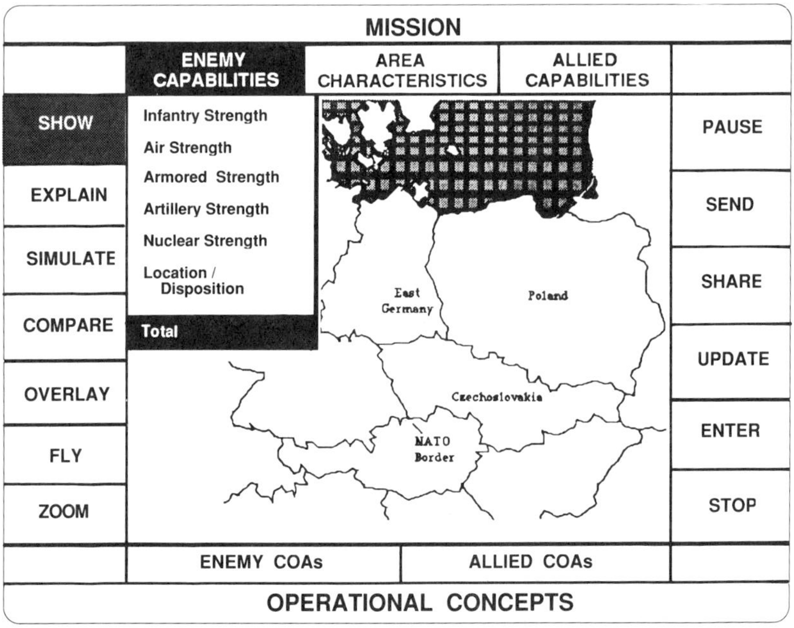
MISSION
ENEMY CAPABILITIES
AREA CHARACTERISTICS
ALLIED CAPABILITIES
SHOW
EXPLAIN
SIMULATE
COMPARE
OVERLAY
FLY
ZOOM
Infantry Strength
Air Strength
Armored Strength
Artillery Strength
Nuclear Strength
Location / Disposition
Total
East Germany
Poland
Czechoslovakia
NATO Border
PAUSE
SEND
SHARE
UPDATE
ENTER
STOP
ENEMY COAs
ALLIED COAs
OPERATIONAL CONCEPTS

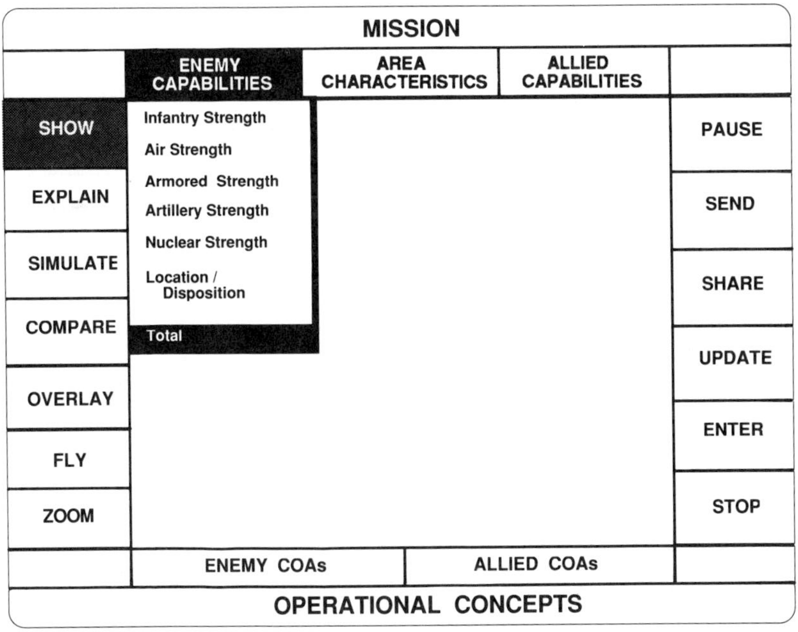
MISSION
ENEMY CAPABILITIES
AREA CHARACTERISTICS
ALLIED CAPABILITIES
SHOW
EXPLAIN
SIMULATE
COMPARE
OVERLAY
FLY
ZOOM
Infantry Strength
Air Strength
Armored Strength
Artillery Strength
Nuclear Strength
Location / Disposition
Total
PAUSE
SEND
SHARE
UPDATE
ENTER
STOP
ENEMY COAs
ALLIED COAs
OPERATIONAL CONCEPTS

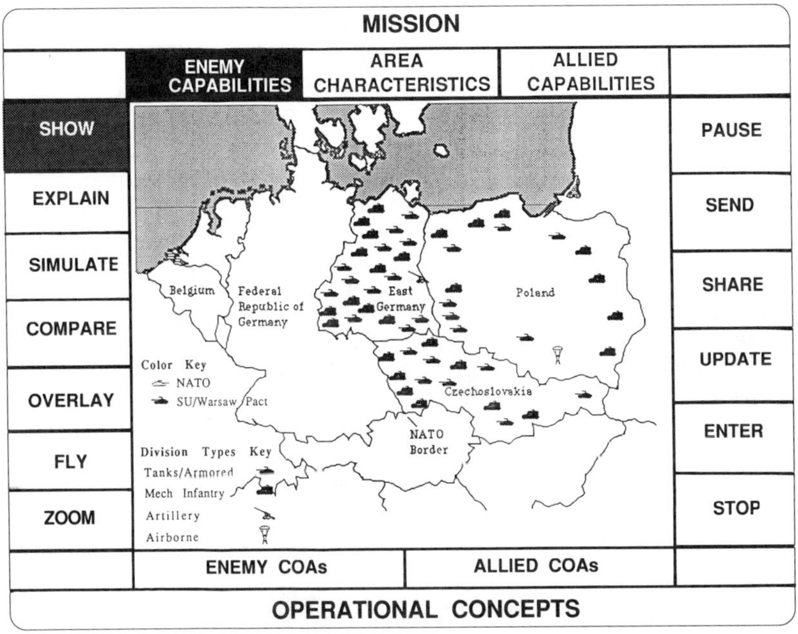
MISSION
ENEMY CAPABILITIES
AREA CHARACTERISTICS
ALLIED CAPABILITIES
SHOW
EXPLAIN
SIMULATE
COMPARE
OVERLAY
FLY
ZOOM
Belgium
Federal Republic of Germany
East Germany
Poland
Czechoslovakia
NATO Border
Color Key
NATO
SU/Warsaw Pact
Division Types Key
Tanks/Armored
Mech Infantry
Artillery
Airborne
PAUSE
SEND
SHARE
UPDATE
ENTER
STOP
ENEMY COAs
ALLIED COAs
OPERATIONAL CONCEPTS

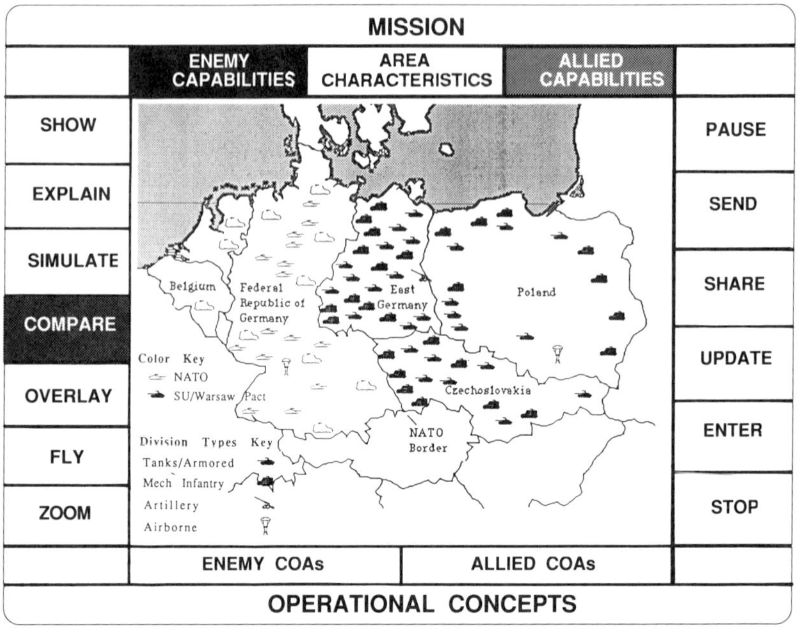
MISSION
ENEMY CAPABILITIES
AREA CHARACTERISTICS
ALLIED CAPABILITIES
SHOW
EXPLAIN
SIMULATE
COMPARE
OVERLAY
FLY
ZOOM
PAUSE
SEND
SHARE
UPDATE
ENTER
STOP
Belgium
Federal Republic of Germany
East Germany
Poland
Czechoslovakia
NATO Border
Color Key
NATO
SU/Warsaw Pact
Division Types Key
Tanks/Armored
Mech Infantry
Artillery
Airborne
ENEMY COAs
ALLIED COAs
OPERATIONAL CONCEPTS

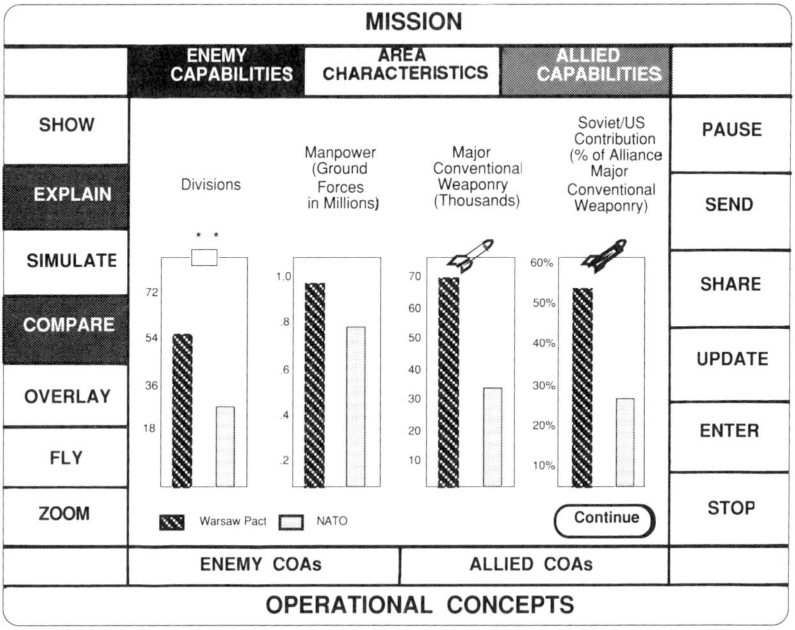
MISSION
ENEMY CAPABILITIES
AREA CHARACTERISTICS
ALLIED CAPABILITIES
SHOW
EXPLAIN
SIMULATE
COMPARE
OVERLAY
FLY
ZOOM
PAUSE
SEND
SHARE
UPDATE
ENTER
STOP
Divisions
Manpower (Ground Forces in Millions)
Major Conventional Weaponry (Thousands)
Soviet/US Contribution (% of Alliance Major Conventional Weaponry)
72
54
36
18
1.0
.8
.6
.4
.2
70
60
50
40
30
20
10
60%
50%
40%
30%
20%
10%
Warsaw Pact
NATO
Continue
ENEMY COAs
ALLIED COAs
OPERATIONAL CONCEPTS

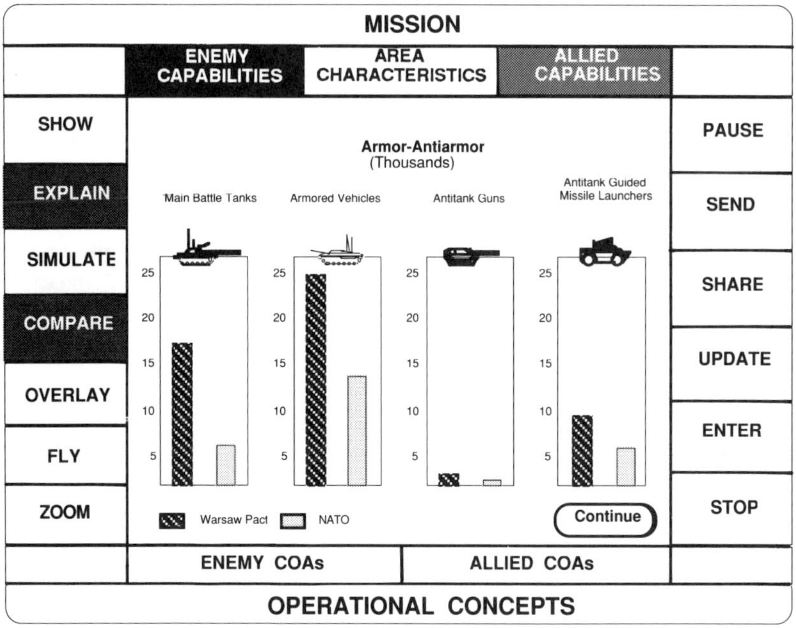
MISSION
ENEMY CAPABILITIES
AREA CHARACTERISTICS
ALLIED CAPABILITIES
SHOW
EXPLAIN
SIMULATE
COMPARE
OVERLAY
FLY
ZOOM
Armor-Antiarmor
(Thousands)
Main Battle Tanks
Armored Vehicles
Antitank Guns
Antitank Guided Missile Launchers
25
20
15
10
5
Warsaw Pact
NATO
Continue
PAUSE
SEND
SHARE
UPDATE
ENTER
STOP
ENEMY COAs
ALLIED COAs
OPERATIONAL CONCEPTS

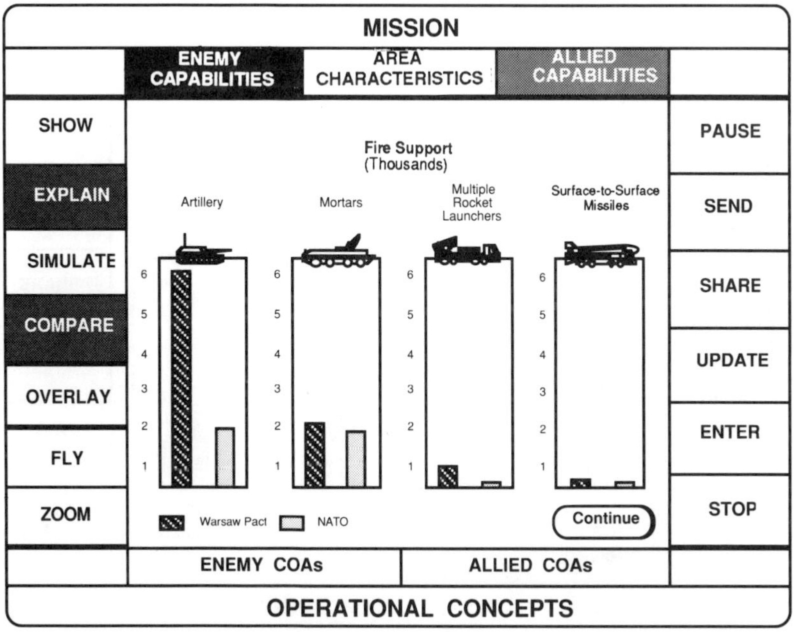
MISSION
ENEMY CAPABILITIES
AREA CHARACTERISTICS
ALLIED CAPABILITIES
SHOW
EXPLAIN
SIMULATE
COMPARE
OVERLAY
FLY
ZOOM
Fire Support
(Thousands)
Artillery
Mortars
Multiple Rocket Launchers
Surface-to-Surface Missiles
6
5
4
3
2
1
Warsaw Pact
NATO
Continue
PAUSE
SEND
SHARE
UPDATE
ENTER
STOP
ENEMY COAs
ALLIED COAs
OPERATIONAL CONCEPTS

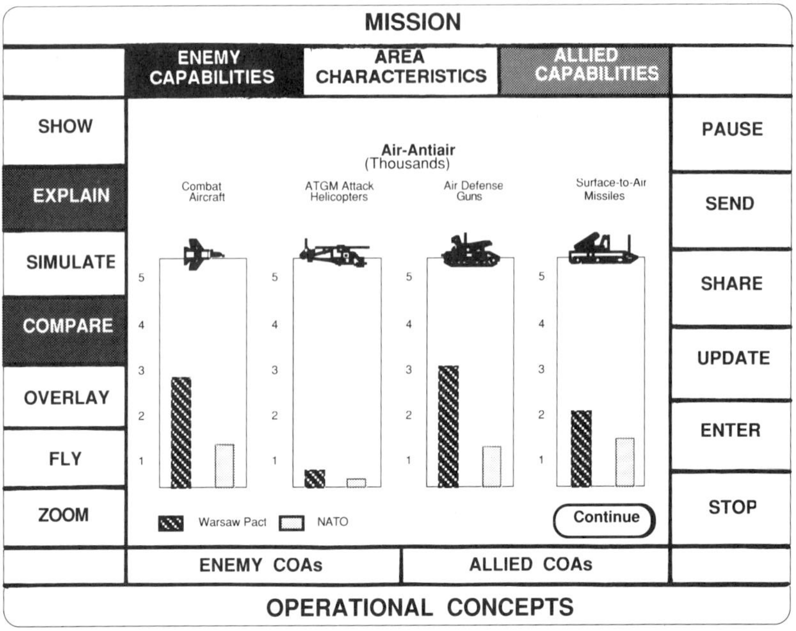
MISSION
ENEMY CAPABILITIES
AREA CHARACTERISTICS
ALLIED CAPABILITIES
SHOW
EXPLAIN
SIMULATE
COMPARE
OVERLAY
FLY
ZOOM
PAUSE
SEND
SHARE
UPDATE
ENTER
STOP
Air-Antiair
(Thousands)
Combat Aircraft
ATGM Attack Helicopters
Air Defense Guns
Surface-to-Air Missiles
Warsaw Pact
NATO
Continue
ENEMY COAs
ALLIED COAs
OPERATIONAL CONCEPTS

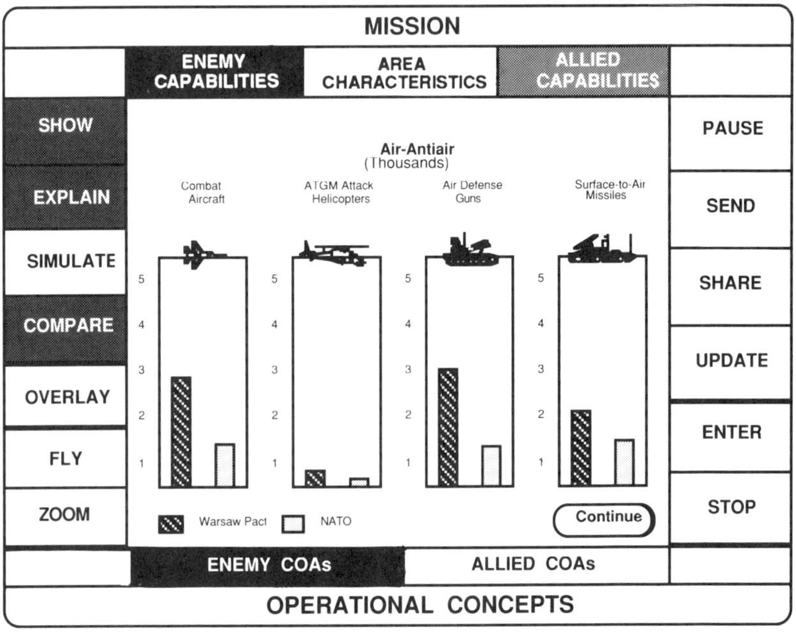
MISSION
ENEMY CAPABILITIES
AREA CHARACTERISTICS
ALLIED CAPABILITIES
SHOW
EXPLAIN
SIMULATE
COMPARE
OVERLAY
FLY
ZOOM
PAUSE
SEND
SHARE
UPDATE
ENTER
STOP
Air-Antiair
(Thousands)
Combat Aircraft
ATGM Attack Helicopters
Air Defense Guns
Surface-to-Air Missiles
Warsaw Pact
NATO
Continue
ENEMY COAs
ALLIED COAs
OPERATIONAL CONCEPTS

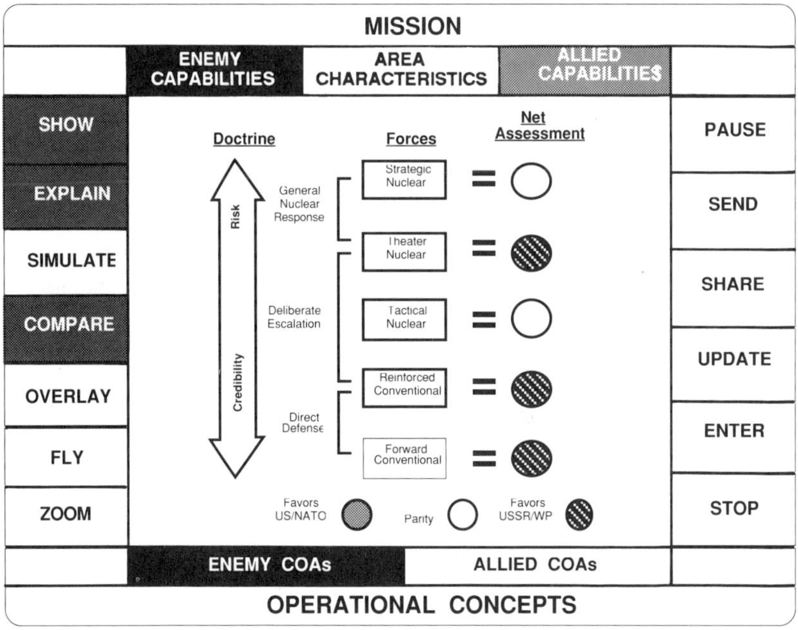
MISSION
ENEMY CAPABILITIES
AREA CHARACTERISTICS
ALLIED CAPABILITIES
SHOW
EXPLAIN
SIMULATE
COMPARE
OVERLAY
FLY
ZOOM
Doctrine
Forces
Net Assessment
Risk
Credibility
General Nuclear Response
Deliberate Escalation
Direct Defense
Strategic Nuclear
Theater Nuclear
Tactical Nuclear
Reinforced Conventional
Forward Conventional
Favors US/NATO
Parity
Favors USSR/WP
PAUSE
SEND
SHARE
UPDATE
ENTER
STOP
ENEMY COAs
ALLIED COAs
OPERATIONAL CONCEPTS

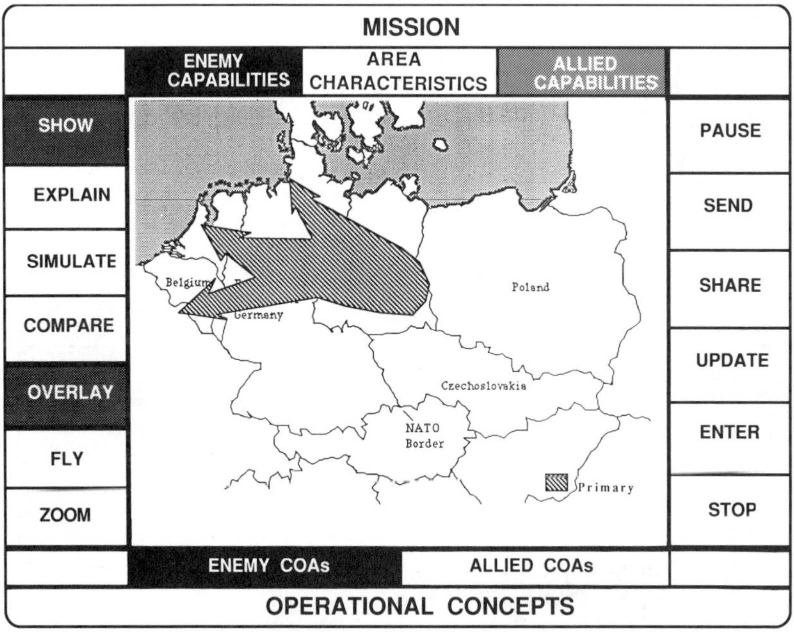
MISSION
ENEMY CAPABILITIES
AREA CHARACTERISTICS
ALLIED CAPABILITIES
SHOW
EXPLAIN
SIMULATE
COMPARE
OVERLAY
FLY
ZOOM
Belgium
Germany
Poland
Czechoslovakia
NATO Border
Primary
PAUSE
SEND
SHARE
UPDATE
ENTER
STOP
ENEMY COAs
ALLIED COAs
OPERATIONAL CONCEPTS

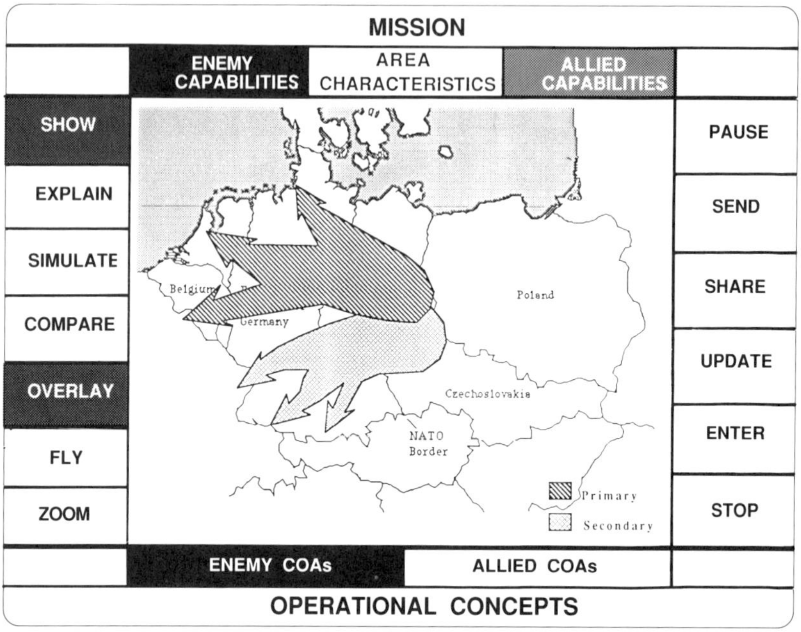
MISSION
ENEMY CAPABILITIES
AREA CHARACTERISTICS
ALLIED CAPABILITIES
SHOW
EXPLAIN
SIMULATE
COMPARE
OVERLAY
FLY
ZOOM
PAUSE
SEND
SHARE
UPDATE
ENTER
STOP
Belgium
Germany
Poland
Czechoslovakia
NATO Border
Primary
Secondary
ENEMY COAs
ALLIED COAs
OPERATIONAL CONCEPTS

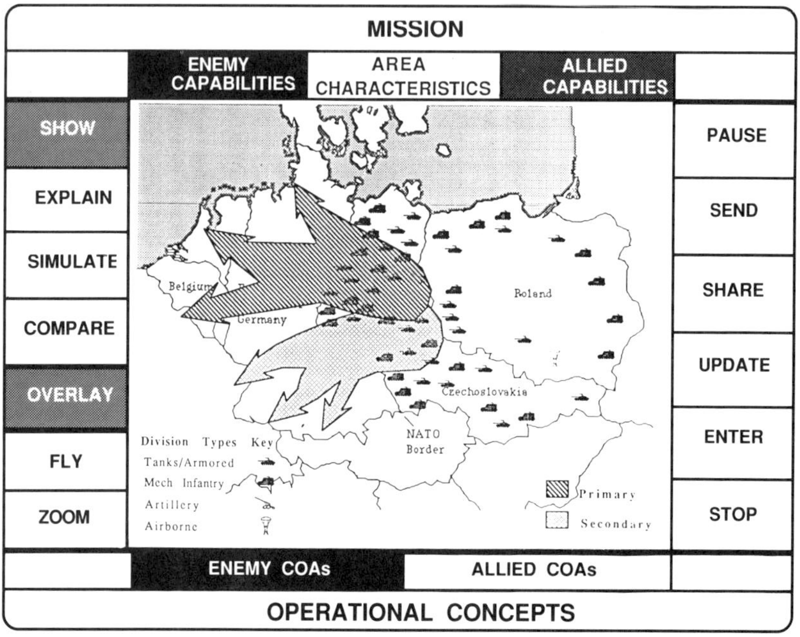
MISSION
ENEMY CAPABILITIES
AREA CHARACTERISTICS
ALLIED CAPABILITIES
SHOW
EXPLAIN
SIMULATE
COMPARE
OVERLAY
FLY
ZOOM
PAUSE
SEND
SHARE
UPDATE
ENTER
STOP
Belgium
Germany
Poland
Czechoslovakia
NATO Border
Division Types Key
Tanks/Armored
Mech Infantry
Artillery
Airborne
Primary
Secondary
ENEMY COAs
ALLIED COAs
OPERATIONAL CONCEPTS

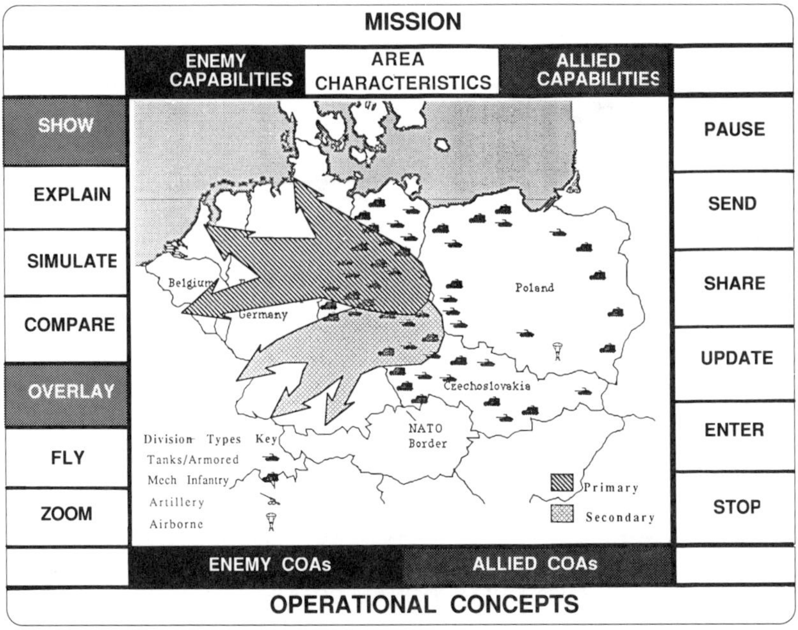
MISSION
ENEMY CAPABILITIES
AREA CHARACTERISTICS
ALLIED CAPABILITIES
SHOW
EXPLAIN
SIMULATE
COMPARE
OVERLAY
FLY
ZOOM
PAUSE
SEND
SHARE
UPDATE
ENTER
STOP
Belgium
Germany
Poland
Czechoslovakia
NATO Border
Division Types Key
Tanks/Armored
Mech Infantry
Artillery
Airborne
Primary
Secondary
ENEMY COAs
ALLIED COAs
OPERATIONAL CONCEPTS

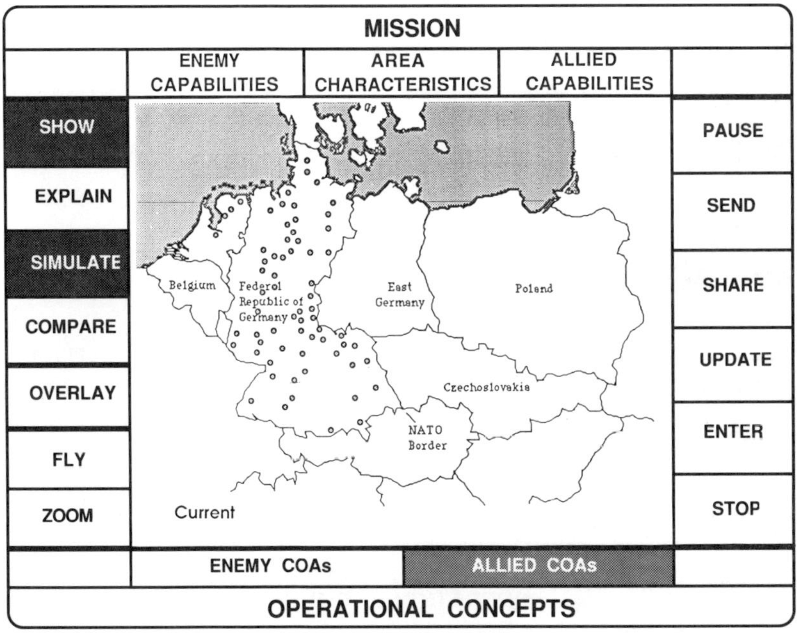
MISSION
ENEMY CAPABILITIES
AREA CHARACTERISTICS
ALLIED CAPABILITIES
SHOW
EXPLAIN
SIMULATE
COMPARE
OVERLAY
FLY
ZOOM
PAUSE
SEND
SHARE
UPDATE
ENTER
STOP
Belgium
Federal Republic of Germany
East Germany
Poland
Czechoslovakia
NATO Border
Current
ENEMY COAs
ALLIED COAs
OPERATIONAL CONCEPTS

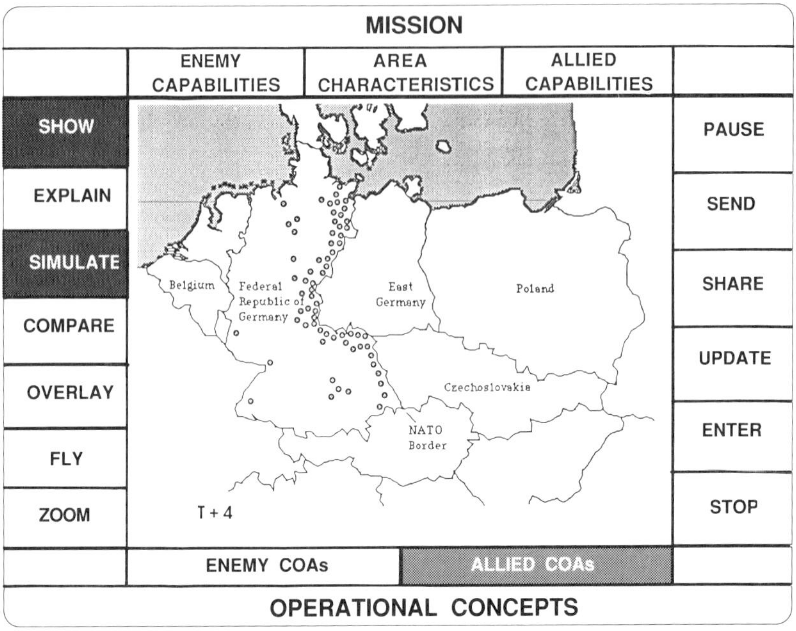
MISSION
ENEMY CAPABILITIES
AREA CHARACTERISTICS
ALLIED CAPABILITIES
SHOW
EXPLAIN
SIMULATE
COMPARE
OVERLAY
FLY
ZOOM
PAUSE
SEND
SHARE
UPDATE
ENTER
STOP
Belgium
Federal Republic of Germany
East Germany
Poland
Czechoslovakia
NATO Border
T + 4
ENEMY COAs
ALLIED COAs
OPERATIONAL CONCEPTS

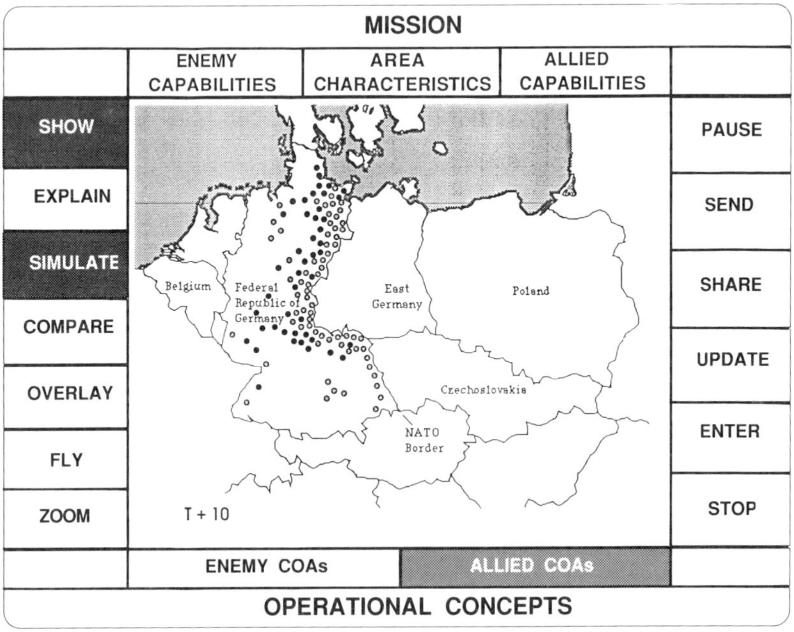
MISSION
ENEMY CAPABILITIES
AREA CHARACTERISTICS
ALLIED CAPABILITIES
SHOW
EXPLAIN
SIMULATE
COMPARE
OVERLAY
FLY
ZOOM
PAUSE
SEND
SHARE
UPDATE
ENTER
STOP
Belgium
Federal Republic of Germany
East Germany
Poland
Czechoslovakia
NATO Border
T + 10
ENEMY COAs
ALLIED COAs
OPERATIONAL CONCEPTS

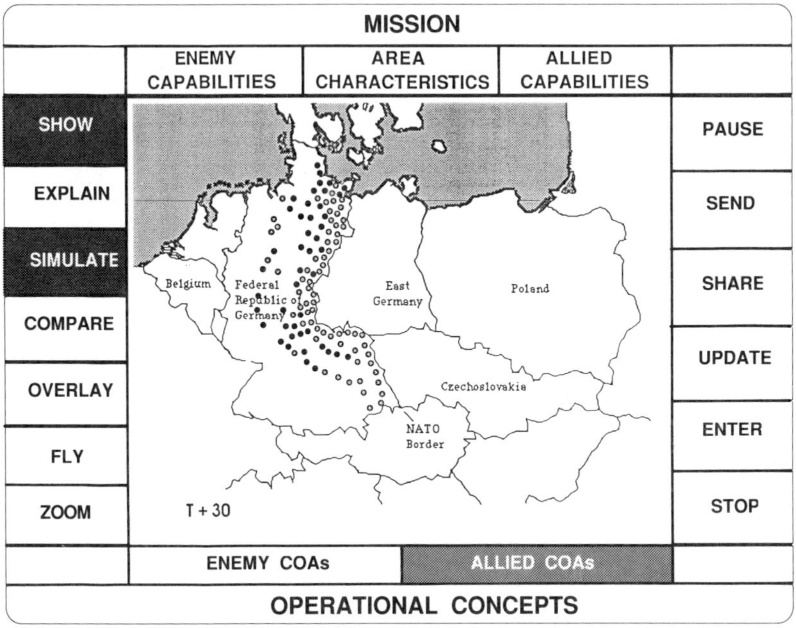
MISSION
ENEMY CAPABILITIES
AREA CHARACTERISTICS
ALLIED CAPABILITIES
SHOW
EXPLAIN
SIMULATE
COMPARE
OVERLAY
FLY
ZOOM
Belgium
Federal Republic of Germany
East Germany
Poland
Czechoslovakia
NATO Border
T + 30
PAUSE
SEND
SHARE
UPDATE
ENTER
STOP
ENEMY COAs
ALLIED COAs
OPERATIONAL CONCEPTS

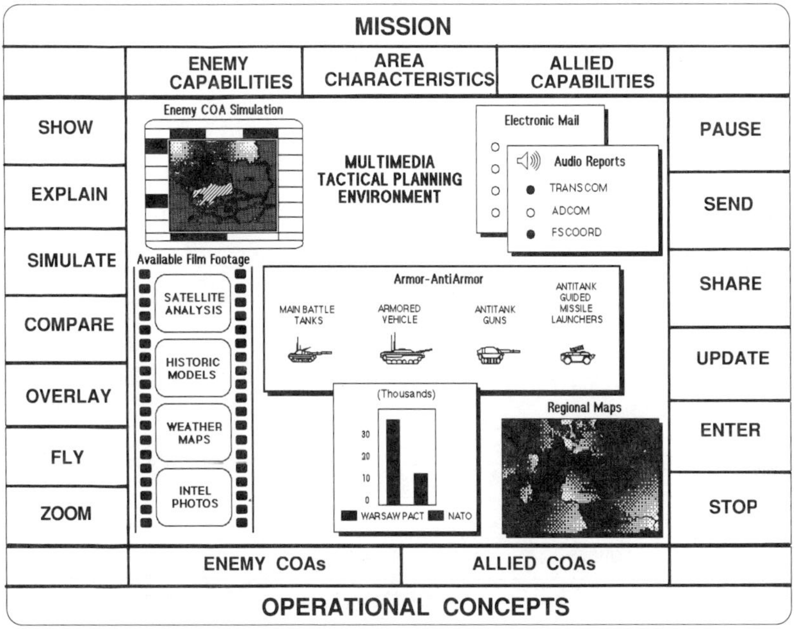
MISSION
ENEMY CAPABILITIES
AREA CHARACTERISTICS
ALLIED CAPABILITIES
SHOW
EXPLAIN
SIMULATE
COMPARE
OVERLAY
FLY
ZOOM
Enemy COA Simulation
MULTIMEDIA TACTICAL PLANNING ENVIRONMENT
Electronic Mail
Audio Reports
TRANSCOM
ADCOM
FSCOORD
Available Film Footage
SATELLITE ANALYSIS
HISTORIC MODELS
WEATHER MAPS
INTEL PHOTOS
Armor-AntiArmor
MAIN BATTLE TANKS
ARMORED VEHICLE
ANTITANK GUNS
ANTITANK GUIDED MISSILE LAUNCHERS
(Thousands)
30
20
10
0
WARSAW PACT
NATO
Regional Maps
PAUSE
SEND
SHARE
UPDATE
ENTER
STOP
ENEMY COAs
ALLIED COAs
OPERATIONAL CONCEPTS

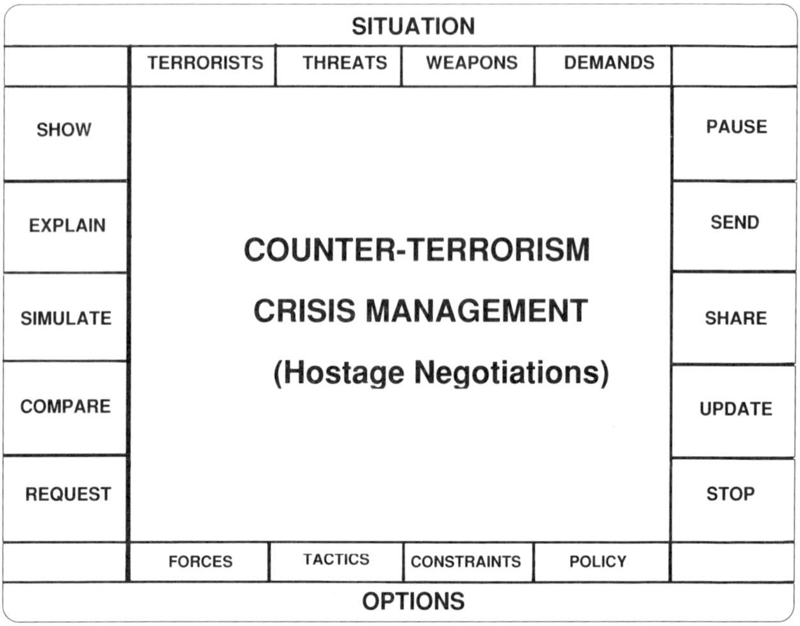
SITUATION
TERRORISTS
THREATS
WEAPONS
DEMANDS
SHOW
EXPLAIN
SIMULATE
COMPARE
REQUEST
COUNTER-TERRORISM
CRISIS MANAGEMENT
(Hostage Negotiations)
PAUSE
SEND
SHARE
UPDATE
STOP
FORCES
TACTICS
CONSTRAINTS
POLICY
OPTIONS

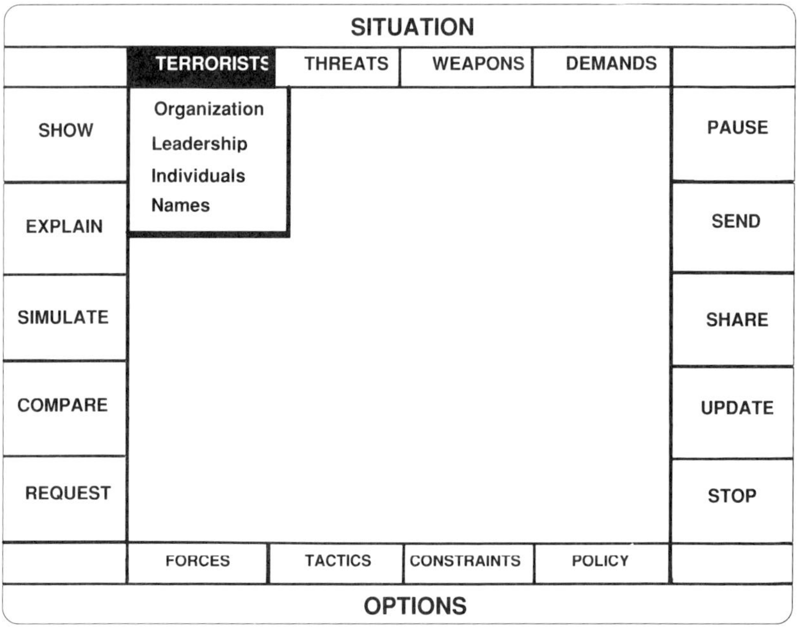
SITUATION
TERRORISTS
THREATS
WEAPONS
DEMANDS
Organization
Leadership
Individuals
Names
SHOW
EXPLAIN
SIMULATE
COMPARE
REQUEST
PAUSE
SEND
SHARE
UPDATE
STOP
FORCES
TACTICS
CONSTRAINTS
POLICY
OPTIONS

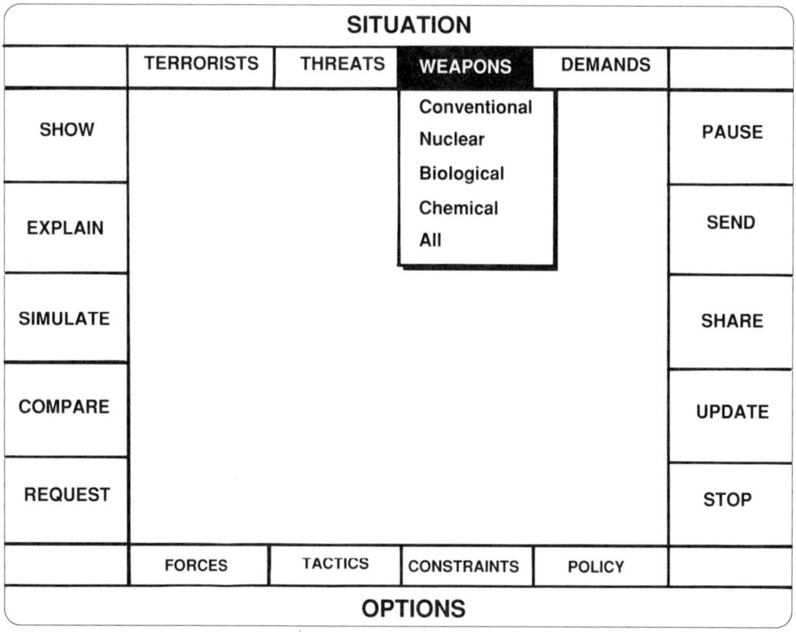
SITUATION
TERRORISTS
THREATS
WEAPONS
DEMANDS
Conventional
Nuclear
Biological
Chemical
All
SHOW
EXPLAIN
SIMULATE
COMPARE
REQUEST
PAUSE
SEND
SHARE
UPDATE
STOP
FORCES
TACTICS
CONSTRAINTS
POLICY
OPTIONS

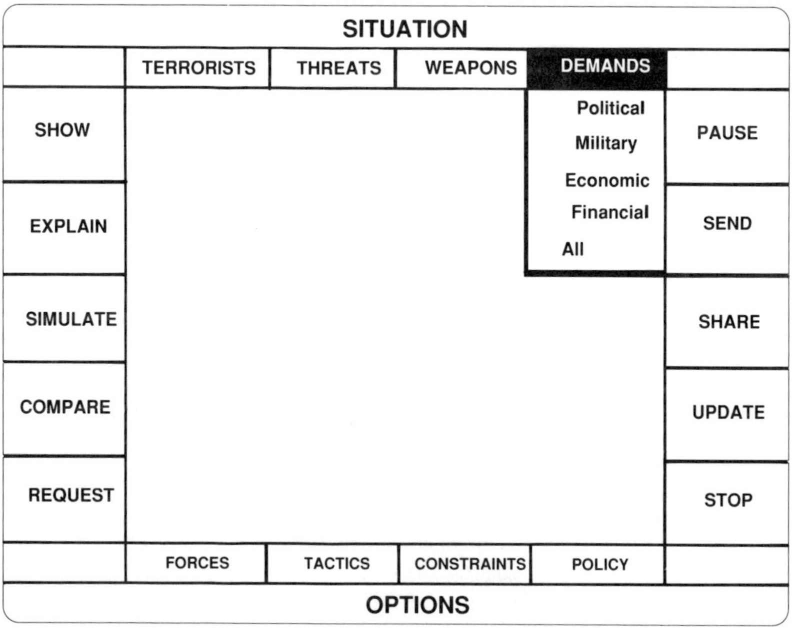
SITUATION
TERRORISTS
THREATS
WEAPONS
DEMANDS
Political
Military
Economic
Financial
All
SHOW
EXPLAIN
SIMULATE
COMPARE
REQUEST
PAUSE
SEND
SHARE
UPDATE
STOP
FORCES
TACTICS
CONSTRAINTS
POLICY
OPTIONS

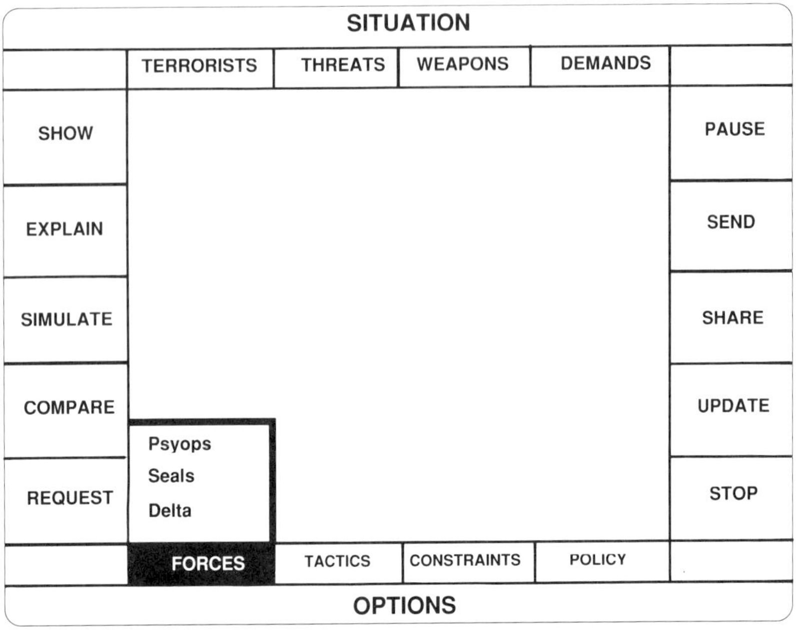
SITUATION
TERRORISTS
THREATS
WEAPONS
DEMANDS
SHOW
EXPLAIN
SIMULATE
COMPARE
REQUEST
PAUSE
SEND
SHARE
UPDATE
STOP
Psyops
Seals
Delta
FORCES
TACTICS
CONSTRAINTS
POLICY
OPTIONS

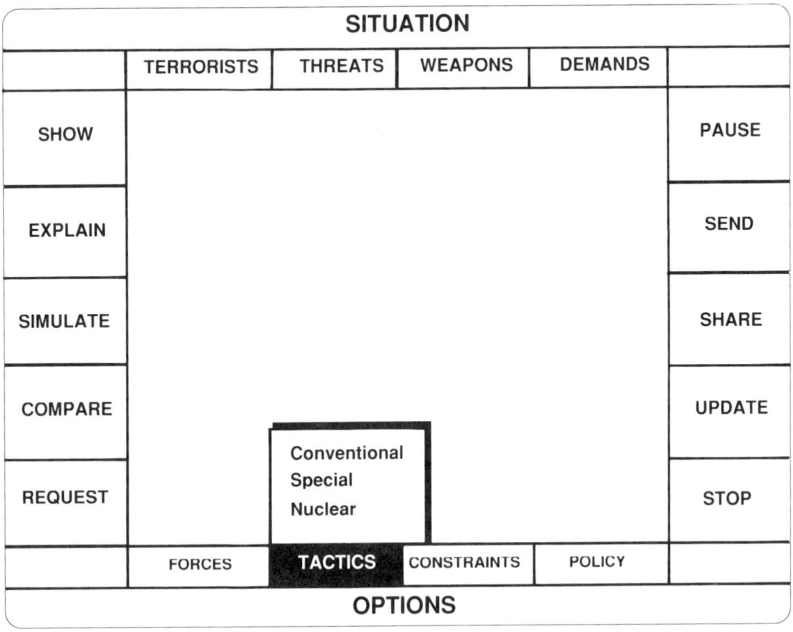
SITUATION
TERRORISTS
THREATS
WEAPONS
DEMANDS
SHOW
EXPLAIN
SIMULATE
COMPARE
REQUEST
PAUSE
SEND
SHARE
UPDATE
STOP
Conventional
Special
Nuclear
FORCES
TACTICS
CONSTRAINTS
POLICY
OPTIONS

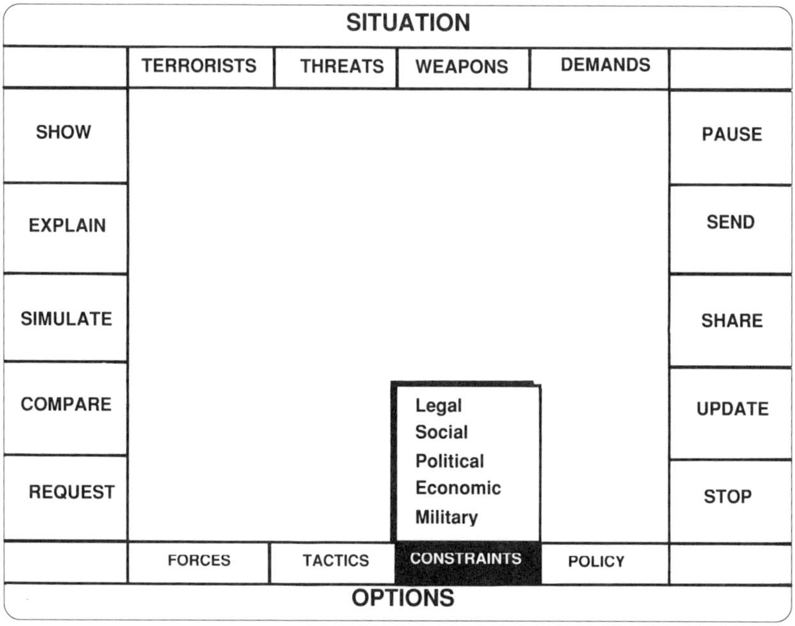
SITUATION
TERRORISTS
THREATS
WEAPONS
DEMANDS
SHOW
EXPLAIN
SIMULATE
COMPARE
REQUEST
PAUSE
SEND
SHARE
UPDATE
STOP
Legal
Social
Political
Economic
Military
FORCES
TACTICS
CONSTRAINTS
POLICY
OPTIONS

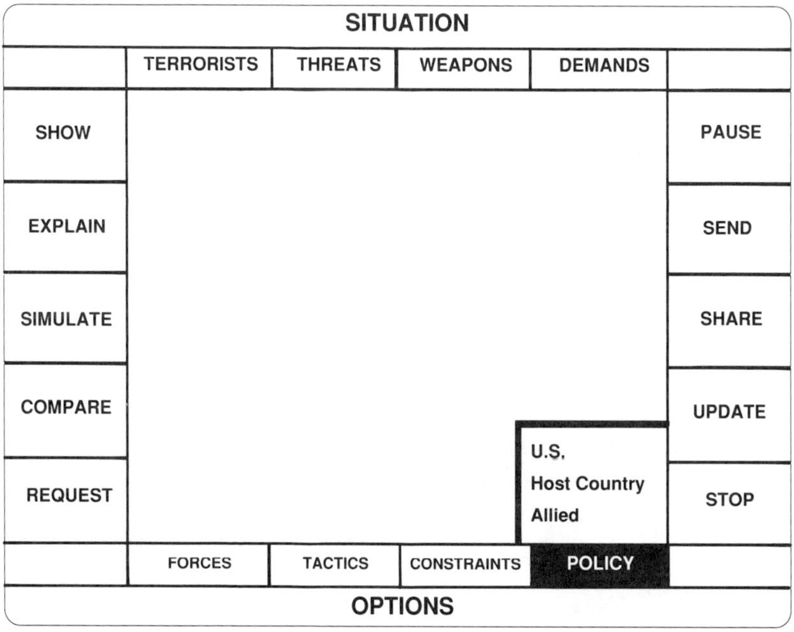
SITUATION
TERRORISTS
THREATS
WEAPONS
DEMANDS
SHOW
EXPLAIN
SIMULATE
COMPARE
REQUEST
PAUSE
SEND
SHARE
UPDATE
STOP
U.S.
Host Country
Allied
FORCES
TACTICS
CONSTRAINTS
POLICY
OPTIONS

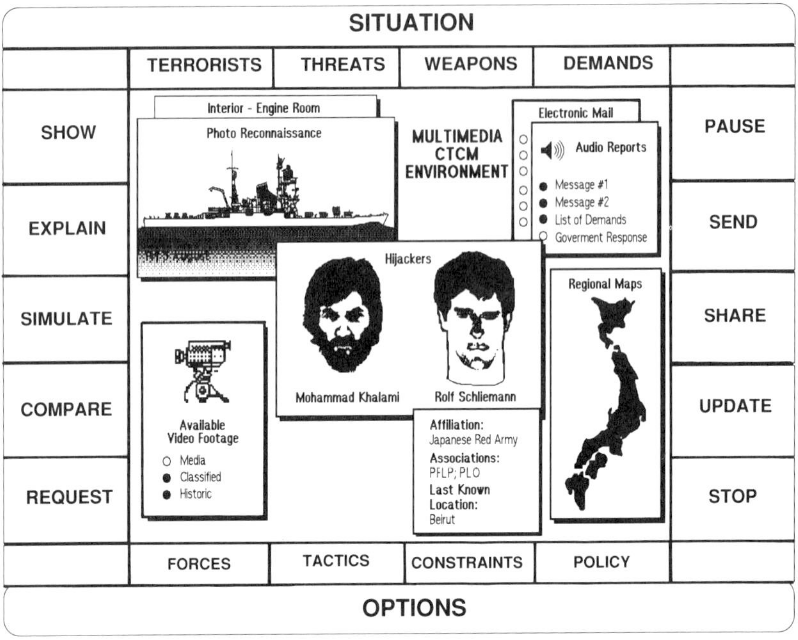
SITUATION
TERRORISTS
THREATS
WEAPONS
DEMANDS
SHOW
EXPLAIN
SIMULATE
COMPARE
REQUEST
PAUSE
SEND
SHARE
UPDATE
STOP
Interior - Engine Room
Photo Reconnaissance
MULTIMEDIA CTCM ENVIRONMENT
Electronic Mail
Audio Reports
Message #1
Message #2
List of Demands
Goverment Response
Hijackers
Mohammad Khalami
Rolf Schliemann
Regional Maps
Available Video Footage
Media
Classified
Historic
Affiliation:
Japanese Red Army
Associations:
PFLP; PLO
Last Known Location:
Beirut
FORCES
TACTICS
CONSTRAINTS
POLICY
OPTIONS

Defense Communications Decision Support Using ISMAUT and Extensions

Bradley S. Stewart, William T. Scherer,
Edward A. Sykes and Chelsea C. White, III

Introduction

We describe a study involving planning and decision making for defense communications systems. An important aspect of this research is the use of imprecisely specified multiattribute utility theory (ISMAUT), and associated software, to assist in various decision making contexts involving defense communications programs and plans. The research provides insight into the problems involved in planning for large defense communications applications, as well as more general perspectives on decision aiding using ISMAUT and related techniques.

Key components of the background of this research are the organizations involved, the problems addressed, and the general approach taken to resolving those problems.

Organizations. The impetus for this effort originated in the Organization of the Joint Chiefs of Staff (OJCS). Personnel working in the area of command, control, and communications (C^3) at OJCS were introduced to the ISMAUT methodology and recognized its potential usefulness for their planning and decision making requirements. The Defense Communications Agency (DCA), and in particular, its Defense Communications and Engineering Center (DCEC), became involved almost immediately. In addition, government contractors became involved in association with their roles supporting DCEC. The party primarily responsible for the work discussed below was Intelligent Systems

Corporation (ISC), the developer of the ISMAUT methodology and owner of the associated software.

Problems. The class of problems addressed in this case included various strategic planning problems for DOD C^3 systems. Specifically, there were two basic problems attacked. In the simpler of the two, it was desired to have an idea of the relative merits of various DOD C^3 programs. The desirability of these programs was defined in terms of a number of criteria arranged in a hierarchy. The difficulty in comparing these programs stemmed from the conflicting and noncommensurate nature of the criteria. A multiobjective approach to assessment was therefore appropriate.

The more complicated problem involved the identification of a collection of programs that would best meet the objectives, with the addition of a budget constraint. Thus the problem was to identify preferred packages of programs, where the cost of each package did not exceed the available funding. There were two major complicating factors. First, the number of feasible packages can be extremely large, even though the number of programs appears reasonable. The number of possible packages is 2 raised to the number of programs. For example, if there are 20 programs, there are over 1 million possible packages. The second complication involves the existence of synergistic effects among the programs, so that preferences among packages were not accurately representable using sums of constituent program scores. Synergy among programs can be either positive or negative, resulting in package scores that are either greater than or less than those produced by a straightforward additive model.

The identification of these two basic problems, even to the level of definition represented by the simple statements included above, was a lengthy interactive process that continued over most of the period during which the work was performed. This problem identification process was complicated by the planning and decision making context, which was closely tied to top level C^3 programming and budgeting activities. The two problems outlined above were only a small part of a very large planning and decision making system that was continuously evolving to keep up with rapidly changing C^3 technology and defense needs.

Approach. Of the many potential approaches to our two basic problems, we chose to pursue solutions based on ISMAUT. For simple comparative assessment of C^3 programs, ISMAUT was applicable with minimal modifications. The packaging problem required a more innovative approach that included a simple rule-based package generation system in conjunction with ISMAUT. To model the synergistic interactions among programs, we used a form of multilinear value function, which is described later along with the other information on the packaging problem.

Prototype software for both problems was produced using FORTRAN, with development work taking place in PROFORT on an AT class personal computer, with ultimate installation on an IBM mainframe.

ISMAUT

Multiattribute utility theory (MAUT) is commonly used to model planning and decision making situations in which there are multiple, conflicting, and non-commensurate objectives. In MAUT based systems, inputs include a list of alternatives, a list of attributes, a precise evaluation of each alternative with respect to each attribute, and a set of precise trade-off weights indicating the relative contribution of the scores on each attribute. A total ordering of the alternatives may then be derived using the simple weighted linear combination of the scores for each alternative. For a complete discussion of MAUT, see Keeney and Raiffa [1].

A major difficulty in MAUT based systems is that a large amount of data must be precisely and completely specified before any results can be obtained. This difficulty was the motivation behind the development of ISMAUT, which will produce results even in cases for which: 1) a complete set of value scores is unavailable, and/or 2) value scores are only imprecisely known, and/or 3) a complete set of trade-off weights is unavailable, and/or 4) trade-off weights are only imprecisely known.

Concepts. ISMAUT requires the following inputs:

- A, a finite set of alternatives, in our case, alternative programs
- I, a finite set of additively independent attributes
- $\{u_i(\alpha)\}$, a set of (possibly imprecise) value scores, giving the value associated with attribute i for selecting program α.

The following information may also be entered in cases for which it is available:

- $\{w_i\}$, a set of (possibly imprecise) attribute trade-off weights, for one or more attributes i
- $\{(\alpha,\alpha')\}$, a set of ordered pairs directly indicating preferences between pairs of alternatives α and α'.

Imprecision in the value scores and trade-off weights is described by set inclusion, a relatively simple, flexible, and behaviorally acceptable approach. The behavioral appeal is based partly upon the way in which natural language evaluative statements may be translated into specific input information. For instance, people will often be able to make statements like, "I don't know which alternative I prefer the most, but among those available, I certainly prefer Alternative X to Alternative Y." Standard MAUT techniques would be unable to incorporate this type of information in any simple way, but ISMAUT accepts it directly in the form of the input "X > Y." As another example of the behavioral attractiveness of the linear inequality representation for imprecision, consider the statement, "In the context of the available alternatives, Attribute 1 is more important to my decision than Attribute 3." This statement also has a simple translation to input information in the form of the linear inequality on the trade-off weights of Attributes 1 and 3 as follows: $w_1 \geq w_3$.

To consider the operation of ISMAUT, let w be a specific row vector of trade-off weights and let u(α) be a specific #I vector of value scores for alternative α, where the notation #X is indicates the cardinality of the set X. Then the *total weighted value* of alternative α is defined as

$$w\, u(\alpha) = \sum_i w_i\, u_i(\alpha)$$

Preferences among alternatives may then be defined in terms of this definition. In particular, an alternative α' is said to be at least as preferred as alternative α if and only if

$$w\, u\, (\alpha') \geq w\, u\, (\alpha).$$

As mentioned above, the utilities may be imprecisely specified by a set of linear inequalities, as may the attribute trade-off weights. The trade-off weights are also assumed to satisfy the following relations

$$w_i \geq 0 \;\; \forall\, i; \qquad \sum_i w_i = 1.$$

Directly-specified pairwise preference comparisons among the alternatives are interpreted to imply inequalities in the space of possible weight vectors according to the following relation for the example of α' preferred to α:

$$\sum_i w_i \max_{u_i \in Ui} v_i(\alpha',\alpha) \geq 0,$$

where $v_i(\alpha',\alpha) = [u_i(\alpha') - u_i(\alpha)]$

and U_i represents the set of all possible values of $u_i(\bullet)$.

The sets U_i are derived from information provided by a decision maker regarding alternative assessments for attribute i. Note that when the U_i are described by finite sets of linear inequalities, as is assumed here, the maximization in the above expression defines a linear program for each attribute i. According to this mathematical interpretation of the implications of a directly-specified pairwise preference relation, the statement that alternative α' is preferred to alternative α means that the trade-off weights are such that the total weighted value of alternative α' is at least as great as that of alternative α' for every value that $u(\alpha')$ and $u(\alpha)$ might assume.

In order to further explain ISMAUT, let the set of inequalities that were specified as input information on trade-off weights be denoted by S_w. Let W be the set of trade-off weight vectors that satisfy all inequalities in Sw and the sum-to-one and nonnegativity constraints. Similarly, let R denote the set of directly-specified pairwise preference relations on the alternatives and S_r be the set of inequalities on the weight space implied by those preference relations in conjunction with the mathematical interpretation provided above. Finally, let $S = S_w \cup S_r$ be the complete set of linear inequalities on the attribute trade-off weights that can be inferred from the input data under the assumptions of the underlying multiattribute decision model.

A set of problem input information is *inconsistent*, according to the underlying multiattribute decision model assumed here, if all vector attribute trade-off weights violate one or more inequalities in S. Otherwise, the problem input information is *consistent*. For the consistent case, we let $W' \in W$ be the set of trade-off weights that violate no inequalities of S and define a new set of preference relations, call it R', with $(\alpha', \alpha) \in R'$ if and only if

$$\min_{w \in W'} \sum_k w_k \min_{u_i \in U_i} v_k(\alpha',\alpha) \geq 0.$$

Since we have assumed the input information is consistent, this set R' will necessarily include the set R, plus any additional preference relations that can be inferred from the input information and the assumed decision model. An alternative α is said to be *dominated* if there exists an alternative α' such that $(\alpha',\alpha) \in R'$ and $(\alpha,\alpha') \in R'$. Otherwise, alternative α is said to be *nondominated*. Note that R' defines a transitive, reflexive, and antisymmetric relation on A.

When used in a standard decision aiding context, the outputs of the ISMAUT process are 1) the set of dominance relations among the alternatives and 2) a description of the allowable weight region, both of which are inferred from the input information under the assumptions of the model. More detail concerning ISMAUT can be found in White, et al. [2], with improvements provided by Scherer, et al. [3]. Related discussions can also be found in [4],through [12]. The use of ISMAUT in assessing the relative merit of DOD C^3 programs is best described using a simple example.

Example. This example illustrates the concepts involved in ISMAUT in a simple fashion, without the complications that are introduced when the attributes are arranged in a hierarchy and the value scores are imprecisly specified. We consider a hypothetical situation in which four C^3 programs are to be evaluated with respect to three criteria. The only difference between the use of ISMAUT and the use of standard MAUT on this example is the use of imprecise specification of the trade-off weight vectors. The objective of the analysis is to determine any dominance relations that can be inferred among the programs based on the input value scores and some imprecise trade-off weight information.

Figure 1 shows four packet-switched data communications networks, the programs and their performance on the attributes dollar cost/month, a "less-is-better" attribute, percent base requirement at 200 ms average packet delay (i.e., the percentage level the base traffic requirements must be scaled to produce a source-to-destination average packet delay of exactly 200 ms.), a "more-is-better" attribute, and maximum link utilization, a "less-is-better" attribute. These performance values represent the raw attribute scores in this example.

For simplicity, we assign a value score of 1 to the best and a value score of 0 to the worst alternative score on each attribute and assign all other attribute value scores in a linear fashion. Table 2 gives the attribute value scores. An alternate approach would be to assess value functions on each attribute according to any of several standard approaches; see, for example, Keeney and Raiffa [1]. The numbers in parentheses in the bottom row of Figure 2 are the differences between the best and worst alternative scores on each attribute.

At this point, one could elicit statements from the user regarding the relative importance of the attributes. For example, the user might feel that the first attribute deserved no more than 20 percent of the total available weight. This statement could be captured by the following linear inequality constraint:

$$0 \leq w_1 \leq .2$$

Program	Cost ($/month)	%base req.	max. link util.
A1	126,000	110	.70
A2	120,000	70	.40
A3	138,000	100	.60
A4	150,000	170	.90

Figure 1. Program Raw Scores.

Program	Cost $v_1(\cdot)$	% base reg. $v_2(\cdot)$	max. link util. $v_3(\cdot)$
A1	0.8	0.4	0.4
A2	1.0	0.0	1.0
A3	0.4	0.3	0.6
A4	0.0	1.0	0.0
	($30K)	(100%)	(.5)

Figure 2. Attribute Value Scores.

Alternatively, or in addition, the user might be able to state that, in the current context, the third attribute was at least twice as important as the second attribute, resulting in the following constraint:

$$w_3 \geq 2w_2.$$

For now, we assume that no such statements are provided in order to illustrate the power of directly specified alternative comparisons. Suppose the user makes the following statement regarding the networks: "I don't know which of the alternatives I prefer the most, but I do know that I prefer A4 to A2."

We shall denote this statement: A4 ≥A2. Under the assumptions inherent in the ISMAUT methodology, the mathematical implications of this statement are as follows:

$$w[v(A4) - v(A2)] \geq 0$$

$$w_1[v_1(A4) - v_1(A2)] + w_2[v_2(A4) - v_2(A2)] + w_3[v_3(A4) - v_3(A2)] \geq 0$$

$$-w_1 + w_2 - w_3 \geq 0 .$$

Since we assume that the trade-off weights sum to one, we also know that

$$-w_1 + w_2 - (1 - w_1 - w_{2)} \geq 0 ,$$

and hence $$w_2 \geq 1/2 .$$

With four (n = 4) alternatives, there are n(n - 1)/2 = 6 possible pairwise comparisons. With three relationships allowed (≥, ≤, and =), there are a total of eighteen possible statements the user could make with implications of the sort described above. We enumerate the weight constraints associated with the ≥ relationships for this example in Figure 3. Note that the ≤ and = relationships may be obtained by replacing the appropriate ≥ sign in the table.

Graphically, we can represent these relationships as in Figure 4. From the graph, it is apparent that A4 ≥ A2 limits the region of admissible w's to the upper portion of the triangle. Looking at the other possible pairwise comparisons, we see that the admissible w region as a result of A4 ≥ A2 is a subset of the admissible W region for A4 ≥ A3. Therefore, from the statement that A4 is preferred to A3, we can infer that A4 is also preferred to A3.

One can easily enumerate graphically the implications of each of the eighteen possible pairwise comparisons. Of them, two yield just one alternative in the nondominated set, seven give two alternatives in the nondominated set, and the remaining nine give three alternatives in the nondominated set. So, in this simple example, with no directly assessed information regarding attribute weights, we can infer at least one additional alternative dominance relation for half of the possible pairwise comparisons the user might make.

Given either trade-off weight constraints or directly specified alternative comparisons, ISMAUT generates a directed graph indicating the dominance relationships that can be inferred among the alternative programs. Continuing this example, assume the user has specified that Attribute 1 should be weighted more than Attribute 2, which in turn should be weighted more than Attribute 3; i.e., $w_1 > w_2 > w_3$. The only dominance this statement implies is that A1 is preferred to A3. Now let's further assume the user specifies that A4 is preferred to

Comparison	Implied Weight Constraint
A1 ≥ A2	$2w_1 + 5w_2 \geq 3$
A1 ≥ A3	$6w_1 + 3w_2 \geq 2$
A1 ≥ A4	$2w_1 - 5w_2 \geq -2$
A2 ≥ A3	$2w_1 - 7w_2 \geq -4$
A2 ≥ A4	$1/2 \geq w_2$
A3 ≥ A4	$6 \geq 2w_1 + 13w_2$

Figure 3. Potential Pairwise Comparisons and the Resulting Weight Constraints.

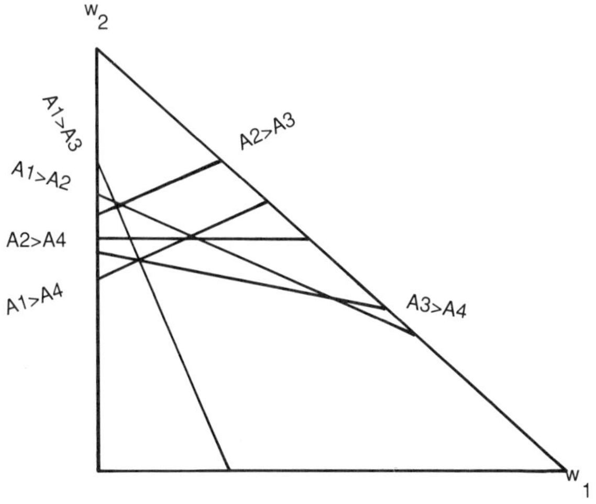

Figure 4. Graphical Representation of Pairwise Comparisons.

A1. These statements together result in the digraph of Figure 5a as an output of ISMAUT. As another example, the statement $w_1 > w_3 > w_2$ by itself results in a total order on the alternatives as shown in Figure 5b.

This example illustrates the use of ISMAUT to comparatively evaluate individual C[3] programs. Extensions that can be added to ISMAUT to address the problem of selecting packages of programs are discussed next.

ISMAUT+

Designed as a preprocessor for ISMAUT, ISMAUT+ assists the decision maker with the problem of generating packages of programs by using information on specific programs over a set of specified attributes. ISMAUT+ takes as input a set of programs and their raw attribute scores, and produces a set of alternative packages along with value scores for each over all attributes. The outputs of ISMAUT+ are produced in a form suitable for direct input to ISMAUT.

ISMAUT+ Concepts. The goal of ISMAUT+ is to generate a tractable number of feasible packages on the basis of the value scores for each package over all attributes, taking into account the synergistic effects between programs within specific attributes. Once the packages have been generated and their value scores determined, they can be considered as alternatives by the ISMAUT program. Thus we have a two-phase process: ISMAUT+ is used for package generation and value score determination, and then ISMAUT is used for comparative evaluation of, and selection from, the resulting packages.

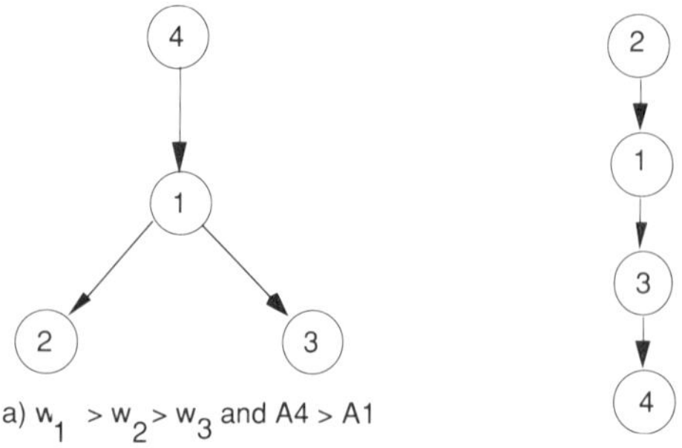

a) $w_1 > w_2 > w_3$ and A4 > A1

b) $w_1 > w_3 > w_2$

Figure 5. Dominance Digraphs for the Example.

Let P be the set of *programs* under consideration and assume P has cardinality #P. As before, let A be the set of *attributes* under consideration, with cardinality #A. Note that the maximum number of packages is equal to the number of possible subsets of P, or $2^{\#P}$. Let $\mathbf{Q} \subseteq \mathbf{P}$ represent a *package*, and assume $R_i(Q)$ is the *raw score* of package Q with respect to attribute i. Below we provide a description of how $R_i(Q)$ is determined from the raw scores of the programs that comprise package Q with respect to attribute i. Assume $V_i(Q)$ is the *value score* of package Q with respect to attribute i. Determination of $V_i(Q)$ from $R_i(Q)$ is also presented below.

Consistent with the assumptions made from ISMAUT, we assume that all attributes are additively independent. Hence, for each attribute i, there is a trade-off weight $w_i \geq 0$, $\Sigma_i w_i = 1$, such that the overall value score of package Q is:

$$V(Q) = \Sigma_i w_i V_i(Q).$$

Given precise trade-off weights, our objective is to choose a $\mathbf{Q}^* \subseteq \mathbf{P}$ such that $V(Q^*) \geq V(Q)$ for all $\mathbf{Q} \subseteq \mathbf{P}$ When the trade-off weight vector $w = \{w_i\}$ is only known to be a member of a given set W, we then seek the relation $R'' \subseteq \mathbf{P} \times P$ such that $(Q, Q') \in R''$ if and only if $V(Q) \geq V(Q')$ for all $w \in W$. Note that $(Q, Q') \in R''$ if and only if

$$\min\{\Sigma_i w_i[V_i(Q) - V_i(Q')]: w \in W\} \geq 0,$$

which is a linear program if W is described by a finite set of linear inequalities. The details for constructing W and hence R were presented in the section on ISMAUT above.

A general model that would allow for capturing the notion of synergy would be the following multilinear form:

$$V_i(Q) = \sum_{j \in Q} k_i(j)\, v_i(j) + \sum_{n,j \in Q}\sum k_i(n,j)\, v_i(n)\, v_i(j) +$$

$$\sum_{m,n,j \in Q}\sum\sum k_i(m,n,j)\, v_i(m)\, v_i(n)\, v_i(j) \cdots,$$

where the k terms capture the synergy between programs. A positive ki(n,j), for example, implies that programs $n \in Q$ and $j \in Q$ have a positive synergy on attribute $i \in A$. One problem with this model is that the series of terms tends to get complex for even small numbers of programs. In practice it is common to truncate the function after the second or third order products. This is reasonable since $v_i(j) \in (0,1)$, which tends to make the higher order terms insignificant unless an extremely large weight is assigned to the product term. Another difficulty with the above value function is the problem of assessing the $k(\bullet,\bullet,\ldots)$ coefficients that capture synergy.

ISMAUT+ addresses the above concerns with the multilinear value function by 1) allowing only second order interactions to be directly assessed, thus assuming that higher order synergistic interactions are negligible, and 2) assessing the synergistic effects directly on the raw attribute scores.

Let $r_i(j)$ be the raw score of program $j \in P$ with respect to attribute i. The ISMAUT+ model is as follows:

$$R_i(Q) = \sum_{j \in Q} r_i(j) + \sum_{\substack{k,j \in Q \\ j<k}} s_i(k,j)$$

where $s_i(k,j)$ is the synergy between programs k and j, on attribute $i \in A$.

The form described above requires that the synergy terms, $s_i(i,j)$, be assessed for all pairs of programs and for all attributes. Thus, the maximum number of synergy terms required is N(N-1)/2. Often, however, many of the synergy terms will be zero and will therefore require only trivial assessment.

The above procedure can be thought of as corresponding to directly assessing the entire k v(n) v(j) term as compared to assessing the k coefficients. The motivation for such a procedure is the inherent difficulty associated with assessing the k coefficients.

ISMAUT+ only calculates the raw value scores for a package, Q, as needed. To show how $V_i(Q)$ is determined from $R_i(Q)$, assume that "more is better" with respect to attribute i. Let

$$M = \max \{R_i(Q):\ Q \subseteq P\}, \text{ and}$$

$$m = \min \{R_i(Q):\ Q \subseteq P\}.$$

Using these quantities, the raw scores are scaled linearly to determine the value scores:

$$V_i(Q) = (R_i(Q) - m)/(M - m).$$

An analogous procedure is used for the "less is better" attributes.

We now consider how packages are generated, and, if possible, eliminated from consideration via user input rules. Let A_F represent the set of all *feasible* packages. ISMAUT+ allows users to indirectly describe A_F through use of the following four types of rules:

1. Minimum and maximum allowable raw scores, for a given attribute, that a package must satisfy. For example:

$$R_i(Q) \geq 60,\ R_i(Q) \leq 90 \text{ for all } Q \in A_F.$$

2. Programs that are specifically required or excluded. For example:

$$\{i,j,k\} \notin Q,\ \{m,n\} \in Q \text{ for all } Q \in A_F.$$

3. Inclusive and exclusive OR rules. For example:

• One and only one program from the set {i,j,k,l} must be a member of Q for all $Q \in A_F$

(Exclusive OR)

• At least one program from the set {m,n,u} must be a member of Q for all Q $\in A_F$

(Inclusive OR)

4. AND Rules. For example:

{l,k,j} $\in$ Q for all Q $\in A_F$. (note: this trivially reduces to rule class 2 - it is only included for accommodation of different behavioral decision making styles).

Given the above allowable rules and package value generation process, ISMAUT+ proceeds as follows:

i. Loop over a restricted subset of the power set of P, where the restriction is based on the rules in rule classes 2 and 4. Note that the maximum cardinality of the set is $2^{\#P - \#E}$, where #E is the number of programs specified by rules 2 and 4. Therefore, each package generated satisfies rule classes 2 and 4. Next,

- For each package generated:

—Verify that the package satisfies rule classes 1 and 3. If any rule is violated, generate next package (return to step i above). Note that rule class 1 requires that the raw value scores be calculated for the generated package.

—If the package does not violate any rules, then:

- Determine if the package dominates or is dominated by any package currently in the nondominated list of feasible packages. Throughout ISMAUT+ we use the simple unweighted vector dominance criterion; i.e., alternative α dominates alternative α' if and only if $v(\alpha) \geq v(\alpha')$ and $v(\alpha) \neq v(\alpha')$.
- Remove any packages from the list that are dominated by the current package.
- If the current package is dominated by a package on the list, do not add it to the nondominated feasible list; otherwise add the package to the list.
- If the list of nondominated feasible packages is equal to the maximum allowed, then stop.

- Generate the next package. Go to step i.

ii. When ISMAUT+ is finished with the generation of feasible, nondominated packages, two tasks are performed:

- The raw attribute scores are linearly scaled between 0 and 1 over all feasible, nondominated packages, where "more is better" is assumed unless the user has stated otherwise. The simple linear procedure described above is used to perform the scaling.

- A data file is written in a form directly readable by ISMAUT.

Example. We now present a simple example of the use of ISMAUT+ in conjunction with ISMAUT to help identify good packages of programs. The process of using these tools is easily described in terms of the following sequence of steps.

Step 1. Define attribute/objective hierarchy. This process is similar to that involved in the use of ISMAUT or any MAUT based technique that accommodates a hierarchy of attributes. For this simple example, we have selected to use the portion of a general C^3I hierarchy for mid-term DOD needs shown in Figure 6.

The specific objectives corresponding to the lowest level attributes in the above hierarchy are as follows:

O1. Provide automated C^2 for SHORAD and airspace management.

O2. Improve interoperable, secure, jam-resistant communications.

O3. Improve intra-base C^3 connectivity for air base survival support.

O4. Increase survivability of radar systems

O5. Integrate intelligence and air surveillance for improved electronic warfare, tracking, and target/threat classification.

The objectives measures corresponding to the above attributes and objectives are as follows:

I1. Level of automated C^2 capability for SHORAD and airspace management.

I2. Level of interoperable, secure, jam-resistant communication capability.

I3. Level of intra-base C^3 connectivity for air base survival support.

I4. Level of survivability of radar systems

I5. Degree of integration between intelligence and air surveillance.

In practice, we would wish to refine these objectives measures to the point where they could be directly measurable in physical terms. For example, the measure for the third objective might be some combination of communications

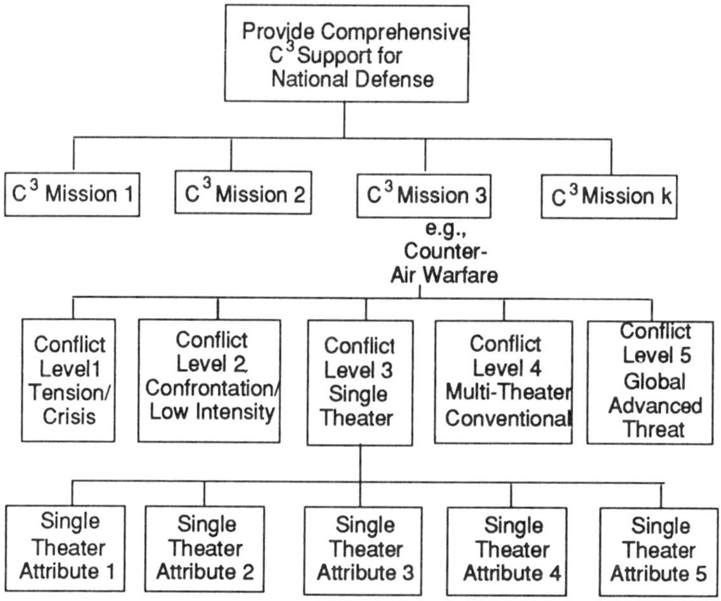

Figure 6. Mid-term C3 attribute/objective hierarchy.

bandwidth available among the installations at the airbase and the graph theoretical connectivity of the underlying network topology. For our simple example, we have avoided many of the complexities of defining these objectives measures in order to concentrate on the use of ISMAUT and ISMAUT+. We will assume that raw scores associated with these measures are assessed subjectively in the form of percentages of maximum attainable levels.

Step 2. Identify programs that may contribute to achieving the objectives. As the first step in creating packages that effectively use the resources available to satisfy the objectives of Step 1, relevant programs are identified. Assume the following programs have been proposed for meeting the counter-air warfare, single theater conflict, mid-term objectives presented above.

P1. Acquire new survivable computer facilities for SHORAD and airspace management.

P2. Acquire hardware for a new interoperable target/threat classification system.

P3. Acquire new radar defense systems.

P4. Acquire new radar with increased hardness.

P5. Acquire new interoperable, secure, jam-resistant communications systems.

P6. Acquire fiber optic communications links for intra-base C^3 connectivity.

P7. Acquire new computer hardware and software for integrating intelligence and air surveillance information.

Step 3. Assess individual raw scores for programs. As mentioned above, we will assume that these raw scores are assessed subjectively on a scale of 0 to 100, roughly indicating percent achievment of each lowest level objective. In this case, each program contibutes to the achievment of at most three objectives. As a preliminary activity, we develop the interaction matrix in Figure 7 that shows which programs contribute to which objectives.

Using this interaction matrix, the programs can be assessed with respect to each of the relevant attributes. In this step we assess the programs individually, ignoring any potential synergistic effects. In the notation of the conceptual presentation above, we assess the $r_i(j)$ terms for each attribute i and program j. These raw scores are shown in Figure 8.

Note that we have added a sixth attribute, cost, represented by the symbol I6. This attribute has the distinctive feature of being the only attribute in this problem for which less is better. It is also the only attribute for which measurable and meaningful raw scores are provided, the numbers entered below being in units of millions of dollars.

Step 4. Assess synergy scores. We now determine the impacts of the synergistic effects among the programs; i.e., we assess the $s_i(j,k)$ terms described earlier. For this hypothetical example, we use the synergy terms shown in Figure 9. Entries are in the form (i) #, where the i in parentheses is the attribute for which the synergy value of # occurs between the two programs indicated by the row and column numbers. For example, in column 4, row 3, the entry (1) -20 indi-

	01	02	03	04	05
P1	X				X
P2	X				X
P3				X	
P4				X	
P5	X	X	X		
P6	X	X	X		
P7	X	X			X

Figure 7. Program/Objective Interaction Matrix

cates a negative interaction between programs 3 and 4 on attribute 1 with magnitude 20.

Step 5. Specify rules for package generation. For this example, we specify the following rules that must be satisfied by all feasible packages Q:

Rule 1 - Budget constraint: $R_6(Q) \leq$ \$225 million

Rule 2 - Minimum coverage of I1: $R_1(Q) \geq 15$

Rule 3 - Require P4: $P4 \in Q$

	11	12	13	14	15	16
P1	75				5	20
P2	5				25	25
P3				65		40
P4				35		60
P5	5	80	30			55
P6	10	15	70			70
P7	5	5			70	35

Figure 8. Raw scores of programs.

P1						(1) 25
	P2					(1) 20 (5) 30
		P3	(4) -20			
			P4		(4) 20	
				P5	(4) 10	
					P6	
						P7

Figure 9. Synergistic effects among programs.

Rule 4 - Require either P2 or P3: $\{P2, P3\} \cap Q \neq \emptyset$

Rule 5 - Require P1 and P6: $\{P1, P6\} \subseteq Q$

These rules are quite restrictive in this problem, but they illustrate most of the possible rule types and they limit the set of feasible packages to a manageable size.

Step 6. Generate feasible packages. Using the ISMAUT+ software, the set of feasible packages for this set of programs and rules can be generated. The feasible packages are shown in Figure 10.

Step 7. Calculate package value scores. Using the linear scaling method discussed above, the ISMAUT+ software determines the value scores for each alternative package on each attribute. The results of this activity are shown in Figure 11.

Step 8. Generate data file for ISMAUT. The ISMAUT+ software generates an appropriately formated file for use by ISMAUT.

Step 9. Run ISMAUT on the set of packages generated. To determine most preferred packages from among those generated by ISMAUT+, use ISMAUT as described previously.

Issues and Extensions

We conclude the paper with a discussion of some of the key issues in defense communications decision support using ISMAUT and the other techniques presented here. In many cases these issues call for extensions of the work reported

Package	Constituent Programs
A1	P1, P3, P4, P6, P7
A2	P1, P2, P4, P6, P7
A3	P1, P2, P3, P4, P6
A4	P1, P3, P4, P6
A5	P1, P2, P4, P6

Figure 10. Feasible Packages.

here, extensions that may be topics for future research.

There are some difficulties with the ISMAUT+ approach to package generation. The most obvious of these is the fact that, unless the generation rules are sufficiently restrictive, there still will be an extremely large set of feasible packages. The ability of the ISMAUT methodology to handle the resulting enormous numbers of alternatives is quite limited, due to the computational demands of solving numerous linear programs. Another promising approach to the packaging problem involves use of probabilistic search methods.

Probabilistic search has recently been the subject of renewed interest in the artificial intelligence and operations research communities due to some notable successes in a variety of computationally complex problem domains. Two probabilistic search algorithms that have received much recent attention are the *genetic algorithm* and *simulated annealing*.

	11	12	13	14	15	16
A1	.55	1.0	.50	1.0	.56	0.0
A2	1.0	1.0	.50	0.0	1.0	.30
A3	.09	0.0	.50	1.0	.20	.20
A4	0.0	0.0	.50	1.0	0.0	.07
A5	.09	0.0	.50	0.0	.20	1.0

Figure 11. Package/Alternative Value Scores.

The basic notion underlying the genetic algorithm is that the process through which biological systems adapt to their environment by co-adapting sets of alleles may be captured in algorithmic form for the optimization of complex, nonlinear systems. Holland [13] is credited as the originator of the mathematical theory underlying the genetic algorithm.

Simulated annealing [14], [15] is also based on an analogy to a physical process, in this case the process of annealing in materials fabrication. The model relies heavily on the use of a temperature coefficient that is slowly reduced during the optimization process, in analogy to the actual physical temperature reduction that takes place in annealing processes.

More recently still, significant success has been achieved using combinations of simulated annealing and genetic algorithms [16]. We have done some preliminary work on the packaging problem using multiobjective versions of simulated annealing and genetic algorithms, with encouraging results.

ISMAUT+ is not the only place in which intractable problems may be encountered; ISMAUT suffers from similar problems under a variety of circumstances. As one example, if the number of trade-off weight constraints and/or directly specified alternative comparisons becomes large; e.g., greater than about 50, then ISMAUT is slowed considerably by the size of the linear programs that must be run.

Computational difficulties also result in ISMAUT when the number of attributes and/or alternatives becomes large. The problem is particularly acute for the common case in which there is a hierarchy of attributes specified as a regular tree with a branching factor of 4 to 7 at each level. We have found it not uncommon that users should wish to use such tree structured hierarchies of depth 4 to 6, which can easily result in hundreds or thousands of lowest level attributes.

A related problem occurs frequently when users can specify the definitions of their desired hierarchy layers, but do not have a specific order as to which layer should appear at what depth in the hierarchy. In many cases we have observed the need for flexibility in dynamically rearranging the hierarchy by exchanging levels.

Some of the tractability problems can be reduced by incorporating additional intelligence into the ISMAUT model. We continue to investigate such possibilities. Various approximation schemes can also be useful in reducing the computational burdens associated with the linear programs.

One key aspect of the ISMAUT methodology that has to date been given little attention relates to various assessment issues. Assessment of utilities is a difficult matter in scalar situations and these difficulties are compounded in the case of multiobjective problems. The further flexibility allowed by imprecise specification of value scores and trade-off weights results in a still more complex assessment problem in the case of ISMAUT. Our experience to date indicates that these assessment issues can be effectively addressed, but they have so far received inadequate attention. Note that these difficulties also interact with the problems caused by having numerous attributes. Assessment must be accomplished on each lowest level attribute, so if there are hundreds of attributes, this may be a prohibitively time-consuming task.

A related assessment complication involves the use of attribute hierarchies. Although it appears that assessment in a hierarchy may potentially provide an easier and more intuitive approach in many cases, the methodology for such assessment is currently undeveloped.

A final issue that deserves mention concerns the fact that we have not yet given adequate attention to the problems associated with introducing ISMAUT and related techniques into existing planning and decision making processes. There invariably is a more or less effective process in existence prior to the introduction of ISMAUT and special care must be given to minimize the disruptions caused by the transition to new technology. At the same time, it appears that ISMAUT and its extensions offer substantial promise for aiding planning and decision making in situations where available information is limited or only imprecisely known. In our experience, the flexibility provided by ISMAUT in contrast to other MAUT based techniques yields substantial improvements in behavioral acceptability.

References

[1] Keeney, Ralph L., and Howard Raiffa. *Decisions with Multiple Objectives: Preferences and Value Tradeoffs.* New York: John Wiley, 1976.

[2] White, Chelsea C., III, Andrew P. Sage, and Shigeru Dozono. "A model of multiattribute decision making and trade-off weight determination under uncertainty. " *IEEE Trans. on Sys., Man, and Cyber.* SMC-14 (Mar/Apr 1984) 223-229.

[3] Scherer, William T., Bradley S. Stewart, Edward A. Sykes, and Chelsea C. White, III. "A new interpretation of alternative pairwise comparisons for a generalization of SMART." *IEEE Trans. on Sys., Man, and Cyber.* SMC-17 (Jul/Aug 1987) 666-670.

[4] Sage, Andrew P., and Chelsea C. White, III. "ARIADNE: A knowledge-based interactive system for planning and decision support." *IEEE Trans. on Sys., Man, and Cyber.* SMC-14 (Jan/Feb 1984) 35-47.

[5] Scherer, William T., Bradley S. Stewart, Edward A. Sykes, and Chelsea C. White, III. *The ISMAUT User's Guide.* Version 1.01, Mar 1986, Charlottesville, VA: Systems Associates.

[6] Scherer, William T., Bradley S. Stewart, Edward A. Sykes, and Chelsea C. White, III. *The ISMAUT Maintenance Manual.* Version 1.01, Mar 1986, Charlottesville, VA: Systems Associates.

[7] Scherer, William T., Bradley S. Stewart, Edward A. Sykes, and Chelsea C. White, III. *The ISMAUT+ User's Guide.* Version 1.01, Dec 1987, Charlottesville, VA: Intelligent Systems Corporation.

[8] Scherer, William T., Bradley S. Stewart, Edward A. Sykes, and Chelsea C. White, III. *The ISMAUT+ Maintenance Manual.* Version 1.01, Dec 1987, Charlottesville, VA: Intelligent Systems Corporation.

[9] Sykes, Edward A. "Multiobjective Intelligent Computer-Aided Design." Ph.D. Diss., Dept. of Sys. Engr., Univ. of Virginia, Charlottesville, VA, Jan 1987.

[10] White, Chelsea C., III, Shigeru Dozono, and William T. Scherer. "An interactive procedure for aiding multiattribute alternative selection." *Omega* 11 (1983) 212-214.

[11] White, Chelsea C., III, and Andrew P. Sage. "A multiple objective optimization-based

approach to choicemaking." *IEEE Trans. on Sys., Man, and Cyber.* SMC-10 (Jun 1980) 315-326.

[12] White, Chelsea C., III, Andrew P. Sage, and William T. Scherer. "Decision support with partially identified parameters." *Large Scale Sys.* 3 (1982) 177-189.

[13] Holland, J., *Adaptation in Natural and Artificial Systems.* Ann Arbor, MI: Univ. of Michigan Pr., 1975.

[14] Bonomi, Ernesto, and Jean-Luc Lutton. "The N-city travelling salesman problem: Statistical mechanics and the metropolis algorithm." *SIAM Rev. 26* (Oct 1984) 551-568.

[15] Vannimenus, J., and M. Mezard. "On the statistical mechanics of optimization problems of the travelling salesman type." *Le Jour. de Physique-Lettres* 45 (1984) L1145-L1153.

[16] Huntley, Christopher L., and Donald E. Brown. "A parallel heuristic for quadratic assignment problems." Tech. Rpt. IPC-TR-89-003, Inst. for Parallel Computation, School of Engr. and App. Sci., Univ. of Virginia, Charlottesville, VA, 16 March 1989, 27 pgs.

Evaluating Decision Support Systems: Toward Integrating Evaluation Methods into the Development Process

Leonard Adelman

Informal evaluations are an inherent part of developing decision support systems (DSSs) because judgment and uncertainty are an inherent part of the DSS development process. The problem and task definitions for which DSS software are developed are typically elusive and unstructured because the strategic and tactical decision making settings within which sponsors of the DSS development efforts typically find themselves are comprised of information input and consequence-of-action uncertainty. The decision makers for whom the DSS is being built are typically unable to accurately describe the wide range of decision processes they use to solve a decision problem, or exactly the type of support they need, thereby adding more judgment and uncertainty to the development process. And the developers themselves are constantly making judgments and decisions about the type of DSS that is "best" to build in order to meet their understanding of the sponsors and decision makers' needs in terms of the system's requirements, interface, analytical methods, software language, hardware configuration, etc., while being constrained by the limiting factors of cost, time, and available personnel. As a result, formal DSS evaluations have to be capable of rigorously addressing the different judgments and decisions and, hence, information needs of different members of the sponsoring and development teams. This requires the application of different evaluation methods, appropriately matched to the information and decision needs of different persons, throughout the DSS development process.

This chapter identifies the applicability of different classes of evaluation methods to different stages in the DSS design and development process. Evaluation is the control mechanism that keeps the development process on track. It is important to realize that the decision to build a DSS is really nothing more than a hypothesis that the DSS will improve the designated decision maker's and, in turn, the organization's decision making and subsequent performance. The evaluator's job from the outset of the development process is to systematically gather, filter, and aggregate data about the DSS in order to test this hypothesis. The application of formal evaluation methods helps members of the sponsoring and development teams monitor the perceived utility of the DSS under development and take corrective action to increase the probability of its use and effectiveness.

Overview

Figure 1 presents Andriole's [1989] "nine step prototyping design blueprint" for developing DSSs. Requirements analysis is the first step, and it has been referred to as "systems targeting" by Andriole. Its goal is to profile the user, the

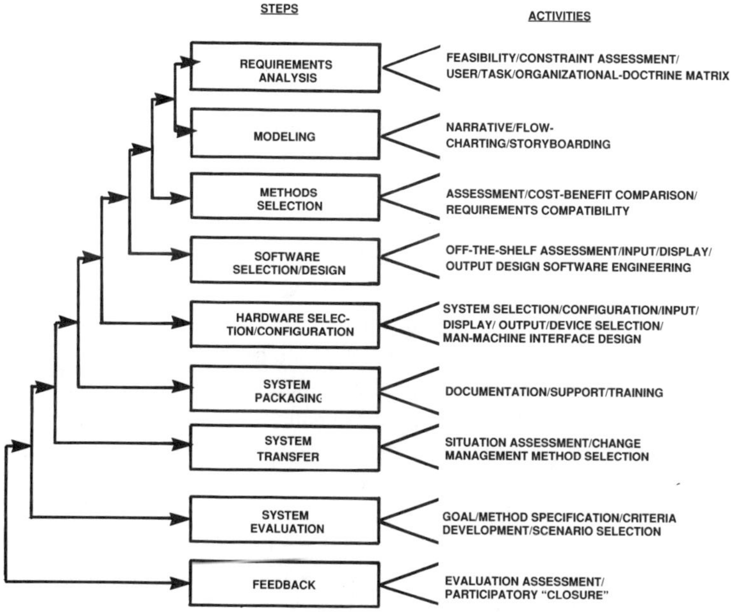

Figure 1. Andriole's Nine-Step Prototyping Design Blueprint.

tasks they perform and the organizational context within which the user and the DSS will operate. The second step is functional modeling. Although there are numerous ways to model a DSS, the goal of each is to show decision makers exactly what the DSS will be capable of doing, and how it will be done. The third step is the selection of analytical methods for the DSS's model-based management system. Analytical methods need to be developed consistent with the requirements analysis and functional models. The fourth and fifth steps are software and hardware selection and the subsequent design and development of the DSS. The sixth and seventh steps are system packaging for and transfer to the host organization. The eighth and ninth steps are system evaluation and feedback, respectively. Consistent with a prototyping perspective, Figure 1 emphasizes iteration between the steps in the DSS design and development blueprint. The use of formal evaluation methods at different steps produces the data required for effective iteration.

Adelman and Donnell [1986] presented a three phased approach for selecting classes of evaluation methods. This evaluation approach is composed of a subjective phase for obtaining users' opinions regarding the DSS' strengths and weaknesses; a technical evaluation phase for "looking inside the black box"; and an empirical evaluation phase for assessing the DSS' impact on performance. Specifically, the subjective evaluation phase focuses on evaluating the DSS from the perspective of potential users. The goal of the subjective evaluation phase is to assess whether the users generally like the DSS; what they consider to be its relative strengths and weaknesses; and what changes they would suggest for improvement. The technical phase evaluates the DSS from both an algorithmic (internal) perspective and a systemic input/output (external) perspective. The goal of the technical evaluation is to identify system problems (e.g., regarding the selected analytical method, the projected costs, the system architecture, the quality of the programming, etc.), and to suggest software modifications to correct these deficiencies. The empirical evaluation phase, in contrast, focuses on obtaining objective measures of the DSS' performance. The goal of this phase is to assess, for example, whether persons make significantly better or faster decisions or use significantly more information working with rather than without the DSS and to identify mechanisms for improving performance.

Evaluation methods can be classified according to the subjective, technical, and empirical phases of this three-phase evaluation approach. Moreover, methods for all three phases are applicable during steps 8 and 9 (evaluation and feedback, respectively) of the development process. In addition, however, specific methods within each classification are more or less applicable to earlier steps in the DSS development cycle. For example, subjective evaluation methods tend to be most applicable early in the cycle (steps 1 and 2) because they represent an explicit means for defining the judgments of members of the sponsoring team and potential users of the DSS. They are, however, also applicable anywhere in the development process where members of the sponsoring and/or development team need to evaluate one option against another. For example, as Andriole [1989] and Hopple [1989] point-out, subjective evaluation methods can be used to systematically address the value and technical judgments inherent in evaluating the utility of available off-the-shelf software and various hardware configurations.

Technical evaluation methods, particularly software testing/verification methods, are most applicable in software design and development (step 4). However, software test and verification methods also have considerable applicability (a) prior to programming code, in terms of verifying both requirements analysis documentation and functional models of the software (steps 1 and 2), and (b) once the development process is well underway during hardware configuration, system packaging, and system transfer (steps 6 and 7). Moreover, there are technical evaluation methods in addition to those for test and verification. In particular, this chapter will also review technical evaluation methods for assessing the adequacy of analytical methods for the model-based management system and for developing estimates of software development costs.

In contrast to technical evaluation procedures, which concentrate on how well the DSS software was developed, empirical evaluation methods focus on how well decision makers can perform the task with (versus without) the DSS. Empirical evaluation methods are most applicable during software design and hardware configuration, and at later steps in the process including after the DSS has actually been developed, packaged, and transferred to the test organization. From an iterative, prototyping perspective, it is anticipated that experiments will be conducted throughout software development (step 4) and hardware configuration (step 5) as a means of objectively measuring the performance of the DSS and testing hypotheses for improvement. During this iterative process, experiments also can be used—although they apparently seldom are—to evaluate system documentation developed in step 6 of the process. Prior to software development, experimental designs can be used in evaluating alternative storyboards for the DSS. After the transfer of the DSS to the test organization, experimental and quasi-experimental designs can be used to evaluate performance. Finally, it is important to note that there are empirical evaluation methods other than experiments. In particular, simulations and statistical analyses of historical data represent two important empirical evaluation methods that are applicable during requirements analysis to assess the potential utility of the DSS.

Subjective Evaluation Methods

The goal of subjective evaluations is to assess the DSS from the perspective of potential users and members of the sponsoring team. This is accomplished by identifying the measures of effectiveness (MOEs) that will provide the information required by members of the sponsoring team in order to assess the utility of the DSS. Identification of explicit MOEs is important at the beginning of the development process because they represent (a) reference points for the development team to use when developing the DSS, and (b) criteria for the evaluators to assess whether the development process is on track. Gaschnig et al. [1983, p. 258] have clearly addressed this point within the context of evaluating expert systems." It is important for system designers to be clear about the nature of their motivations for building an expert system. The long-range goals must also be outlined explicitly. *Thus stage 1 of a system's development, the initial design, should be accompanied by explicit statements of what the measures of the program's success will be and how failure or success will be evaluated.* (italics theirs) It is not uncommon for system designers to ignore this issue at the outset. If the evaluation stages and long-range goals are explicitly stated, however, they will necessarily have an impact on the early design of the expert system."

Riedel and Pitz [1986, p. 986], as well as others [e.g., Adelman and Donnell, 1986; Andriole, 1989; Hopple, 1989; Ulvila et al. 1987], have pointed out that "Multiattribute Utility Technology (MAUT) provides a formal structure for conceptualizing MOE's, a mechanism for both decomposing the global MOE into its component dimensions and for reintegrating them to yield one summary measure of value." When applying MAUT to the evaluation of DSSs, the DSS is conceptually decomposed into attributes that can be defined well enough so that one can obtain either subjective or objective measures (MOEs) of how well the DSS performs on each attribute. This decomposition typically proceeds through the creation of a value hierarchy, such that the global attribute entitled "the overall utility (or value) of the DSS" is decomposed into major categories of attributes, which are further decomposed, and so forth, until one is reasonably confident that one can define and obtain precise, reliable, and valid measures (or scores) of the DSS on each attribute. Table 1 presents the value hierarchy developed by Adelman and Donnell [1986].

Reintegration typically occurs within MAUT through the application of utility functions and relative importance weights. A DSS is typically evaluated on many different attributes, all of which should be defined as precisely as possible. The natural measurement scale for an attribute depends on the nature of the attribute. For example, the scale for an attribute could be in objective units (e.g., minutes for time) or subjective units (e.g., answers on a questionnaire) depending on the nature of the attribute. A common scale is required to compare scores on one attribute with scores on another, that is, "apples with oranges," and, by so doing, obtain an overall score for the item (e.g., DSS) being evaluated by the decision maker. A utility scale, which conceptually measures psychological value or satisfaction, meets this requirement. Utility (or value) functions are used to translate the DSS's performance on an attribute into a utility score on that attribute. Then, relative importance weights (or other forms of decision rules) are used to assess the relative value of a utility scores and, thereby, obtain an overall utility score for the DSS.

0.0 OVERALL UTILITY

- 1.0 Aid/User Interface
  - 1.1 Match with Personnel
    - 1.1.1 Training & Technical Background
    - 1.1.2 Work Style, Workload & Interest
    - 1.1.3 Operational Needs
  - 1.2 Aid Characteristics
    - 1.2.1 General
      - 1.2.1.1 Ease of Use
      - 1.2.1.2 Understanding Aiding Process
      - 1.2.1.3 Ease of Training
      - 1.2.1.4 Response Time
    - 1.2.2 Specific
      - 1.2.2.1 User Interface
      - 1.2.2.2 Completeness of Data Files
      - 1.2.2.3 Accuracy of Expert Judgements
      - 1.2.2.4 Ability to Modify Judgements
      - 1.2.2.5 Understanding of Aid's Algorithms
      - 1.2.2.6 Utility of Graphs
      - 1.2.2.7 Utility of Print-outs
      - 1.2.2.8 Understanding of Text

- 2.0 User/Aid Organization
  - 2.1 Efficiency Factors
    - 2.1.1 Acceptability of Time for:
      - 2.1.1.1 Task Accomplishment
      - 2.1.1.2 Data Management
      - 2.1.1.3 Set-Up Requirements
    - 2.1.2 Perceived Reliability Under Average Battle Conditions
      - 2.1.2.1 Skill availability
      - 2.1.2.2. Hardware Availability
  - 2.2 Match with Organizational Factors
    - 2.2.1 Effect on Organizational Procedure and Structure
    - 2.2.2 Effect on Other People's Position in the Organization
    - 2.2.3 Effect on Information Flow
    - 2.2.4 Side Effects
      - 2.2.4.1 Value in Performing Other Tasks
      - 2.2.4.2 Training Value

- 3.0 Organization/Environment
  - 3.1 Decision Accuracy
  - 3.2 Match Between Aid's Technical Approach and Problem's Requirements
  - 3.3 Decision Process Quality
    - 3.3.1 Quality of Framework for Incorporating Judgement
    - 3.3.2 Range of Alternatives
    - 3.3.3 Range of Objectives
    - 3.3.4 Weighting of Consequences
    - 3.3.5 Assessment of Consequences of Alternatives
    - 3.3.6 Re-examination of decision making process
    - 3.3.7 Use of Information
    - 3.3.8 Consideration of Implementation and Contingency Plans
    - 3.3.9 Effect on Group Discussions
    - 3.3.10 Effect on Decision Makers' Confidence

Table 1. Adelman and Donnell's MAUT DSS Evaluation Hierarchy.

There are other subjective evaluation methods besides MAUT. For example, Adelman [in press] also discusses traditional cost-benefit analysis, the dollar-equivalent technique, decision analysis, and a MAUT-based cost-benefit analysis. It is important to emphasize that in all cases these methods use personal judgments. This initially might disturb and be difficult for members of both the sponsoring and developing teams, for it emphasizes the subjective process decision makers go through when evaluating DSSs. To quote Riedel and Pitz [1986, pp. 987-988], "There is no way to avoid the fact that the overall MOE must be based on such judgments, or the fact that no mechanical procedure can replace this subjective assessment..." Through the explicit identification of MOEs and procedures for converting performance scores into a global MOE, subjective evaluation methods provide reference points for the development team to use when developing the DSS and criteria for the evaluator to monitor in order to assess whether the development process is on track.

Technical Evaluation Methods

Three classes of technical evaluation methods are overviewed in this chapter: (a) methods for selecting analytical methods for the model-based management system (MBMS), (b) methods for estimating software development costs, and (c) formal software testing/verification methods. The latter category is the one typically thought of when one asks the question, "How well was the software built?" Obviously, the answer can affect the results of the empirical and subjective evaluations, for a poorly programmed DSS will affect how well users can perform desired tasks with the DSS and, in turn, what they think of it. However, the selected analytical method(s) and the estimated software development costs also can affect the empirical and subjective evaluation phases. DSS performance and preference is particularly sensitive to the analytical methods because failure to select the best method(s) means the software will not effectively support the user's decision making process. Similarly, given the highly competitive nature under which many DSS development contracts are issued, inaccurate estimation of software costs increases the probability that there will not be enough money for a high quality technical product and, in turn, the probability of poor performance in the empirical and subjective evaluations.

The third step in Andriole's [1988] DSS development approach is the selection of analytical methods that are as consistent as possible with the requirements definitions for and functional model of the DSS. There is a wide range of analytical modeling methods. Andriole, for example, has classified them into four categories: decision analysis, operations research, management science, and computer science. Moreover, each of these categories contains many methods. Decision analysis includes utility/value modeling methods, probability modeling methods, and mixed (value-probability) modeling methods. Operations research includes descriptive and inferential statistical methods and modeling and optimization techniques. Management science includes program management methods. Computer science includes "conventional" algorithmic methods and artificial intelligence.

The evaluation of analytical methods for the DSS is a critical and often difficult decision for the design team for three reasons. First, analytical methods differ in their appropriateness for different stages in the decision making process. Second, there are tradeoffs inherent in using one method over another even when the methods are appropriate for a given stage in the process. Third, developers are seldom experts in every analytical method, let alone in matching methods to requirements definitions. Research has begun to develop taxonomies for matching analytical methods to requirements. These can be used in step 8 to look inside the "black box" and evaluate the method embodied in the transferred DSS prototype. However, they also can be used early in the DSS development process when the consequences of an error in method selection are less costly. Table 2 presents the "requirements/methods matrix" developed by Andriole [1989].

The dependent variable for software cost estimation models is some quantification of "effort." Effort is typically expressed as the number of person-months required to develop a system in order to generate a metric that is not subject to organizational variations in salary structures. [An actual dollar cost can, of course, be generated by multiplying the effort estimate by the developer's aver-

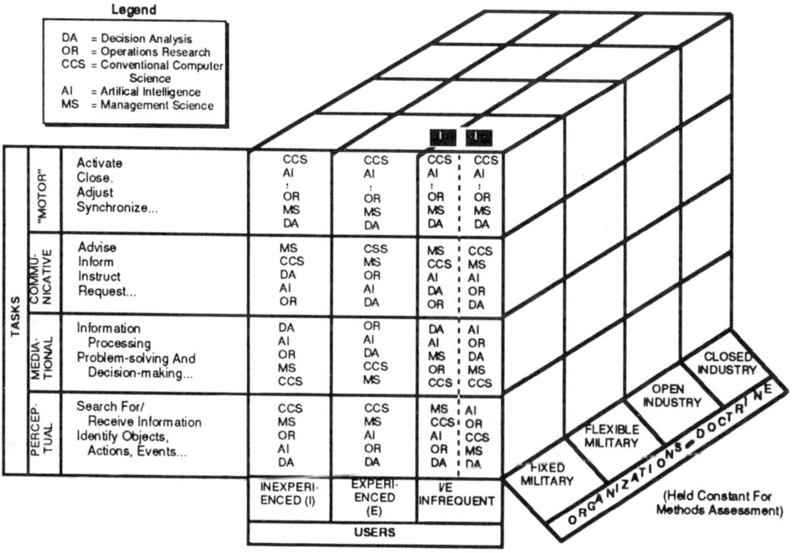

Table 2. Andriole's Requirements/Methods Matrix.

age "fully loaded" cost (i.e., salary, fringe benefits, overhead, etc.) or by estimating the percentage involvement, and, thus, person-months for different personnel categories, and then using the average category costs to estimate the cumulative costs.] Models differ, however, in both the cost factors used as the independent variables to predict effort and the presumed relationship between these factors and effort.

Pfleeger [1989] classifies software cost estimation models into four categories: experiential, static, dynamic, and hybrids. Experiential models rely on the "expert judgment" of key people who have the necessary experience, background, and business sense to make the cost estimation. Experiential models can take a number of different forms, and a number of different procedures exist for obtaining their estimates, such as asking experts for their global judgments, using the Delphi technique (e.g., see Fairley [1985], or using Albrecht and Gaffney's [1983] function point model. In contrast, static cost estimation models [e.g., Bailey and Basili, 1981; Halstead, 1977] typically use regression analysis to generate equations relating cost factors to effort. The idea is that one can develop a "best fit" equation to predict effort (the dependent variable) by using a data base of previous projects for data on cost-relevant factors (the independent variables). Dynamic estimation models incorporate time as a parameter of the equation. For example, Putnam [1977] used the Rayleigh equation and curve to estimate the number of full-time personnel required at different times in a development project. Putnam's model is significantly different from static models because it emphasizes the trade-off between development time and total development effort. Finally, there are hybrids that combine aspects of two categories. For example, Boehm's [1981] well-known Constructive Cost Model (COCOMO) uses a static, nonlinear equation for predicting effort. But, the coefficient and exponent are based on expert judgment, not regression analysis. The interested reader is referred to Conte et al. [1986] and Kemerer [1987] for empirical studies comparing the predictive accuracy of different cost estimation models.

Verification testing should be systematically performed for DSSs, just like any other software products. Fagan and Miller [as reported in DeMillo et al. 1987] have identified four phases for software testing. The first phase is manual analysis in which the requirements specification, the design and implementation plan, and the program itself is analyzed by experienced software engineers for problems. The second phase is static analysis—which may be manual or automated—in which requirements and design documents and software are analyzed, but without code execution. The third phase is dynamic analysis in which software is executed with a set of test data. The fourth phase, which Fagan and Miller consider to be optional, is attempting to prove the program as being correct. According to DeMillo et al. [1987], most testing techniques focus on either static or dynamic analysis. Detailed discussions of these and other methods can be found in, for example, DeMillo et al. [1987], Fairley [1985], Pressman [1982], and Rushby [1988].

Four points are important about software testing methods. First, as Hamlet [1988, p. 666] points out, each software testing method has strengths and weaknesses and represent "imperfect test methods." As with methods for analytical method selection and software cost estimation, evaluators need to use multiple methods to obtain accurate feedback. Second, the intent of testing is to find errors. As Fairley [1985, p. 268] points out, "... one has most confidence in programs with no detected bugs after thorough testing and least confidence in a program with a long history of fixes." Third, the best way to minimize the number of errors in a system, and the amount of time, effort, and money required to fix them, is to eliminate errors early in the development process. Consequently, as Gelperin and Hetzel [1988] point out, software development life cycles are becoming "preventive" through the application of technical software testing methods early in the development process. And, fourth, dynamic testing methods rely heavily on empirical analysis via test data. However, the reader should maintain a clear distinction between the empirical results of software testing methods and those of empirical evaluation methods. The former concern how well the DSS software was developed; the latter how well decision makers can perform the task with and without the DSS.

Empirical Evaluation Methods

Empirical evaluation methods can be classified into four categories: experiments, quasi-experiments, simulations, and statistical analyses of historical data. Experiments are by far the most common and commonly thought of empirical evaluation method. Moreover, they are, particularly appropriate when a number of people would actually use the developed DSS, for experiments are designed to help generalize from a test sample to the larger population.

When one thinks of an experiment, one typically thinks of a factorial design where (a) one or more factors are systematically varied as the independent variable(s), and (b) the dependent variable(s) are quantitative, objective measures of system performance. One of the independent variables is almost always "the degree of decision support." This independent variable typically has two levels. The first level is an "unaided condition;" it is a control group that is supposed to represent the degree of support the decision maker has without the DSS. The second level is an "aided" condition, which is the support level provided by the DSS. Although the statistical tests are designed to test the null hypothesis that there is no difference between performance for the two levels, there is often the implicit, if not explicit hypothesis that the aided condition will lead to significantly better performance, otherwise the DSS would not have been built. Consequently, evaluators must take steps to remove potential bias from the experimental test.

For example, effort should be directed toward accurately representing the unaided condition. This can be facilitated by pretesting the condition with representative users or technical personnel. Second, the aided condition should be pretested too. Pretesting the aided condition can help assess the adequacy of training and whether minor changes should be made to the DSS to prepare it for a formal evaluation. Third, effort should be given to developing two (or more) problem scenarios. Although the DSS may be a prototype with many limitations, it still has to be capable of handling at least some of the complexities of the real world task. Fourth, participants should be randomly assigned to experimental conditions to ensure a true experiment. Fifth, when domain experts are used to evaluate solution quality, they must be blind as to whether the solution was produced with or without the DSS. As Gaschnig et al. [1983] point out, the initial evaluations of MYCIN, an early (and now famous) expert system for medical diagnosis, were biased against the system because the participating physicians knew that they were examining the output of a computer program.

Consistent with an iterative, prototyping development approach, the representativeness of the experimental setting and the level of the DSS's performance requirements should advance throughout the development cycle. Although the latter iteration is routinely acknowledged, the former is not. It must be remembered that most information and decision technology fails to be successfully implemented for organizational, not technical reasons [Adelman, 1982; Huber, 1986]. Consequently, increasing the fidelity of the organizational and environmental interfaces between the DSS and its users is essential in generalizing the laboratory performance results to the real world.

Ideally, field experimentation would be used to assess whether the DSS significantly improved performance over the pre-DSS (i.e., unaided) organizational setting. For example, organizational units (e.g., divisions/sections in a large company or governmental agency) would be randomly assigned to the "with DSS" and "without DSS" conditions, and their performance measured over a long enough time period for stabilization (i.e., reach an asymptote). If possible, a "placebo" condition should be included. Organizational units in this condition would be given some "treatment" that was not hypothesized to have any effect on performance. This condition is analogous to giving patients sugar pills when evaluating new drugs, and is oriented to controlling for any positive effects in the "with DSS" condition that is the result of being given special treatment. It is important to note that the unit of analysis is the performance of the organizational unit; consequently, a large enough sample of units would be required for performing statistical tests.

The sample size and randomization requirements of true experiments is typically not possible in many organizations. Quasi-experimental designs should be used in such situations. To quote Campbell and Stanley [1966, p. 34], "There are many social settings in which the research person can introduce something like experimental design into his scheduling of data collection procedures (e.g., the *when* and *to whom* of measurement), even though he lacks the full control over the scheduling of experimental stimuli (the *when* and *to whom* of exposure and the ability to randomize exposures) which make a true experiment possible. Collectively, such situations can be regarded as quasi-experimental designs." [itallics theirs]

There are a number of different types of quasi-experimental designs. Campbell and Stanley [1966], for example, identify 14 types. These include: (a) time series designs, where the organizational unit is measured for a long period before and after receiving the DSS; (b) multiple time series designs that do not use randomization, but do use a control group that does not receive the DSS; and (c) non equivalent (and non randomized) control group designs that rely on analysis of covariance to assess whether the pretest and post test difference for the DSS group is significantly better than that of the control group.

The last two types of empirical evaluation methods considered here are simulations and statistical analyses of historical data. Both represent an alternative to subjective evaluation methods prior to or during step 1 of the DSS development process. As an illustration of the former, consider the simulation done by Leverich et al. [1987] of the logistics network for moving ammunition to and around Western Europe. As Shank [1989] points out, the various nodes in this network represent decision points for considering the potential value of DSSs to, for example, better assess the ammunition needs of other nodes or how to better move and/or store different types of ammunition. That is, the simulation provides an analytic (versus subjective) means for testing the potential effect of alternative DSSs over a range of conditions representing the uncertainty in the environment. Moreover, by providing an overall systemic prediction of a DSS's effect, it helps eliminate the possibility of building DSSs that generate only localized (versus systemic) improvements.

The statistical analysis of historical data also represents an empirical evaluation method for use in requirements analysis. In addition, it can be used in certain circumstances to help select analytical methods for the DSS. For example, Girill et al. [1988] have shown how monitoring historical usage statistics for the National Magnetic Fusion Energy Computer's document-delivery software "...has not only prompted specific refinements to existing programs, but has also yielded general design constraints for the next-generation system and for automated data-reduction tools as well."

Summary

Table 3 indicates the relative applicability of each of the three classes of evaluation methods to each of the nine steps in the DSS development process. The ratings were created for steps 1 through 7 by (a) indicating which class of methods had the highest applicability for that step, and then (b) scoring the applicability of the remaining two methods relative to it. As can be seen, all three classes of evaluation methods are applicable to different degrees throughout the

Table 3

The Relative Applicability of Evaluation Methods to Different Steps in The DSS Development Process

Step in Development Process	Subjective Methods	Technical Methods	Empirical Methods
1) Requirements Analysis	High	Medium	Medium
2) Modeling	Medium	High	Medium
3) Methods Selection	Low	High	Low
4) Software Selection/ Design/Development	Medium	High	High
5) Hardware Selection/ Configuration	Medium	High	Medium
6) System Packaging	Low	High	Medium
7) System Transfer	Low	High	Low
8) System Evaluation	High	High	High
9) Feedback	High	High	High

development process, as well as being highly applicable in steps 8 and 9.

It is important to note that, although the assessment presented in Table 3 is based on the evaluation literature and my experience, it still represents only my judgment. To say it differently, it is not based on the results of objective data collected during empirical evaluations of the DSS development process for a large number of projects. Consequently, others may disagree with various aspects of the matching based on their experience. The goal is not to present a definitive assessment but, rather, to demonstrate the potential applicability of different evaluation methods to different steps in the DSS development process and, thereby, facilitate the use of evaluation methods throughout the DSS design and development process. As with all scientific endeavors, such evaluation activities will, in time, lead to cumulative empirical data to replace the subjective data in Table 3.

Evaluation represents the control mechanism in the DSS design and development process. It provides the feedback required to keep the development effort on track, for the application of formal evaluation methods helps members of the sponsoring and development teams monitor the perceived utility of the DSS under development and take corrective action to increase the probability of its use and effectiveness. For this reason, effective evaluations are an essential ingredient to the successful application of a requirements driven, prototyping approach to DSS development. Although some may argue that increasing the emphasis on evaluation activities will increase software development costs, research reviewed herein suggests that it will actually decrease costs because of the cost savings achieved by correcting errors earlier in the development process. More importantly, it will increase the proportion of DSSs that are used effectively.

Acknowledgements

This paper is based on research performed under a subcontract to George Mason University from Science Applications International Corporation (SAIC) under a contract from the Fort. Leavenworth Field Office of the U.S. Army Research Institute for the Behavioral and Social Sciences (ARI). The author would like to thank Dr. Stan Halpin of ARI and Mr. Paul McKeown of SAIC for their support. The views and conclusions expressed in this paper are those of the author and not those of the above persons or organizations.

References

1. Adelman, L. "Involving Users in the Design of Decision-Analytic Aids: The Principal Factor in Successful Implementation," *Journal of the Operational Research Society*, vol. 33, 1982, pp. 333-342.
2. Adelman, L. *Evaluating Decision Support Systems*, QED Information Sciences, in press.
3. Adelman, L., and Donnell, M. L. "Evaluating Decision Support Systems: A General Framework and Case Study," in *Microcomputer Decision Support Systems: Design, Implementation, and Evaluation*, S. J. Andriole (ed.), QED Information Sciences, 1986.
4. Albrecht, A. J., and Gaffney, J. E. "Software Function, Source Lines of Code, and Development Effort Prediction," *IEEE Transactions on Software Engineering*, vol. SE-9, 1983, pp. 639-648.
5. Andriole, S. J. *Handbook for the Design, Development, Evaluation, and Application of Interactive Military Decision Support Systems*, Petrocelli Books, Inc., 1989.
6. Bailey, J. W., and Basili, V. R. "A Meta-Model for Software Development Resource Expenditures," *Proceedings of the 5th International Conference on Software Engineering*, 1981, 107-116.
7. Boehm, B. *Software Engineering Economics*, Prentice-Hall, 1981.
8. Campbell, D. T., and Stanley, J.C. *Experimental and Quasi-Experimental Designs for Research*, Rand McNally, 1966.
9. Conte, S.D., Dunsmore, H. E., and Shen, V. Y. *Software Engineering Metrics and Models*, Benjamin/Cummings, 1986.
10. DeMillo, R. A., McCracken, W. M., Martin, R. J., and Passafiume, J. F. *Software Testing and Evaluation*, The Benjamin/Cummings Publishing Co., 1987.
11. Fairley, R. E. *Software Engineering Concepts, McGraw-Hill, 1985.*
12. Gaschnig, J., Klahr, P., Pople, H., Shortliffe, E., and Terry, A. "Evaluation of Expert Systems: Issues and Case Studies," in *Building Expert Systems*, F. Hayes-Roth, D. A. Waterman, and D.

B. Lenat (eds.), Addison-Wesley, 1983.

13. Gelperin, D. and Hetzel, B. "The Growth of Software Testing," *Communications of the ACM*, vol. 31, 1988. pp. 687-695.
14. Girill, T. R., Luk, C. H., and Norton, S. "The Impact of Usage Monitoring on the Evolution of an Online-Documentation System: A Case Study," *IEEE Transactions on Systems, Man, and Cybernetics*, vol. SMC-18, 1988, pp. 326-332.
15. Halstead, M. *Elements of Software Science*, Elsevier, 1977.
16. Hamlet, R. "Special Section on Software Testing," *Communications of the ACM*, vol. 31, 1988, pp. 662-667.
17. Hopple, G. W. *The State-of-the-Art in Decision Support Systems*, QED Information Sciences, 1989.
18. Huber, G. P. "The Decision-Making Paradigm of Organizational Design," *Management Science*, vol. 32, 1986, pp. 572-589.
19. Kemerer, C. F. "An Empirical Validation Software Cost Estimation Models," *Communications of the ACM*, vol. 30, 1987, pp. 416-429.
20. Leverich, B., Bankes, S., Berman, M., and Miller, L. *The Wartime Theater Ammunition Distribution System: A Preliminary Analysis*. (Working Draft.) Santa Monica, CA: RAND Corporation, 1987.
21. Pfleeger, S. L. *An Effort Estimation Model for Object-Oriented Development*. (Ph.D. Dissertation.) Fairfax, VA: School of Information Technology and Engineering, George Mason University, 1989.
22. Pressman, R. S. *Software Engineering: A Practitioner's Approach*. McGraw-Hill, 1982.
23. Putnam, L. H. "A General Empirical Solution to the Macro Software Sizing and Estimating Problem," *IEEE Transactions on Software Engineering*, vol. SE-4, 1978, pp. 345-361.
24. Riedel, S. L. and Pitz, G. F. "Utilization-Oriented Evaluation of Decision Support Systems," *IEEE Transactions on Systems, Man, and Cybernetics*, vol. SMC-16, 1986, pp. 980-996.
25. Rushby, J. *Quality Measures and Assurance for AI Software*. (NASA Contractor Report 4187) Washington, DC: National Aeronautics and Space Administration (Code NTT-4), 1988.
26. Shank, J. *Pre-Development Evaluation of Decision Aids in Large Complex Systems*. (Ph.D. Dissertation Proposal.) Fairfax, VA: School of Information Technology and Engineering, George Mason University, 1989.
27. Ulvila, J. W., Lehner, P. E., Bresnick, T. A., Chinnis, J.O., and Gumula, J. D. E. *Testing and Evaluating C^3I Systems That Employ Artificial Intelligence*. Reston, VA: Decision Sciences Consortium, Inc., 1987.

Advanced Information Technology for Next Generation Decision Support

Stephen J. Andriole

Introduction

The design, development and use of decision support systems (DSSs) will change dramatically as we approach the twenty-first century. Our expectations about what DSSs should do are rising as rapidly as the requisite technology is evolving. By the year 2000 problem-solvers will use DSSs to deal with all sorts of simple and complex problems. They will also benefit from systems capable of providing much more than database support and low-level inference making. Just as importantly, the distribution of DSS power will be expanded beyond specialized professionals; decision support systems will be available to us all, on and off the job.

This article examines the new trends, and attempts to describe how DSSs will be designed, developed and used a decade or so from now. It is thus speculative, though anchored in trends definable today.

The article attempts to address the following questions:

- How will definitions of decision support evolve over time?
- How will emerging methods, models and technologies affect the design, development and use of next generation decision support systems?; and
- What role will future decision support systems play in the larger information systems world?

The Range of Decision Support

It is safe to say that most DSSs today support decision-making only indirectly. There are DSSs that manage projects, provide easy access to operational

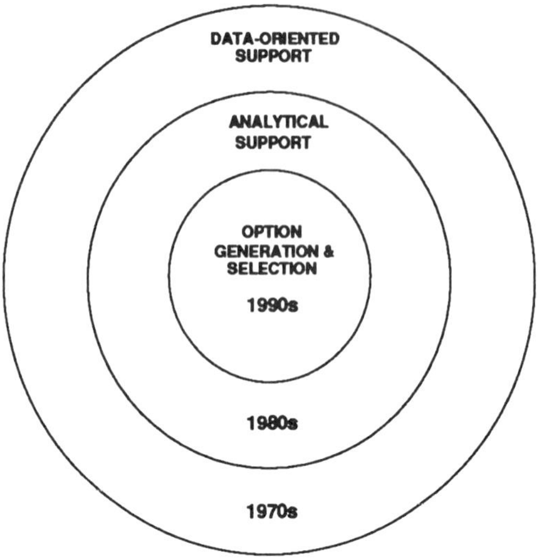

Figure 1. The Range of Decision Support.

data, and otherwise deal with relatively structured problems. This distinction between "structured" and "unstructured" targets of opportunity is important to understanding the range of today's DSSs and the promise of tomorrow's. Early proponents of DSS's technology hoped that their systems would help decision makers generate, compare, and implement decision options, but most DSSs only indirectly support these activities. Real-time option generation and evaluation has evaded designers—except in some rare instances where a special (single) purpose system was developed to address a very well-bounded problem (Andriole, 1989).

Figure 1 suggests the range of decision support today and where it might be in the future. Note that early DSSs were data oriented. The field migrated slowly toward model oriented applications, and over the next five to ten years should move toward systems that synthesize data, models, knowledge, and interfaces into some creative wholes.

Figure 1 also suggests that future decision support systems applications will grow considerably. Next generation systems will be capable of addressing operational, tactical and (some) strategic structured and unstructured problems via the application of data, knowledge, and models that exploit their integration.

Figure 2 suggests where the action will be. The technology in applied analytical methodology, user-computer interface (UCI) techniques, and display technology, will grow considerably over time as the nature and depth of application domains expands.

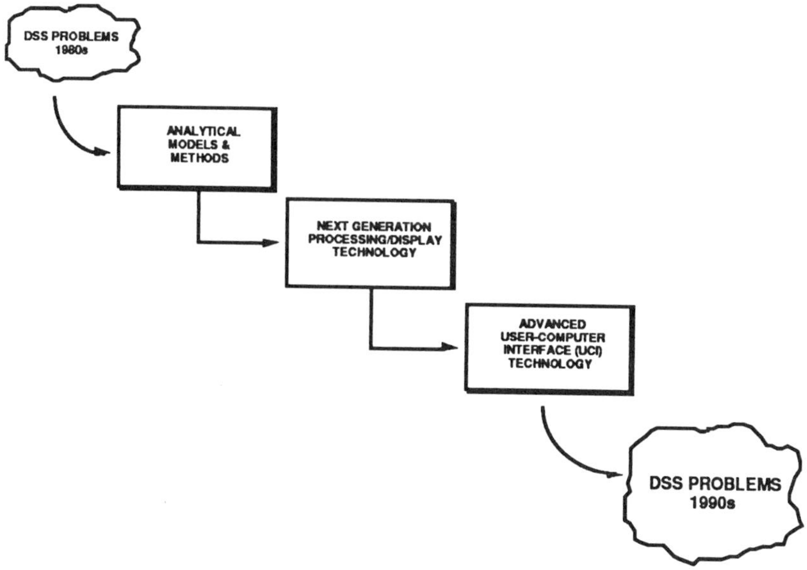

Figure 2. Research and Development Opportunities.

The new applications perspective on decision support will be extremely broad, reflecting the capabilities of new systems that will be embedded (in larger information systems) and functional on many levels. Future decision support systems will improve the efficiency of decision makers and information managers, resource allocators and administrators, strategic planners and inventory controllers.

The broad perspective will be permitted by the new technology (see below) that will emerge over the next several years, and a centralization in the corporate and governmental workplace, a centralization that will not be "ideological" but rather driven by the same technology that will permit the design and development of more powerful DSSs. In all likelihood, the movement toward technological networking and system integration will translate into new imperatives for the management of information and decision making. In short, offices (and homes) of the future will transform themselves because new technology will permit the change and, significantly, because it will demand it.

What will be driving what? Will new corporate and governmental requirements suggest new DSS requirements, or will new DSSs suggest new requirements? Would next generation DSSs look the way we expect them to look if they were conceived in an applications vacuum, or is the interpretation and anticipation of applications driving the "form" that future DSSs will take?

The questions are relevant to all parts of the information and manufacturing economy. In the automobile industry, for example, robots are solving specific requirements because robotic technology has finally evolved into a cost-effective alternative to human labor. Word processing is now cheap and efficient. But has computer-based word processing changed, eliminated, or reduced requirements? It is interesting that many social theorists have argued that the development of the office copier has dramatically changed the way information is produced and distributed. Some argue that copiers have triggered a paper explosion that has over-satisfied the need for information. In other words, copiers—just because they exist—now satisfy far more requirements than can actually be identified. Will DSSs suffer a similar "success"?

While definitions of decision support will grow, so too will our understanding of computer-based problem solving. Decision support, while very broad in concept and application, will nevertheless be subsumed under the general rubric of computer-based problem solving which, over time, will also experience radical change. Expectations about what computers can do for users will continue to rise. Consistent with the evolution in expectations about the power of DSSs, computer-based problem solving systems of all kinds must satisfy analytical requirements.

Advanced Information Technologies

There are a variety of tools, methods, techniques, devices and architectures available to the DSS designer; many more will emerge as we move toward the 21st century. The challenge—as always—lies in the extent to which designers can match the right tool or method with the appropriate problem. This section looks at a number of technology options now available to the designer, options that will evolve quite dramatically during the next five to ten years.

Models and Methods

Figure 3 suggests the range of methods and models available to the DSS designer today (Hopple, 1986). The taxonomy is by no means complete, though it is representative of the way methods, tools and techniques can be categorized and assessed. Figures 4, 5, and 6 from Sage and Rouse (1986) show how several of the major methods classes can be described and assessed. "Assessment," of course, is the key. DSS designers must know precisely what method (or methods) to apply to which decision support requirements. Figure 7 from Andriole (1989) indicates how methods can be rank-ordered against a set of requirements.

Over the past few years the DSS community has seen the preeminence of knowledge-based tools and techniques, though the range of problems to which heuristic solutions apply is much narrower than first assumed. It is now generally recognized that artificial intelligence (AI) can provide knowledge-based

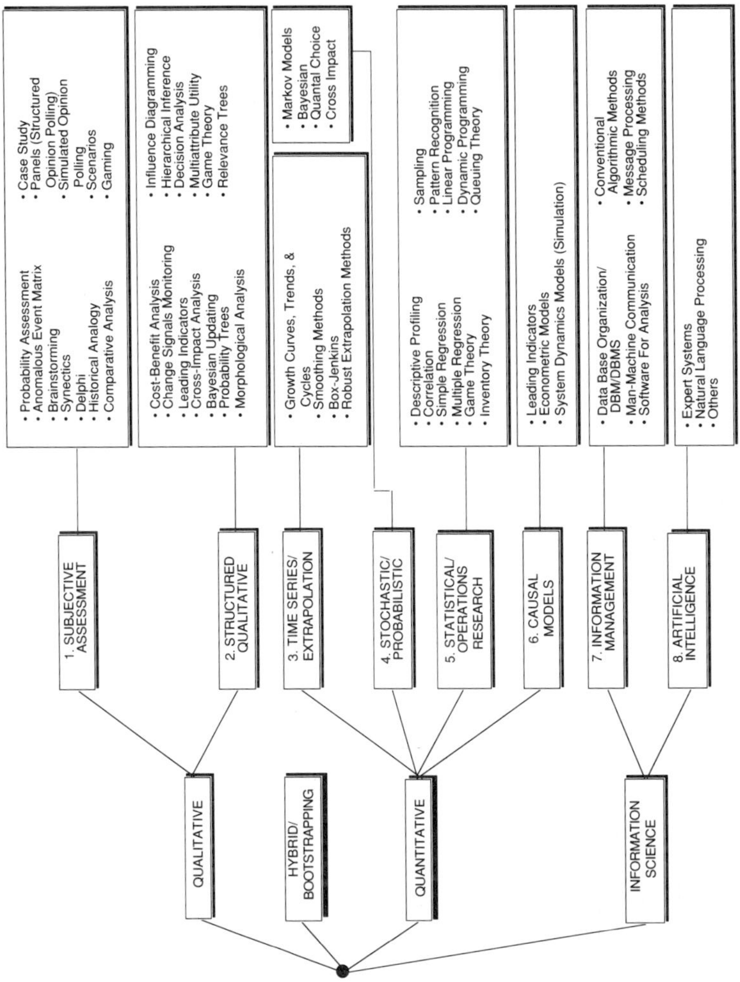

Figure 3. A Taxonomy of Methods and Models (Hopple, 1986).

CRITERIA \ METHOD	Artificial Intelligence	Cognitive Science	Decision Analysis	Operations Research and Control Theory
Objectives	Software Performance	Explanation Of Cognition	Aiding Decisions	
Methods	Symbolic; Powerful Software Tools	Symbolic; Computation-Oriented	Numerical; Axiomatic Dec. Rules	Models
Products	Software and Tools	Theories; Training and Aiding Princ.	Decision Prescriptions; Software/Tools	Optimal Solutions
Strength	Exploitation of Context; Software Tools	Emphasis on Process Over Product of Behavior	Well-Founded and Structured	Breadth and Rigor
Weaknesses	Often Ad Hoc; Avoids Engineering Methodology	Often Ad Hoc; Avoids Psychological Methodology	Biased Toward Choice; Not Judgement in Context	Avoids Context; Assumption Laden

Figure 4. Methods/Models Descriptions (Sage and Rouse, 1986).

CRITERIA / METHOD	AI & Expert Systems	Decision Analysis & Decision Support	Human/System Interaction
Objectives/ Expectations	Design of "Intelligent" Systems (KBS) Human-like Capacities Heuristics	Normative Modeling, Aids to Option Generation, choice & selection	Effective Interface - (Adaptive, intelligent?) Between Humans and Decision Tools - Transparent Interface
Typical Analytical Concern or Question to Subject	"Models" of Human Processes, Search & Representation, Symbolic Processing	Decomposition, structure - elicitation of Subjective Judgements, Evaluation of Cost/Benefit MAU, Bayes Policy Capturing	Prototyping; Performance Modeling
Products	LISP Machines, Shells	Process Aids (Computer, Other)	I/O Devices, Configurations
Strengths: Input		Structure	
Process	Explicit Model, Ad Hoc	Normative Models, Face Validity of Models	
Output	Explanation Facility	Audit Trail Sensitivity Analysis	
Weaknesses: Input	Primitive	Often Data Intensive	
Process	Comprehensibility, May Use Unrealistic Belief Structures	Normative Models	
Output	Comprehensibility	Sensitivity Analysis	

Figure 5. Multi-Criteria Methods/Models Assessment (Sage and Rouse, 1986).

METHOD / CRITERIA	OR & Control Engineering	Data Base Management	Cognitive Science/ Psychology
Objectives/ Expectations	Provide Results of Quantitative Analysis to Clients	Facilitate Storage, Retrieval, Manipulation of Data (Information)	Understanding & Describing Human Cognition
Typical Analytical Concern or Question to Subject	Linear Programs, Math Modeling, Optimization, Dynamic Systems Analysis, Statistics	Relational, Hierarchical, Spatial Structuring, Retrieval Strategies	Modeling Theories Lens Model Attribution Theory Empirical/Experimental
Products	System Model "Autopilot," etc. (Control Systems) Aids Modeling/Simulation Languages	DBMS Software, Database Designs	Human Models/Decision Heuristics (Validated?) Capability and/or Limitation Assessments
Strengths: Input	Can Use Massive Real Data		
Process	Rigorous; Engineering Methods		Experimental/Empirical
Output	Quantification		
Weaknesses: Input	Requires Volumes of Data	Inflexible	
Process	"Artificial" Lack of Face Validity (GIGO)		Mind Reading, Laboratory Bound
Output	Explanation Facility		Narrow Domains

Figure 6. Multi-Criteria Methods/Models Assessment (Sage and Rouse, 1986).

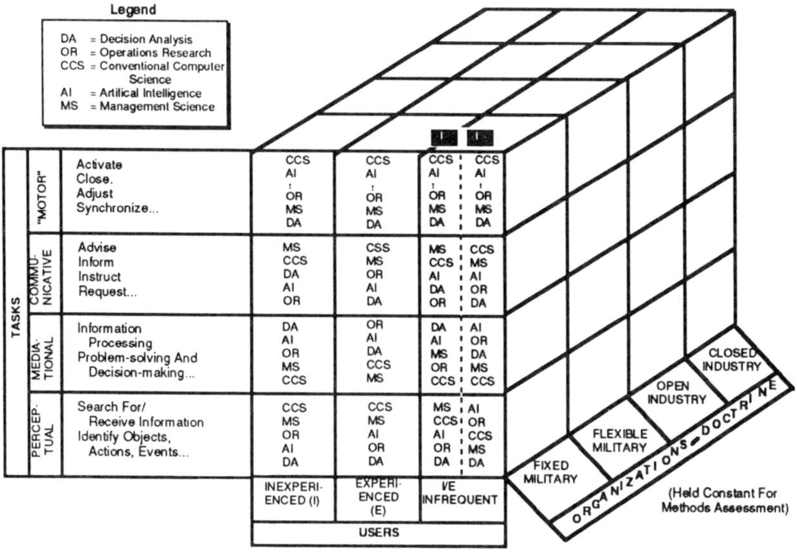

Figure 7. Some Models/Methods Rankings.

support to well-bounded problems where deductive inference is required (Andriole, 1990). We now know that AI performs less impressively in situations with characteristics (expressed in software as stimuli) that are unpredictable. Unpredictable stimuli prevent designers from identifying sets of responses, and therefore limit the applicability of "if - then" solutions. We now know, for example, that so-called expert systems can solve low-level diagnostic problems, but cannot predict Soviet intentions toward Poland in 1995. While there were many who felt from the outset that such problems were beyond the applied potential of AI, there were just as many sanguine about the possibility of complex inductive problem solving.

The latest methodology to attract attention is neural network-based models of inference making and problem solving. As seen in Figure 8, neural networks are applicable to problems with characteristics that are quite different from those best suited to AI. Neural networks are—according to Hecht-Nielsen (1988)—"computing systems made up of a number of simple, highly interconnected processing elements which process information by their dynamic state response to external inputs." Neural nets are non-sequential, non-deterministic processing systems with no separate memory arrays. Neural networks, as stated by Hecht-Nielsen, comprise many simple processors that take a weighted sum of all inputs. Neural nets do not execute a series of instructions, but rather respond to sensed inputs. "Knowledge" is stored in connections of processing elements

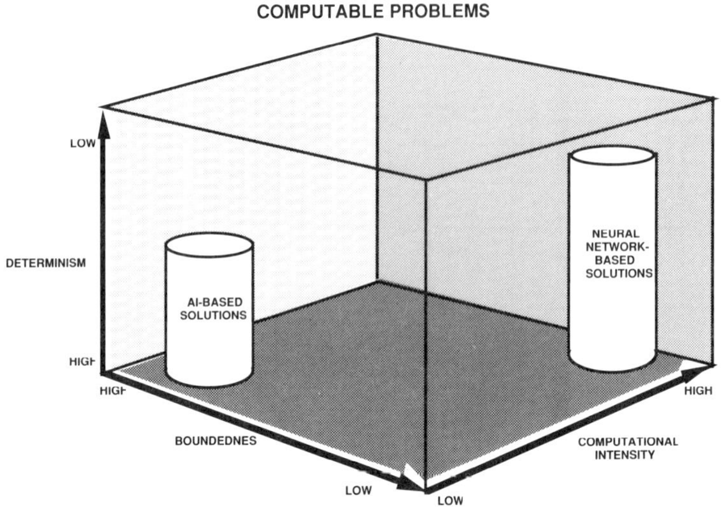

Figure 8. The Applicability of Artificial Intelligence and Neural Network-Based Models and Methods.

and in the importance (or weight) of each input to the processing elements. Neural networks are allegedly non-deterministic, non-algorithmic, adaptive, self-organizing, naturally parallel, and naturally fault tolerant. They are expected to be powerful additions to the DSS methodology arsenal, especially for data-rich, computationally intensive problems. The "intelligence" in conventional expert systems is pre-programmed from human expertise, while neural networks receive their "intelligence" via training. Expert systems can respond to finite sets of event stimuli (with finite sets of responses), while neural networks are expected to adapt to infinite sets of stimuli (with infinite sets of responses). It is alleged that conventional expert systems can never learn, while neural networks "learn" via processing. Proponents of neural network research and development have identified the kinds of problems to which their technology is best suited: computationally intensive; non-deterministic; non-linear; abductive; intuitive; real-time; unstructured/imprecise; and non-numeric (DARPA/MIT, 1988).

It remains to be seen if neural networks constitute the problem solving panacea that many believe they represent. The jury is still out on many aspects of the technology. But like AI, it is likely that neural nets will make a measured contribution to our inventory of DSS models and methods.

What does the future hold? Where will the methodological leverage lie?

In spite of the over selling of AI, the field still holds great promise for the design and development of DSSs. Natural language processing systems—systems that permit free-form English interaction—will enhance decision support efficiency and contribute to the wide distribution of DSSs. The Artificial Intelligence Corporation's INTELLECT natural language processing system, for example, permits users to interact freely with a variety of data base management systems. Cognitive Systems, Inc. BROKER system permits much the same kind of interaction with the Dow Jones' databases. These systems reveal how natural language interfaces will evolve over time, of how users will be able to communicate with databases and knowledge bases in ways that are compatible with the way they address human and paper data, information and knowledge bases. When users are able to type or ask questions of their DSSs in much the same way they converse with human colleagues, then the way DSSs will be used will be changed forever.

Of particular interest is the disproportionate attention that natural language interfaces have received vis-à-vis expert systems. This imbalance will be redressed by the year 2000.

Expert systems will make routine many decision making processes. Rules about investment, management, resource allocation, and office administration will be embedded in expert DSSs. Problems that now have to be re-solved whenever a slight variation appears will be autonomously solved.

Smart data base managers will develop necessary data bases long before decision support problems are identified. DSSs of the 1990s will be capable of adapting from their interaction with specific users. They will be able to anticipate problem solving "style," and the problem solving process most preferred by the user. They will be adaptive in real-time, and capable of responding to changes in the environment, such as a shortage of time.

The kinds of problems that will benefit the most from AI will be well bounded, deductive inference problems about which a great deal of accessible and articulate problem solving expertise exists. The DSS community will abandon its goals of endowing computer programs with true inductive or abductive capabilities in the 1990s, and the dollars saved will be plowed back into so called "low level" AI.

In terms of the range of decision support that we can expect in the 1990s and beyond, AI tools and techniques are much more likely to solve problems in the outer concentric circles of Figure 1 than the inner ones, save situations where option generation and selection occur in well-bounded, deductive inference areas.

DSS designers in the 1990s will also benefit from a growing understanding of how humans make inferences and decisions. The cognitive sciences are amassing evidence about perception, biasing, option generation, and a variety of additional phenomena directly related to DSSs modeling and problem solving. The world of technology will be informed by new findings; resultant DSSs will be "cognitively compatible" with their users.

Next generation DSSs will also respond to the situational and psychophysiological environment. They will alter their behavior if their user is making many mistakes, taking too long to respond to queries, etc. They will slow down or accelerate the pace, depending on this input and behavior. The field of cognitive engineering—which will inform situational and psychophysiological system design strategies—will become increasingly credible as we approach the 21st century. The traditional engineering developmental paradigm will give way to a broader perspective that will define the decision making process more from the vantage point of requirements and users than computer chips and algorithms. Principles of cognitive engineering will also inform the design and human computer interfaces (see below).

Some future decision support software will be generic and some will be problem-specific. Vendors will design and market generic accounting, inventory control, and option selection software. These models will be converted into templates that can be inserted directly into general purpose systems. The template market will grow dramatically over the next five to ten years.

It is extremely important to note the appearance of DSS development tools. Already there are packages that permit the development of rule-based expert DSSs. There are now fourth generation tools that are surprisingly powerful and affordable. These so-called "end-user" systems will permit on-site design and development of DSSs that may only be used for a while by a few people. As the cost of developing such systems falls, more and more throw-away DSSs will be developed. This will change the way we now view the role of decision support in any organization, not unlike the way the notion of rapid application prototyping has changed the way application programs should be developed.

Hybrid models and methods drawn from many disciplines and fields will emerge as preferable to single model-based solutions largely because developers will finally accept diverse requirements specifications. Methods and tools drawn from the social, behavioral, mathematical, managerial, engineering, and computer sciences will be combined into solutions driven by requirements and not by methodological preferences or biases. This prediction is based in large part upon the maturation of the larger design process, which today is far too vulnerable to methodological fads. Hybrid modeling for DSS design and development also presumes the rise of multidisciplinary education and training, which is only now beginning to receive serious attention in academia and industry.

User-Computer Interface (UCI) Technology

Twenty years ago no one paid much attention to user interface technology. This is understandable given the history of computing, but no longer excusable. Since the revolution in microcomputing—and the emerging one in workstation-based computing—software designers have had to devote more attention to the process by which data, information and knowledge are exchanged between the system and its operator. There are now millions of users who have absolutely no sense of how a computer actually works, but rely upon its capabilities for their very professional survival. A community of "third party" software vendors is sensitive to both the size of this market and its relatively new need for unambiguous, self-paced, flexible computing. It is safe to trace the evolution of well-designed human-computer interfaces to some early work in places like the University of Illinois, the Massachusetts Institute of Technology, (in what was then the Architecture Machine Group, now the Media Lab), Xerox's Palo Alto Research Center (Xerox/Parc), and, of course, Apple Computer, Inc. The "desktop" metaphor, icon-based navigational aids, direct manipulation interfaces, and user guided/controlled interactive graphics—among other innovations—can all be traced to these and other organizations.

Where did all these ideas come from? The field of cognitive science and now "cognitive engineering" is now—justifiably—taking credit for the progress in UCI technology, since its proponents were the (only) ones asking why the user-computer interaction process could not be modeled after some validated cognitive information processing processes. UCI models were built and tested, and concepts like "spatial database management" (from MIT's Architecture Machine Group [Bolt, 1984]), hierarchical data storage, and hypertext were developed. It is no accident that much UCI progress can be traced to findings in behavioral psychology and cognitive science; it is indeed amazing that the cross-fertilization took so long.

UCI progress has had a profound impact upon the design, development and use of DSSs. Because many of the newer tools and techniques are now affordable (because computing costs have declined dramatically), it is now possible to satisfy complex UCI requirements even on personal computer-based DSSs. Early data-oriented DSSs displayed rows and rows (and columns and columns) of numbers to users; modern systems now project graphic relationships among data in high resolution color. DSS designers are now capable of satisfying many more substantive and interface requirements because of what we have learned about cognitive information processing and the affordability of modern computing technology.

The most recent progress in UCI technology is multimedia, or the ability to store, display, manipulate and integrate sound, graphics, video and good old fashioned alphanumeric data (Ragland, 1989; Ambron and Hooper, 1988; Aiken, 1989). It is now possible to display to users photographic, textual, numerical, and video data on the same screen, as Figure 9—from Aiken (1989)—shows. It is possible to permit users to select (and de-select) different displays of the same data. It is possible to animate and simulate in real time—and cost-effectively. Many of these capabilities were just too expensive a decade ago and much too computationally intensive for the hardware architectures of the 1970s and early 1980s. Progress has been made in the design and execution of applications software and in the use of storage devices (such as videodisks and compact disks [CDs]). Apple Computer's Hypercard software actually provides drivers for CD players through a common UCI (the now famous "stack"). DSS designers can

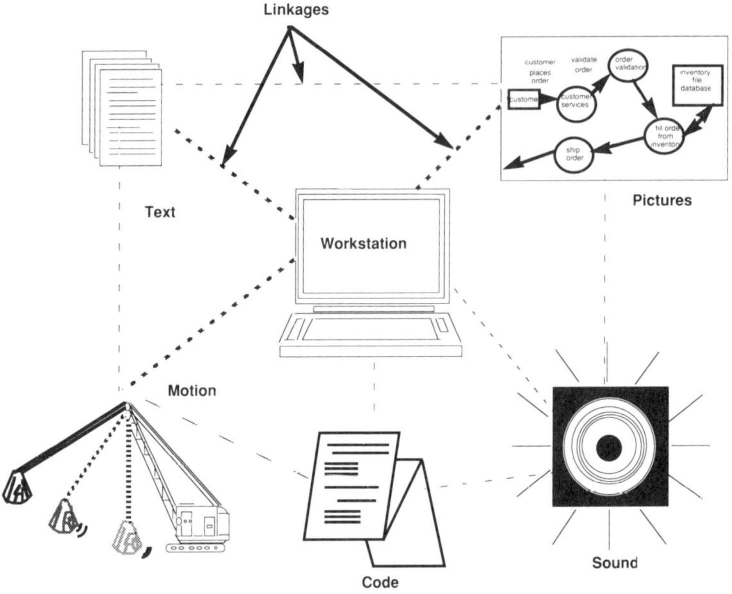

Figure 9. Multimedia Technology (Aiken, 1989).

exploit this progress to fabricate systems that are consistent with the way their users think about problems. There is no question that multimedia technology will affect the way future DSSs are designed and used. The gap between the way humans "see" and structure problems will be narrowed considerably via the application of multimedia technology.

Direct manipulation interfaces (DMIs) such as trackballs, mice and touch-screens have also matured in recent years and show every likelihood of playing important roles in next generation DSS UCI design and development. While there is some growing evidence that use of the mouse can actually degrade human performance in certain situations, there are countless others where the payoff is empirically clear (Ramsey and Atwood, 1979; Ledgard, Singer and Whiteside, 1981; Bice and Lewis, 1989). Touch screens are growing in popularity when keyboard entry is inappropriate and for rapid template-based problem solving (Smith and Mosier, 1984).

The use of graphical displays will dominate future UCI applications. Growing evidence in visual cognition research (Pinker, 1985) suggests how powerful the visual mind is. It is interesting that many problem-solvers—professionals who might otherwise use a DSS—are trained graphically not alphanumerically. Military planners receive map-based training; corporate strategists use graphical trend data to extrapolate and devise graphic scenarios; and a variety of educators use case studies laden with pictures, icons, and graphics of all kinds. Complicated concepts are often easily communicated graphically, and it is possible to convert complex problems from alphanumeric to graphic form. There is no question that future DSSs will exploit hypermedia, multimedia, and interactive graphics of all kinds.

Speech input and output should also emerge over the next five to ten years as a viable UCI technology. While predictions about the arrival of "voice activated text processors" have been optimistic to date, progress toward continuous speech input and output should be steady. Once the technology is perfected there are a number of special purpose applications that will benefit greatly from keyboard- and mouse-less interaction.

The use of advanced UCI technology will foster a wider distribution of DSS technology. Early DSSs were used most productively by those familiar with the method or model driving the system as well as interactive computing itself. In other words, in order to exploit DSS technology one had to have considerable computing expertise. Advanced UCI technology reduces the level of necessary computing expertise. Evidence suggests that training costs on the Apple Macintosh, for example, are lower because of the common user interface. Pull-down and pop-up menus, windows, icons, and direct manipulation via a mouse or trackball are all standard interface equipment regardless of the application program (and vendor). If you know how to use one Macintosh program chances are you can use them all to some extent. Such interface uniformity is unheard of in other than Macintosh-based software systems, yet illustrates the enormous leverage that lies with the creative application of advanced UCI technology.

UCI technology will also permit the use of more methods and models, especially those driven by complex—yet often inexplicable—analytical procedures. For example, the concept of optimization as manifest in a simplex program is difficult to communicate to the typical DSS user. Advanced UCI technology can be used to illustrate the optimization calculus graphically and permit users to understand the relationships among variables in an optimization equation. Similarly, probabilistic forecasting methods and models anchored in Bayes' Theorem of conditional probabilities while computationally quite simple are conceptually convoluted to the average DSS user. Log odds and other graphic charts can be used to illustrate how new evidence impacts prior probabilities. In fact, a creative cognitive engineer might use any number of impact metaphors (like thermometers and graphical weights) to present the impact of new evidence on the likelihood of events.

Finally, advanced UCI technology will also permit the range of decision support to expand. Anytime communications bandwidth between system and user is increased, the range of applied opportunities grows. UCI technology permits designers to attempt more complex system designs due to the natural transparency of complexity that good UCI design fosters.

Some argue that the interface may actually become "the system" for DSS users. The innards of the system—like the innards of the internal combustion engine—will become irrelevant to the operator. The UCI will orchestrate process, organize DSS contents and capabilities, and otherwise shield users from unfriendly interaction with complex data, knowledge, and algorithmic structures.

Hardware

The hardware that supports DSS's technology today is "conventional." There are turnkey systems as well as generic hardware configurations that support the use of DSSs. CPUs, disk drives, keyboards, light pens, touch screens, and the like can be found in a variety of DSSs. There are also microcomputer systems, as well as systems that require larger (minicomputer) hardware configurations.

Next generation DSSs will be smaller and cheaper and therefore more widely distributed. They will be networked and capable of up-loading and down-loading to larger and smaller systems. Input devices will vary from application to application as well as the preferences of the user. As suggested above, voice input will dramatically change the way a small set of DSSs are used in the future; voice activated text processing will expand the capabilities of DSSs by linking decision support to word processing and report preparation in a "natural" unobtrusive way, though it is likely that robust voice activated systems will not appear until the late 1990s.

Many DSSs will have embedded communications links to databases and knowledge bases, other systems on decision support networks and the outside world via conventional communication systems. IBM's acquisition of Rolm suggests that the merger between computing and voice systems is well underway. Future DSSs will have (some selected) voice input capabilities, conventional headset communications, deep data base linkages, and a "place" on a much larger decision support network.

Briefcase- and smaller-sized DSSs will become widespread. The embedding of spreadsheets in popular portable microcomputers suggests that decision support chips will be developed and embedded in future hardware configurations. In fact, not unlike some of the more powerful calculators of the 1970s, future DSSs will permit users to mix-and-match decision support chips within a single decision processor. The chips will be sold independently by decision support vendors.

Future DSSs will also be integrated with video display systems of several genres. There will be video disk-based DSSs as well as packaged systems that integrate powerful computer-generated imagery capabilities. The cost of both video options is falling rapidly, and the DSS consumer of the future will be able to select the one that best serves their needs.

It is safe to say that video will become integral to future decision support. But it will be integrated directly into the system, not an optional, expensive afterthought.

Behavioral scientists have just about convinced system architects—via the amassing of tons of evidence—that information, concepts, and many ideas can be communicated much more effectively via graphic, symbolic, and iconic displays (Smith and Mosier, 1984; Schneiderman, 1987). DSSs that do not have these and related capabilities will fail.

The revolution in high resolution display technology will exert a profound impact upon next generation DSS design and use. Many UCI technologies will exploit high resolution displays thereby accelerating the movement toward graphic computing.

Processor technology is also evolving rapidly. Just a decade ago, most of us computed on Intel 8088 microprocessors, while today everyone is waiting for the 486. Processors like the Motorola 68030 (and next generation 68040) have placed enormous power not only in the hands of users, but—perhaps more importantly—DSS designers as well. It is sad to say that applications software is today lagging the capabilities of such chips; at the same time, even assuming a consistent lag, DSSs in the 1990s and beyond will benefit from applications software that exploits the revolution in microprocessor design.

The issue of power, however, does beg the larger requirements question. In other words, it is safe to assume that raw computing power will be ready for next generation DSS concepts. The challenge—as always—will lie in the application of the power to validated user requirements. If the truth be told, there are many successful DSSs that today require less than 20 percent of available computational power; many future DSSs may well find themselves with abundant power—and nowhere to go! Regardless of available computing power, DSS designers will have to adhere to sound information systems engineering principles well into the 1990s (Andriole, 1990).

We are witnessing the demise of the distinction among mainframe, mini and microcomputers. Tomorrow there will be "workstations." Some will be more powerful than others, but nearly all will be available to individuals at reasonable prices. The balance between capability and price will continue to perplex vendors, since users will demand more and more capabilities for less and less money. Pricing strategies will determine how much power becomes "affordable." Future DSS design and use will work within this changing marketplace and, because of some new usage strategies (see below), will remain largely unaffected by the instability of the workstation marketplace.

Decision Support Systems as Organizational and Executive Information Systems

DSSs will be used very differently in the future than they are today. They may function as clearinghouses for our professional problems. They may prioritize problems for us, and automatically solve some of them. They will become problem solving partners, helping us in much the same way colleagues now do. The notions of DSSs as software or hardware, and users as operators, will give way to a cooperative sense of function which will direct the design, development, and application of the best DSSs.

They will also be deployed at all levels in the organization. Today, decision support is targeted at mid-level management; tomorrow all levels will be supported by powerful interactive, adaptive systems. The distribution of DSSs will permit decision support networking, the sharing of decision support data, and the propagation of decision support problem solving experience (through the development of a computer-based institutional memory of useful decision support "cases" that might be called upon to help structure especially recalcitrant decision problems). Efficient organizations will actually develop an inventory of problem/solution combinations that will be plugged into their decision support networks.

DSSs will also communicate with systems in other organizations in other parts of the world. Falling communications costs will permit global linkages and contact with databases, expert systems, inventories, and the like, thereby multiplying the capabilities of in-house DSSs by orders of magnitude. This global networking is not in the distant future, but only five to ten years away.

The most important change will occur in the way DSSs interface with other information systems. Most contemporary DSSs are "disembodied," that is, distinct from larger corporate, government or military information systems. Actual use of many DSSs involves leaving one system to activate another. It is common in the military application of decision support for users to work alternately with mini and microcomputers, manually feeding the output from one system into the other. A good deal of this can be explained by acquisition and procurement craziness, but just as much can be traced to obsolete concepts of how DSSs should be used. As the range of DSS problems and capabilities increases, fewer and fewer systems will be disembodied; to the contrary, the most successful systems will be embedded in larger organizational and executive information systems.

Future executive information systems will provide "portals" for users to explore. It will be possible to perform all sorts of tasks via myriad application programs (that ideally will have common user-computer interfaces). Figure 10 shows a hierarchy of requirements that today are addressed by a variety of disembodied application programs; Figure 10 also suggests the role that next generation DSSs will play in the larger problem solving process.

Finally, next generation DSSs will bridge the gap between our professional and personal worlds. Because they will have capabilities to manage our professional lives, they will also be capable of managing our personal lives as well. Since future decision support "delivery systems" will be expandable and inexpensive, the integration of personal management modules will be inevitable. This blurring of the traditional lines between our professional and personal

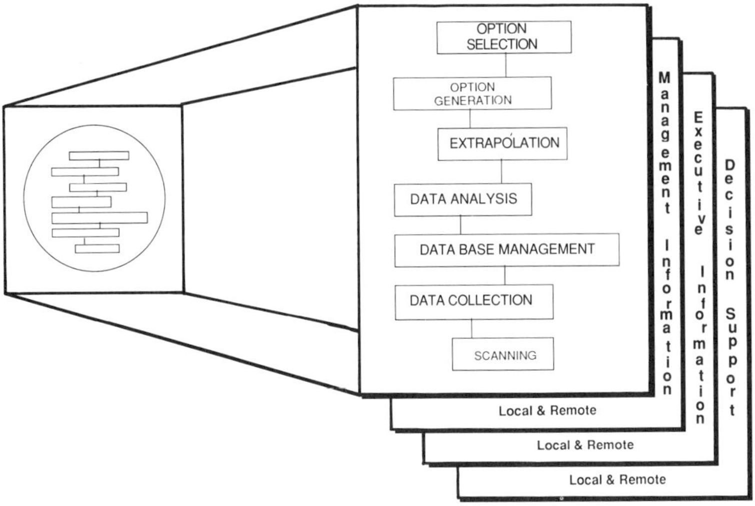

Figure 10. Integrated Problem-Solving System Concept.

worlds may not be desirable, but it is likely. Again, the technology will drive such changes to a significant extent, technology that will trigger changes because of its capabilities not necessarily in response to real requirements. In other words, there is a technological destiny built into our forecasts about the role of future DSSs.

Summary and Conclusions

The net effect is staggering. Decision support in the 1990s will be enormously broad and powerful. It will be distributed and networked. It will be "intelligent" and inexpensive. The effects of this reality are difficult to predict precisely, though a number of the ideas expressed above are inevitable. Have we given enough thought to the direction in which decision support technology is taking us? Have we assessed the desirability of the direction? Have we determined the impact that the next generation DSS will have upon the office of the future? Will they define the office of the future? Or will the office of the future shape decision support? We have avoided these and other tough questions. If we are to manage and exploit the new technology, we must address all of the pertinent questions as soon as possible. When we do, we will be able to inform and direct progress so that the next generation is as capable and responsive to real needs as it can possibly be.

The technology tour presented in this article suggests that DSS designers will leverage emerging technology to fundamentally change the nature of interactive decision support. While the recent past enjoyed data-oriented decision support, next generation systems will provide analytical support of all kinds. Of particular value will be the speed with which routine problems will be solved and the advisory support that DSSs will provide users in areas as complex as option generation, evaluation and selection. Interface technology will play a pivotal role in the introduction of new and hybrid models and methods; the hardware community will make sure that more than enough computational power is available.

The whole concept of "decision support" will evolve to accommodate changes in the larger corporate, governmental and military information systems structure. Networking and advanced communications technology will permit linkages to databases and knowledge bases—and the routines to exercise them. Not only will distinctions among mainframe, mini- and microcomputing fade, but distinctions among management information, executive information, and DSSs will also cloud. Ironically, the concept of centralization may re-appear, not with reference to central computing facilities but with regard to enormous systems conceived functionally as hierarchies of capabilities. Users may well find themselves within huge computing spaces capable of supporting all kinds of problem solving. Advanced communications technology will make all this possible; users will be able to travel within what will feel like the world's largest mainframe, which conceptually is precisely what a global network of data, knowledge, and algorithms is.

The same users will be able to disengage the network and go off-line to solve specific problems. This freedom will expand the realm of analytical computing in much the same way microcomputing expanded the general DSS user community.

Finally, all this technology will permit designers to fulfill user requirements in some new and creative ways. Up until quite recently, technology was incapable of satisfying a variety of user requirements simply because it was too immature or too expensive. We have crossed the capability/cost threshold; now designers can dig into a growing tool bag for just the right methods, models, and interfaces. By the year 2000 this tool bag will have grown considerably. Talented DSS designers should be able to match the right tools with the right requirements to produce systems that are user-oriented and cost-effective.

The future of DSS design, development and use is bright. While some major changes in technology and application concepts are in the wind, next generation DSSs will provide enormous analytical support to their users. We can expect the range of decision support to grow in concert with advances in information technology.

References

Aiken, PH (1989) *A Hypermedia Workstation for Software Engineering.* Fairfax, VA: George Mason University, School of Information Technology and Engineering.

Ambron, R and Hooper, C (1987) *Interactive Multimedia: Visions of Multimedia for Developers, Educators and Information Providers.* Redmond, WA: Microsoft Publishers, Inc.

Andriole, S. J. (1990) *Information System Design Principles for the 90s: Getting it Right!* Fairfax, VA: AFCEA International Press.

__________ (1989) *Decision Support Systems: A Handbook for Design and Development.* Princeton, NJ: Petrocelli Books, Inc.

Bice, K and Lewis, C (1989) *Wings for the Mind: Conference Proceedings: Computer Human Interaction.* Reading, MA: Addison-Wesley Publishing Co.

Bolt, R. A .(1984) *The Human Interface: Where People and Computers Meet.* Belmont, CA: Lifetime Learning Publications.

Defense Advanced Research Projects Agency (DARPA)/Massachusetts Institute of Technology (MIT) (1988) *DARPA Neural Network Study.* Fairfax, VA: AFCEA International Press.

Hopple, G. W. (1986) "Decision Aiding Dangers: The Law of the Hammer and Other Maxims." *IEEE Transactions on Systems, Man and Cybernetics.* SMC-16. 6. November-December.

Ledgard, H, Singer, A, and Whiteside, J. (1981) *Directions in Human Factors for Interactive Systems.* New York, NY: Springer-Verlag.

North, R. L. (1988) "Neurocomputing: Its Impact on the Future of Defense Systems." *Defense Computing.* January-February.

Pinker, S., ed. (1985) *Visual Cognition.* Cambridge, MA: MIT/Bradford Books.

Ragland, C. (1989) "Hypermedia: The Multiple Message." *MacTech Quarterly.* Spring.

Ramsey, H. R. and Atwood, M. E. (1979) *Human Factors in Computer Systems: A Review of the Literature.* Englewood, CO: Science Applications, Inc.

Sage, A. P. and Rouse, W. B. (1986) "Aiding the Decision-Maker Through the Knowledge-Based Sciences." *IEEE Transactions on Systems, Man and Cybernetics.* SMC-16. 4. July/August.

Schneiderman, B. (1987) *Designing the User Interface.* New York, NY: Addison-Wesley.

Smith, S. L. and Mosier, D. (1984) *Design Guidelines for the User Interface to Computer-Based Information Systems.* Bedford, MA: The MITRE Corporation.

PART 5
C^2 Requirements Analysis and Modeling

MECCA: Multi-Faceted Evaluation of Command and Control Assets

Dr. Richard E. Hayes

Problem

Command and control (C^2) assets cannot escape evaluation, nor should they. They are expensive, both in dollars and the demands they make on the key people in military organizations. Expenditures for systems that support command and control must be scrutinized carefully, particularly when budgets are tight.

C^2 systems must compete with weapons, logistics, personnel, and other programs for funds. Credible, valid measures of their performance and impact are necessary if they are to compete effectively and receive an appropriate share of available resources.

Moreover, command and control is a vital component of military effectiveness. The longstanding debate about whether command and control is a force multiplier need not be resolved to see this.

It is adequate to assume that C^2 is a force multiplier, but ranges between 0 and 1.00: that is, perfect command and control allows the unit to function to its full military potential, degraded command and control limits its effectiveness.

Ultimately, command and control assets form a system that includes people, hardware, software, and procedures. However, those systems include a variety of subsystems (management information systems, displays, decision support systems, etc.) that can be identified and evaluated both individually and in the larger context of overall command and control.

Evaluation normally focuses on these subsystems. As a consequence, it is often restricted to rather narrow technical reviews that determine whether the system meets "requirements" rather than broader issues of functional capacity and impact. Military organizations routinely recommend development on the basis of technology push (new aircraft that fly faster, carry more payload, turn more steeply; submarines that dive deeper, run more quietly, go faster; tanks with better armor, more firepower, etc.) They recognize that military professionals will be able to develop strategies and tactics that take advantage of these improved capabilities in a variety of missions and situations.

However, a different standard has been applied to command and control systems. They have been designed for particular environments and threat situations. Unfortunately, all military systems (including C^2 systems) will have to operate under unforeseen circumstances and in the face of threats designed specifically to confound them. When this happens, their inherent capabilities and the existence of reserve capacities permit the military professionals using them to overcome obstacles and accomplish missions. Hence, it is essential that C^2 systems be designed in terms of a comprehensive view of command and control and the range of environments within which they must be capable of operating.

The movement to evolutionary acquisition, rapid prototyping, and the heavy user involvement in the design, development and implementation of C^2 systems is a recognition of this need; and a partial solution. However, it has the negative consequence of making the opinions of current users and the needs of the particular war-fighting environments they foresee the only standards. While essential, these standards are too narrow.

There is a need, therefore, for a comprehensive approach to the evaluation and testing of command and control systems that will encourage development of generic capabilities that apply across a wide range of military situations and provide the data necessary to (a) demonstrate the potential impact of a C^2 system on military performance, (b) support development and refinement during evolutionary acquisition, and (c) document the impact of the system so it can compete effectively for the funding it deserves and requires. MECCA, Multi-faceted Evaluation of Command and Control Assets, is designed to meet this need.

Background: What Exists

While there have been a variety of efforts to develop approaches and tools for evaluation of command and control, most of them have dealt with only a part of the problem.

In 1982 the Defense Communications Agency (DCA) undertook a research and development program to identify the determinants of effective command and control at the theater level and to provide both guidelines for the design of such systems and ways to measure their performance (Hayes, 1983). The resulting approach, the Headquarters Effectiveness Assessment Tool (HEAT) has since been applied primarily as a measurement tool for evaluating specific headquarters (two joint exercises, several fleet exercises, a variety of Army divisions command post and field exercises, several laboratory experiments and war games) and secondarily as a research tool.

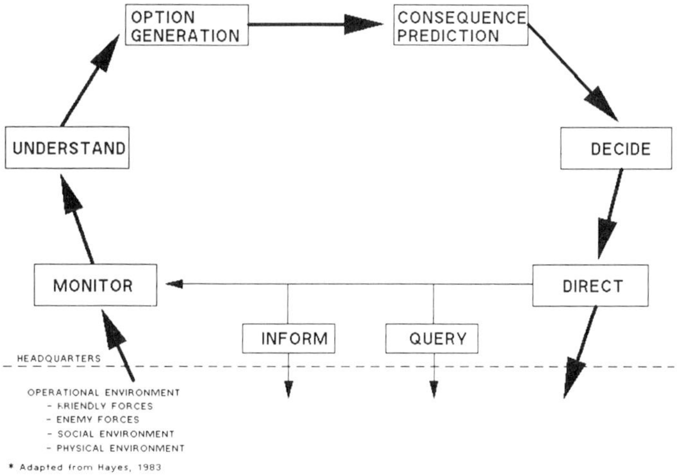

Figure 1. Overview of the Headquarters Effectiveness Assessment Tool.

HEAT has been used more widely than any other approach partly because it introduced some important ideas. First, it established the idea that there is a difference between the quality of the command and control processes present and the overall performance of the C^2 system. Second, it specified the goal of higher level headquarters as establishing "adaptive control" over an environment (friendly forces, enemy forces and physical elements). Third, it identified the key functions that a command and control system performs (Figure 1). In addition, it linked the performance of command and control systems to the crisis management and small group decision making literatures; providing an empirical and theoretical basis for analysis. HEAT also stresses the need for user involvement in specifying the standards against which performance is measured.

HEAT represents important progress, but its best developed elements are the theoretical structure that underlies it and the measurement system. The theoretical structure is not widely understood. Further, HEAT focuses on senior headquarters: ideally theater level. This leads to an emphasis on planning that is not appropriate for other levels. Moreover, it has never been employed in the design phase of a C^2 system. Finally, its original version has proven manpower intensive and therefore expensive.

DCA refocused the HEAT program in 1987 on the design of command and control systems and began research on ways to recognize "optimum" C^2 performance. However, progress to date has been more theoretical than practical. Review at a workshop convened by DCA in December, 1988, indicated that the program needed more field application of the new ideas. That process began in the Spring of 1989 with an application to TRANSCOM performance in a joint exercise.

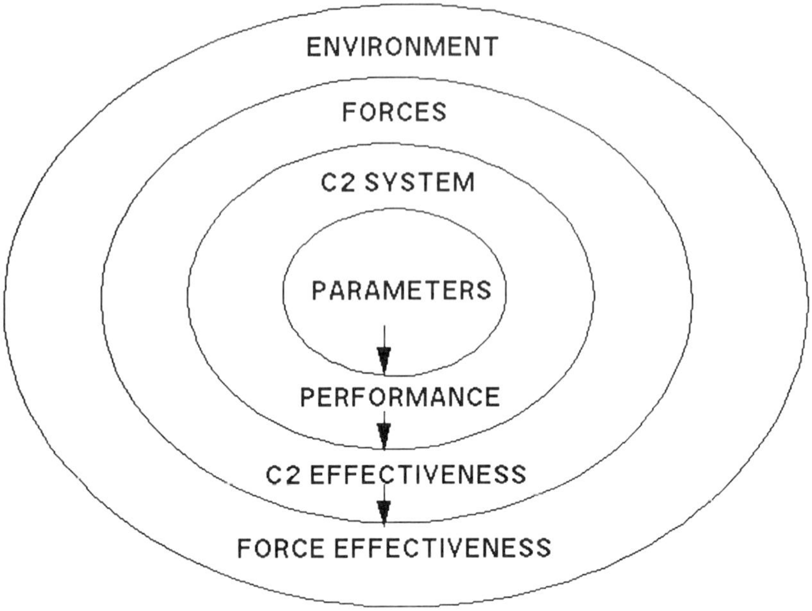

Figure 2. MORS Workshop view of "Layers of Performance."

In 1983 the Military Operations Research Society convened a Workshop on Command and Control Evaluation (Sweet, Metersky and Sovereign; 1985). Drawing on the whole community, the Workshop explicitly identified four levels of measurement necessary to evaluate command and control: (a) Measures of Force Effectiveness (impact of the force); (b) Measures of Effectiveness of the command and control system (performance of C^2 processes, such as timeliness or knowledge of enemy order of battle); (c) Measures of Performance (ability of the system to perform technical tasks such as sense accurately) and (d) Parameters, which deal with characteristics of the system (luminosity of screen, etc.) These distinctions are based on the level of analysis that applies (see Figure 2).

The MORS Workshop resulted in the *Modular Command and Control Evaluation Structure* (MCES). This approach organizes the process of evaluation into the six steps necessary to organize any quasi-experimental design: (1) identify the problem as the decision maker understands it, (2) bound the system of interest, (3) identify the processes of interest, (4) specify the measures to be used, (5) collect the data needed, and (6) aggregate the results into meaningful patterns. MCES also focuses attention directly on the "decision maker" or user, who is both the source of the problem statement and the judge of the results. It leaves the analyst free to select a variety of measures and functional approaches.

MCES weakness results from this flexibility. There are no specific standards or measures of performance. Rather, they are specified for each application based on the general categories set up by the MORS workshop. Indeed, the first publicly reported application (Sovereign and Sweet, 1988) was funded by DCA as an experiment using HEAT. In addition, the heavy involvement of the user means that MCES is vulnerable to the error of overemphasizing immediate needs and current missions. While MCES provides an approach more easily applied to systems design, it is inadequate because it lacks comprehensive and specific guidelines as to what to measure.

The need for lower level measures of performance was addressed for the Naval Air Development Center (NADC) by Defense Systems, Inc., in a system entitled DAME (Decision Aid Measurement and Evaluation) in the mid-1980s. DAME recognizes explicitly that different information will be available at the design, development, and implementation stages of a decision aid.

As a consequence, DAME specifies unique testing to be performed at different stages. For example, at the design stage information is collected on the allocation of functions to humans and machines to ensure that humans are not asked to do things machines do better (remember facts, sort by known criteria, etc.) and examines design issues parametrically. In the laboratory, where prototypes are tested, DAME applies a variety of capacity and human factors tests intended to catch problems before they reach the field. Field tests are like HEAT applications, with detailed focus on the quality of information handling and the interfaces of the machines to humans.

In 1986, the Army Research Institute began adapting HEAT to its need for a way to evaluate staff processes (Hayes, et al., 1987). The resulting tool, called ACCES (Army Command and Control Evaluation System) translated HEAT into Army language, adjusted some of its structure to reflect ARMY doctrine, and simplified the measurement approach to reduce the manpower required for field exercises. Both the Battle Command Training Program (focused at corps and division) and the Army's l essons learned program have been influenced by it.

A parallel effort, MAPP (Method of Assessing Planning Performance) was developed by ARI to evaluate military decision making. It employs videotape as well as direct observation to compare actual behaviors with those expected and desired (Metlay, et al., 1985). Its focus on planning to the exclusion of other command and control functions limits its utility.

The Army Operational Test and Evaluation Agency (OTEA) has also done some pioneering work. Beginning with funding from the HEAT program and later on their own initiative, they applied HEAT to specify a baseline of division C^2 performance and as a part of their field test of the first module of the Maneuver Control System. Their explicit recognition of the need for baseline performance was a significant development.

More important, they undertook this effort as part of a new concept that recognizes the need for a different approach to the evaluation of C^2 systems. In

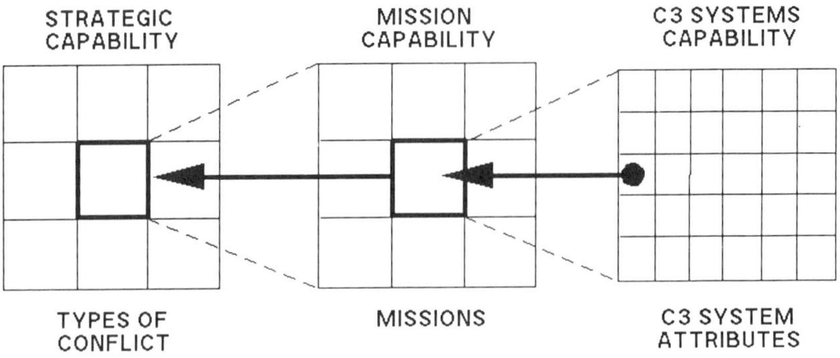

Figure 3. Logic of Mission Oriented Analysis.

1985 they indicated dissatisfaction with traditional methods that tend to focus on the communications aspects of C^2 systems and to measure them against technical specifications. At the same time, they recognized that waiting until C^2 systems were ready for field testing before becoming involved was an error. Evolutionary acquisition, rapid prototyping and heavy user involvement imply early testing. Hence, they evolved a doctrine of "continuous and comprehensive evaluation" that seeks to link designers, developers and testers into a coherent team and perform meaningful evaluations early so their results can be fed back into the development process.

The Center for Command and Control Communications Systems (C4S) at DCA has developed and applied a tool they call Mission Oriented Analysis (MOA) that plots alternative packages of C^2 systems against the capabilities they create and their impact of mission accomplishment in a variety of levels of crisis and conflict (Figure 3). The weakness of this approach is that it assumes that the level of performance of C^2 systems is known and that their impact on operational capability are understood. Its strengths are (1) it asks the correct questions (what capabilities are provided under what conditions) and (2) it relates systems capacity to expenditures.

While a variety of relevant initiatives can be identified, none has the breadth to support the design, development, implementation and improvement of C^2 systems. Yet evaluation is critical to both quality and cost-effective development. Hence, a more comprehensive system is needed.

Solution

The solution lies in combining the existing approaches and experience into a comprehensive system that is valid, reliable, capable of objective measurement of the quality of command and control, and applies to all C^2 systems. The approach must be multidimensional and multidisciplinary, yet simple enough to be understood by the variety of actors that will be impacted.

MECCA, the Multi-faceted Evaluation of Command and Control Assets, is such a system. It includes explicit recognition that eleven different dimensions will need to be evaluated to measure the quality of a command and control system:

- Force Effectiveness (mission accomplishment, casualty ratios, territory, etc.);
- Command and Control Effectiveness (quality of decisions, quality of planning, decision timeliness, etc.);
- Quality of Command and Control Processes (understanding, option generation, etc.);
- Quality of Command and Control Products (situation reports, intelligence summaries, databases, etc.);
- Responsiveness to User Criteria (commanders and staff);
- Acceptance by Users;
- Availability/Survivability (mean time between failures, meantime to repair, redundant capability, degraded performance, etc.);
- Quality of Information Provided (completeness, accuracy, age, consistency throughout the system, etc.);
- Interoperability (within headquarters, across levels, with allies, etc.);
- Time Consumed (training required to develop and maintain proficiency, demands during crisis and combat use, etc.); and,
- Efficiency (resources required to perform effectively, preferably in life cycle terms).

MECCA also incorporates important distinctions that have prevented generalization of some of its predecessors. For example, it specifies different types of measures for the four different stages of system development: systems design, prototype development, laboratory testing, and operational testing. Like DAME, therefore, it asks the same set of questions at each level, but recognizes that the information available and the level of precision with which they can and should be answered will be different.

Because it deals directly with all phases of system development, MECCA provides the ideal tool around which to organize information flows between designers, developers, users and testers.

MECCA also distinguishes between levels of command when defining effective command and control. As Jaques and Jacobs (1987) have shown, different strata of military organizations are presented with qualitatively different challenges, types of information, and standards of performance. Hence, the specific measures of C^2 process, products and effectiveness will differ across levels, as will their relative importance.

In addition, MECCA introduces the ideas of "appropriate cognitive complexity" and "connectedness" as measures of command and control effectiveness.

Rather than relying, as HEAT must, on the actual testing of a set of options in the context of a simulation or exercise to judge command and control effectiveness, the work of Streufert (1981) can be adapted to assess the match between the nature of the command and control challenge presented and the nature of the response (options chosen). Where cognitive complexity is appropriate, the ability of the command and control system to identify actions in one phase of a battle that will contribute to effective performance later and create a more favorable military situation in the future are stressed.

Following both the work of Olmstead et al. (1978) and the experience in applying HEAT, MECCA employs stress testing to evaluate command and control. The issue is not whether a particular mission can be accomplished, but rather the capacity of the system. Conscious manipulation of the pace of battle and the quantity and quality of decision making required are used to determine breaking points. This allows escape from scenario driven testing.

MECCA also uses an explicit strategy, termed Management Objectives, to ensure that the commander and senior staff are both actively involved in the specification of the system and postured to take full advantage of the potential of any new or revised system. This approach recognizes that many C^2 systems fail because their objectives are not clear, become confused over time, or are transformed into technical specifications that do not reflect functional needs.

MECCA has also been designed to minimize the cost of evaluation. It calls for the use of the computers and communications systems in the C^2 system to collect and process a great deal of the information needed to evaluate command and control. Most of this information ought to be preserved and used for systems management. For example, structured message traffic like status and position reports, patterns of usage for data bases, resource consumption and times of key events such as detection of hostile force elements are important for a variety of purposes.

Other information, such as "ground truth" location of red and blue forces are available from the exercise or simulation drivers or controllers, but need to be preserved in electronic form to support rapid analysis.

The analytic tools implied are quite simple (spreadsheet calculations) that are no real burden to the types of computer systems needed to drive exercises or support command and control. The calculations are repetitive; measuring differences between ideal and actual performance, measuring time elapsed between critical events, and checking consistency of information across different locations and functions.

Overview of an Application

The process of applying MECCA to a command and control system is illustrated in Figure 4. In a typical case, a decision support system is to be introduced into a CINC headquarters. The initial requirement has developed from a complaint by the CINC that the information systems employed in his headquarters provide information primarily of interest to his subordinates and that the elements of information are not combined.

For example, he notes that he is given current expenditure rates for munitions, while his task is to ensure an adequate supply beginning six to twelve days in the future. Moreover, he is offered information on enemy forces in and potentially entering his area of interest and on munitions available, but the two data sets are unrelated, forcing him to estimate the relationship between them. He also points out that supplies are being reordered on the basis of contingency plans developed for the outbreak of a conflict. This assumes that the pace of battle and the quality of the threat opposing him will not change. These seem like poor assumptions.

The flow of the application is shown in Figure 4. The potential requirement is first subjected to Management Objectives review, in which the initial idea is stated in terms of functional requirements, which are then systematically reviewed with (a) the commander, (b) senior staff, (c) mid-level managers who would be involved, (d) potential users, (e) skeptics, (f) specialists in systems design and (g) resource allocators. The results of this review are fed back to the commander and senior staff along with recommendations for altering the requirements, scrapping the idea, and/or systems design guidelines. These include estimates of the resources and time needed and potential problems and opportunities to review progress.

Assuming the idea is approved, there are now a set of explicit goals endorsed by management, a budget and a schedule. Based on these the areas where performance should be measured (C on Figure 4) can be specified. All eleven of the areas identified in MECCA are candidates, but some aspects of them may not apply. In this case, the commander is not worried about interoperability with allies, although direct connection with other systems in his headquarters are needed. Linkages for subordinate commands to provide information directly are also needed to make the system cost effective.

Given the areas of performance, the project leader can also specify the particular measures of interest (D) for the design, prototype, laboratory, and operational versions of the system. As Table 1 shows, these reflect the different nature of the information available at each level. For designs, the major issues are explicit consideration of each goal specified. For example, interoperability requires that the design identify interfaces with other systems, but it is not yet possible to know whether these will work unless obvious errors have been made. Similarly, provision should be made for system checks to ensure inputs are within realistic ranges (no more than 360 degrees for direction, no negative values for force strength, etc.)

Prototypes can be tested for performance against both baselines and technical specifications. However, they are typically not operated by the eventual users, so detailed tests are not appropriate.

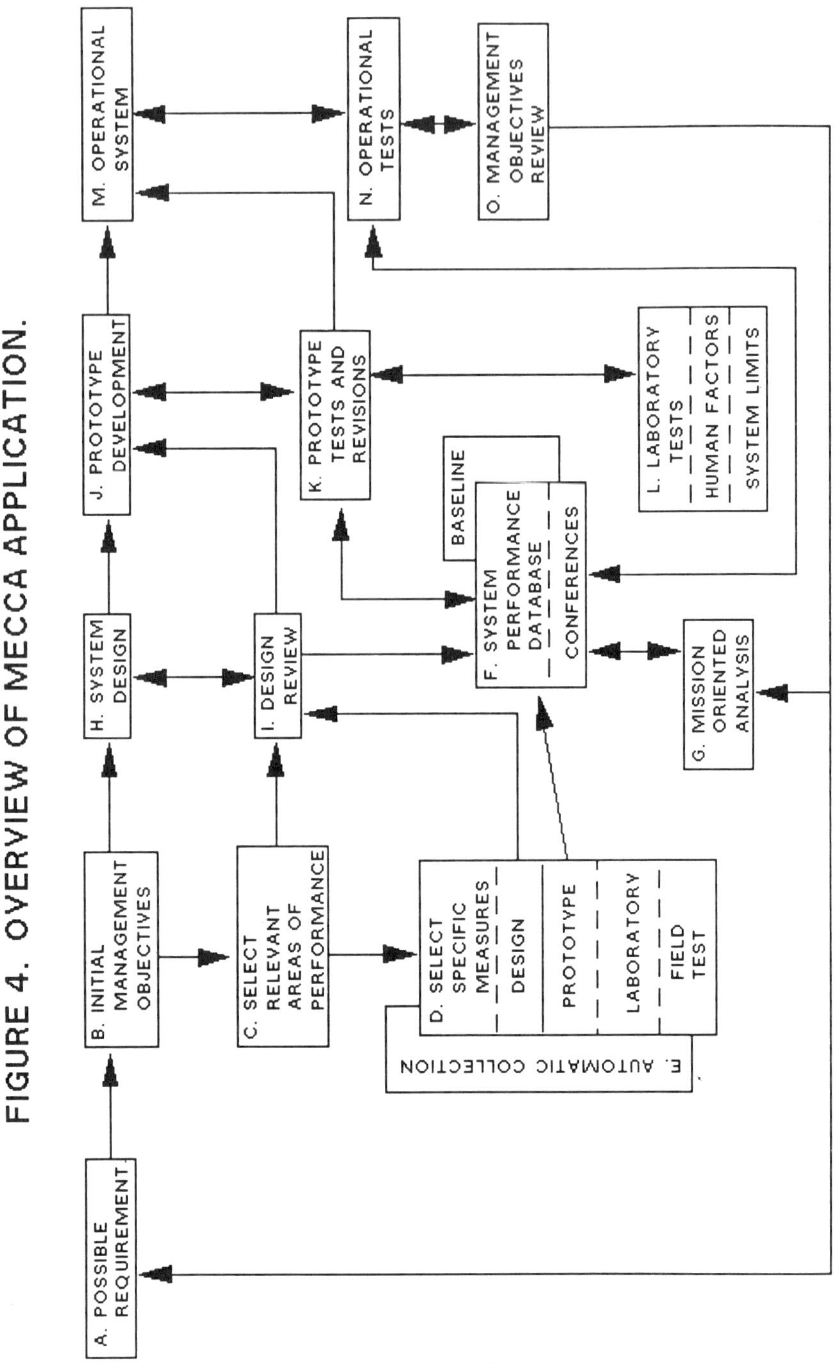

Figure 4. Overview of MECCA Application.

PERFORMANCE ARENA	SYSTEM DESIGN	PROTOTYPE (PERFORMANCE VS BASELINE & GOALS)	LABORATORY (STRESS TESTS)	OPERATIONAL (STRESS TESTS)
FORCE EFFECTIVENESS	o COVERAGE OF MANAGEMENT OBJECTIVES	MISSION ACCOMPLISHMENT CASUALTY RATIOS ENGAGED, DESTROYED OR TIED DOWN	SAME	SAME
COMMAND AND CONTROL EFFECTIVENESS	o "VISION" INTO FUTURE o LINKAGE OF INTEL, OPS, LOGISTICS, AND PERSONNEL o DEGREE OF CONTROL SOUGHT	SAME, PLUS CONNECTEDNESS APPROPRIATE COMPLEXITY		
COMMAND AND CONTROL PROCESS QUALITY - DEVIATION FROM IDEAL - SPEED - CONSISTENCY	TREATED o MONITOR o UNDERSTAND o OPTION GENERATION o PREDICTION	PERFORMANCE	SAME	SAME
COMMAND AND CONTROL PRODUCT QUALITY	GOALS FOR o COMPLETENESS o PRECISION o ACCURACY o AGE OF INFORMATION o CONSISTENCY o TIMELINESS	PERFORMANCE ON THE SAME ISSUES	SAME, PLUS PERCEPTIONS OF USERS	SAME
RESPONSIVENESS	GOALS FOR o COMPLETENESS o TIMELINESS	SURVEY OF USERS	SAME, PLUS USER ATTITUDES	REVIEW BY COMMAND AND STAFF

Table 1. Example MECCA Measures: CINC Decision Support System.

PERFORMANCE ARENA	SYSTEM DESIGN	PROTOTYPE (PERFORMANCE VS BASELINE & GOALS)	LABORATORY (STRESS TESTS)	OPERATIONAL (STRESS TESTS)
ACCEPTANCE BY USERS	o USER INVOLVEMENT	SURVEY OF USERS	SAME	SAME
AVAILABILITY/ SURVIVABILITY	DESIGN GOALS FOR o MTBF o MTR o DEGRADED PATTERNS o REDUNDANCY	PERFORMANCE DATA	SAME	SAME
INFORMATION QUALITY	SELF-CHECKS; DESIGN GOALS FOR o ACCURACY o COMPLETENESS o AGE o CONSISTENCY	PERFORMANCE DATA	SAME	SAME
INTEROPERABILITY	GOALS FOR o HQ o LEVELS o ALLIES	PERFORMANCE (INTERFACES)	SAME	SAME
TIME CONSUMED	DESIGN GOALS - TRAINING - NOVICE USERS - EXPERIENCED USERS	PERFORMANCE DATA	TRAINING/PROFICIENCY CURVE TESTS OF DIVIDED ATTENTION, NOISE, ETC.	TRAINING/PROFICIENCY CURVE PERFORMANCE DATA
EFFICIENCY	DESIGN GOALS ARTICULATED	RESOURCES CONSUMED	SAME	SAME

Table 1. Example MECCA Measures: (Cont.)

Laboratory testing allows in depth evaluation of human factors such as training time to achieve proficiency and operations in the presence of noise or when distracted by other tasks. Laboratory tests also provide the first meaningful venue for stress testing: examining the capabilities of a C^2 system across the range of likely demands and finding the upper bounds of the pace of battle it serves effectively.

Stress testing also contributes to operational evaluation, but the types of exercises implied force variation of the pace of battle over time and within scenarios realistic enough to provide useful training. Operational tests provide the only valid data on responsiveness and user acceptance and the most meaningful values on other dimensions.

As the measures are specified, the project team can also identify opportunities for automatic data collection (E) and the design of the system performance database (F). As this is being done the needs for baseline data are clarified, the schedule of conferences needed to keep the designers, developers, users and testers in communication is articulated and (G) a Mission Oriented Analysis of the system that establishes the capabilities to be created and the range of conflict to be supported (which also helps define the range of scenarios for testing). The MOA is modified whenever performance data indicate the systems capabilities or impact differ from this original estimate.

The automatic collection requirements become part of the system design (H) and are examined, along with the other design measures, at the formal design review (I). The other key driver of this review is the Management Objectives Review.

A successful design initiates the prototype phase (J) which includes tests (K) and demonstrations that feed the Performance Database. Baseline information becomes relevant when the prototype works and is ready for laboratory testing (L).

Operational testing (M) should not begin before human factors testing is completed (although prototype tests may actually be done with users under evolutionary acquisition) and system acceptance should not occur until stress testing in the laboratory is complete. Rushing a system to the field only to discover that it cannot be operated by the people it is intended for or when the noise levels common in the field occur not only wastes testing money, but also gives the system a bad reputation for factors that might be readily fixed if identified in the laboratory. When this occurs, user acceptance may be impacted with severe long-term impact on the C^2 system's utility.

After operational testing and before system acceptance, the Management Objectives Review and Mission Oriented Analysis should be revisited. These top level examinations of the system ensure that it meets real needs and contributes to military effectiveness and cost-effective performance.

Conclusion

Evaluation is essential for the development of effective C^2 systems. Properly done, it will save money and provide the evidence necessary to justify funding of important C^2 initiatives.

While there have been a variety of efforts to develop tools for C^2 evaluation, some of which have been useful, no comprehensive system has been generated. By blending the useful parts of earlier efforts with the lessons learned from their application and recognizing the multidimensional and multidisciplinary nature of the problem, however, a coherent solution can be shaped without massive new investments. MECCA has been designed to meet this need.

It can be used by those developing new C^2 systems or interested in improving existing ones.

References

Hayes, R. E. (1983) *Theater Headquarters Effectiveness: Its Measurement and Relationship to Size, Structure, Functions, and Linkages; Vol.. I, Measures of Effectiveness and the Headquarters Effectiveness Assessment Tool.* Defense Systems, Inc., McLean VA

Hayes, R. E., et al. (eight others). (1987) *Methodology for the Performance Measurement of Army Command Staff Operations.* Defense Systems, Inc., McLean VA

Jacobs, T. O. and E. Jaques, (1987) "Leadership in complex Systems" in *Human Productivity Enhancement. Vol II Organizations and Personnel.* Praeger, NY

Metlay, W. et al. (5 others). (1985) *Methodology for the Assessment of the Command Group Planning Process.* Applied Research and Evaluation Program, Hofstra University, Hempstead NY

Olmsted, J. A., H. E. Christiansen and L. L. Lackey. (1978) *Research on Training for Brigade Command Groups: Factors Contributing to Unit Combat Readiness.* HRRO, Alexandria VA

Sovereign, M. and R. Sweet, (1988) "Evaluating Command and Control: A Modular Structure," *The C^3I Handbook*, Edition 3. EW Communications, Inc.

Streufert, S. and Susan C. Streufert. (1981) "Stress and Information Search in Complex Decision Making: Effects of Load and Time Urgency," Hershey Medical Center, Hershey, PA

Sweet, R., M. Metersky and M. Sovereign. (1985) Command and Control Evaluation Workshop. U.S. Naval Postgraduate School, Monterey, CA

Military Planning and Technology

Frederick W. Kulik

Introduction

Much time and money is being spent on developing expert systems or artificial intelligence (AI) components to assist military planners. The military has taken a half-hearted approach to supporting this research. They furnish the funds but do not always furnish the domain expertise. At best the military is not sure AI and advanced technologies can improve military planning but there is the fear that it might and the military will be left behind. Thus programs are funded on the anticipation that something new and grand may surface that will be of benefit to the planner.

What is ignored is that the military, and the Army specifically, hasn't paid sufficient attention to how it teaches or how it does military planning. In peacetime military planning is really only done for military exercises or for contingency plans which are seldom, if ever, executed. These military plans usually are developed over a period of months or even years. This is a luxury that would not exist in a wartime environment. Since most plans are not executed, the military never gets a grade on how well they were conceived. The few plans which are executed such as the aborted Iranian rescue mission or the Grenada invasion are replete with problems that raise serious questions about planning.

It is against this backdrop that new technologies are emerging which have promise to aid in military planning. Will these technologies ever see the light of day in the military planning process? Can these technologies help overcome some of the shortcomings in today's planning process?

PLANNING FINDINGS

- INADEQUATE COMMANDER'S GUIDANCE
- INADEQUATE EDUCATION OF PLANNERS
- PLANNERS WERE NOT REWARDED
- PLANNERS ASSIGNED TO OTHER DUTIES
- DIFFERENT PERCEPTIONS OF US AND NATO PLANNERS

Figure 1. A number of problems in planning were common to a majority of the headquarters that were visited.

What Is Wrong With Military Planning?

Numerous formal and informal studies have been done in the past ten years which have addressed the problems in military planning. These studies have been performed by such diverse groups as the Strategic Studies Institute of the Army War College, the Army Staff, the Command and General Staff College as well as by a number of civilian analysts. Not surprisingly, the results of these studies conclude to varying degrees on a number of findings and with my own personal views on the state of military planning.

In the fall of 1982, the Chief of Staff of the Army established a special group of 10 officers at the Army War College at Carlisle Barracks, Pennsylvania with the charter to assess the state of military planning in the Army. This group spent two years visiting and analyzing planning at all Army major commands as well as at some joint and combined headquarters. As a member of this group I had the opportunity to discuss planning with U.S. and NATO planners at all echelons from SHAPE headquarters to division level. Though, invariably we would start with planners stating, "We don't have any problem with planning at *this* headquarter." We would find after a few hours of discussion and promises of non-attribution that some problems were prevalent. Figure 1 lists some of the common planning findings uncovered in the staff visits.

Inadequate Commander's Guidance

The lack of commander's involvement in the early stages of the planning process was especially disturbing. The commander's conceptual understanding of the nature of the problem is key to successful planning. The commander is normally the most experienced officer at the headquarters and should have the best understanding of what his superior headquarters wants him to accomplish. This experience and understanding of the higher headquarter's mission is fundamental to starting the planning process in the right direction. In some cases, the Chief of Staff attempted to fill the commanders role for initial guidance only to have the plan "shot down" at the decision brief because the commander wanted another approach. This, of course, is frustrating for planners who can spend weeks or months on a plan only to have to redo it overnight because the commander wanted a different course of action. When confronted with this finding, a number of commanders agreed that they did not spend enough time in the initial phases of the planning process, but blamed their behavior on the press of business.

In those cases where planners did receive adequate guidance a different dynamic appeared to be working. The commander would call in his Chief of Staff, G-3, the Plans Officer and possibly a trusted, bright staff officer to discuss the task or mission that had to be planned. This meeting did not normally take much more than an hour and all participants were able to speak as equals about the task at hand. Out of this discussion would come the commander's view of the situation, what he intends and hopes to accomplish, what resources he would allocate to the task and any specific course of action he wanted explored. After such a meeting, all participated in initial formulation of ideas or courses of action from the junior staff officer doing the actual plan writing to the officers in his approval chain. As a result, all key players understood the commander's intent. Much work would still have to be done within the planning process such as staff estimates, or more detailed mission analysis and exploration of courses of action prior to a commanders decision: but a good planning foundation was established.

Inadequate Education of Planners

Almost without exception, all of the planners interviewed became planners through on-the-job training. In the late 1970s, the Command and General Staff College (CGSC) had dropped planning as a course of instruction due to the demand of other more important curricula such as Survival, Escape, Evasion and Resistance (SEER) or some other current "hot button" topic. So, for a period of about 10 years, the logic of planning and the disciplined approach of the Military Decision Making Process (Figure 2) to include the Estimate of the Situation (Figure 3) were not formally taught at CGSC or, for that matter, at the sister institutions such as the Armed Forces Staff College (AFSC). With a lack of education in the fundamentals even innovative thinkers would have a difficult task putting their ideas into a cohesive and integrated plan which units could then implement.

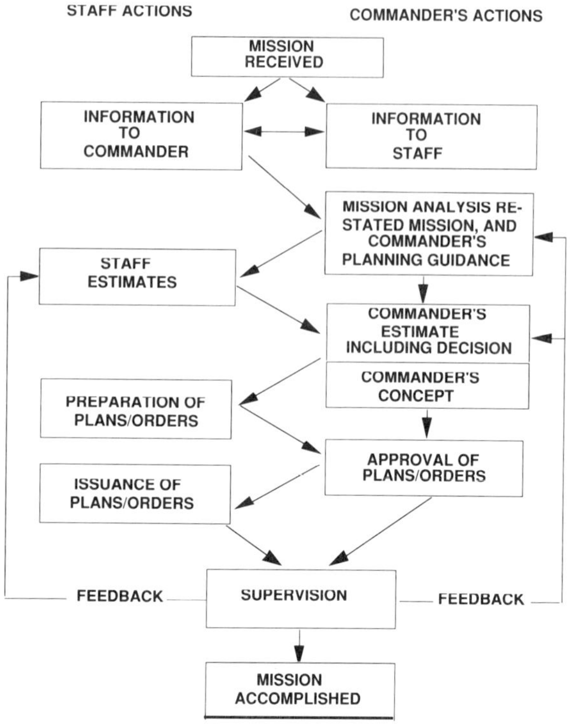

Figure 2. Military Decision Making Process.

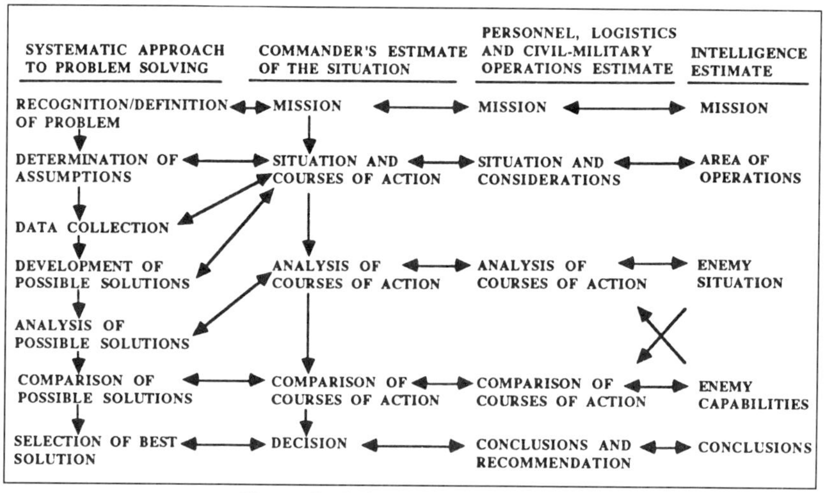

Figure 3. Estimate of the Situation.

In the mid 1980s, the Army realized some of the shortcomings in the teaching of planning subjects and the Combined Arms and Services Staff School (CAS^3) at Fort Leavenworth was a first step in coming to grips with the task of developing planners. A special elective for planners as well as more emphasis on planning in the basic CGSC curricula was attempted. However, there are currently about 10 years of graduates of CGSC some of who are now generals and colonels in senior management positions who probably have not received any formal education in the planning process.

Planners Were Not Rewarded

In almost all of the U.S. headquarters that were visited planners viewed themselves as unappreciated. Many hoped to leave the plans shop to become staff officers in the current operations shop. The glamor job was to be an "ops officer" who would deal with the current situation. It was the operations officers who usually were selected for early promotion to battalion and brigade command. These were the officers who were clearly on the path to high rank and responsibility. If by chance a "hard charger" was assigned as a planner, he would work mightily to get reassigned to a more prestigious and career-enhancing position. The irony was that many of the planners enjoyed the job and only desired to be recognized for their contributions.

Planners Assigned to Other Duties

Some planners at higher headquarters were pulled from their primary war planning duties and used to backstop other activities. In one case, they planned the monthly retirement reviews, monthly retreat parades and ceremonies for visiting dignitaries. Substantial time was consumed with these protocol and ceremonies issues instead of developing or enhancing the General Defensive Plan (GDP) or other contingency plans.

Perceptions of NATO and U.S. Planners

One of the most illuminating findings of the staff visits was the difference in perception of non-U.S. NATO officers and their U.S. counterparts at combined headquarters. The multinational staffs at a combined headquarters bring their own national perception of the role planners play in the decision making process. One German lieutenant colonel (who worked for a USAF colonel who reported to a Belgian colonel who in turn reported to a U.S. general who worked for a German general) told of his anxiety about the planning process at his

NATO headquarters. When given a job to plan an operation with a minimum of guidance by his USAF colonel he quickly realized he didn't have all the facts or assumptions needed to make a coherent plan. He voiced his concerns to the USAF colonel who did not have the information that was needed. Instead of going "up-the-chain" to get the information the USAF colonel instructed the German to come up with a "strawman" plan. The German found this very distressing because German staff procedures require the planner to obtain the information required before constructing the plan. Germans are taught not to guess what the commander might want but to ask. The flying of "trial balloons" or preparing a "strawman" is a cop out on the staffs responsibilities to obtain the most accurate guidance on which to build the plan. This particular German officer finally solved his dilemma in a novel way. He waited for his USAF colonel and Belgian colonel to be away on travel and then he was free to go up the chain-of-command to obtain the information he required.

Vietnam Experience

The Vietnam War profoundly influences the U.S. military's view on planning. It is an old saw but it probably has an element of truth that we prepare and train for future wars based on our experiences in the last war.

The battalion and company commanders of the Vietnam era are now the general officers and senior staff officers of the Army. To a great extent, the operationally permissive environment of Vietnam has conditioned an entire generation of Army leadership. Resources in Vietnam were relatively unconstrained and detailed planning was not essential. In general, plans and planning ended when a unit was inserted on the landing zone (LZ). Subsequent actions were more responsive to what the enemy might do or what the commander flying in a helicopter overhead might direct than in following any detailed plan. As a result, most Army officers are viewed as quite effective in conducting operations and responding to unexpected events. This may be acceptable in situations like Vietnam when different unit's actions are independent and need not be coordinated. But this approach in large-scale coordinated operations can be disastrous.

There is now a substantial body of Army leadership who succeeded in Vietnam and must believe that if their methods of operating led to success then they must be correct. This may account for some of the lack of personnel involvement in planning that was encountered at some of the highest levels of command. The whole framework of planning and the development of logical thought has been undermined by the Vietnam experience.

Innovative Thinking

What is the source of innovation in military planning? Where do the strategists really come from? Are they born or are they developed? Innovation is grounded in a thorough grasp of the fundamentals with a perception of where opportunities might be. You can teach the fundamentals, but how does one see the opportunities. Opportunities are by nature transient and fleeting. One must seize the moment boldly but not recklessly.

I would contend that those who would be innovative have at least two characteristics. One of these is a sense and study of history and the other is time to reflect. The study of history in the military is a sometime thing. More attention is being paid to military history in the Army school system but individual study is required to achieve any real sense of the subject.

Time to reflect on military tactical situations and to postulate potential actions is essential. But to expect this creative process to take place without some stimulus is wishful thinking. Short of actual battle, wargames provide the best vehicle for generating new tactical situations. Wargames invariably do not allow planning to be done very well. Moreover, they do not model concepts well. Traditionally, wargames tend to be attrition oriented and cannot easily be re-run in near real time to explore other outcomes. It is this facility to raise questions during the conduct of operations and allow the wargame to be rapidly (on the order of one or two minutes) rerun that make insights and innovation possible.

Technology and Institutional Fears

Much of current Army leadership is not knowledgeable of the computer revolution. They appreciate the power of computers, but lack hands-on experience. As a result, they find it difficult to embrace or to really understand the opportunities that computers present to military planners. However, through fear that computers might revolutionize military planning, they give half-hearted support to prototype decision aids.

The captain and major level of the military have a much better appreciation of the technology and are more receptive to new ways to approach planning through decision aids. In time this generation of the military leadership will be in decision making positions and will insure that computers will be used and make a contribution to military planning.

The military has been using a form of decision aids since before the days of Napoleon. In fact, the decision aids have not materially improved in the interim. The primary decision aid has been the map. Commanders rely on 1:50,000 or 1:100,000 scale maps pasted together to portray the units area of operations. Clear acetate overlays which depict the intelligence situation or the operational scheme are dropped over the map and become the defacto method of focusing and displaying data for the commander. It is in that environment that the commander and his staff conduct mental wargames using rules of thumb (i.e. don't attack unless you have at least a 3:1 ratio of combat power) modified by varying degrees of information about the enemy, his own force, logistics, weather, terrain, morale, etc. The rules of thumb are straightforward but it is the commanders grasp of such intangible tasks as morale or uncertainty that decision aids cannot quite capture.

In the near term, it is unlikely that computers will handle the entire military planning problem. However, technology exists today that can facilitate the correlation and display of data to assist the commander in his planning process. Large screen displays are approaching the fidelity of 1:50,000 scale maps and expert systems exist that can fuse and present data from multiple sources in a graphic and coherent manner.

Obviously, computers can process large amounts of data much better than humans. For example, the intelligence acetate overlay mentioned earlier, typically has grease pencil annotations of all intelligence spot reports of the last 24 or 36 hours, with some coding as to reliability and credibility. More than a few days reports rapidly clutter the overlay and it becomes too complicated to be of use. A computer generated overlay would allow the spot reports to be presented in a more usable format. For example, you could request the display of all reports in the last 6 hours, or all reports with a credibility rating greater than some coded value in some geographic area that involved armor units. This ability to manipulate the information would be extremely useful in analyzing potential enemy actions. The analysis or deduction is still done by the human but the data is presented to the human rapidly by the computer.

Overdirection

One of the enduring premises of military command and control is the use of mission-type orders. In other words "tell me what you want done but don't tell me how to do it." The advent of improved communications that can transmit large amounts of information rapidly and accurately is having its effect on relatively autonomous operations. With developments in packet switching and interconnectivity between tactical and strategic systems it is now possible for the National Command Authorities to be in direct contact with the soldier in the foxhole. This raises to new heights the phenomena of the division commander in a helicopter in Vietnam giving directions to his company commander on the ground.

This desire for information at higher headquarters can assume ludicrous dimensions. During one of the Marine incursions into Beirut, Lebanon, an overzealous OJCS staff officer was requesting the anticipated forward trace of Marine positions in Beirut before the Marines had left their landing ships. The

OJCS staff officer wanted the information for the morning briefing for the JCS. An exasperated EUCOM officer asked if JCS would like to know where every Marine is standing. The OJCS staff officer asked, "Can you get that?"

This fear of overdirection or of having someone looking over your shoulder is very inhibiting. If you feel your every action is being evaluated, your actions will tend to be pro forma. This may be very desirable in a delicate situation but may not be what is desired when boldness and innovation are called for. Technology is hardly useful if it inhibits personal initiative and ingenuity.

Models, Wargames and Simulation

We have progressed a long way from the sand tables and toy soldier models to computer assisted wargames and simulations. At the lower echelons, battalion level rehearsals of operations in SIMNET-like simulations provide surprising realism and fidelity. At the division and corps level, computer wargames are still ponderous, tedious, time consuming and do not assist in real-time planning. There isn't time for these large involved wargames to explore multiple courses of action and help the planner. The JANUS, McClintic, JTELS, etc., simulations are adequate for training or exploring doctrine over time, but are just not rapid enough for a planning regime.

There are, however, emerging expert system rule-based simulations which are surprisingly rapid (on the order of a minute or so) and are very powerful planning tools. Their power lies in their transparency. That is, the user can easily see why a particular course of action resulted and furthermore can modify the operative parameters to see an alternative. Additionally, it is possible to ask "what if?" questions and obtain a whole tree of alternative solutions along with supporting rationale. This simulation approach allows exploration of many courses of action and the relative merits of each in less than 10 minutes.

Knowledge Acquisition and System Development

Developers of expert systems require access to true domain experts on a long-term and continuing basis. As we have seen, there is a dearth of truly expert military planners. The Army school system attempts to fill this void by making a few of its faculty members available as domain experts. Unfortunately, most of these officers do not have real world planning experience. This is further compounded by the fact that real world planners tend to deviate from the school approach.

The military is reluctant to assign the best and brightest to the development of expert systems. There are just too many other more important problems to solve. How then can this problem be approached? There are two possibilities:

- Senior Champion. The enthusiastic support of a senior military officer can do much to ensure availability of the right domain experts and access to the proper levels of command. This approach has been especially effective in the Navy's development of the Joint Operational Tactical System (JOTS). The continuous support of an admiral in JOTS has kept the program moving in the face of much bureaucratic opposition and detractors. Without this support, the program would have foundered long ago.
- Use the war colleges. Each year the military's best and brightest are sent to the service war colleges where they are expected to complete a research project. These projects take many forms and many students would welcome an opportunity to contribute to expert system development. They have the requisite knowledge, the time and the inclination.

Conclusions

Military planning has been slowly recovering from a period of neglect, some of which can be attributed to Vietnam and to the lack of command interest. Technology exists that can make a substantial improvement in military planning. Any technology must be transparent to the users until they develop confidence in its utility. Of and by itself this requires a man-in-the-loop approach for all decision aids.

There are some exciting opportunities for both the military planner and the expert system developer. The field is ripe for improvement and the technology is here to make the transition.

The President or the Computer: Who Will Authorize Employment of SDI?

Murray R. Berkowitz

Strategic Defense System

"Strategic Defense" today refers to a strategic ballistic missile defense (BMD) system. Strategic ballistic missile defense systems can be classified according to the phase in which they intercept the attacking missile. The flight path of a missile is composed of three phases: (1) boost phase, (2) mid-course phase, and (3) terminal phase. The initial boost phase is defined as the time the missile is launched until it leaves the atmosphere. This phase is typically four minutes for ICBMs. This is followed by the mid-course phase which covers the period during which the missile is outside the atmosphere. This accounts for over 80 percent—that is, approximately 25 to 30 minutes—of the flight time for an ICBM.[SCH80] The final, terminal phase is defined as the time when the missile reenters the atmosphere and closes on the target. This phase is typically 15 seconds. There are advantages and disadvantages attendant to interception during each of these three phases of missile flight.

Interception of ballistic missiles during boost phase has several advantages. The attacking missile is (described as) large, slow, and "soft." It has not yet deployed multiple reentry vehicles or penetration aids (e.g., decoys or chaff).[SCH80] The missile case and engine are particularly vulnerable during this highly stressed powered portion of a missile's flight trajectory.[SME78] Creating a small velocity or directional error in the attacking missile during this phase results in a much greater error (miss distance) at the far end of the missile's trajectory.[GAR81] Considerable "leverage" (i.e., the ratio of reentry vehicles to defensive units that is required for the attacker to overcome the defensive system) is achieved by destroying the missile before it can deploy its Multiple Independently Targeted Reentry Vehicles (MIRVs). This can have the effect of destroying as many as 10 reentry vehicles in later portions of the missile's flight path.[SCH80] The distinct disadvantage regarding boost phase defense systems is that they would almost certainly operate from space. This entails the considerable cost of lifting the components of the system into space.

The basic advantage of mid-course phase interception of ballistic missiles is that due to the relatively long time for interception, the threat "cloud" of attacking missiles can be analyzed. This extended time frame also allows human directed control centers to launch successive waves of interceptor missiles in response to the structure of the attack.[DAV79] Conversely, the disadvantage is that the attacking missiles have time to make use of penetration aids such as chaff and decoy projectiles to confuse the defense.

Terminal phase interception of ballistic missiles has several advantages. Decoys and chaff are "filtered out" as they fall through the atmosphere at a slower rate than the attacking reentry vehicles. The reentry vehicles falling through the atmosphere provide observable wakes that aid in discrimination (i.e., "signatures"). The velocity of the attacking reentry vehicles is slowed as they penetrate the atmosphere and are thus easier to track and intercept. There are several disadvantages to terminal phase interception. This is very much a "high risk" interception. The computers that would direct the defense system must operate in a compressed time frame (15 seconds), directing a complex network of radar and defensive missiles in a nuclear environment. The reentry of the vehicles into the atmosphere will result in the breakup of their fuel tanks which will provide radar clutter. This clutter will make discrimination of enemy reentry vehicles more difficult. If the system fails at this point, no human-directed actions are possible in the time frame of this phase.[SCH80]

Strategic ballistic missile defense systems also can be classified according to the environment in which they intercept the attacking missiles and reentry vehicles. Endoatmospheric systems intercept attacking reentry vehicles within the atmosphere while exoatmospheric systems intercept outside the atmosphere. Terminal defense systems are endoatmospheric and mid-course defense systems are exoatmospheric. Boost phase defense systems (e.g., satellite-based lasers) would fire from an exoatmospheric position and "kill" the attacking missiles in both endoatmospheric and exoatmospheric environments. In general, endoatmospheric systems defend only relatively small areas (e.g., an ICBM field, a bomber base, or a city). Thus, endoatmospheric defense systems would not be able to concurrently provide defense of both cities and strategic forces unless the different "target sets" overlapped geographically. Exoatmospheric defense systems are indiscriminate in terms of their target defense potential because of the large areas they protect.[SCH80]

The different targets Strategic Ballistic Missile Defense Systems can defend differ in the levels of performance they require of the system. A key element in the performance requirements is "leakage," the percentage of incoming reentry vehicles that penetrate the defensive system. The lower the level of leakage required, the greater the technical demands on the system and the greater the cost. The following are some example targets such a system would have to defend and their associated leakage requirements. For "single non-replaceable hardened point targets" (e.g., NORAD) a low to nil leakage would be required since the target cannot be replaced. For "multiple hardened point targets" (e.g., the Minuteman ICBM silos) variable levels of leakage would be acceptable depending upon what percentage of the target set is required to survive. For "single non-replaceable area/point soft targets" (e.g., Washington, D.C., key radar installations, etc.) low to nil leakage is required. For "multiple soft area/point targets (e.g., cities, SAC bomber bases, radars, etc.) variable levels of leakage can be acceptable depending upon what percentage of target losses is considered "acceptable." [SCH80] Based upon the technical requirements imposed by the hard/soft nature of the defended targets and the preferred level of leakage, these target sets can be ranked with respect to the technical sophistication they require of the defensive system as follows: (1) multiple hardened point targets (technically least demanding); (2) multiple soft/area point targets; (3) single non-replaceable hardened point target; and (4) single non-replaceable soft area/point targets (technically most demanding).

Finally, strategic ballistic missile defense systems must deal with three general problems: (1) defending the target set and successfully coping with penetration aids and countermeasures; (2) defending the defensive system components themselves so as to prevent a total collapse of the system; and (3) gaining favorable cost-exchange ratios in relation to offensive systems.[SCH80]

Application of AI Techniques for BM/C^3

To better understand the role of BM/C^3 a broad definition is useful. One such definition may be obtained by paraphrasing a quotation given in Levis and Athans paper. [LEV87]

> Battle management/command, control, and communications (BM/C^3) is the military organizational structure which exercises authority and direction over the available forces in the accomplishment of a specific mission. Battle management functions are performed through an arrangement of equipment, communications, facilities, and subsystem interactions which are employed in directing, coordinating and controlling forces and operations in the accomplishment of the mission.

While this definition is vague in some respects, it captures the ubiquitous nature of BM/C^3, and focuses attention on the distinction between the all-encompassing strategic defense system, and its numerous but subordinate parts. An epigrammatic definition is that BM/C^3 is the "mapping" of assets into actions. When this is interpreted in the context of a strategic defense system, it is apparent that BM/C^3 is manifested by the system architecture. The arguments regarding the extent of BM/C^3 may be seen as merely disputes about the domain and range of the MAPPING, rather than about the fundamental nature of BM/C^3.

The need for a broader view of BM/C^3 is made clear in Van Trees' paper.[VAN80] In discussing command and control systems, he points out that "the utility of communications, intelligence, warning, surveillance and reconnaissance cannot be determined without casting these systems in a command and control systems context and understanding ... the dependencies and interrelationships which exist among these contributing subsystems. The ability to evaluate and understand command and control systems which are working properly as designed in situations that are correctly anticipated is not enough. The increasing investment which our adversaries are making in developing countermeasures and in their own command and control systems heightens the need to understand how our systems would operate at diminished levels of capacity and capability and the impact of this reduced performance on our fighting capability."

The study of BM/C^3 naturally leads to investigation of different architectures. The overall system architecture is the direct consequence of a particular BM/C^3 concept selected. The possible architectural forms range from those that are highly centralized to those that are nearly autonomous. The former are very vulnerable to countermeasures and introduce possibly unacceptable time delays, but they are highly efficient in resource allocation. The latter are more survivable, but they are rather inefficient in their utilization of scarce resources.[ATH87] Obviously, there is a spectrum of architectural choices between these two extremes.

Communications technology is fundamental to all BM/C^3 functions. Any communications architecture must simultaneously support a sensor architecture, a weapons architecture, and a command and control architecture. These communications links should be transparent. While communications by themselves do not convert the various processing algorithms into command and control functions, critical battle management problems arise when the communications are not transparent to all users. They become conspicuous when the links associated with the NCA are disturbed or interrupted. Clearly then, a primary architectural issue in strategic defense BM/C^3 is that of the organization of the available communications assets. This subject should be phrased in terms of desirable topological properties of the communications network.

Thus, the processing algorithms at each node of the strategic defense system network cannot be designed without a careful delineation of the communication network that supplies the node. This depends upon the functional requirements to be met at the nodes. That is, on the degree of autonomy with which the elements of the system perform their functions. The current preference is for the system components to perform "implicit coordination," which has a highly autonomous structure. This requires the development of a class of "active" algorithms which will clarify anomalous data and seek missing data by active search. To accomplish this, an algorithm at a given processing node must be able to infer the network state and then choose how to best obtain the requisite data—that is, it must be imbued with some (artificial) "intelligence."

The question of "what AI can do for battle management" is of concern to a wide range of systems, AI, and BM/C³ specialists, as well as "operators." The first AAAI Workshop on AI Applications to Battle Management was held on 16 July 1987 at the University of Washington. A synopsis of the findings has been reported.[BON88] This workshop came up with three lists:

[1] what problem-solving battle-management tasks the AI community knows how to do; [2] what tasks we cannot do and have little hope of doing before 1997; and [3] what tasks we think we might know how to do, but it is still too soon to tell.

Very briefly summarizing, the report says that AI technology can be applied to do hierarchical planning, stereotypical planning, sensor fusion, intelligent decision aids and user interfaces, and better simulations in support of battle management. In addition, it was stated that "(James) Allen's interval formulation for time is well-suited for temporal representation in battle management." Areas where AI technology will not be able to support battle management in the near term (i.e., less than 10 years) were natural language understanding of messages due to the inability to read free text; building deep-knowledge systems; and the inability to manage uncertainty and reason in real time. Areas where additional efforts might show some nearer term payoffs are in the realms of replanning, cooperating systems, situation assessment, wargaming, and terrain reasoning.

The aforementioned workshop discussed the application of artificial intelligence across the breadth of battle management; no attempt was made to focus on the application of AI technology to the problems of BM/C³ for a strategic defense system. Looking at the above summary and the earlier discussion of the strategic defense system, one can see that relatively few of the areas identified as "key needs in battle management of AI research topics for years to come" are significant to the strategic defense BM/C³ problem. The domain is comparatively circumscribed when compared with the classic tactical AirLand Battle Management problem domain (or the naval equivalent). These "key needs" then include managing uncertainty and reasoning in real time, as well as cooperating systems. The ability or inability to read free text is of no consequence for a strategic defense BM/C³ system. The battle to be managed will take place far above the Earth's surface making terrain reasoning, likewise, of no consequence. The system will not have to be imbued with deep-knowledge, nor will there be sufficient time during an engagement for replanning and wargaming. "Situation assessment" as commonly discussed in the BM/C³ community will not be a need; however, it is most likely that some sort of "adaptive control system" will be needed. The reason for this is that ballistic missiles, both ICBMs and submarine launched, have calculable, predictable trajectories, thus making the problem more tractable. The cruise missile problem will require more "adaptive control" than "situation assessment" during an engagement.

SDI Employment Authorization

To understand the BM/C^3 problem in context, it is necessary to review the requirements of the system as relates to subordination of the automation to human control. The following paragraphs are excerpted from the *Final Report of the Organizing Committee of the First Workshop on SDI:BM/*C^3. [DOD88]

To be effective, a strategic defense system must have the capacity to evolve over time. The system will probably be established with a basic set of elements. During the system's life, these component elements will be modified or replaced, and their roles will be expanded or contracted as changing demands are placed upon the system. Thus, a flexible (phased) system implementation, subject to reevaluation on the basis of technological improvements and changing world conditions is an indispensable requirement.

It is envisioned that system control and management during deployment, during long-term quiescent operation, and during brief periods of partial/total engagement will be exercised by the President, the Secretary of Defense and duly authorized successors. Collectively they are the National Command Authorities (NCA), and the system must be such that this primary human command and control function can be exercised expeditiously and unambiguously.

System readiness and the capability of the system to respond properly depends upon the ability of the communication network to supply the requisite data to each of the nodes in an expeditious manner. This is true whether the system architecture is hierarchical or decentralized. The flexibility inherent in an architecture derives in large part from the adaptability of the communication network. Because of its fundamental importance, the communications architecture should be responsive to direct control of the NCA. Research is recommended on methodologies which can be used to compare and select appropriate command and implementation structures which yield a robust anthropocentric system.

Human control over the strategic defense system is to be exercised at several levels, for diverse purposes and on disparate time scales. The NCA must be constantly apprised of system readiness, threat magnitudes, and current world conditions during deployment, modification, and protracted periods of quiescent operation. The NCA must make prompt decisions and issue commands in response to perceived threats. They must authorize modification of defense doctrine in an appropriate dynamically varying strategic situation which may include a nuclear exchange.

In the overall strategic defense system there are numerous interactions and trade-offs to be made in the allocation of BM/C^3 functions to the different levels within the hierarchy and to different nodes within a level. This allocation is made on the basis of considerations of cost, reduced sensitivity to countermeasures, and uncertainties in the way that the encounter will evolve. A flexible architecture is essential to permit a rational maturation through the inclusion of advanced subsystems and components.

The role of the human decision maker in the strategic defense system does not alter this fundamental view of BM/C^3. Many of the BM/C^3 functions can be performed either by a human or by an algorithmic surrogate. Clearly, any task involving the expeditious processing of large quantities of data is of necessity outside the human purview. A judicious aggregation of the data can enable a human to adroitly make those decisions upon which the success of the strategic defense system depends. Changes in rules-of-engagement, demonstrations of national will and resolve, and placing the strategic defense system on alert status are inherently human decisions that must be referred to the NCA acting in a direct management role.

The need for a human presence is a controversial theme in discussions of complex systems, both military and civil. For example, the space shuttle is under human control except for phases of ascent and descent. In NASA, there is such a lack of trust in the ability of computers to automatically detect off nominal operating conditions, that computers are not permitted to turn off other computers. This can only be done by a person in authority.[ROG86]

The NORAD system provides the closest operational analog to at least part of the strategic defense system and the principal operating decisions are all made by humans working individually or in teams. Although the anecdotal evidence suggests that some of the most untoward system errors have been caused by misguided human intervention, there is no indication of any intent to reduce the human presence in the future. Instead, the proposals from both NASA and NORAD would increase the human presence by providing aids that would enhance human effectiveness.

Rigorous reasons for this fixation on giving a human an online role have not been clearly articulated. In many discussions of complex systems, the human is not thought of as an integral part of the system architecture, but is rather given an external position as a "user" of data or an "input" to the rest of the system. This extrinsic view of the human has led some investigators to relegate the human to a very limited position in the system. It has been suggested that "the tactical system will have to operate in an automated mode simply because there is no time for humans to evaluate the huge amounts of sensor information and to arrive at superior weapon engagement in the short time available." [ATH87]

There are significant practical limits on the role that can be played by humans. The basic issue is how to definc and support the role of human decision making, given the characteristics and limitations of human information processing capabilities. To clearly display the unique contribution that can be played by the human, a clear understanding of his areas of proficiency is essential. Some of the areas of particular human competence are "...decision making in semantically rich problem domains, ... analogical reasoning and problem structuring, ...(and) information processing and application of heuristics." [WEL80]

What is sought is a system that utilizes the best combination of both automated and human decision making to accomplish the mission requirements while at all times recognizing the fundamental need for a *political decision maker* to direct the strategic posture of the nation in an ongoing manner. While there is a continuum of different possible roles for the human decision maker, two distinct categories of responsibility are evident. Thus, it is for the NCA to determine the appropriate response to any given strategic situation and all possible assistance should be given to them in order that their choice of action and their changes in rules-of-engagement be judiciously made. At a lower level in the strategic defense system hierarchy, a decision maker must allocate specific resources and formulate defensive strategy in a timely manner.

In the final analysis, if the system is to have the highest "probability of kills" and thereby destroy the greatest number of incoming enemy missiles, then clearly the strategic defense system must be employed with the first two to four minutes—that is, during boost phase. It is extremely unlikely that the NCA can be contacted, a decision made, and authorization of system employment granted within that time frame. Thus the system would have to be imbued with the ability to "authorize" its own employment. On the other hand, fifteen or twenty minutes is a much more realistic time frame for the NCA to be contacted and authorize employment of the strategic defense system. This means that the system will have to deal with the problems associated with late mid-course/early terminal phase interception. The overall "probability of kill" is greatly reduced; many more enemy missiles would reach their targets. In essence, the selection of this time-frame option would make the system requirements the same as those of the Anti-Ballistic Missile (ABM) systems being considered during the early 1970s (prior to the ABM treaties).

President Reagan's Strategic Defense Initiative (SDI) sought to design and develop a system vastly superior to the capabilities of the 1970s ABM systems. To achieve this superior capability effectively means boost phase interception is a systems requirement. As has been shown, due to the severe time constraint, for the strategic defense system to intercept during boost phase would require autonomy of the system—"the system" would have to be imbued with and delegated the Presidential authority to make the decision and employ the strategic defense system. It ultimately remains for the human to make the decision whether the strategic defense system will require direct man-in-the-loop control over system employment and thereby select the time frame for interception. This very human decision (aided or unaided) must be made before final system requirements become system design specifications.

References

[ATH87] Athans, M., "Command and Control (C^2) Theory: A Challenge to Control Science," *IEEE Trans. on Automatic Control,* Vol. AC-32, April 1987.

[BON88] Bonasso, Peter R., "What AI Can Do for Battle Management," *AI Magazine*, Fall 1988.

[DAV79] Davis, William A., Jr., "Ballistic Missile Defense into the Eighties," *National Defense* (September-October) 1979.

[DOD88] Department of Defense, "Final Report of the Organizing Committee of the First Workshop on SDI:BM/C^3," 1988.

[GAR81] Garwin, Richard L., "Are We on the Verge of an Arms Race in Space?," *Bulletin of Atomic Scientists*, May 1981.

[LEV87] Levis, A. H. and M. Athans, "The Quest for a C^3 Theory: Dreams and Realities," 1987 Symposium on C^2 Research, August 1987.

[ROG86] Rogers, William P., et al., Report to the President by the Presidential Commission on the Space Shuttle Challenger Accident, Washington, DC, 6 June 1986.

[SCH80] Schneider, William, Jr., et al., *US Strategic Nuclear Policy and Ballistic Missile Defense the 1980s and Beyond,* Cambridge: Institute for Foreign Policy Analysis (IFPA), 1980.

[SME78] Smeroff, Barry J., "Strategic and Arms Control Implications of Laser Weapons: A Preliminary Assessment," *Air University Review* (January-February) 1978.

[VAN80] Van Trees, H. L., "Keynote Address," Proceeding for Quantitative Assessment of Utility of Command and Control Systems, National Defense University, January 1980.

[WEL80] Welch, J. A., "State of the Art of C^2 Assessment," Proceeding for Quantitative Assessment of Utility of Command and Control Systems, National Defense University, January 1980.

Toward Analytically Aiding the Cost-Effective Allocation of Training Resources

Leonard Adelman, John F. Patterson,
Robert F. Gibson and Michael L. Donnell

Multi-Attribute Utility Theory (MAUT) represents an axiomatic based approach to quantifying the judgments inherent in evaluating whether various alternatives satisfy decision making objectives [Keeney and Raiffa, 1976; Von Winterfeldt and Edwards, 1986; and Watson and Buede, 1987]. MAUT and the methods developed for implementing it have been applied in numerous settings, including the military. For example, MAUT has been used by various procurement agencies to evaluate prospective systems [e.g., Bresnick, 1982; O'Connor, 1982]. A MAUT model is an integral part of the U.S. Marine Corps standard combat evaluation procedure [Barsily, Catalogne, and Marlow, 1981]. And Thompson et al. [1988] are currently using MAUT concepts and procedures to develop a measurement instrument for assessing the value of tactical intelligence output to different C^2 users.

This paper will describe how an extension of MAUT—MAUT based cost-benefit analysis—can be used to analytically aid the cost-effective allocation of C^2 training resources. In general, MAUT based cost-benefit analysis is used to help design the most beneficial option packages for different levels of cost. Within the military, it has been used to develop advanced helicopter designs [Adelman, 1984], critical aspects of the U.S. Marine Corps annual budget [Watson and Buede, 1987], and a development plan for decision support system (DSS) technology [Rockmore et al., 1982]. In this chapter, we describe how MAUT based cost-benefit analysis was embedded in a DSS prototype for assisting in the cost-effective allocation of training resources to units in the Marine Air Command and Control System (MACCS). The prototype is called TRACES for Training Requirements and Cost Evaluation System.

It is important to note at the outset that, although TRACES was well received by the personnel at the Operations Division of the Plans, Policies, and Operations Department at Marine Corps Headquarters for whom it was developed, it has not been converted from an R&D to an operational system. Although there are reasons, it is not our purpose to discuss them here. Rather, our purpose is to highlight the MAUT based cost-benefit modeling approach embedded in TRACES because that approach has potential utility in numerous other domains.

The remainder of this chapter is divided into three major sections. The first section overviews the MACCS allocation problem as defined during the time of the project. [Note: The Marine Corps organizational types, their configurations, and the benefit and cost assessments were formulated in the 1981 time period.] The second section describes the technical approach and provides an example of how TRACES is used to advise the user of cost-effective allocations of training resources of the MACCS. The third section discusses the distinctive features of the approach, with a focus on the generality of the technology to other C^2 training environments.

The MACCS Allocation Problem

The Marine Air Command and Control System (MACCS) supports ground units in eleven types of major training exercises. A field exercise with a Marine Amphibious Force (MAF) is the largest and most complex training exercise and will be used as our example throughout this paper. The basic decision facing MACCS personnel who are supporting a field exercise with a MAF is how to deploy the six agencies within the MACCS so they receive the greatest increase in readiness for a specific cost. This decision problem is represented in Table 1. The six rows represent the six agencies in the MACCS and the columns represent the possible deployment configurations for these agencies. As indicated in Table 1, MACCS personnel could select:

- a nondeployed or deployed Tactical Air Command Center (TAAC),
- a nondeployed, an Early-Warning, or a deployed Tactical Air Operations Center (TAOC),
- a full Direct Air Support Center (DASC) or a full-DASC plus a mini-DASC,
- up to three batteries of Light Anti-Aircraft Missiles (LAAM),
- up to three Air Support Radar Teams (ASRTs), and
- up to two platoons of Forward Area Air Defense (FAAD) with ordnance.

The MACCS chosen to support the field exercise with the MAF would be defined by the one deployment configuration for each of the six agencies.

AGENCIES	DEPLOYMENT CONFIGURATIONS							
	1	2	3	4	5	7	8	9
TACC	Non-Deployed	Deployed						
TAOC	Non-Deployed	Early Warning	Deployed					
DASC	Full	Full + Mini						
LAAM	None	1 BTRY	2 BTRY	3 BTRY				
ASRT	None	1 ASRT No Ord-nance	1 ASRT Ord-nance	2 ASRT No Ord-nance	2 ASRT Ord-nance	3 ASRT No Ord-nance	3 ASRT Ord-nance	
FAAD	None	1 PLATOON No Ord-nanc	1 PLA-TOON Dummy Ordnance	2 PLA-TOONS No Ord-nance	2 PLA-TOONS Dummy Ordnance			

Table 1. MACCS deployment configuration choices for field exercise with a MAF.

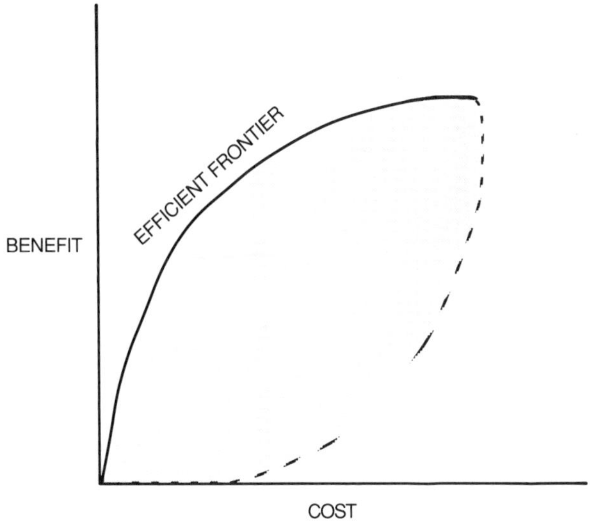

Figure 1. Benefit-cost representation of possible deployment configurations.

Each successive deployment configuration within each agency provides more training benefit, but for increased cost. The problem is knowing which deployment configuration for each of the six agencies will provide the most cost-effective training for the MACCS overall. Considering the present example, there are 1,120 (2x2x2x4x7x5) possible deployment configurations for a MACCS supporting an integrated field exercise with a MAF. The 1,120 possible deployment configurations are uniquely defined by the combination of configurations for the six agencies. Each of these 1,120 configurations results in a potentially different degree of overall readiness improvement for the MACCS for a different amount of overall dollar cost.

The above situation is depicted in Figure 1. Each dot represents one of the 1,120 possible MACCS deployment configurations; its location in Figure 1 depends on its level of total benefit (i.e., readiness return) and total cost. The set of configurations tends to form a football-shaped region bounded above by what is called "the efficient frontier." The MACCS deployment configurations on this frontier are the most cost-beneficial configurations because they provide the most benefit for a given level of cost. These are the deployment configurations that the decision aiding method was designed to identify.

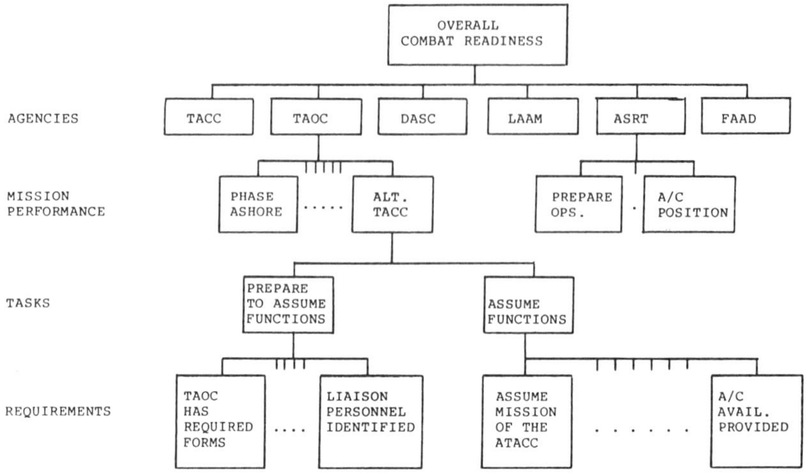

Figure 2. Schematic MAUA model of MCCRES for Marine air command and control.

Technical Approach and Illustrative Example

The MAUT based cost-benefit modeling approach is composed of two major components: (1) An evaluation model that specifies how well each MACCS agency (as well as the MACCS as a whole) is performing each of its primary tasks, and (2) an allocation model that specifies the most beneficial major training option and associated agency deployment configuration for a specific level of cost. Each component is considered below, in turn.

MCCRES: Evaluating Combat Readiness

The Marine Corps Combat Evaluation System (MCCRES) is the evaluation component of the system. It incorporates a MAUT model that permits the systematic assessment of a unit's combat readiness. Figure 2 is a schematic of the MAUT model of MCCRES for the MACCS. The top-level factor is the overall combat readiness of the MACCS. At the next level down are the factors representing the individual combat readiness of each agency in the MACCS. The individual agencies are, in turn, decomposed into separate categories of tasks that represent the appropriate mission performance standards (MPSs) for the MCCRES evaluation for each agency. These MPSs are decomposed into specific tasks which are then decomposed into specific requirements for each task that represent observable activities. Thus, the scores on different activities are systematically aggregated (in a weighted additive fashion) to provide evaluation scores for individual performance areas and thereby, yield an overall performance score for the MACCS.

The MAUT model is used to provide an overall combat readiness score for the MACCS. First Marine Corps evaluators rate in a simple "yes" or "no" manner whether the agencies satisfied each of their requirements during the MCCRES evaluation. An agency's score of each of its tasks is calculated by multiplying the relative importance weights assigned to the requirements comprising a task by the ratings on those requirements and summing the products. An agency that failed to satisfy important requirements on a task would receive a low score on that task. In a similar fashion, the agency's score on each MPS is calculated by multiplying the relative importance weights assigned to the tasks by the agencies' scores on those tasks and summing the products; a low score on an MPS implies that the agency performed poorly on important tasks within that MPS. The MPSs are then assigned relative importance weights which are used in deriving an overall combat readiness score for the agency. The more combat-ready the agency, the higher the overall score produced by the MCCRES evaluation. Thus, poor overall agency performance can be readily attributed to poor performance on specific mission performance standards, tasks, and requirements. Finally, if so desired, the agency's scores can be weighted and summed to provide an overall combat readiness score for the MACCS.

TRACES: Matching Readiness Return and Training Resources

TRACES is the allocation portion of the modeling framework and portable computer software developed to implement it. As inputs, TRACES uses the MCCRES combat readiness scores for the MACCS. In its memory, TRACES has cost and benefit assessments for all possible deployment configurations of the six MACCS agencies for each of 11 major training exercise options. The cost assessments represent a constructive guess at the proportional costs for the operational deployment costs for each of eight possible variations of a five-day field exercise, and each of three possible variations of a three-day command post exercise, for each deployment configuration of each MACCS agency. (These costs were not based upon empirical cost data and were created only to test the model.) The benefit assessments, again made for each deployment configuration of each MACCS agency for each training exercise, represent experts' best estimates of the expected increase in combat readiness for each MPS. As outputs, TRACES specifies the training exercise and deployment configuration that provides the greatest increase in the MACCS' overall combat readiness for different, specified levels of cost.

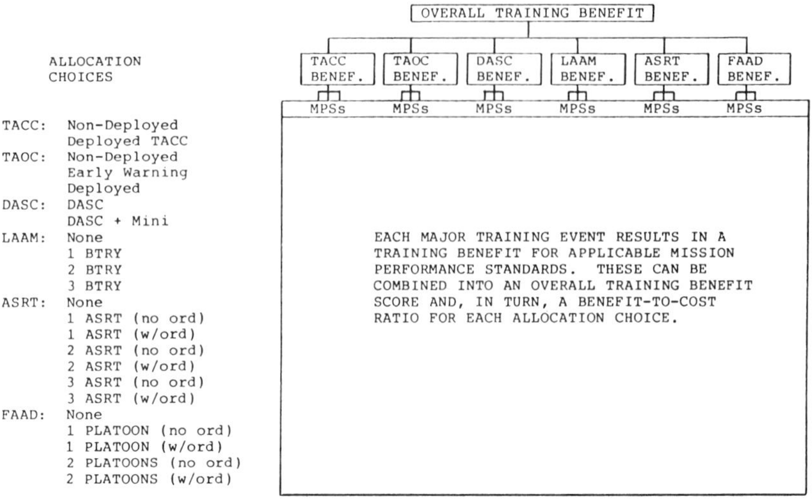

Figure 3. Schematic representation of the benefit model for a field exercise with a Marine Amphibious Force.

Figure 3 presents a schematic of the benefit model within TRACES for the Field Exercise with a MAF (one of the 11 possible major training exercises). The benefit model includes the same MAUT hierarchy as in MCCRES to ensure the explicit integration of the evaluation and training components of the overall system. The top-level factor is the overall benefit to the MACCS that is produced by any of the 1,120 possible deployment configurations that might be selected to support a field exercise with the MAF. Overall benefit is decomposed into the benefits obtained for each of the MPS for each of the six agencies. (Marine Corps personnel thought the MPSs accurately reflected MACCS strengths and weaknesses; therefore, tasks and requirements were not included in the model.) The TRACES software contains similar benefit models for each of the other 10 major training exercises.

Just as the MACCS hierarchy is used in MCCRES to calculate an overall combat readiness score during an actual evaluation, the same hierarchy is used in TRACES to predict the number of MCCRES points one can expect to make up with a particular training event for a specified exercise, where an event is comprised of one deployment configuration for each agency. By moving up the hierarchy, one estimates the training event's benefit for each agency's MPSs, for the agency overall, and then, if so desired, for the entire MACCS. If cost were not a factor in allocating training resources, one would simply deploy the configurations that most effectively train those MPSs and agencies that performed poorly in the last MCCRES evaluation. Since cost is a factor in allocating training resources, the TRACES software identifies the training events that have the

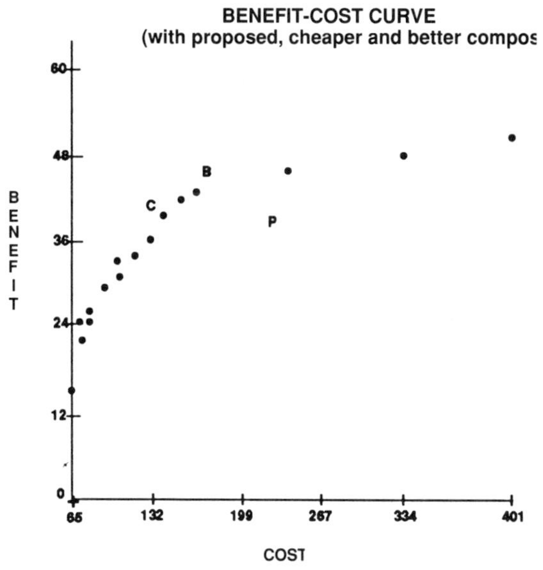

PROPOSED EXERCISE
FX; WITH MAF; BOTH
AD+AS

Agency	Deployment
TACC	Deploy
TAOC	Deploy
DASC	Full DASC
LAAM	1 (NO MI)
ASRT	1 (NO ORD)
FAAD	1 PL(-ORD)

COST	BENEFIT
241	38

CHEAPER EXERCISE
FX; WITH MAF;
BOTH AD+AS

Agency	Deployment
TACC	Nondeploy
TAOC	Nondeploy
DASC	Full DASC
LAAM	1 (NO MI)
ASRT	1 (WI ORD)
FAAD	2 PL(-ORD)

COST	BENEFIT
122	38

BETTER EXERCISE
FX; WITH MAF;
BOTH AD+AS

Agency	Deployment
TACC	Nondeploy
TAOC	Nondeploy
DASC	Full + mini
LAAM	1 (NO MI)
ASRT	3 (WI ORD)
FAAD	2 PL(+ DUM)

COST	BENEFIT
191	43

Figure 4. Using traces to identify more cost-effective MACCS deployment configurations than the one proposed.

highest overall benefit-to-cost ratios for different levels of cost. These events represent the deployment configurations that lie on the efficient curve, as illustrated in Figure 1.

TRACES also identifies the number of MCCRES points that the MACCS can be expected to make up by implementing each of the training events. In addition, the software helps MACCS decision makers evaluate the overall adequacy of their initially proposed training event. This is accomplished by identifying whether the total benefit-to-cost ratio for the proposed training event (deployment configuration for the six agencies) lies on the efficient frontier. This capability, as well as those mentioned in the above paragraph, is illustrated in Figure 4.

The proposed event is identified by the P. Based upon the assessments made by Marine Corps experts in MACCS training, a hypothetical set of scores for MACCS performance and rough cost estimates, it can be seen that the initially proposed training event does not lie on the efficient frontier in this example. The TRACES software does, however, identify two alternative training events that do lie on the efficient frontier. The C (for cheaper) alternative provides the same or more benefit for less money; the B (for better) alternative provides more benefit for the same or less cost.

The table below the graph in Figure 4 identifies the three training events. The proposed event contains the following: a deployed TACC, a deployed TAOC, a full DASC, one LAAM battery, one ASRT with no ordnance, and one FAAD platoon with no ordnance. In our hypothetical example, the MACCS could receive the same benefit, an additional 38 points on the MACCS' next MCCRES evaluation, for about $120,000 less by not deploying the TACC or TAOC and, instead, deploying a second platoon of FAAD. In contrast, the MACCS could receive approximately 16 percent more benefit (i.e., an additional six points on the MACCS' next MCCRES evaluation) than the proposed training event for $50,000 less if it did not deploy the TACC, did not deploy the TAOC, used both a full and mini-DASC, kept the one LAAM battery, had three ASRTs with ordnance, and had two FAAD platoons with dummy ordnance. These two training event deployment configurations have higher benefit-to-cost ratios than the proposed one because, in the judgment of experts in Marine Air Command and Control, these configurations will exercise important performance areas requiring remediation for either less money, more benefit, or both less money and more benefit than the proposed exercise.

The expert judgments relating training resources to readiness are an important distinctive characteristic of the modeling approach. The experts quantitatively defined benefit as the percentage of the deficit made up (PDMU) on a MPS by the chosen deployment configuration. Consequently, the more effective the configuration in training a given MPS, the higher the PDMU.

Table 2 illustrates the benefit assessments (PDMUs) for the alternative agency deployment configurations for a field exercise with a MAF for each of the five TACC MPSs. The table shows that, in the opinion of the experts, a nondeployed TACC can make up, on the average, 20% of the deficit on the "phase control ashore" MPS, 25% of the deficit on the "display" MPS, 60% of the deficit on the "management of aircraft" MPS, 10% of the deficit on the "external agencies" MPS, and 35% of the deficit on the "succession of command and control" MPS. In contrast, a deployed TACC can be expected to make up 50%, 25%, 60%, 30%, and 45%, respectively, on the five MPS, and so forth. As can be seen, the readiness return for the more expensive deployed TACC, over the less expensive nondeployed TACC, depends on the MPS. For example, the deployed TACC is two and one-half times more effective than the nondeployed TACC on "phase control ashore," yet no more effective on "man-

	PHASE CTL	DSPLY	MNG. A/C	EXT. AG.	SUCC OF C^2
TACC: Nondeployed	20	25	60	10	35
Deployed TACC	50	25	60	30	45
TAOC: Nondeployed	20	44	20	50	20
Early Warning	10	22	10	10	20
Deployed TAOC	22	44	20	70	30
DASC: DASC	20	25	20	0	15
DASC + Mini	22	25	20	0	15
LAAM: None	0	0	0	0	0
1 BTRY	2	2	0	0	5
2 BTRY	2	2	0	0	5
3 BTRY	2	2	0	0	5
ASRT: None	0	0	0	0	0
1 ASRT (no ord)	2	2	0	0	0
1 ASRT (w/ord)	2	2	0	0	0
2 ASRT (no ord)	2	2	0	0	0
2 ASRT (w/ord)	2	2	0	0	0
3 ASRT (no ord)	2	2	0	0	0
3 ASRT (w/ord)	2	2	0	0	0
FAAD: None	0	0	0	0	0
1 PLATOON (no ord)	2	2	0	0	5
1 PLATOON (w/ord)	2	2	0	0	5
2 PLATOONS (no ord)	2	2	0	0	5
2 PLATOONS (w/ord)	2	2	0	0	5

Table 2. Relative effectiveness scores for TACC MPSs for a field exercise with a MAF.

agement of aircraft," an MPS for which a nondeployed TACC is extremely effective.

It must be emphasized that these and the other benefit and cost assessments should not be considered to be objective data, but rather, to be the best subjective estimates of officers knowledgeable in Marine Air Command and Control. In many operational training environments, expert judgment is the only feasible means of relating readiness and resources; consequently, it is relied upon daily. By using experts' subjective assessments of cost and benefit, the computer program implementing the MAUA modeling framework assumes the role of an adviser whose duty is to recommend a particular course of action and supply the rationale for it.

Others, of course, may disagree with this advice and with the cost and benefit assessments upon which it is based. Consequently, the computer program implementing the modeling framework permits users to enter their own estimates of the costs (fixed and operational) and benefits, as well as agency and

MPS weights. Users can, in turn, determine whether their judgments result in the selection of a different set of cost-beneficial training events. In this way, users can perform sensitivity analyses to evaluate the implications of differences in opinion with the MACCS experts who made the initial assessments for the aid. By doing so, the modeling approach expands the users' ability to evaluate the adequacy of their own as well as their advisors' judgments on the relationship between resources and readiness.

Conclusion

We close this paper by discussing three points about the MAUT based cost-benefit analysis approach to allocating training resources, as implemented in TRACES. First, it emphasizes expert judgment. Abell [1981] noted that, in general, there were two approaches for estimating readiness as a function of resources. "The first includes econometric models that depend on regression analysis of data reflecting historical or empirical observations of resource expenditures." (p. 13) "The second kind of approach postulates mathematical models of processes, relationships, and resources, typically at a detailed level." (p. 15) Although Abell prefers the second approach, he notes that, "Detailed data are required and the models employed tend to be tedious, cumbersome, and costly." (p. 15)

TRACES uses a third approach, expert judgment, to relate readiness and resources. In most training environments, expert judgment is the only feasible means of relating readiness and resources; consequently, it is relied on daily. Notwithstanding the potential benefit of expert system technology, system developers should not forget that MAUT provides an axiomatic approach to representing knowledge that has had a long and successful history. In developing TRACES, for example, tried-and-true paired comparison procedures were used to help the two participating MACCS personnel judge the relative effectiveness of each training resource on each performance measure. As a result, benefit-to-cost ratios were obtained in a coherent, explicit fashion to represent the judgment of military experts on the relation between readiness and training resources.

The second point is that TRACES' technical approach is applicable to most training environments. Conceptually, the DSS has only two components: an evaluation and a remediation component. The evaluation component measures the unit's level of performance at any given time. The remediation component then predicts the relative effectiveness and cost of different training efforts given the unit's performance. Since all training programs have both an evaluation and a remediation component, although they may not be as explicit and retraceable as the Marine Corps', the potential exists for adapting the DSS to many, if not all, training environments.

Third, TRACES, in terms of a metaphor to describe its approach to computer-aided problem solving, represents a knowledgeable subordinate [see

Andriole et al., 1986; Patterson, 1981]. In contrast to expert system technology, however, the goal is to provide an alternative, normative versus descriptive perspective that decision makers can match against their own. The goal was to organize (i.e., engineer) expert judgment within the framework of a MAUT based cost-benefit analysis, not represent it in terms of problem solving heuristics. Whether this approach is better or not is an empirical question beyond the scope of this chapter. However, the technical approach does permit one to take full advantage of the theoretical and computational power provided by MAUT and cost benefit analysis and, thus, to expand one's ability to think about the problem at hand from a different perspective. Moreover, a MAUT based cost-benefit analysis readily permits decision makers to perform sensitivity analysis in order to assess the implications of differences (or modifications) in judgment. The DSS does not, however, produce an answer; that is the job of the decision maker, not the subordinate.

Endnote

The aid described in this paper was developed under Contract M00027-81-C0052 from the U.S. Marine Corps to Decisions and Designs, Inc. The views expressed in this paper represent those of the authors and are not an official position of the U.S. Marine Corps or any other organization.

References

Abell, J. B., "Resources to Readiness: The Tactical-Fighter Case," *Defense Management Journal*, Second Quarter 1981.

Adelman, L., "Real-Time Computer Support for Decision Analysis in a Group Setting: Another Class of Decision Support Systems," *Interfaces*, vol. 14, 1984, pp. 75-83.

Andriole, S. J., Black, H. H., Hopple, G. W., and Thompson, J. R., "Intelligent Aids for Tactical Planning," *IEEE Transactions on Systems, Man, and Cybernetics*, vol. SMC-16, 1986, pp. 854-864.

Barsily, Z., Catalogne, P.R., and Marlow, W. H., "Assessing Marine Corps Readiness," *Defense Management Journal*, First Quarter, 1981.

Bresnick, T. A., "Source Selection for the 80's: Light Armored Vehicle (LAV) Case Study," *49th Military Operations Research Symposium (MORS)*, Kirtland AFB, New Mexico, June 1982.

Keeney, R. L. and Raiffa, H., *Decisions with Multiple Objectives*, Wiley, 1976.

O'Connor, M. F., "Relating Promised Performance to Military Worth: An Evaluation Mechanism," *Defense Management Journal*, 3 October 1977, pp. 36-46.

Patterson, J. R., Randall, L. S., and Stewart, R.R., *Advisory Decision Aids: A Prototype*, Decision and Designs, Inc., 1981.

Rockmore, A. J., Hemphill, R. A., Riemenschneider, R. A., Donnell, M. L., and Lehner, P. E., *Decision Aids for Target Aggregation: Technology Review and Decision Aid Selection*,

PAR Technology Corporation, 1982.

Thompson, J. R., Landee-Thompson, B., Fichtl, T. C., and Adelman, L., *Measurement and Evaluation of Military Intelligence (MI) Unit Information Processing Performance: Year 1 Technical Report*, Technical Report 88/6504, Science Applications International Corporation, McLean, VA, March 1988.

von Winterfeldt, D. and Edwards, W., *Decision Analysis and Behavioral Research*, Cambridge Univ. Press, 1986.

Watson, S. and Buede, D. M., *Decision Synthesis: The Principles and Practice of Decision Analyis*, Cambridge University Press, 1987.

PART 6
Advanced Communications Technology

ATCCS: An Integrated C^3 Environment

R.J. Rechter; D.F. Bender; W. A. Holdrege;
Maj. J. Walbeck, USA; and C. Pizzutelli, Jr.

The concept of integrated battlefield command, control and communications (C^3) is not new and has always been an important objective. Success on the modern battlefield cannot be achieved in the absence of integrated C^3. The U.S. Army's requirements and development communities must ensure that the user battlefield automated systems (BASs) and the supporting tactical communications systems evolve compatibly over the next decade toward the primary goal: an integrated C^3 environment.

Performance requirements for Army Tactical Command and Control Systems (ATCCS) have been developed based on the operational needs of the five battlefield functional areas (BFAs)—Maneuver, Fire Support, Air Defense, Intelligence/Electronic Warfare (EW) and Combat Service Support. The BASs supporting each BFA were designed to meet the degree of distributed functionality required and to provide corresponding distributed processing, data storage and display capabilities. The tactical communications systems—Area Common User System (ACUS), Army Data Distribution System (ADDS) and Combat Net Radios (CNRs)—must provide the communications backbone for each BAS within its intended operational area. They also must provide organic support in their integrated force level support roles throughout ATCCS.

The fluidity, tempo and tight coordination required of maneuver forces under the Army's AirLand Battle concept and future Army 21 doctrine require that tactical communications systems be developed in parallel with the command and control (C^2) systems they service. Not only must C^2 and communications systems evolve compatibly in their technical response to evolving user requirements and a changing threat environment, they also must make maximum use of existing, fielded assets.

For tactical C^2 systems, this means a carefully controlled evolution of applications software in a standardized, nondevelopmental item (NDI) platform environment (for example, the Army Command and Control System [ACCS] Common Hardware Software [CHS]). For tactical communications, this reuse and capability extension will include deployment of packet switch appliques to circuit switched bearer systems (for example, Mobile Subscriber Equipment [MSE]). (The Program Executive Offices [PEOs] for C^2 Systems and Communications currently are working in conjunction with the Communications-Electronics Command [CECOM] Research and Development [R&D] Center and the Signal School to ascertain the best, most economical solution to the packet switch issue for their automated systems.) The development of new techniques to manage multiple, interconnected, tactical communications media effectively (for example, Integrated System Control) will aid in servicing future C^2 system user needs.

The successful integration of the C^2 and communications communities must maximize interface compatibility among BFAs and minimize C^3 system breakage as C^2 user requirements and organic communications capabilities evolve over time. Any major changes in digital data support capability, jam resistance, channelization and automated user system interface capabilities will affect overall C^3 system performance dramatically.

Automated methods for adaptively interconnecting the available media to provide appropriate overall functional connectivity should be the goal of any future tactical communications systems development. Communications media assets should be presented to the C^2 user in a thoroughly transparent manner. Ideally, any C^2 user should be able to gain required voice and data access to supporting systems, regardless of the communications environment, in response to the fluidity of the AirLand Battle.

ATCCS as Battlefield C^3 Integrator

The ATCCS shown in Figure 1 is an integrated family of automated, interoperable C^2 and communications systems. This federation of systems provides C^3 support to the five BFAs. The support is provided via the ACUS, in which the new MSE is replacing the old Tri-Service Tactical (TRI-TAC) architecture; the ADDS, which consists of the Enhanced Position Location Reporting System (EPLRS) and the Joint Tactical Information Distribution System (JTIDS); and the CNRs in which the Single Channel Ground and Airborne Radio System (SINCGARS) and Improved High Frequency Radio (IHFR) are replacing the old VRC-12 and GRC-106 radios.

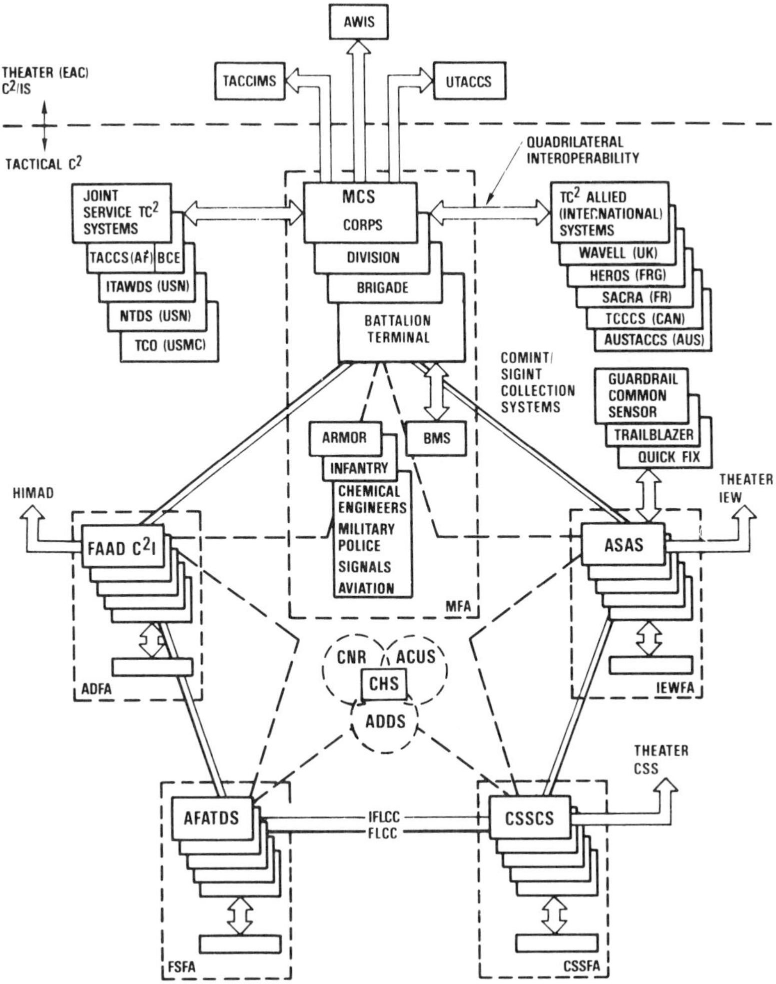

Figure 1. Army Tactical Command and Control System (ATCCS) architecture and its theater interfaces.

Partial functional C[2] integration currently is being achieved within ATCCS through the Maneuver Control System (MCS), followed by Segment 11 implementation of the Initial Force Level Control Capability (IFLCC). The ultimate ATCCS C[3] integration objective hinges on the future implementation of the Force Level Control Capability (FLCC) and fully integrated digital data transport in the supporting tactical communications systems.

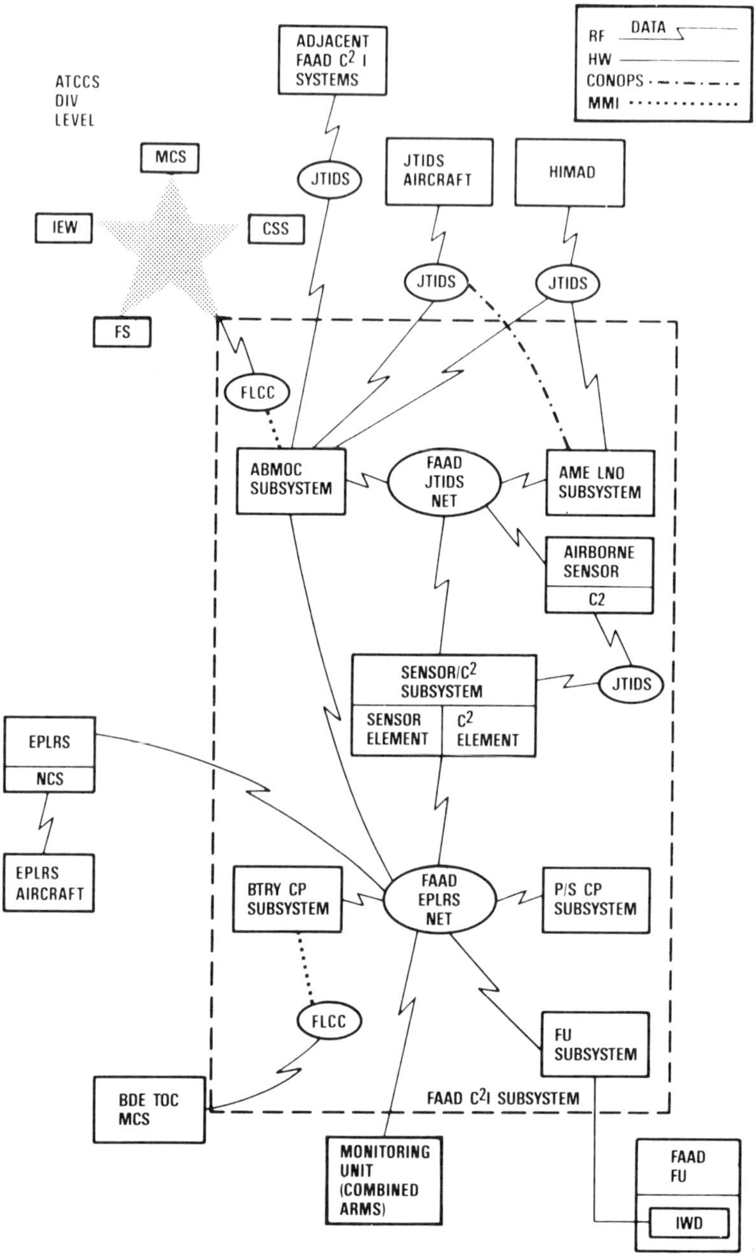

Figure 2. Overall Forward Area Air Defense Command, Control and Intelligence (FAADC2) functional connectivity.

The following describes how the various organic tactical communications systems function to meet the top-level communications requirements of two ATCCS BFAs—Air Defense and Maneuver.

Forward Area Air Defense Command, Control and Intelligence (FAAD C^2I)

The BAS for Air Defense has the most rigorous response time and data quality requirements of all the BFAs. It also is characterized by broadly distributed functional area assets, distributed processing and data base management and hierarchical and joint service interfaces. The FAAD C^2I system architecture, shown in Figure 2, has both internal and external requirements, communicating with High and Medium Altitude Air Defense (HIMAD), Air Force JTIDS-equipped aircraft, adjacent FAAD C^2I systems and other ATCCS BFAs. The ADDS supports these communications needs.

Air Defense internal communications (shown in Figure 2 within the dashed box) are serviced by EPLRS. EPLRS interconnects the air defense command post elements, the Air Battle Management Operations Center, battlefield sensors and fire unit subsystems. The EPLRS architectures feature spread-spectrum, synchronous time division multiple access (TDMA) nets operating in the 420-450 megahertz (MHz) range with integral relaying. The EPLRS Network Control System provides network management for the EPLRS net, as well as an interface to EPLRS equipped aircraft.

All internal BFA functional connections, such as Airspace Management Element Liaison Officer, airborne sensors, HIMAD, selected Air Force aircraft and adjacent FAAD C^2I systems, are provided by JTIDS nets. JTIDS is also a spread-spectrum, synchronous TDMA system operating in a line of sight mode from 960 to 1,215 MHz, offering a higher data rate capability than EPLRS. The deterministic dissemination characteristic of the EPLRS and JTIDS nets supports the speed-of-service requirements imposed by weapons systems whose response times are measured in seconds. Both systems feature high-rate, hybrid pseudo-noise phase modulation (spread-spectrum)/frequency hopping radio frequency modulation modes. Therefore, both provide jam resistance and reduced probability of intercept and system features to enhance interoperability.

The Maneuver Control System

The Maneuver Functional Area (MFA) has highly distributed, widely integrated mobile C^2 system automation, with the broadest requirements for hierarchical and lateral interoperability of any ATCCS BAS. MCS is served by all three types of organic tactical communications systems, providing intra- and

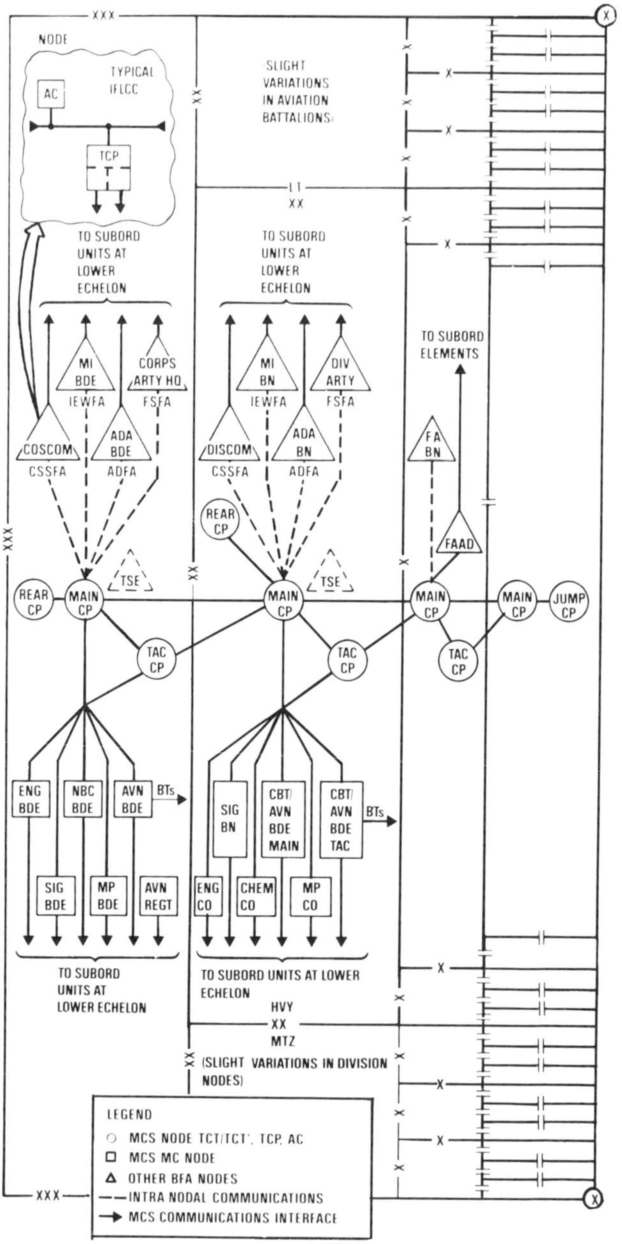

Figure 3. Maneuver Control System (MCS) architecture and its Army Tactical Command and Control System (ATCCS) interfaces.

inter-nodal connectivity between operational facilities, maneuver functional units, allied C^2 systems, other BFA nodes and adjacent echelons. The MCS IFLCC software segment provides functional integration of all ATCCS C^2 nodal systems. MCS employment range is corps to battalion. Figure 3 shows the overall functional architecture with various BFA elements and operational facilities.

The required communications interfaces for MCS include Automatic Digital Network, MSE, EPLRS, VRC-12/SINCGARS, circuit switches (TTC-39), message switches (TYC-39), four wire and various secure data (communications security [COMSEC]) devices (KY-57, KY-68, KG-84). Protocols such as X.25, Quadrilateral Protocol/Packet Switch and Standard NATO Agreements, U.S. Message Text Formatting and ACCS message standards also are required.

Because MCS is an evolutionary development program, each successive software segment has increased or enhanced communications interface capability. The QLP/PSN capability, which is a longer term interoperability feature of MCS, started with the X.25 network layer packetizing in MCS Segment 9 and will continue to grow as successive MCS software segments are fielded.

The Impact of ATCCS Requirements on Organic Tactical Communications

Optimal C^3 integration will require certain tradeoffs between C^2 system design drivers and supporting communications systems features and utilities. Typical nodal C^2 systems design criteria include data acquisition and dissemination requirements, mobility, security policy, distribution of assets, survivability, affordability and functional performance. Sunk cost investments in processing and data base management system hardware/software and NDI use further constrain the C^2 design process.

The technical requirements for tactical communications systems typically include security (bulk link encryption with validated secure data devices, integrated COMSEC for SINCGARS and other physical security means); appropriate speed of service; appropriate quality of service; protocol and format compatibility; survivability and jam resistance (to minimize vulnerability to physical and EW threats); coverage and connectivity; and low probability of intercept and non-uniqueness of electronic signature.

In general, those C^2 system features that enhance battlefield performance and survivability (such as distributed architectures), increase stress on supporting communications systems. Widely distributed user data bases, for example, which involve high levels of concurrency and centralized system configuration management, require far more battlefield communications capability than a loosely federated, dispersed and autonomous user concept. Communications requirements also depend on the software processing architecture employed. The tactical communications burden resulting from a highly centralized C^2 system architecture would differ widely from that of a highly distributed C^2 system in terms of capacity, connectivity, speed of service and robustness of the supporting communications systems. The ATCCS user's performance and survivability requirements in both benign and stressed environments will be met in an affordable manner only when a careful balancing of C^2 and communications system design and implementation has been achieved.

Currently, battlefield C^2 users have highly structured and limited access to the various organic tactical communications media. A battalion at the Front Line of Troops (FLOT), for example, has access primarily to CNR, and perhaps EPLRS; operational facilities at division or brigade can access MSE; corps also has MSE for internal requirements and employs this system to achieve external connectivity. It obviously would be beneficial if these media assets were interconnected adaptively and transparently and independent of the user's echelon, Electronic Order of Battle, operational unit, distance from the FLOT, and so forth. An effectively integrated C^2 system would use powerful, standardized computing elements in its C^2 systems (such as the ACCS CHS); would incorporate reusable software; and would employ a flexible C^3 architecture that could be tailored easily to enhance system performance.

Achieving an Integrated C^3 Environment

Achieving the sought after integration between the C^2 and communications elements will necessitate extensive systems engineering and integration efforts, such as those ongoing in MFA, as well as software modularization and architectural layering. Some form of communications network management overlay could be used to integrate the various tactical C^2 systems into the C^3 environment while maximizing use of fielded communications assets such as MSE, SINCGARS, EPLRS, JTIDS and satellite communications. This could be done via C^2 system-driven communications system control; by making the integrated communications systems independent of and transparent to C^2 users through their own internal management structure; or through a combination of both approaches.

ATCCS will use CHS to integrate BASs while optimizing the use of NDI technology. The current contract with the MILTOPE Corporation using Hewlett-Packard as the prime subcontractor will provide a family of standardized, interoperable computer processors and associated peripherals on a Unix based operating system with other system support software. The use of these common computers and system software will facilitate interoperability, continuity of operations, training, maintenance, logistics, interchangeability and configuration management for both hardware and software. A recent addition to the CHS family is the Standard Integrated Command Post System program. By this initiative, the BFA control systems will receive a common facility for housing the BFA Operational Facilities. (Although the CHS is targeted to BFA automa-

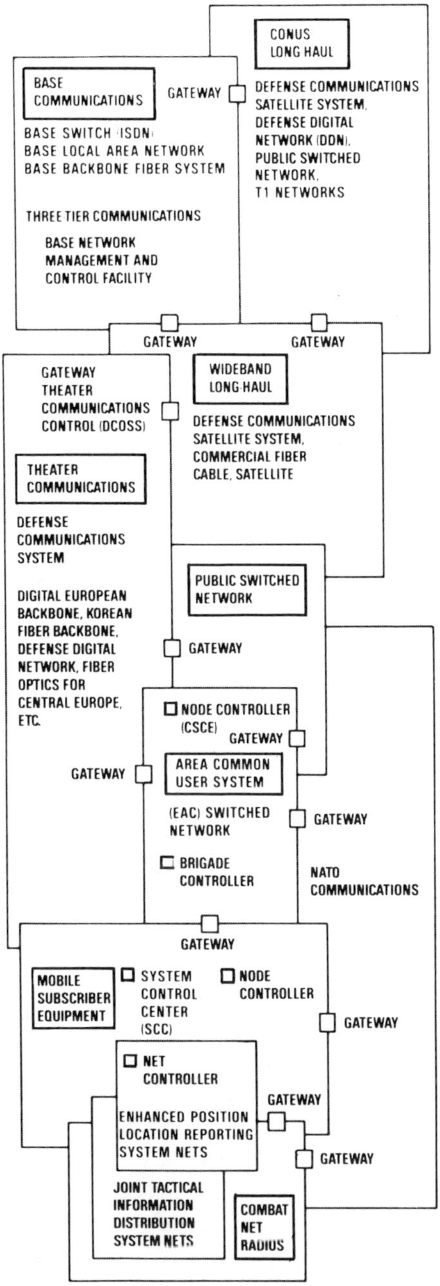

Figure 4. Integrated global network communications concept.

tion systems, it also could be employed in other tactical roles such as the organic communications systems.)

A C^2 user/host/terminal should be able to communicate through CNR, ADDS, MSE or multichannel media, as C^2 system performance requirements dictate. A global network management function, or appropriate subsets of such organic communications network management, is called for here. For example, combinations of integral communications systems management, such as the EPLRS net, could be gatewayed to other media (for example, MSE). The global gateway network would then be controlled by the C^2 system users or by global communications network managers. Figure 4 provides a current view of global network communications, from the sustaining base to the tactical networks. The gateway representations are of functional, possibly nonautomated, exchange points in the near term.

Artificial intelligence techniques are envisioned effectively to plan and manage multiple, interconnected, tactical communications media. Some examples include making data system assignments (for example, MSE, EPLRS, SINCGARS); routing via each system control function; allocating adequate channel capacity based on constraints of terrain, availability of facilities, user statistics, dynamic rerouting, CONOPS and C^2 system/user commands and connectivity requirements. Tactical systems inherently require constant reconfiguration to satisfy user needs (for example, mobility, mission changes) and to respond to enemy action (for example, jamming, battle damage) and random equipment failures. The mission driven needs of the various C^2 user systems (such as FAAD C^2's use of EPLRS) mandate real time communications control actions. Using expert system technology to manage tactical networks would reduce troop planning and operations burdens and would promote high levels of performance. Figure 5 illustrates a network management concept that uses expert systems techniques to interconnect and manage MSE, EPLRS and echelons above corps area communications systems globally.

Expert systems technology also could be used to add intelligent flexibility to data distribution, formatting and routing in future interconnected tactical networks incorporating gateways and packet switching appliques. Figure 6 shows a packet switched overlay on an illustrative slice of a division level fire support communications network.

Finally, substantive communications and C^2 systems changes have to be tested thoroughly before fielding. This would require not only developmental testing of system capabilities in an engineering test bed at the ATCCS Experimentation Site and the Army Interoperability Network, as well as appropriate milestone operational tests, but also a Total Systems Tactical Validation at III Corps. Experience has shown that all too often, what works in the laboratory or test bed significantly degrades in the field once the system is subjected to the rigors and uncertainties of battlefield use.

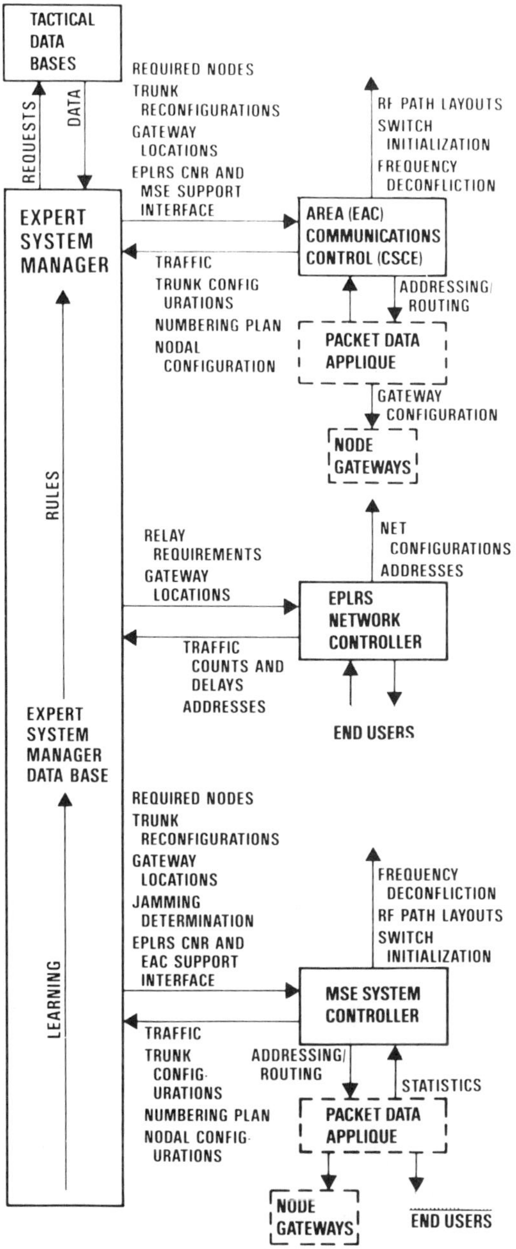

Figure 5. Network management data flow.

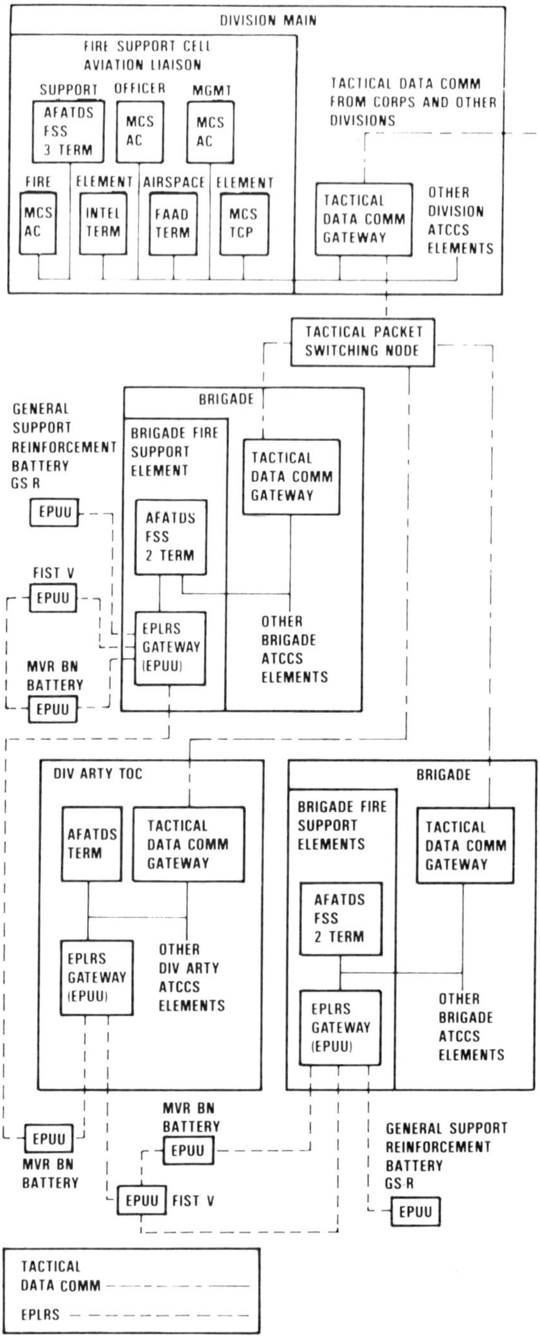

Figure 6. Typical Tactical Data Communications Division support slice.

Summary

Tactical communications must be able to support C^2 system requirements in a lethal, complex and dynamic battlefield environment. Effective integration of C^2 user requirements and communications design can provide a cohesive C^3 concept to help U.S. forces cope with the complexity and tempo of modern warfare. The establishment of the PEO for C^2 Systems and PEO for Communications Systems in May 1987 was a major step toward bringing the C^2 materiel developer community together paralleling the central user focus and coordination provided by Training and Doctrine Command's Combined Arms Center. Now the requirements and user communities must strive to abandon the old ways of conventionally partitioning the C^2 and communications functions. Army organizational structures must be modified to foster a C^3 view of BASs.

The goal of an integrated ATCCS environment must be afforded the same priority as any advanced weapon system today. Without integrated C^3, tomorrow's battle commanders will be unable to implement rapid decisions in the overwhelming environment of tomorrow's AirLand conflict. The substantial investments in weapons and sensors simply will not pay the necessary dividends to achieve victory. Extensive systems engineering and integration efforts must be supported by Congress and the Army to maximize communications connectivity and interoperability.

This article originally appeared in the June 1989 issue of SIGNAL.

Integrated Distributed Computing Using Heterogeneous Systems

Stephen T. Vinter

The increasing sophistication and the standardization of communication protocols are making the construction of large and complex distributed applications feasible. However, several remaining factors contribute to the complexity of building and maintaining these applications. First, the different types of hosts that populate a network make interconnection difficult. The chaos created by the continuing lack of operating systems standards and the need for specialized hosts, such as Symbolic processing hosts and multiprocessors, is particularly difficult to handle. Second, conventional facilities used to build distributed applications, such as remote file systems and remote procedure calls, are not adequately rich to make the complexity of large applications manageable. Third, the parallelism, partial failure modes and difficulty of obtaining a consistent global view make distributed systems inherently more difficult to understand.

One distributed computing environment for building and operating distributed applications is Cronus. Funded primarily by the U.S. Air Force Rome Air Development Center to support command and control (C^2) applications, Cronus addresses each of the sources of complexity that confronts the distributed application developer. Cronus offers a comprehensive and integrated approach to building distributed applications that will reduce the cost of application development and maintenance and prolong application lifetimes. It has been used to build a range of applications, including prototype C^2, office automation and simulation applications. Figure 1 shows the resources of a wide area Cronus test bed, currently in use for distributed software development and experimentation.

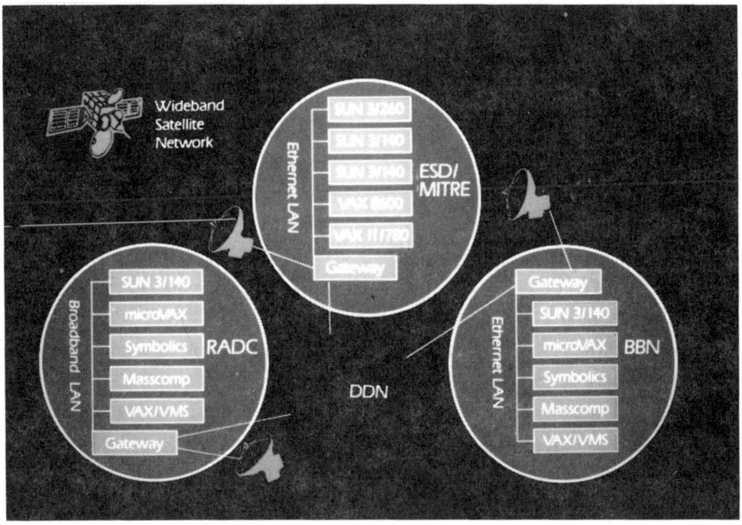

Figure 1. Cronus, currently in use in wide area test beds spanning multiple cities, integrates software and hardware resources from different organizations.

Characterizing Distributed Applications

Large distributed applications generally have long lifetimes because their build time is long, and their replacement cost is high. As a result, an evolutionary approach is needed in planning and designing them. Such an approach advocates modularity, an open architecture and uniformity.

Modularity makes it possible to upgrade or replace software components (hereafter simply referred to as components) with little or no impact on the rest of the system. A system is modular if it can be decomposed into a set of components with clear functions and well-defined interfaces. The appropriateness of the granularity of decomposition determines how easy the system is to understand and to maintain. For example, procedures have too fine a granularity to be used as the primary structural unit in defining large applications.

An open architecture[1] makes possible the inclusion of new components and technologies as they become available. An architecture is open if it allows easy interconnection of new components without modifying the existing system. The Open Systems Interconnection (OSI) protocol standardization efforts have drawn attention to the importance of open architectures.

A system is conceptually uniform if its structure can be understood by understanding a small set of concepts applied throughout the system. The implementation of a system exhibits uniformity if functions are used commonly by a variety of different components (such as with libraries). This makes it possible to modify a single module and affect all of the components that use it.

Uniformity is the most difficult of these properties to achieve. Distributed systems commonly are built using an amalgamation of software and hardware supplied from different sources. Sites commonly have multiple incompatible protocols running over the same network and occasionally different networks for different types of machines. Different machines commonly have different operating systems, and software frequently is developed using different programming languages. Interconnecting software components dispersed throughout this environment is extremely difficult.

Imposing uniformity, such as requiring the use of UNIX, has its own problems. It is difficult or administratively impossible to enforce. For example, many flavors of UNIX remain despite progress toward its standardization. And, imposing uniformity can lead to inappropriate hardware and software solutions; for example, non-UNIX symbolic processing machines would be excluded from sites requiring UNIX.

A Modular Solution

Cronus provides a wide range of capabilities for distributed application development and execution, as illustrated in Figure 2. The most basic mechanism in Cronus to support distributed computing is a network transparent, host-

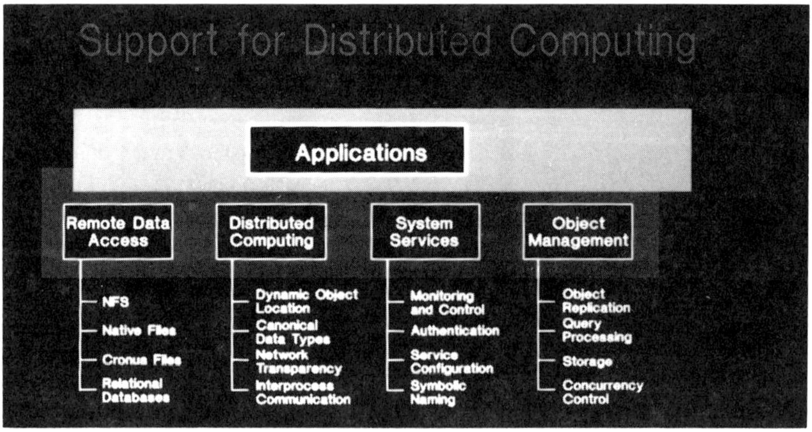

Figure 2. Cronus proves a wide range of capabilities for distributed application development and execution.

independent, interprocess communication facility. Because Cronus defines an object-oriented programming paradigm[2], all interprocess communication is framed as invocations of operations on objects.

Using the object-oriented programming paradigm, each application is structured as a set of types. A type defines a set of objects and the operations used to manipulate them. Cronus provides a set of system types, such as hosts, processes, files, directories (for organizing object names), principals (synonymous with users) and groups of principals. The object types created during application building correspond to the basic resources maintained by the application. For example, an electronic mail application might define mailbox, mail message and post office object types. The distinction between system types and application types is useful only to designate the commonly reused functions provided by system types. Except in rare instances, Cronus treats system types in precisely the same manner as it treats application types. This makes it possible to easily modify individual system types and create new ones.

How an object is implemented is largely independent of the definition of the object's type. Some objects are data structures, some are devices (for example, printers), and others are programs or functions. Some objects that contain data are stored on secondary storage; others are maintained in memory.

An object's location and the way the object is implemented also are hidden from the component accessing it. Objects are accessed by referencing them by name, specifying the operation used to manipulate them and passing the parameters used by the operation. When an operation is invoked, Cronus dynamically determines the object's location and routes the invocation request to it. This dynamic object location makes possible the movement of objects between hosts without modifying the programs that access them.

The decomposition of each application into a set of types and their associated objects results in highly modular applications. The strict definition of object interfaces (for example, operations) ensures well-defined boundaries between components. Additionally, because types correspond to the resources used by applications, their granularity is entirely under the control of the developer. As the granularity of objects increases and the operations used to access it become more complex, the cost of using the object (such as communication overhead in accessing it) decreases. Generally, it is possible to understand a great deal about an application simply by understanding its structure (that is, the set of types used to implement it).

The modularity of Cronus applications also contributes to their maintainability. With modularity, it becomes easier to isolate problems and make incremental replacements. Similarly, application decomposition provides a clear delineation of functionality and clean, easily documented interfaces between components, thus facilitating application development by multiple developers.

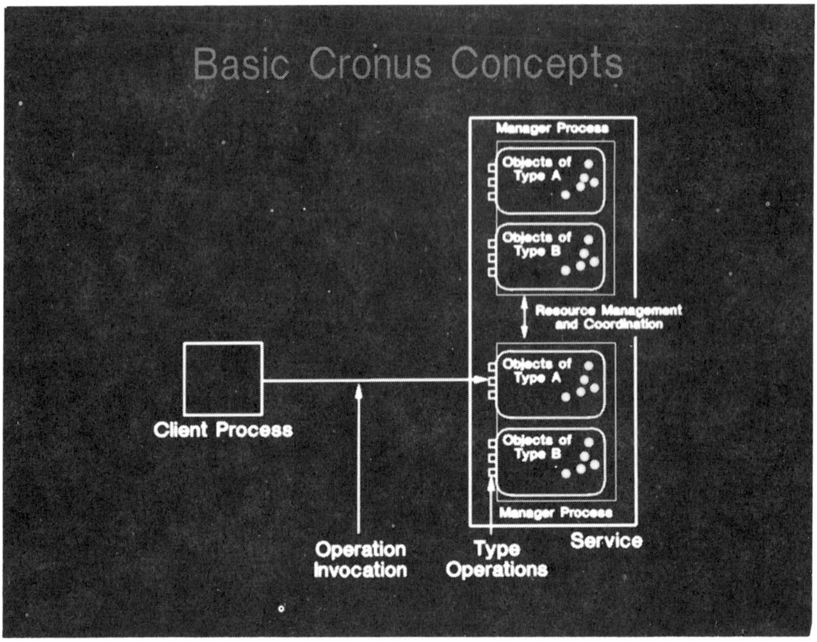

Figure 3. The Cronus object model hides the details of application implementation, facilitating evolution, scalability, survivability and flexible resource management.

Open Architecture

Objects and the components that use them are implemented in a variety of ways, but their implementations can be described using a basic set of concepts, shown in Figure 3. The component accessing an object is a client. When a client invokes an operation on an object, Cronus delivers the invocation request to the object's manager. Clients may be short-living or long-living processes (such as user interfaces and managers). A manager generally is implemented as a long-living process; examples of managers include file managers and authentication managers. On receiving an invocation request, a manager manipulates the representation of the referenced object and returns the results to the client.

A manager may manage more than one object type, and several managers may manage the same type. When the managers of a type cooperate in implementing the type, the group of managers is called a service. Examples of system services include Authentication, File and Monitoring and Control.

The architecture of Cronus and its applications is open because of the ease with which new types can be created, new managers of existing types can be started and can interact with existing ones in cooperating to provide a service. Consider objects that represent radar detections: Each detection is implemented as a data structure containing the approximate location of the detection, its direction, its rate of change and its hypothesized identity. Cronus provides object-oriented data storage within managers that is well-integrated with the object-oriented interprocess communication. When an invocation request is received, the manager reads the object from secondary storage into a record and begins processing.

As the number of detections being managed increases, the application operator may choose to add more detection managers to balance the processing load and use the secondary storage on other hosts. After starting the new managers, the operator may migrate a subset of objects to the new managers. When an operation is invoked on an object, Cronus dynamically will locate the objects at their new location. The implementer of managers is provided protocols to forward requests between managers for load management. The ability to add new managers and reassign the workload allows the application defining the detection objects to scale, thereby handling the changing numbers of detections and processing requests. New hosts also can be added easily to a site.

Facilities also are provided to support survivability, or the ability of an application to continue operation despite failures to a subset of its components (for example, host failures and communication failures). A replication mechanism is available to application developers that automatically will maintain copies of objects at the managers of the type. When the manager of a frequently used object becomes inaccessible, the dynamic object location mechanism will locate a new copy of the object. The replication mechanism also will create copies of an object automatically when the object is created; allow the application developer to control the trade-off between the availability and consistency of objects using a parameterized voting scheme; and detect inconsistent copies of objects should the developer permit them.

Selective Uniformity

Cronus runs on top of existing operating systems (such as various flavors of UNIX, VMS and Genera) on various hardware platforms (including Suns, VAXes, PC/ATs and Symbolics). Distributed applications have been developed in C, FORTRAN and Commonlisp. Figure 4 is an example of how Cronus has been used to integrate an existing expert system and data base with customized application processing components. The only current single requirement of Cronus is the availability of TCP/IP for transport-level communication.

Uniformity across this diversity is imposed by the object-oriented programming paradigm. Several aspects of uniformity relate to the use of the object

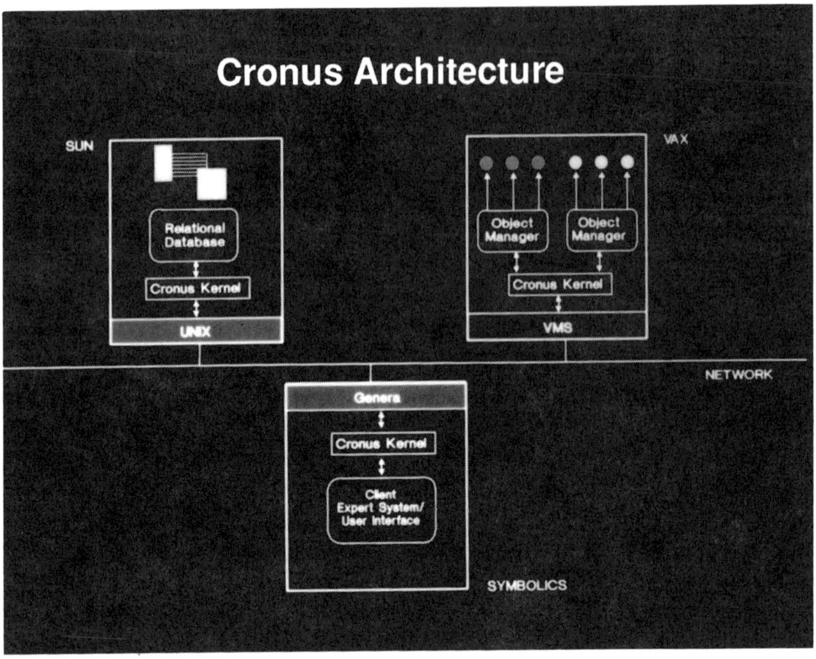

Figure 4. Cronus supplements host operating systems and software, integrating complex existing systems and data bases with customized application components.

model. Conceptual uniformity exists because the abstractions object, type, client, manager and service are used to define both applications and Cronus itself, regardless of the hardware and software used in an implementation.

The implementation of object types often varies, however, to exploit the characteristics of specialized processors. For example, it is possible (though uncommon) to implement two managers of a single type in radically different ways for different host types. And while it is possible to implement a type that can be ported to several different host types and operating systems, a developer also is free to use whatever native operating system services are available on a host. Thus, the application developer may trade off the portability of the application against the use of special purpose features provided by a host, operating system or language.

A set of software development tools offers the ability to create a degree of implementation uniformity. When a new type is created, the developer is encouraged to define the interface of the type in a special specification language for defining types. This type definition language, independent of any programming language, is used to define the operations, access control constraints and concurrency control constraints of the type.

Once the type interface is defined, code in several languages can be generated automatically to implement the low-level, routine functions found in virtually all managers (including message parsing, parameter validation, data conversion, access control and remote procedure call stubs). This allows developers to concentrate their energies on the application-specific functions rather than on mundane tasks common to most distributed applications. The interfaces also are placed in a data base and can be accessed remotely to assist in debugging and browsing.

A second means of supporting implementation uniformity is through the support of type inheritance. All types developed with the tools are organized into a type hierarchy used during the development process. When a new type is implemented, it becomes the descendant of one or more types when placed in the hierarchy. The type is able to inherit, or reuse, operations and other information defined by its parent and ancestor types. In this way, operations supporting replication, associative access, access control and administrative functions can be reused automatically by any application. This relieves the application developer of the burden of reimplementing these functions, and it ensures that the operation interfaces for these functions are uniform across all types using them.

Uses and Experiences

Cronus has been used to support a wide range of applications, including C^2 prototypes, distributed simulation, office automation, real-time processing of tracking data and decision aids. Simple applications have been built using the interprocess communication facility to remotely access commercial relational data base systems. More complex applications combine sophisticated user interfaces, artificial intelligence components, data bases and simulations. Figure 5 illustrates a few of the host systems and services currently in use in Cronus applications. All applications have been built using the type specification language and the software development tools.

Cronus is in use at 12 sites, each consisting of from two to 50 hosts, to build small and medium size applications (for example, four to 15 types). Cronus only recently has been available, and most uses of it to date have been in test beds. Cronus releases are generated approximately every eight months; release 1.4 is scheduled for September 1989.

One of the common misconceptions about Cronus is that it is too big and contains unneeded functions. In fact, developers are free to use the off-the-shelf services provided by Cronus, or they may use facilities customized for their application. For example, an application may opt to store data in native files or a commercial relational data base instead of the object-oriented data storage facility provided by Cronus. Similarly, objects can be assigned symbolic names using the hierarchical catalog maintained by the Catalog system service. The objects

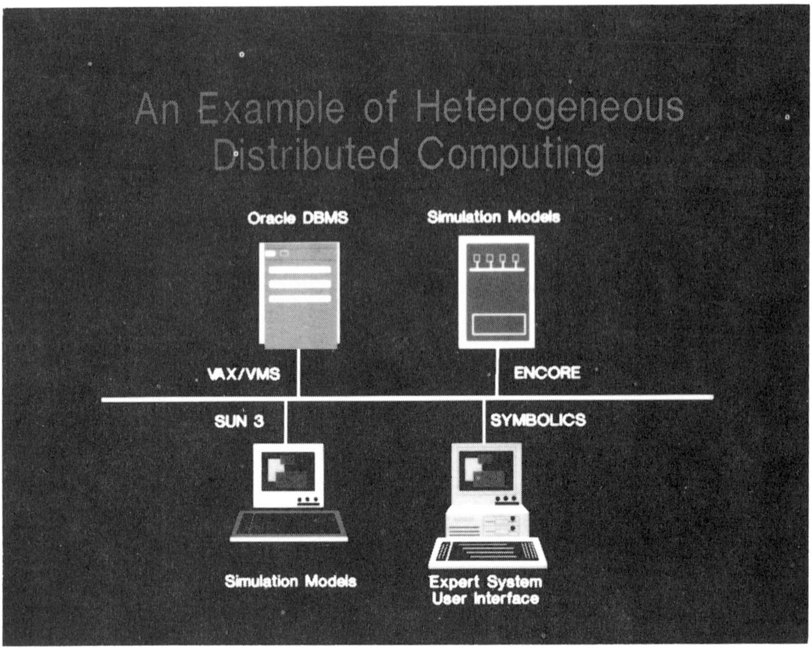

Figure 5. Existing Cronus applications are distributed across a variety of work stations and time sharing, symbolic and parallel processing systems.

can be self-naming, or developers can implement their own naming service.

Cronus attempts to provide a variety of choices in an off-the-shelf fashion, in addition to allowing developers to implement their own solutions. For example, Cronus allows operations to be invoked using conventional remote procedure calls, where the client blocks until the operation completes; unacknowledged invocations, for low-effort invocations that do not have return parameters; and asynchronous invocations, which allow client processes to execute in parallel with the processing of one or more invocation requests.[3]

Many distributed applications built today are implemented using rudimentary interprocess communication facilities and simple files to store data. Experience has indicated that more sophisticated general purpose mechanisms and services that provide functions commonly needed by application developers significantly save development time and satisfy the operational requirements of applications. General-purpose solutions also may reduce the cost of software maintenance.

Surprisingly, Cronus has turned out to be effective in quickly integrating and distributing existing software previously incompatible or implemented on a single host. Examples include supporting remote access to relational data bases from AI applications and distributing FORTRAN simulation components that were run serially on a single host.[4]

Conclusion

Cronus addresses the complexity of building distributed applications by providing flexible, off-the-shelf solutions to the problems commonly confronted by application developers. The diversity of hardware and software found in distributed systems is handled by imposing a degree of uniformity that allows, but hides, the customization of components on each machine. The Cronus architecture provides a rich, object-oriented programming paradigm that can express the modules composing a distributed application in ways that more closely correspond to how developers and users think of them. In addition, a variety of approaches is provided to assist developers in handling the complexity of the distributed environment, including software development tools, debugging tools and monitoring and control services.

Cronus has proved useful for rapid software development, integrating existing applications and integrating different technologies. It will be useful for supporting a wide range of applications, including C^2 and other complex distributed applications.

Cronus continues to evolve. Current areas of development include support for Ada, porting Cronus to several new hosts (such as multiprocessors) and implementation in a wide area network (WAN) environment. WAN operation will provide configuration, resource management and access control mechanisms to support efficient application operation and administrative policies for controlling the interactions between sites.

Endnotes

[1] A. Caston, "Are Open Systems Coming of Age?" *SIGNAL*, January 1989, pp. 61-64.

[2] A. Jones, The Object Model: A Conceptual Tool for Structuring Software (Berlin: Springer-Verlag, 1978).

[3] K. J. Schroder, J. G. Cole and C. Eide, "Cronus TVE Final Technical Report," BBN Systems and Technologies Corporation, *Technical Report 6841*, January 1989.

[4] E. F. Walker and P. Neves, "Asynchronous Remote Operation Execution in Object Oriented Distributed Systems," submitted to OOPSLA 1989.

The author is grateful for the assistance and support of Dr. Richard E. Schantz and Kenneth J. Schroder of BBN and Richard Metzger and Tom Lawrence of RADC.

This article originally appeared in the June 1989 issue of SIGNAL.

Videoteleconferencing: Applications and Trends

G.A. Redding

This paper describes technologies available for integration in video teleconferencing (VTC) facilities. Video teleconferencing does not require the use of leading edge technologies. Instead it integrates commercially available technologies. But, while hardware is readily available, organizations that set telecommunications standards have yet to establish acceptable protocols providing the "holy grail" of systems interoperability.

It is important that VTC technologies be cost-effective to the site sponsor, easy to use by the facilitator, and unobtrusive to the user. Accomplishing this technical engineering and integration feat, while ensuring other VTC sites have compatible technologies is a difficult process.

Video teleconferencing activities in the DOD covers the full range of defense agencies, military departments and their major commands. VTC is a subset of teleconferencing, a family of communications activities that do not normally involve video transmission techniques. For the majority of VTC sponsors and users adding video may be a new start, although it is a natural extension of traditional voice and data services. This video extension is potentially resource intensive, involving significant costs for equipment, facilities, and dedicated communications links and personnel and is, therefore, a prime candidate for a systems engineering approach.

VTC is a large functional area without an office of primary responsibility in DOD: there is no sponsoring command or agency and there is no single user community. These facts alone present significant challenges to successfully employing VTC technologies.

VTC is a hybrid technology. VTC Program Managers are faced with coordinating their requirements and introducing new technologies into established functional areas such as: information resources management, telecommunications management, audiovisual management, and security management.

Television cameras and microphones, and user "friendly" data management systems have been around the DOD for a long while, and that's a trap! Sponsors and users tend to underestimate the scope of the VTC project by assuming they already have a working knowledge of all of the components.

Differing approaches by DOD components have resulted in costly and uncoordinated applications of VTC technologies, ranging from desktop VTC workstations costing as little as $6,000, to TEMPEST secured facilities costing up to $4,000,000 per site. Large permanent facilities often bristle with proprietary hardware and software systems that tend to drive up capital costs for other VTC users even though they are not affiliated with a network sponsored by a major command. Within the DOD this stifles interoperability, and ultimately jeopardizes effective communications, the basic requirement for VTC in the first place. This scenario is further compounded when DOD activities attempt to interact with other federal agencies and with private industry.

The most complex teleconferencing applications employ video technologies. The integration of analog video components with less critical audio components raises the system base line for communications circuitry. VTC networks require large bandwidths to process useable transmission signals. In addition to dedicated line charges, costs factors spill over into the construction of special facilities, CODEC equipment, and security measures (physical, computer and communications). While VTC events may be fully interactive, the digitized video signal may diminish screen resolution and become a communications detractor when recomposed into an NTSC format. Communications problems are exponential when networking (especially secured) multipoint to multipoint sites since the audio, video, and control signals may be traveling on completely different circuits.

There are several reasons for including video with aural communication. There is an intrinsic value in being seen and an ability to assess both language and expression presents better reinforcement. Adding a video component to an event requires the participants to immerse themselves in the communications process to a greater degree than in a simple audio conference. Being on the receiving end of a conversation is a passive activity. Taking a simple phone conversation as an example, communication lacks full expression. In a VTC environment, some of the abstracts of aural communication can be tempered by concurrent visual confirmation.

Background

The growth of DOD VTC facilities and systems has been incremental and fragmented. Ad hoc events dominated VTC applications through the early

1980s. Event sponsors usually sought the services of an audiovisual studio and made transmission arrangements through the local telephone company. This was more a business television approach using analog video. A commercial NTSC broadcast signal was send to the receiving location(s). Full video interaction was not the order of the day, except for an optional audio return loop.

The largest consolidated DOD VTC program has been established through the Defense Commercial Telecommunications Network (DCTN). This provides a single vendor solution for digital transmission services and VTC facilities, keyed to dedicated facilities with communications leased at 1.544 MHz or T-1 (near full motion service). As with most turnkey systems, the DCTN service provides an elegant solution for a specific application. But DCTN installation shortcomings include unfavorable response times, limited standard interface for transmission rates less than T-1, limited provisions for multipoint VTC sessions (especially secured), and backup capability. The DCTN service does not readily integrate with base cable systems, local area networks (LAN) or individual workstations. Another major drawback is that most DCTN systems are leased. The current AT&T lease arrangement (running through 1996) for DCTN facilities poses problems in obtaining and integrating other than AT&T endorsed telecommunication services. The DCTN approach is a communication network solution for leased VTC services and systems with yearly sustaining costs for each site approaching $400K, excluding in-house overhead. Since the Defense Switched Network (DSN) is the primary communications vehicle, exceptions to its use are handled on a case-by-case basis by DCA.

Within the DOD, permanent VTC facilities are often managed as Visual Information (VI) activities since most of the hardware components fall into traditional audiovisual categories. This is a carryover from the use of TV studios in ad hoc VTC situations.

Security of VTC systems deserves attention. Each Federal Agency sets up its own security procedures, based on general guidance from the National Security Agency. There is a lack of working procedures that would satisfy most security officers. What may be a *secured* facility (physical and communications) to the Army may not be acceptable to the Navy. Who determines the security baseline? Further, if a session is intentionally mixing secured and unsecured facilities, what happens when the discussions become sensitive? As mentioned above, security devices and procedures are a major component of all VTC scenarios, but they are not yet being adequately addressed in detail.

Application of New Technologies

Video teleconferencing enjoys the benefits of emerging technologies without having to pay for development. VTC is still a requirements-driven system, sprinkled with traditionally reliable components utilizing three basic categories of signals: image, sound, and data. Imagery includes still and motion optical film and video and digitized graphics. Sound elements are either acoustic or electronic. Data is provided by electronic (computer) emanation and from the output of video compression techniques.

Desktop Systems

There are relatively few commercially available desktop VTC components. They were initially bypassed by the major DOD VTC effort, specifically the DCTN project, but they offer functionality that often surpasses larger systems. Initiating a VTC session from a desktop device requires little more effort than making a standard phone call. The call up procedure on a local area net can be managed from a keypad and the interconnections with external communications are relatively straightforward actions.

Image teleconferencing from a desktop system offers an informality that appeals to a large user base. It is much easier to start an ad hoc conference than it is to work through formal networks which require lead times, coordination and scheduling. Desktop systems offer the opportunity to share images through a LAN, and can readily integrate NTSC voice and video signals, along with PC text and graphics. Cost factors should not be overlooked. Since desktop systems share many PC components and are portable, they are a viable choice over fixed facilities costing upwards of $750,000 each.

Within a LAN environment, communications can be fully interactive with point-to-point and multipoint-to-multipoint. Over 30 workstations can be supported on a broadband system, mixing analog and digital signals to produce high quality images. The workstations can be interconnected through a cluster server and the cluster server can be used as a gateway to other cluster servers or any number of external communications links. Computer chip technology now offers board-size CODEC processors that can be inserted in a PC slot, increasing ad hoc networking independence. A single desktop VTC workstation ($5500) with cluster server ($9000) and a 56kb/s-384kb/s data rate CODEC ($30,000) provides a basic interoperable configuration. By using a standard interface, linking to a DCA approved CODEC may permit a tie-in to full DCTN service.

In using either a hard wire or fiber system, desktop VTC workstations also enable the integration of remote locations into a more formal VTC environment. Senior officials can monitor and interact from their own offices

Desktop systems provide a network solution that could prove extremely useful during emergency or contingency operations. Desktop VTC applications appear to have a ready home in military command centers. Action or desk officers would be able to access computer based data, retrieve graphics from optical storage devices and be able to interact with colleagues or subject matter experts through the same workstation: all in real time. *All* of the information, including the imagery, can be shared. That is the real advantage of desktop video teleconferencing. This functionality is relevant to computer based command post exercises or real-time decision making. In an emergency situation, supervisors would be able to evaluate the speed of problem solving and perhaps perform analysis on the quality of the decisions given.

The DOD has also experimented with systems larger than the desktop, but which provide the same flexibility. Fly away or "Rapid Deployment" systems have been field tested and are commercially available.

Combat Camera

Visual documentation of military operations has been a longstanding requirement. From Matthew Brady (a contract photographer) to the present, the military has been committed to documenting its operations. Some of the more obvious reasons are: visual history and archives, intelligence gathering, public affairs, research and development documentation, and medical evaluation.

As traditional film based media is replaced by video, and as analog video moves toward digital processing, the ability exists for combat camera teams to provide imagery in *real time*. The DOD has fielded still video camera systems that enable a combat photographer to input images directly into a transceiver for relay over communication links. This is a relatively low cost technology that would enable decision makers in remote locations to, for instance, evaluate a battle scene and to direct countermeasures. DOD testing of still video cameras have already allowed video images to be sent from remote locations to the Pentagon within 10 minutes: not real time, but a step toward teleconferencing.

A scenario might be a missile attack on a naval vessel. Video images of the damage could be relayed in near real time to senior military commands and to repair and rescue activities, more quickly and accurately marshalling resources to assist the stricken ship. Pentagon staffs could assume oversight positions in their video teleconferencing rooms; subject matter experts could be networked using desktop VTC systems; and shipbuilding technical staffs could overlay video images onto structural computer aided design (CAD) graphics.

There is a downside to such concepts and military history provides examples of the risks of skip-echelon transactions and too much real-time communications to remote echelons.

Distance Learning (or Teletraining)

The readiness of Reserve Components (RC) is an increasing concern to the Congress. A number of studies have been initiated on how fast the RC forces could be marshalled. Each of the Services directs its own RC training program, and each views *distance learning* differently. Most of the programs are modeled after what secondary schools, colleges and universities are currently providing. Each of the military departments has their own version of a "School of the Air."

Additionally, the military supports specialized distance learning activities such as courses offered by the Army Logistics Management College, which provides training to thousands of DOD employees each year. Aside from the real problems of funding and command support, there are also a few technological restrictions. The Services are not disposed to indiscriminately uplinking military courses. Even though the course material may not be sensitive, military trainers take care to ensure that only the intended audience is serviced. While there is yet no firm DOD guidance, capabilities are being studied by the military depart ments for encoding or encrypting (or both) all satellite communications for training applications.

There are considerable logistical problems in both encoding and encryption. Encoding is more than an analog to digital compression algorithm. Both encoding and encryption are equipment intensive, with encoding systems being the least expensive. Encoded signals provides a measure of privacy through controlled access, but would not meet National Security Agency (NSA) guidelines for communications security. CODECS provides only a measure of "privacy." Encryption provides *security*, but perhaps at a higher security level than is available at the training sites. Encryption devices are "key" controlled, and staff trainers in remote locations might not have the required control mechanisms to store and safeguard keys.

The economics of distance learning are a driving force to exploit this technology. Using a Direct Satellite Broadcasting (DSB) methodology, military school houses are able to uplink a full broadcast signal which is retrieved by television receive only (TVRO) satellite dishes. Instructor travel time could be significantly reduced and service members would also benefit from reduced travel costs. Since only the instructor's image is broadcast, interactivity is achieved by back looping an audio channel over normal phone lines. Using a full broadcast bandwidth also enables the linking of computer terminals, further enhancing interactivity. However, if the training application calls for full duplex service among all participants, costs would be prohibitive using full broadcast (not to mention availability) and even compressed video is still expensive. This is just one application of an integrated service digital network (ISDN). In lieu of dedicated communications, another major obstacle for ISDN for the DOD is getting ad hoc satellite access.

The technology of distance learning via VTC techniques also enables courses to be downlinked to unattended sites. This is a distribution technology and not interactive. A decoding device could be installed with a video tape recorder and a timing device to receive course material overnight when satellite transmission tariffs are lower. This technique currently is being used in the private sector by such firms as Federal Express which produces a (very) early morning show (FXTV) covering that day's business volume and other (training) segments. These programs are beamed via satellite to selected locations. Employees replay the tapes and are able to incorporate current information into their daily routine.

As a training tool FXTV claims to be three times as effective as conventional methods. Given the military's proclivity for briefings, DSB provides a fairly simple mechanism for getting out the message.

Encoding transmissions can also provide internal control of courseware distribution. The encoding device can be set to allow access only to authorized decoders on a transmission-by-transmission basis, if required. This technique is in daily use for many DSB proponents.

High Definition Television (HDTV)

Within the past few months there has been considerable attention focused on HDTV. While the Defense Advanced Research Projects Agency (DARPA) has fueled the search for a technical standard for HDTV by dangling $30 million in research funding, other federal agencies and the Congress also have entered the picture. For the most part, the race for HDTV standards is focused on economic and political issues rather than on operational applications. Given that the use of standard TV is primarily for commercial entertainment, it's reasonable to assume future uses of HDTV will be along the same line.

There are three basic applications for HDTV under study: imaging/production, distribution, and broadcast. VTC could be involved in all aspects, if there were an application. Is there a niche for HDTV within DOD VTC operations? It depends. So long as "talking heads" are the dominant image, HDTV seems an expensive way to achieve visual interaction. One of the main benefits in face-to-face communications is the premise that dialogue is enhanced. It is tougher to say "no" in a face-to-face dialogue, and easier to get a point across. If visual cognition is used as the touchstone or confirmation tool, then HDTV technologies do not appear to provide significant added value.

One of the better applications of HDTV technologies lies in its use as a professional quality production tool. The quality of the image offers a richness and perception of depth that is desirable in the entertainment industry, but VTC events are not intended to be production showcases. Most of the HDTV audio enhancements would not be applicable to VTC either. VTC events are intended to be interactive, relatively informal, and free of technological distractions. At this time HDTV does not appear to enhance that process.

Within the DOD, there is considerable controversy over acceptable transmission rates within the conventional broadcast bandwidth. Cost has forced most DOD VTC users to settle for less than full motion service through compression techniques. If it is tough to finance T-1 service representing a 60:1 compression, introducing a HDTV technology requiring larger initial bandwidths for transmission would make HDTV cost prohibitive.

Digitized products are now available for VTC applications which offer much greater resolution than that promised by HDTV. Assuming HDTV will settle on approximately 1250 lines of resolution, this is well short of 2000 line resolution routinely available to current VTC users for still frame and graphics transmissions. Commercially available digital scanning devices offer full color renditions with at least 1800 lines of resolution. The principle technological push is toward improving digital transmission techniques for VTC. This is true of all imaging devices. For VTC perhaps more attention should be directed toward developing less costly D1 and D2 specifications to better handle and store digitized data on tape media and then use that digital technology to *push* VTC applications.

HDTV's application to VTC is severely limited in an analog medium. Assume analysis of a moving image (such as an explosion) on HDTV. A freeze frame of such an image would have interleaved and overlapping information, which is still less satisfactory than digital data. A collection of single frames processed in rapid succession provides only an illusion of motion, whether on standard television or HDTV.

As HDTV becomes available through a distributive cable system, it may achieve some utility provided a VTC facility uses local area networks as part of its communications pipe. But all this ignores the capital investment costs of HDTV imaging and processing equipment. VTC systems do not currently operate using industry accepted "broadcast standards" and a leap into HDTV seems improbable.

As described above, with the possible exception of on line compression algorithms, video teleconferencing does not use leading edge technologies. It provides opportunities for expression through real-time access that have been too costly to pursue in the past. The basic audio and imaging components are the same; it's just a matter of signal processing. While digital processing may solve teleconferencing distribution problems, it will not enhance the basic message. Discipline is still required for effective communications. When the novelty of viewing talking heads passes, video teleconferencing will come of age.

Definitions

Conferencing. Programs and meetings for the purpose of presenting and exchanging information, comparing views, learning, planning, and decision making. Conferences may be held in one location or conducted simultaneously at multiple locations and be linked by telecommunications systems.

Teleconferencing. The use of telecommunications systems by groups of three or more, at two or more locations for the purpose of conferencing with one another; or, two-way communications between two or more groups, or three or more individuals, remote from each other, using a telecommunications medium; or interactive group communication through an electronic medium. It includes audio/aural/video and audio/aural/graphics conferencing.

Video teleconferencing. Video teleconferencing is a subset of conferencing involving both analog and digital imagery, with audio. It excludes telephone conferencing, secure voice, and secure data or computer conferencing when no audio is used.

Bibliography

OMB Circular A-130 "Management of Federal Information Resources."

DOD Directive 5000.1 "Major and Non-Major Defense Acquisition Programs," 1 September 1987.

DOD Directive XXXX.X "Teleconferencing Networks and Activities," unpublished draft.

Defense Audiovisual Policy Office, American Forces Information Service, OASD)Public Affairs," unpublished draft.

This paper reflects the personal views of the author and are not to be construed as official or representing the opinion of the U.S. Department of Defense.

PART 7
Issues in C^2 Systems Engineering

Computer-Aided Systems Engineering Applications to Multidisciplinary Command and Control

Howard Eisner

Introduction

The overall subject of this book is the nature and improvement of command and control (C^2) systems engineering and the focus of this chapter is advanced tools for multidisciplinary systems engineering. The key elements are:

- Command and control
- Multidisciplinary features
- Advanced tools for systems engineering

With respect to the latter, we address computer aids as representing the class of advanced tools that we apply to improve our ability to perform the tasks and activities of systems engineering. The implication is simply that we can improve C^2 systems by improving systems engineering, and both can be better accomplished through the utilization of current and projected computer-based tools.

Command and Control

Command and control is a vital part of any large-scale defense system. The broad model for describing the role of C^2 is shown in Figure 1. Sensor systems, whether they be hardware or otherwise, feed information through communication links to the C^2 system. The C^2 element evaluates this information, using it to decide where and how to employ forces against the enemy. C^2 instructions are provided, also through communication links, and these forces are activated. Battle

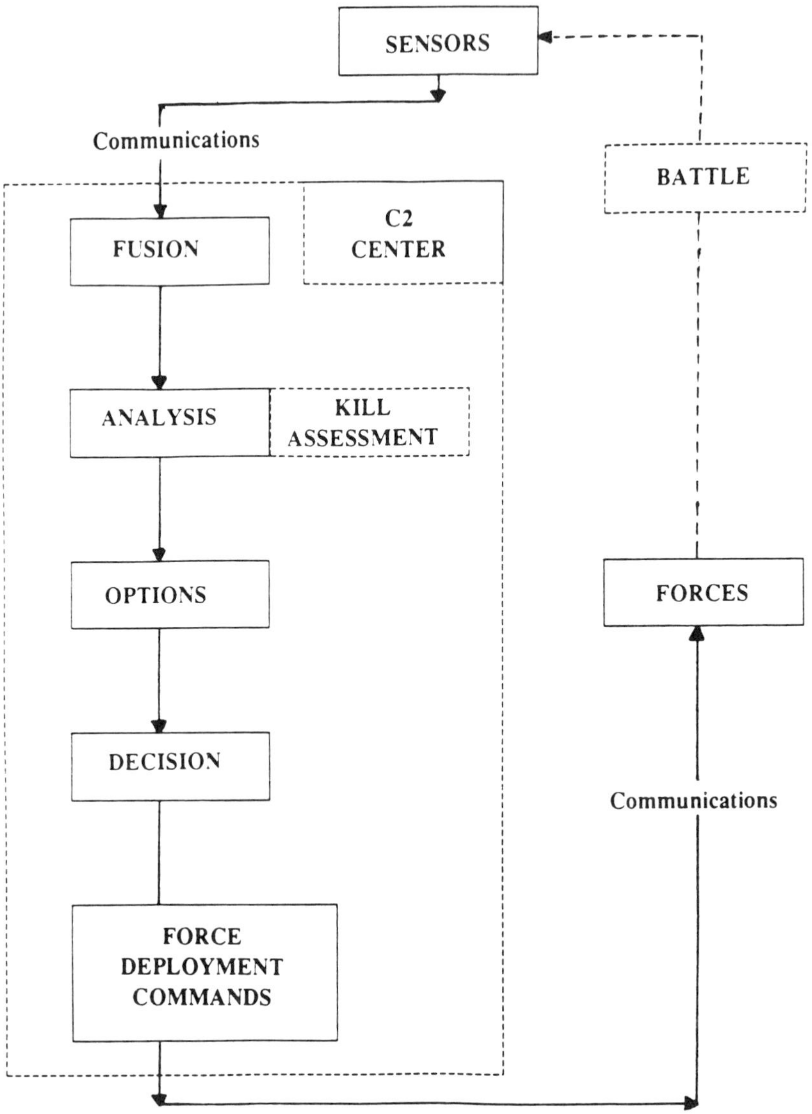

Figure 1. Simplified command and control.

results are observed by the sensors, fed back to and evaluated again by the C2 element, and new instructions issued. This process continues until the battle is over.

In the real world, the above simply-scribed process is an extremely complex one, as we can illustrate by briefly citing some of the systems we have been working on to carry out C2 functions.

The Strategic Defense Initiative (SDI)

This very large-scale strategic initiative focuses on defense against ballistic missiles. The C^2 system has been thought of as embedded within the Battle Management/C^3 function [1]. This "system of systems" is in an early stage of technology development and systems engineering and represents perhaps the most complex defense system challenge ever undertaken, with corresponding challenges evident in the subordinate C^2 element.

The Army Tactical Command and Control System (ATCCS)

In the tactical arena, the Army's ATCCS is also a "system of systems," incorporating five nodal C^3I systems: Forward Area Air Defense Systems Command and Control (FAADS C^2), the Advanced Field Artillery Tactical Data System (AFATDS), the All Source Analysis System (ASAS), the Maneuver Control System (MCS) and the Combat Service Support Control Systems (CSSCS). The five star ATCCS also incorporates a variety of weapons command and control devices and communications systems [2]. This massive system also utilizes leading-edge technologies as part of the systems and also to carry out systems engineering to develop the system.

Airborne Command and Control

The airborne command post (ABNCP) mission has required continuous (24 hours a day, 365 days a year) operation as an alternative mobile and survivable platform from which command and control of U.S. forces can be maintained [3]. Modernization into a network of worldwide command post resources has been extremely challenging even in this relatively small system (in comparison to those cited above), incorporating significant technology upgrades and challenging interoperability, survivability and reliability attributes.

The Command and Control Improvement Program (CCIP)

The Command and Control Improvement Program (CCIP) was established "to coordinate the application of state-of-the-art technology and to ensure commonality, interoperability and the efficient application of resources" [4]. Innovative approaches to these improvements have included: the Experimentation and Evaluation Facility, the Command Center Design Handbook and a Model Command Center (MCC) concept. All have helped to facilitate the identification of promising new technologies, promulgation of standards and the development of new implementation methods.

Air Traffic Control

Air traffic control (ATC) systems are longstanding capabilities that have been designed and constructed for both the civil and military sectors. The Federal Aviation Administration is responsible for the civil system which assures safety in the terminal area as well as for enroute flights. Upgrades for this system are in process utilizing our most up-to-date radar and computer technologies. Command and control ATC for civil applications has been based upon ground-based systems, but various satellite support concepts have been explored over the years.

With the complexity represented by the above illustrative real world systems, it is clear that we must put ourselves in a position to utilize our best technology assets to assist us in carrying out C^2 systems engineering in the most effective and efficient manner.

The Multidisciplinary Nature of C^2

There are a variety of functions to be executed within the command and control element of a system. Some of these include:

- threat warning
- attack assessment
- situation display
- situation projection
- communication with command authorities
- coordination of assets
- weapons assignments
- decision to engage
- damage assignment (own and enemy)
- reallocation/deployment of resources

Although this is not an exhaustive set of mutually exclusive functions, they serve to indicate the numerous concerns and needs of a commander or battle manager. Subordinate subsystems are part of the C^2 element in order to execute each of these functions. Maximum use is made of advanced technology, with particular emphasis on computer-related assets, to provide the commander with the necessary tools to do the job.

Of particular importance in the C^2 element are the time lines under which the commander must operate. Such time lines may be extremely short as illustrated by the SDI for which one has of the order of five minutes during the ICBM boost phase to determine when and how to engage targets. Inability to effect damage on enemy weapons during this phase makes the problem considerably more difficult during subsequent phases of the defense scenario.

The human capacity in the C^2 element is fundamentally a mismatch to the capacities represented by the communications and computer processing capabilities that support command and control. That is, we are able to process and analyze input data (to the command center) at rates that are orders of magnitude higher than the capacity of the human being in the loop. This well-known problem [5] has forced us to work harder at understanding how to maximize the man-machine interface capability. The man-in-the-loop has also greatly expanded the multidisciplinary nature of the problem. Without a person as part of the process, we can number-crunch at multi-megabit rates with little problem. With a person in the loop we have to be concerned as well with such aspects as interactive graphics and displays, rapid generation of alternatives and a broad class of human decision aids. Thus, a short list of disciplines needed to systems engineer, develop and operate modern C^2 centers would include, in addition to classical engineering and computer science disciplines:

- algorithm development
- cost-effectiveness analysis
- decision theory and analysis
- decomposition theory
- diagramming and plotting
- dynamic programming
- forecasting and prediction
- game theory
- heuristics
- man-machine interfacing
- modeling and simulation
- network analysis
- optimization theory
- probability theory
- psychological decision theory
- queuing theory
- risk analysis
- search theory
- structured analysis
- training methods
- visualization techniques

Although some of the above are partly covered in the fields of engineering and computer science, many are not. Thus, we have a considerable challenge when we tackle the design and operation of a C^2 center. Much of that challenge requires that we understand the precise role of the man-in-the-loop and the best way to match the human capacity to that of the information sources that surround him.

From these and similar considerations, we can clearly see the multidisciplinary nature of command and control. This characteristic is demanded by the fact that we generally require that the human element remain in the command and control loop, fearing to give up control of critical elements exclusively to machines. Drawing the proper boundary between machine and human functions, including the design of appropriate interfaces, will remain critical issues for the foreseeable future.

C^2 Systems Engineering

We can approach the matter of what it is that constitutes C^2 systems engineering by first examining the broad elements of systems engineering in general. Thirty such elements are listed in Table 1 [5]. Although these apply to all large-scale systems, if we focus on command and control more specifically, the following sub-elements of systems engineering appear to be of special significance:

- requirements analysis
- system and subsystem design
- trade-off/alternatives evaluation
- interface definition
- program and decision analysis
- risk analysis
- specialty engineering
- integration and test
- training

From the above, together with an examination of the previously cited functional elements and disciplines of command and control, we narrow down to the following very specific and critical systems engineering concerns with respect to either constructing or operating a modern command and control system:

- requirements analysis/threat assessment
- decision and risk analysis
- man-machine interaction
- interfacing and interoperability
- modeling and simulation
- training

Requirements Analysis/Threat Assessment

A most important part of requirements analysis is threat assessment as an element of the system itself as well as a key driver in terms of designing the system in order to be able to handle the threat as it may evolve over time. A C^2 system that is able to deal with expanded levels of threat is said to be a "robust" system. Thus, particular focus for requirements analysis is to assure that the system is responsive to appropriate levels of threat.

Table 1
Elements of Systems Engineering

1. Definition of Needs/Goals/Objectives
2. Requirements Analysis
3. Requirements Allocation
4. Functional Analysis
5. Functional Allocation
6. Specification Development
7. Systems and Subsystem Design
8. System and Subsystem Analysis
9. Trade-off/Alternatives Evaluation
10. Software Development and Analysis
11. Interface Definition
12. Schedule Development
13. Life-Cycle Costing
14. Technical Performance Measurement
15 Program and Decision Analysis
16 Risk Analysis
17. Integrated Logistics Support (ILS)
18. Pre-Planned Product Improvement (P^3I)
19. Reliability-Maintainability-Availability (RMA)
20. Specialty Engineering
21. Integration
22. Test and Evaluation
23. Configuration Management
24. Quality Assurance
25. Training
26. Production
27. Installation
28. Operations and Maintenance
29. Operations Evaluation/Modification

Decision and Risk Analysis

In the operation of the real system, threat data is sent to the C^2 element and decisions made as to if, when and how to respond to the threat with force deployment. Therefore, in this sub-element we must be concerned with supplying the proper decision aids to assist the commander in making the very best choices, particularly in an attack situation. An implicit part of such critical decisions is an assessment of the risk of alternative courses of action, some of which are related to the two classical errors:

- failure to act when it is necessary to do so
- acting when it is not necessary or indeed desirable to do so

As indicated previously, often the commander has precious little time to make force deployment decisions, emphasizing the need to supply appropriate displays and rapid simulation results.

Man-Machine Interaction (MMI)

We cite MMI as a particularly important aspect of "specialty engineering" in a command and control setting. The reasons have been cited above—we must find better ways of matching the capacity of the human decision maker to the very rapid time lines of the battle and its generation of enormous amounts and rates of information.

Interfacing and Interoperability

In addition to the above MMI matter, the broader issues of interfacing and interoperability are critical to command and control. Even if the C^2 elements are internally interoperable, interface and interoperability problems with the sensor, communications and force projection systems are often plentiful. Thus the commander must assess the trustworthiness (e.g., reliability, availability, accuracy) of these and other systems in making command and control decisions. Internal incompatibilities within the C^2 element will only add further degradation to the operation of the overall system.

Modeling and Simulation

This capability is extremely important in that it supports virtually all the other elements. We rely on modeling and simulation to test how well our system will operate against a real world set of threats. In a large-scale system (e.g., the SDI scenario), there is no other way than to resort to extensive modeling and simulation. Often, our confidence in the operation of the system as a whole is predicated upon extensive modeling and simulation in a test bed.

Training

Training has now been accepted as a critical aspect of modern warfare, whether it be in a tactical or strategic arena. To achieve superiority, training generally requires simulating warfare conditions over a wide range of options such that human behavior patterns under stress are accurate and repeatable. Here again we observe the need for a simulation capability.

Computer-Aided Systems Engineering (CASE) Applications

Computer tools have been used increasingly to support systems engineering for command and control systems. These have included aids in order to "systems engineer" the system as well as to be part of the system to continually test its command and control capabilities.

These tools have been associated with all levels of computers, but in recent years have proliferated in desktop (personal computer) and workstation configurations. As examples, Table 2 lists some of the categories of computer aids that have been brought to systems engineering tasks and activities for both personal computers and workstations. This technology is extremely fast moving, representing a "technology push" that has been invaluable to today's systems engineers.

As another specific example, it is reported that microcomputers have revolutionized communications analysis and planning [6]. In this particular application, a microcomputer model was developed, under sponsorship of the Defense Communications Agency, to assess RF propagation in a nuclear disturbed environment. Literally hundreds of other microcomputer applications have been reported in the literature as usage has increased with improvements in the price-performance of this class of machine.

Having narrowed the elements of systems engineering in general (Table 1) down to a subset that appears to be of critical importance with respect to command and control systems, we are now in a position to do the same with computer-based tools by selecting a subset of such tools that are of special relevance:

- requirements languages/tools
- decision support tools
- interactive graphics/displays
- expert systems
- database management systems
- simulation packages
- computer-aided instruction (CAI)
- integrated computer-aided software engineering environments
- integrated computer-aided systems engineering environments

Each of these computer aids/tools is discussed below in relation to the critical C^2 systems engineering concerns, using Figure 2 as a guide for mapping one against the other.

Requirements Languages/Tools

This class of tools provides the capability of analyzing system requirements, including traceability, relationships with specifications, performance and the test and evaluation program. Through its ability to track requirements across the full

Table 2.
Selected Categories of Computer Tools Available for Microcomputers and Workstations

A. Microcomputers
- Spreadsheets
- Integrated Packages
- Database Management Systems (DBMSs)
- Decision Support Systems
- Graphics Displays and Interfaces
- Structured Analysis Tools
- Computer-Aided Software Engineering (CASE) packages
- Computer-Aided Engineering (CAE)
- Simulations
- Project Management Aids
- Statistical Analysis Packages
- Linear Programming Tools
- Expert Systems

B. Workstations
- Artificial Intelligence
- Communications and Networking
- Data Base Management Systems
- Electrical Engineering
- Graphics and Imaging
- Manufacturing
- Mathematics and Statistics
- Mechanical Engineering
- Professional Services (e.g., Training)
- Software Engineering
    - -Cross Development Tools
    - -Editors and Utilities
    - -Languages
    - -Real-Time Processing
    - -Software Development Tools

system life cycle, it can also display interfaces and interoperability needs. Such tools range from a complex and expensive package such as PSL/PSA to an overlay built upon a microcomputer-based database management system.

Decision Support Tools

Many commercial decision support systems apply directly to command and control decision and risk analysis. In addition, numerous such tools have been built by contractors under government sponsorship to respond to particular C^2

Selected Computer Aids/Tools \ *Critical C2 Systems Engineering Concerns*	*Requirements Analysis/Threat Assessment*	*Decision and Risk Analysis*	*Man-Machine Interaction*	*Interfacing/Interoperability*	*Modeling and Simulation*	*Training*
Requirements Languages/Tools	●			○		○
Decision Support Tools		●	●	○	○	●
Interactive Graphics/Displays	○	●	●	○	○	●
Expert Systems	○	○	○	○	○	○
Database Management Systems	●	○	○	●	○	○
Simulation Packages		●	●	○	●	●
Computer-Aided Instruction (CAI)	○	●	●	○	○	●
Integrated Computer-Aided Software Engineering Environment	○	○	○	○	○	○
Integrated Computer-Aided Systems Engineering Environment	●	●	●	●	●	●

● Computer Aid/Tool Directly Applicable
○ Computer Aid/Tool Indirectly Applicable

Figure 2. Relationship between selected computer aids/tools and critical C² systems engineering concerns.

requirements. In some cases, rudimentary tools exist in the form of hard copy documents such as the Command Center Design Handbook referred to above in the Command and Control Improvement Program [4]. The latter can be brought into the world of modern computer aids through expert system or computer-aided instruction applications.

Interactive Graphics/Displays

A central theme in a C² setting is that of providing the appropriate interactive graphics and displays that the commander will use directly in making critical battle decisions. Great strides have been made in this technology, particularly in the form of high resolution large-scale displays and workstation graphics.

Expert Systems

Expert systems apply across the board (see Figure 2) in at least the sense that we wish to capture expert knowledge in C^2 systems engineering as well as the operation of the C^2 system in an operational setting. This technology continues to move along at a rapid pace, finding applications in the world of command and control as well as related surveillance, communications and weapons deployment systems [8,9,10].

Database Management Systems (DBMSs)

DBMSs are the underpinnings of many computer tools and are available at all levels of computers. In terms of our critical C^2 systems engineering concerns, they apply especially to requirements analysis and keeping track of system and subsystem interfaces. We are now also at a point at which such systems can incorporate multi-level security (MLS) capabilities. Architectures have evolved from flat file, network and hierarchical types to the generally accepted relational type of model.

Simulation Packages

Simulation packages, as shown in Figure 2, are key elements of decision and risk analysis, man-machine interaction, modeling and training. In a command and control environment, we must be in a position to simulate battle conditions and test the responsiveness of our C^2 capability to alternative scenarios. Often, this is done in a "test bed," allowing us to go off-line to explore the boundaries of our ability to handle a wide range of threat and response options. Perhaps the largest such system considered to date is the National Test Bed (NTB), designed to support the SDI program.

Computer-Aided Instruction (CAI)

CAI clearly supports decision and risk analysis, man-machine interaction and training. An integrated environment for CAI in the context of C^2 also includes simulation, expert systems, interactive graphics/displays and decision support tools.

Integrated Computer-Aided Systems Engineering Environment

Paralleling the above, we are also now in a position to construct integrated environments to support both the systems engineer as well as the builders and operators of C^2 centers. The goal is to provide an integrated set of tools to allow our systems engineers to carry out the full range of systems engineering tasks and activities (see Table 1) in a fully integrated manner [5].

Summary

We have explored the multidisciplinary nature of command and control, several illustrative real world C^2 systems, systems engineering in relation to command and control and important computer tools that can be and have been used to support critical C^2 systems engineering. In the ever-expanding world of computer technology and related software, it is clear that computer aids will continue to be developed and applied to the elements of C^2 systems engineering. As this occurs, we will bring these tools together in an integrated environment to support both the systems engineer and the activities at the command and control center. This will provide continuing improvements, especially in areas related to matching man and machine capacities, C^2 decision making and performing in domains with short time lines requiring essentially error-free operations.

References

1. J. L. Boyes and D. Shore, "Focusing on SDI BM/C^3," *SIGNAL* , AFCEA, June 1987.

2. Gen. L. C. Wagner, Jr. "Modernizing the Army's C^3I," *SIGNAL* , AFCEA, January 1989.

3. Col. D. Wilson and Maj. W. Calcatrra, "Airborne Command and Control Modernization," *SIGNAL* , AFCEA, March 1987.

4. D. T. Signori, Jr., "The Command Center Improvement Program: An Innovative Approach," *SIGNAL* , AFCEA, March 1987.

5. H. Eisner, "Computer-Aided Systems Engineering," Prentice-Hall, Inc., 1988.

6. R. B. Henriques and J. A. Roth, "Microcomputers Revolutionize Communications Analysis and Planning," *SIGNAL* , AFCEA, June 1987.

7. "Catalyst: A Catalog of Third-Party Software and Hardware from Sun Microsystems, Inc.," Sun Microsystems, Inc., Mountain View, CA, January 1988.

8. Issue on Artificial Intelligence, *SIGNAL* , AFCEA, June 1986.

9. N. Fowler III, "Artificial Intelligence and Air Force C^3I Systems" and D. C. Walden, "Successful Applications of Artificial Intelligence," *SIGNAL* , AFCEA, May 1987.

10. Issue on Supercomputers and Artificial Intelligence, *SIGNAL*, AFCEA, April 1987.

A Change in System Design Emphasis: From Machine to Human

M. L. Metersky, J. M. Ryder and M. A. Leonardo

Introduction

Even though the majority of decision support systems have met their specifications, they have not exhibited predicted performance. We believe this occurs because the role played by the human in the system, particularly in terms of decision making, has not been adequately considered in the design process. The inability to achieve expected performance is prevalent in weapons systems and command and control systems. This is because system performance is tempered by human capability, especially where decision making is involved. Recent trends include more and better sensors, improved communication links and advances in data processing that combine to produce more data. However, large amounts of data cannot be assimilated by a human rapidly enough to underwrite quality decisions. In addition, operational situations have become more complex and fast-paced, greatly impacting human decision making behavior. In spite of these factors, current design philosophy and practice tends to ignore the human role in system performance. Metersky (1986), indicates that system performance approaching theoretical limits may be attained by reorienting the system design approach to stress the decision making function. Proper emphasis of the human as an integral part of a system could result in maximizing system performance.

In the past, and to a large extent even today, the emphasis in system design has been on defining hardware requirements. In many cases, advances in hardware technology, not the ability to meet mission requirements, was the driving factor that determined the need for upgrades to or replacement of a system. Even though software has increased in importance and percentage of cost in system development, it is still philosophically considered as a means to facilitate hardware performance. This thinking has had an adverse effect on system performance by relegating decision requirements, which are derived directly from mission requirements, to a minor role in system design. It is the premise of this paper that system design should be dictated by decision requirements, since human decisions determine how well and to what degree a system's inherent capabilities will be utilized.

In the decision oriented approach proposed, the system is considered not as a package of hardware and software, integrated to fulfill well-defined mission functions, but rather as a decision making system composed of three interacting elements: hardware, software and humans. The design approach based on this viewpoint begins with mission requirements. From this, decisions necessary to perform specified mission functions—and decision augmentation software and hardware needed to support them—are then defined. This design approach focuses on maximizing the decision making ability of an entire system by viewing the three system components as complementary and by integrating their capabilities to provide synergistic performance.

Decision Oriented System Design Approach

The utility of any system is ultimately a reflection of the philosophy used in its design. The most commonly applied system design approach[1] begins with a detailed statement of platform mission requirements. Mission requirements are then used to derive functional requirements that will support successful mission accomplishment. The functional requirements are allocated to either hardware or software components of the system. In practice, this approach emphasizes hardware considerations, with software being designed primarily to facilitate hardware use. In the approach described herein, a system is viewed as being composed of two interdependent parts: a *process* consisting of people, procedures and organizations, in essence *the way* a system reaches a decision; and *physical components*, (platforms sensors, computers, communications networks, and weapons), the *means* of implementing decisions that derive from the *process*.

Because an informed, timely and "correct" *decision* is the key factor in system effectiveness, system design should be based on *decision requirements*. System hardware and decision augmentation software should be designed to support decision requirements that derive from mission requirements. This helps to ensure that the final system is constructed in a cost-effective manner that should ultimately result in improved mission performance. The aim of mission requirements is to delineate the system's operational performance objectives. Moreover, mission requirements provide preliminary estimates for required crew capabilities, as well as an initial description of how the system will operate in conjunction with other systems. Typically, mission requirements are developed after significant preliminary analysis of the anticipated threat and consideration of the availability of new technology during the life of the system being designed.

The design process begins with several preliminary studies which highlight the operational requirements for a new system and develop a more comprehensive understanding of how it would be integrated into the overall command and control structure of existing systems. Integration is considered at the macroscopic level and would identify all communication links and their content, both input and output, which must be established between the new system and other organizations/units.

Figure 1 defines a system design approach which can be applied to develop a C^3 system (or any system) on a decision-oriented basis that has been initiated with a mission requirements study. After mission requirements have been defined and macro-organizational studies completed, the decision oriented design approach continues with an identification of decision requirements. The focus of this effort is to determine all the decisions which the human elements of the system must make and identify operational sources that can provide needed data/information input. Decision implementation and execution are also critical concerns and are identified as part of the macro-organizational study. In addition, the available time is defined to make each decision in the context of the operational environment. Decision time is an important aspect of system design, since it has a major impact on the concept of system operation.

After decision requirements have been specified, each decision is decomposed into a sequence of decision tasks. A task is defined as part of a decision making process that is initiated either by receipt of new data—used by a change in the operational situation and requiring a response—or to accommodate previously established mission planning considerations. Decision tasks are analyzed in terms of complexity, cognitive processing requirements and algorithmic techniques. Task identification can be effected by the organization environment in which the system will operate. This is the principal reason a detailed macro-organizational analysis is required early in system design. Attention is given to such factors as inter/intraorganization human communications links, organization structure that might impact information/data flow and inter/intrapersonnel relationships. Only after the organizational framework is defined can the required time be specified to perform each decision task and to stipulate the timely availability of decision specific data to support the task.

The results of task analysis provide a direct input into the formulation of the system concept of operations. The concept of operations provides a top level specification of the system's functioning within the operational environment and its integration into the command and control structure. Questions such as how many operators are required, operator responsibilities, etc., would be addressed. In addition, decision task allocation to hardware, software or human would be

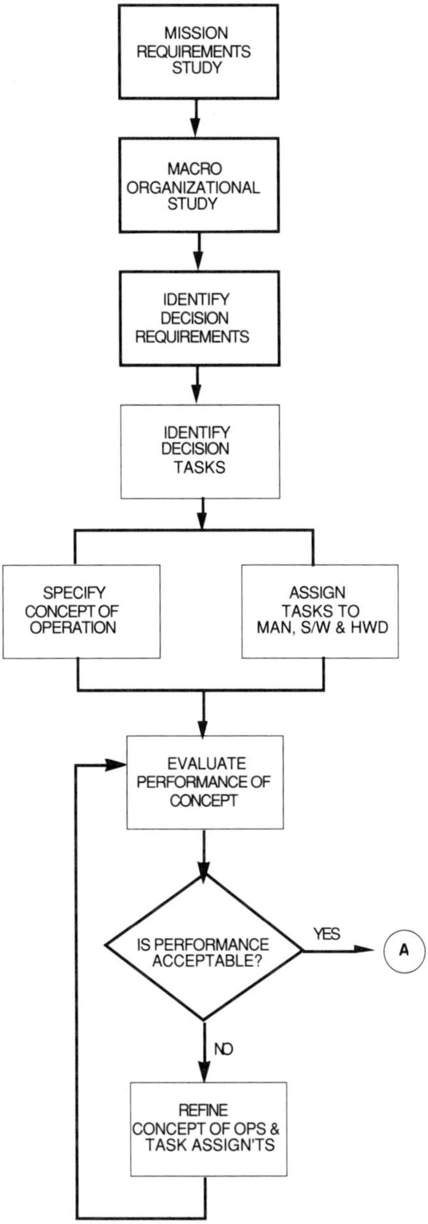

Figure 1a. Decision Oriented Design Approach.

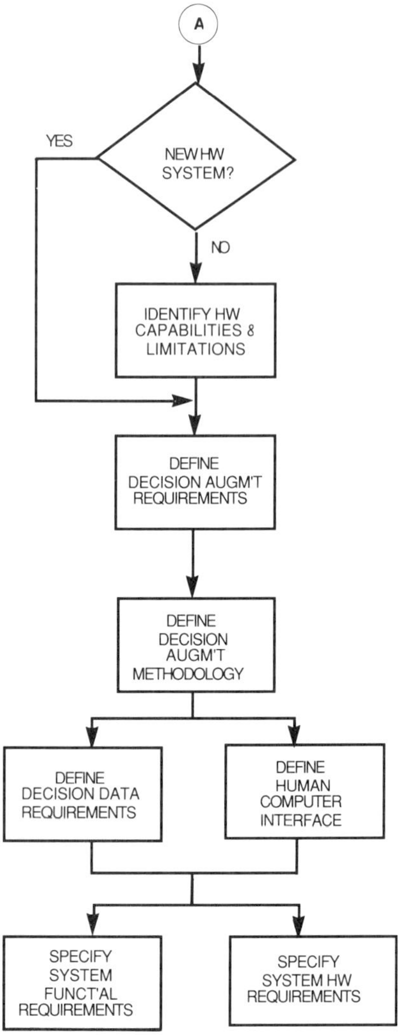

Figure 1b. Decision Oriented Design Approach (cont.)

examined at a macro-level. For example, tasks which are computationally intensive, highly repetitious, and stable are likely candidates for hardware (firmware) implementation. Those which are computationally intensive but subject to modification, as well as simple computational tasks, would be candidates for software implementation. Tasks which require judgment, intuition, or require the use of subjective input, would be assigned to the human.

Once the concept of operations is complete, an evaluation of system performance is conducted. The emphasis in this design phase is to identify the most satisficing concept of system operations. Performance is evaluated in terms of ability to meet mission requirements, as well as adherence to prespecified constraints such as cost, weight, etc.

There is more latitude with respect to system design when a new system is being developed as compared to modifying or improving one that already exists. When enhancements are being made to an old system, the design of new functions is constrained by existing hardware capabilities as well as by the existing software architecture. In order to ensure that new functions are capable of being integrated into the present system, a study must first be performed to define the capabilities and limitations of the existing hardware and software. Results of this study must be taken into account in all subsequent system design efforts.

Once a concept of operations is resolved that satisfies design requirements, decision augmentation requirements are determined. They are based on the need to provide the information necessary to support specified operational decisions. Technical methodologies must be selected which would provide the required information at the requisite fidelity. Choices range from use of heuristics to mathematical and/or statistical algorithms, or expert systems. The decision as to which methodology would be most appropriate would be based on problem complexity and the ability of the selected methodology(s) to effectively model operationally meaningful measures of performance that meet information requirements.

Providing decision specific information is only one factor in ascertaining whether a high quality decision will be made. Since a human makes decisions based on information provided, its content and proper presentation also play a large role. Specification of human computer interface (HCI) is one the most important aspects of the decision oriented system design. The HCI represents the end user's concept of the system—it is the only visible part of the system. If the users cannot readily relate and adapt to the output or feel comfortable with the format, they may not accept and will not use the system. Considerable effort must be given to this phase of the design. Much of it involves working with personnel knowledgeable of the operational environment to gain insight into their cognitive tasks.

Once the decision augmentation methodology has been defined, data needed to provide decision specific information can be ascertained. This can include operational data extracted from communication links interfaced directly to the decision augmentation system or that which must be manually entered. In addition, requirements include "nontraditional" data such as parametrics needed by a mathematical model, or rules used by experts. Collecting this type of data can itself be a formidable and time consuming task.

System hardware requirements should be based on the need to provide the previously specified data,; that is, the ability of the hardware to provide accurate, timely, complete and contextual data to develop the information for a decision maker. Decision augmentation, HCI and hardware requirements become the major inputs to a total system functional specification. When developing this specification, the system should be viewed from *both* an overall system perspective and the environment in which it must operate.

Measuring Performance

One technique that may be used to determine performance is to simulate the system through use of an object oriented programming language. Object oriented languages, which form a relatively new class of programming languages, provide a natural mechanism for representing general C^3 systems, in that they permit a system representation in the form of a network of communicating nodes, or *objects*. Each object has the form of an autonomous processor which has associated with it class ancestry, local data, message templates and methods. Objects inherit data and message templates via a hierarchical class structure. Objects communicate with one another via messages. An object can be human or hardware, thus permitting organizational communications to be modeled and decision process variables to be examined. Methods, which are operations on local data, are executed upon the receipt of specific messages. Simulation would be used to represent relevant BLUE and RED architectures as collections of objects. The simulation would involve definition of scenarios (selection of objects and their local data), C^3 architecture definition (selection of message templates and methods), and a high-level simulation event driver or clock. Measures of Effectiveness (MOE) can be collected at appropriate nodes within the network. This object oriented simulation would allow a user to examine and obtain quantitative statistics on a broad class of C^3 related issues. It would allow the study of alternative BLUE and RED architectures involving communication protocols, command and control procedures, organizational structure, organizational communications and decision process variables, intelligence processing capabilities, platform and sensor performance, and tactical doctrine. Performance evaluation, using an object oriented technique, or some other appropriate methodology, could provide results that would require modifying the concept of operations to meet mission requirements and/or satisfy external constraints. This iterative procedure would continue until an acceptable concept of operations is identified.

Macro-Organizational Analysis

Systems operate within the structure of an organization. Any organization is a systematic arrangement of people and technology whose aim is to accomplish some set of objectives. It is important to realize than an organization exists within the confines of a larger environment. In order to survive and operate effectively, both the environment and its boundary must be completely defined and their implications on organizational effectiveness both understood and accounted for in system design. In an adversarial environment, it is one set of forces and its associated command elements—including C^3—operating against an adversary with similar functional elements. This dual environment is characterized by multiple elements operating against multiple elements, and is similar in nature to commercial organizations operating in a competitive environment. Awareness of and attention to the impact of organizational factors is important to the survival of a commercial organization and becomes even more important when the operating environment is conflict oriented rather than competitive. Organizational effectiveness becomes the major factor impacting performance in many systems. When system performance is examined within the context of an environment the result is operational performance. In order to determine the influence of organizational factors on operational performance, analyses are required that are not normally conducted as part of current system design efforts. These analyses are qualitative and concentrate on factors that relate to the decision process within an organization. Figure 2 indicates the task-source relationship for analyses that should be conducted for a decision oriented system design effort. This top-down analysis is initiated with a determination of mission requirements. Next, decision requirements are established that specifically address mission requirements and are obtained directly from decision makers who operate in the environment under consideration. Behavioral/social science methods such as questionnaires, interviews, verbal protocols and observation of the decision maker in the operational environment would be employed. All tasks shown, including the decision analysis in the broadest sense, would be included in an organizational analysis. Integral to these tasks are identification and evaluation of the following:

- Organizational decision process and its impact on operational performance.
- Decision making responsibility assignments to components of the decision process environment.
- Informal and formal interpersonal and inter/intragroup communication links.
- Relationships among organizations and groups within an organization.
- Need and/or availability for decision augmentation.
- Organizational structure design compatibility with the environment in which it must operate.
- Organizational and procedural problems that negatively impact the decision process.
- Obstacles that impede timely, accurate and complete information transfer between and among people, groups and organizations.

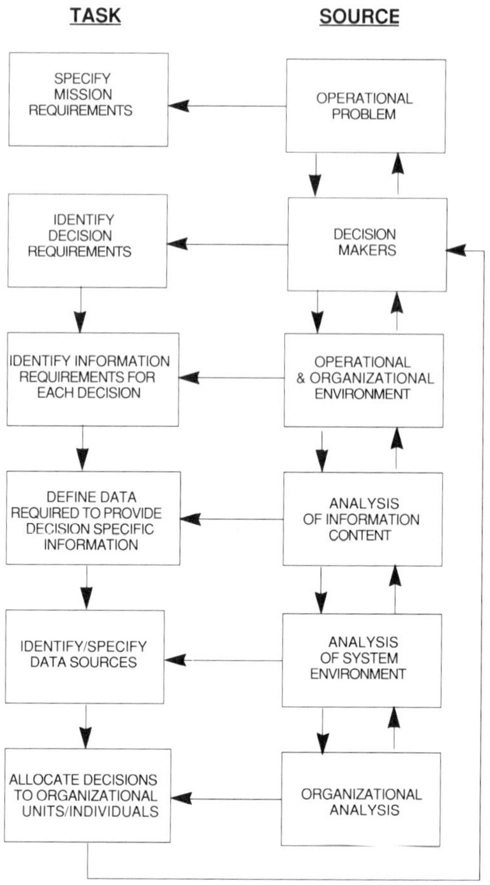

Figure 2. Analysis Methodology.

The organizational areas cited above are not currently considered part of system design tasks. Since the major task of command and control is decision making, organizational factors can and do influence system performance.

Decision Task Analysis

Decision oriented system design includes an analysis of the decision tasks the system is to support. Traditionally, task analysis has focused on the behavioral aspects of task performance to support control and display design and placement, as well as human/machine function allocation. However, operation of C^3

systems involves primarily cognitive processes with little overt performance, requiring a new approach to task analysis. Cognitive task analysis (Card, Moran & Newell, 1983; Rasmussen, 1986a, 1986b), also referred to as cognitive engineering (Norman, 1986) or cognitive systems engineering (Hollnagel & Woods, 1983; Woods & Roth, 1988), provides methods for analysis of the information processes comprising decision tasks. This analysis not only supports allocation of tasks to human, machine, cooperative or adaptive (Rouse, 1988), but also provides the basis for determining the type of decision augmentation that will enhance total system performance.

The goal of cognitive task analysis is to understand the decisionmaking process, including the mental activities involved, the information/knowledge required, the mental models used to support reasoning about the task domain, and the types of errors that are made in the decision making operational environment. Methods for this analysis include interviewing subject matter experts, analyses of thinking aloud protocols, human performance modeling, psychological scaling techniques (i. e., psychometric methods which derive structure from judgments about concepts in a domain), and human-in-the-loop simulation experiments. As technological capabilities for automation and intelligent decision support expand, the need for cognitive task analysis will increase in order to furnish the basis for building systems that effectively integrate human and machine to perform complex tasks.

The Human Element

The most important element of a C^3 system is the human. Often ignored as an integral part of the design, the human is the major determinant of C^3 system effectiveness. Human decisions determine how well and to what degree a system's inherent capabilities will be utilized. The human may present the most difficult system design problem. Humans have limited working memory (Miller, 1956: Simon, 1974) and are not proficient at performing unaided complex numerical calculations. Furthermore, in evaluating alternatives, humans have difficulty integrating input from multiple sources (Dawes, 1979), are biased towards considering solutions from past experience or initial evidence (Tversky and Kahneman, 1974) and usually do not simultaneously contemplate more than three hypotheses (Lusted, 1976; Mehle, 1982; Rasmussen, 1981). In general, when confronted with excess data, humans resort to heuristics which simplify the problem and reduce the amount of input that must be processed (Kahneman, Slovic and Tversky, 1982; Tversky and Kahneman, 1974). All of these actions are taken to reduce the cognitive complexity of making decisions.

The human is a complex organism. Place him/her in an organizational environment and these complexities are manifested in many different ways. In spite of the individual personality, prejudices, preferences or goals, each must interact cooperatively with people of diverse character. The organizational structure can and often does impede interpersonal and intergroup data/information exchange. Even informal communication links are most likely inadequate to meet the communication requirements of an organization. This facet of the organization is just one of many superimposed on the individual that effects performance. Others of major importance are the factors of power and conflict. Input from a source in a position of power is more likely to be accepted in spite of conflicting input or the affected individual's (or organizational unit's) own convictions (Ference, 1970). In an organizational environment, where conflict is considered undesirable, decision makers may be inhibited from making the most appropriate decision simply to avoid conflict. Of equal importance are psychological factors: Once an individual has formulated an hypothesis, disconfirming input would be both ignored and not sought. Thus, it is not only organizational factors that must be considered in system design, but the psychological factors that influence both human behavior and decision making.

Decision Augmentation

The system design approach proposed is facilitated by introducing decision augmentation. Decision augmentation software is designed to reduce human cognitive overload and all its associated problems. In addition, it will satisfy human information requirements by performing, at some level, data-to-information transformation. Information is defined herein as data processed to match the cognitive capability of the human and related to a specific decision task. Decision augmentation software would also be designed to account for information uncertainty in the process of maximizing decision quality. The decision augmentation orientation focuses on the decision requirements necessary to successfully accomplish mission requirements.

A decision augmentation orientation is most important when system performance is highly dependent on decision quality—the goal for most systems. One of the reasons that systems are not attaining their theoretical performance is that little attention was given to the decision making process. Figure 3 shows a simplified schematic of a C^3 system with and without decision augmentation orientation. In either case, data—including intelligence, environment, sensor, threat, and own force—comes into the system and would be processed. Without a decision augmentation orientation, the processed data is presented directly to the decision maker. The two shaded boxes are added within the dotted box to provide a decision oriented design. Properly designed, these two additional elements would provide decision specific information, displayed to allow maximum operator understanding and ease of cognitive processing. This consideration of the human decision making system component can help to overcome some cognitive processing limitations, biases and psychological fac-

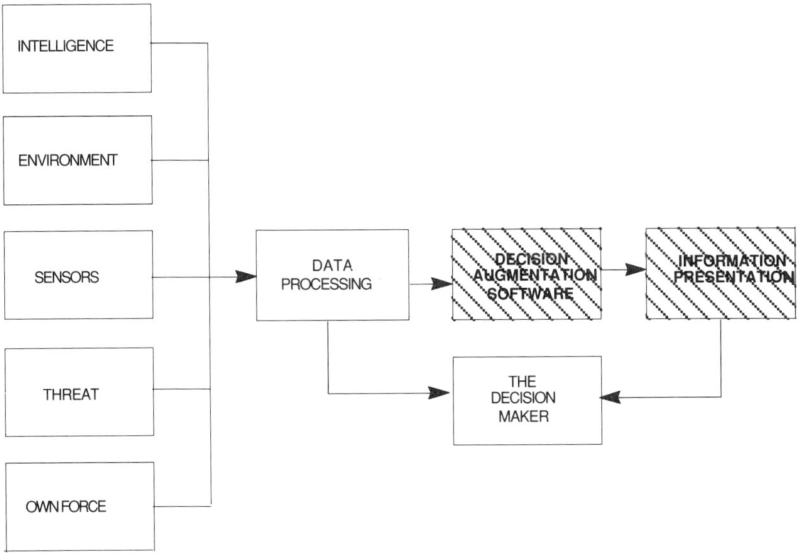

Figure 3. Decision Function Emphasis.

tors that affect quality of decisions. Decision augmentation allows more complex rules to be used, more alternatives to be investigated, uncertainty to be considered and more accurate calculations to be made. It would also provide the ability to predict action consequences and evaluate their associated utility. Information presentation would be based on an understanding of human cognition. Mechanisms of information presentation directly effect the ease with which a human can assimilate and use what is presented. For effective task performance, the display content must be limited to that relevant to the current decision task. Since the displays present only relevant information, easy access of supporting data must be provided; especially data useful in resolving conflicts and historical data for trend analysis.

Decision augmentation systems vary in level of automation involved. They range from completely *automatic*, in which no human action is involved, to *manual*, in which only computational aid is provided. Intermediate levels include *semiautomatic*, in which the system provides evaluated decision alternatives and/or recommended courses of action which the operator can accept or override, and *interactive*, in which the human and software work as a team and symbiotically support each other. Figure 4 indicates the interrelationship between decision augmentation level, information processing requirements, implementation technique and HCI requirements. The decision augmentation technique should match the decision situation environment. While expert sys-

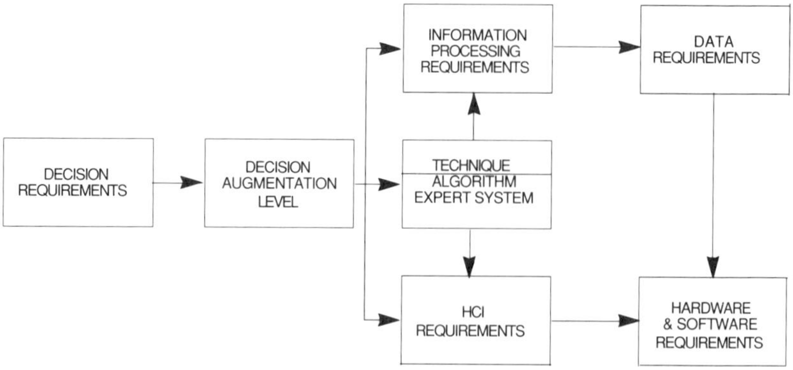

Figure 4. Impact of Decsion Augmentation Level.

tem techniques may be effective in many decision situations, other techniques are also useful. A number of taxonomies for decision situations and decision augmentation techniques have been proposed, (Wohl, 1981; Zachary, 1986), which provide guidance in determining the type of decision augmentation technique to use. This has to be envisioned within the context of both the organizational and operational environments in which the system would be employed.

Time is one of the major parameters in both environments that would impact decision augmentation level. Three time variables must be minimized to maintain control over the operational situation.

$$T_{control} = T_{collect} + T_{decide} + T_{xmt/rcv}$$

In order to maintain operational control:

$$T_{control} < T_{crit} - T_{op}$$

where: T_{crit} = critical time (time within which the operation must be executed in order to have the intended effect)

T_{op} = time required to execute the operation

If time were the only factor affecting decision level, their inverse relationship would suggest that the less time available for decision making, the higher the level of automation. However, the level of decision augmentation that would increase decision effectiveness depends on more than just time. It is also based on, at a minimum, the following:

- Operational situation
- Data load
- Decision importance
- Intuition likelihood
- User acceptance (of decision level)

Most of the above are self-explanatory, but intuition likelihood is not immediately obvious as an indicator of decision augmentation level. The decision maker would use prior knowledge and experience to manage input overloads. In operational situations where the human feels comfortable, intuition (other terms are hunches, gut feel, etc.) plays a large role in decision making. Because the decision maker can be affected by biases, which may be erroneous or not related to the current situation, intuition could produce an unsatisfactory decision, especially in a time critical, high data rate situation. Table 1 shows the relationship among the above five conditions and decision augmentation (DA) levels in terms of an idealistic environment.

The real world is neither idealistic nor cooperative. It is easy to envision a realistic operational environment where conflict among the conditions listed is not only possible but very likely. If a decision is to have the effect intended—such as timely intercept in a space defense system—it may be necessary to have certain decisions made automatically (based on thresholds). This is not what would be done in an ideal environment.

Technology Implications

Emphasis on decision making functions has a number of implications for the design and implementation of C^3 systems. These include understanding the role of cognition, the importance of human computer interface, the requirement for

Table 1

LEVEL OF DA	OPERATIONAL ENVIRONMENT	INPUT LOAD	DECISION IMPORTANCE	USER ACCEPT	INTUITION LIKELIHOOD
Automatic	$T_{crit} = \Delta$	High	Low	*	Low
Semi-Auto	$T_{crit} > \Delta$	High	Mod	*	Low
Interactive	$T_{crit} >> \Delta$	Mod	Mod	*	Mod
Manual	$T_{crit} >>> \Delta$	Low	High	*	High

Δ = Short time

* Variable, a function of user

planned innovation acceptance, the implications of new and improving technology and the need for adaptive capability.

A basic understanding of human cognitive processing is required to fully realize decision augmentation potential. Decision augmentation systems should be designed to provide assistance in those areas where human capability is limited, while still capitalizing on human strengths. For example, humans have attention and memory limitations; inherent heuristics they employ in input processing and limited ability to process any input, but especially numerical data, in complex, stressful and input overload situations. Decision augmentation has the potential for providing significant improvements over unaided decision making.

HCI design can adversely affect decision behavior, especially in decisions involving time pressure and high data volume. This can be mitigated by carefully structuring the interface and exploiting innovative HCI techniques. Information transmission can be improved and user learning time and processing demands reduced. In order to maximize performance in a decision augmentation system, attention must be given to the relationship between the system and the user; that is, the amount and kind of user-augmentation interaction, dialogue style and output formats (Schwartz and Jamar, 1983).

Improved understanding of and attention to acceptance of innovation is needed. Whether a decision system is used or not depends both on the design and how it is introduced (Mackie and Wylie, 1985). The system must be designed so that it meets the user's need for reduced cognitive and physical workload. Perception is a major factor in user acceptance. When decision augmentation software is actively involved in the decision making process—in some cases without human intervention—it should not *appear* to erode individual control or authority. User authority can be preserved by making automated systems understandable (Fitter and Sime, 1980) and by being able to "explain its behavior" in understandable terms. Before accepting results, the user must develop confidence in the system. This requires that the automated system accept user ideas and provide an evaluation for comparison with the automated system's solution. Decision augmentation system operation should require only minimum knowledge of computers and should not involve complex operating procedures. The user community should be involved throughout the development process to facilitate the introduction of new decision augmentation systems (Adelman, 1982). They should also be involved early in the development cycle, beginning with decision requirements analysis. Bringing the user community into the development process from its inception guarantees user design inputs and familiarity with its capability. This translates not only to user acceptance, but also to the ability to take full advantage of the system when it is fielded.

Expert system techniques provide a major tool for implementing decision augmentation systems. They allow, for example, the use of complex decision rules and the ability to consider multiple hypotheses. Furthermore, the knowledge of highly experienced personnel can be encoded and applied by all levels of personnel. Using expert systems, corporate knowledge—a major determinant of organizational decision making performance—could be added to instead of lost, as is currently the situation in military C^3 systems.

An additional design problem exacerbated by the military environment is the need for the decision augmentation system to adapt to the user and the environment. For maximum flexibility and performance, the system must be able to adapt to a dynamic operational environment. In addition, there are individual differences among decision makers in decision style, risk taking and the importance placed on different criteria. The ability of neural net technology to learn may provide this needed adaptability. Having the capacity to adapt to both the decision maker and the environment would not only have a major impact on system performance, but also on user acceptance.

Summary

In the past, and to a large extent today, the emphasis in system design has been on defining hardware requirements. In many cases it is advances in hardware technology rather than ability to meet mission requirements that provide the impetus for upgrades to C^3 systems (or any system). Even though software has increased in importance and percentage of cost in system development, it is still philosophically considered as a means to facilitate hardware performance and not as a way to augment decision makers. This thinking has had an adverse effect on system performance by relegating decision requirements—which derive directly from mission requirements—to a minor role in the system design process. System design should be dictated by decision requirements, since the decisions humans make determine how well and to what degree a system's inherent capabilities will be utilized and performance realized.

References

Adelman, L. "Involving users in the development of decision-analytic aids: the principle factor in successful implementation." *Journal of the Operational Research Society,* 1982, 33, 333-342.

Card, S. K., Moran, T., and A. Newell. *The Psychology of Human-computer Interaction.* Hillsdale, N. J.: Erlbaum, 1983.

Dawes, R. M. "The robust beauty of improper linear models in decision making." *American Psychologist*, 1979, 34, 571-582.

Ference, T. P. "Organizational communications systems and the decision process," *Management Science*, 1970, 883-96.

Fitter, M. J., and M. E. Sime, "Creating responsive computers: Responsibility and shared decision making." In H. T. Smith and T. R. G. Green (Eds.), *Human Interaction With Computers*, 1980. London: Academic Press.

Hollnagel, E., and D. D. Woods, "Cognitive systems engineering: New wine in new bottles." *International Journal of Man-Machine Studies*, 1983, 18, 583-600.

Kahneman, D., P. Slovic and A. Tversky, (Eds.) *Judgment under Uncertainty: Heuristics and Biases.* 1982, New York: Cambridge University Press.

Lusted, L. B. "Clinical decision making." In D. Dombal and J. Grevy (Eds.), *Decision Making and Medical Care.* 1976, Amsterdam: North Holland.

Mackie, R. R. and C. D. Wylie, "Technology Transfer and Artificial Intelligence: User Considerations in the Acceptance and Use of AI Decision Aids." Human Factors Research Division of Essex Corporation, *Technical Report TR51231-1*, November, 1985: Goleta, CA.

Meehl, P. C. *Clinical Versus Statistical Prediction.* 1954, Minneapolis: University of Minnesota Press.

Metersky, M. L. "A C^2 Process and an Approach to Design and Evaluation," *IEEE Transactions on Systems, Man and Cybernetics.* 1986, SMC-16, 6, 880-889.

Miller, G. A. "The Magic Number Seven, Plus or Minus Two: Some Limits on our Capacity for Processing Information." *Psychological Review*, 1956, 63, 81-87.

Norman, D. A. "Cognitive Engineering." In D. A. Norman and S. W. Draper (eds.), *User Centered System Design.* 1986, Hillsdale, N. J Erlbaum.

Rasmussen, J. "A Framework for Cognitive Task Analysis in Systems Design." In E. Hollnagel, G. Mancini, and D. D. Woods (eds.), *Intelligent Decision Support in Process Environments.* 1986a, Berlin: Springer-Verlag.

Rasmussen, J. "Information Processing and Human-Machine Interaction." 1986b, New York: North-Holland.

Rasmussen, J. "Models of Mental Strategies in Process Control." In J. Rasmussen and W. Rouse (eds.), *Human Detection and Diagnosis of System Failures.* 1981, New York: Plenum Press.

Rouse, W. B. "Adaptive Aiding for Human/Computer Control." *Human Factors*, 1988, 30, 431-443.

Schwartz, J. P. and P. Jamar. "Lack of Guidance for Decision Aid Interface Design." *Association for Computing Machinery SIGCHI Bulletin*, 1983, 15, 13-17.

Simon, H. A. "How Big is a Chunk?" *Science*, 1974, 183, 482-488.

Tversky, A., and D. Kahneman. "Judgment Under Uncertainty: Heuristics and Biases." *Science*, 1974, 185, 1124-1131.

Wohl, J. G. "Force Management Decision Requirements For Air Force Tactical Command and Control." *IEEE Transactions on Systems, Man and Cybernetics*, 1981, SMC-11, 618-639.

Woods, D. D., and E. M. Roth. "Cognitive Systems Engineering." In M. Helander (Ed.), *Handbook of Human-Computer Interaction.* 1988, Amsterdam: Elsevier.

Zachary, W. "A Cognitively Based Functional Taxonomy of Decision Support Techniques." *Human-Computer Interaction*, 1986, 2, 25-63.

[1]System Engineering Management Guide. Contract MDA903-82-C-0339 Defense Management College, Fort Belvoir, VA.

Prospects for Cognitive Systems Engineering

Stephen J. Andriole and Leonard Adelman

Introduction

Our working premise is as follows:

Recent advances in cognitive engineering and advanced information technology have provided new opportunities for C^2 systems designers; opportunities that have, by and large, gone unnoticed by the "major" C^2 systems houses. It is now possible to dramatically enhance human performance via the application of the principles of cognitive engineering and the tools and techniques of advanced information technology to C^2 problems. Cognitive engineering is a relatively new field of inquiry that calls for the design and development of computer-based information systems consistent with the way humans store, retrieve, "display" and process information. The advanced information technologies available to us now include hypermedia, multimedia, simulation, animation, cost-effective color, sound, multi-dimensionality, and true multi-tasking, among others. The principles of cognitive engineering and the power of advanced information technology can be synthesized to yield interactive systems that can greatly enhance human performance in C^2.

This article examines a set of "generic" inference- and decision-making requirements, assesses the literature in the cognitive sciences that pertains to inference- and decision-making, and then presents some information technology-inspired prototype system concepts that attempt to leverage the findings in cognitive science. The process is represented below:

Principles of Cognitive Engineering	+	Advanced Information Technology	+	Command & Control Requirements	—>	Prototypes

Requirements

C^2 requirements occur at many levels and in many substantive areas. The role of information systems in C^2 is extensive. The number of problems these systems have faced over the years is legendary. While "cognitive compatibility" is by no means a panacea, it does represent a goal consistent with the general thrust in enhanced user-computer interaction (UCI), a thrust responsible for "user friendly" design approaches and—in some cases—operational systems.

This article suggests that there is enormous leverage in the design, development and deployment of well cognitively engineered systems. In order to demonstrate the power of cognitive systems engineering, a set of "generic" inference- and decision-making requirements—representative of large classes of C^2 problems—are examined. Some examples of "generic" inference- and decision-making requirements are presented below. However, note that this short list is followed by a list of "display" requirements, since nearly all substantive requirements have display correlates:

Inference-Making Requirements

Substantive Requirements

- Assessment of events and conditions; inferences about likely event outcomes;
- Identification of high probability crisis situations;
- Mechanisms for discriminating between "signals" and "noise" in the target environment(s);
- Techniques for minimizing "false alarms" and maximizing "hits";
- Inter- and intra-group communications; and
- Heuristics for adaptive monitoring; methods for quasi-automated monitoring and warning . . .

Display Requirements

- Distilled displays of threat;
- Displays of geographical vulnerability;
- Historical displays of comparative cases; and
- "What-if" displays . . .

Decision-Making Requirements

Substantive Requirements

- (Tactical and strategic) option generation;
- Option evaluation;
- Assessment of operational constraints;
- Information/knowledge processing;
- Search for historical precedents (successful and unsuccessful);
- "What-if" analyses;
- Calculation of decision option outcomes; and
- Inter- and intra-group communications . . .

Display Requirements

- Displays of costs and benefits;
- Displays of alternative options;
- Displays of option rank-orders;
- Displays of communications networks;
- Displays of decision option outcomes; and
- Methods and techniques for rapid information/knowledge processing . . .

These and other functions and tasks are representative of the kinds of activities that define command and control. The challenge remains how to design and develop interactive systems that will satisfy these requirements while simultaneously minimizing operator workload. One way to accomplish these goals is through cognitive systems engineering.

Cognitive Systems Engineering

Recently a number of applied psychologists, information scientists, and systems engineers have joined forces to create a new field of inquiry currently know as "cognitive systems engineering." The key word in the label is *engineering*, because it reflects an intention to *apply* what we know about human cognitive information processing to larger systems engineering efforts. Had "cognitive engineers" been around (in name, at least!) in the mid-1970s, they would have "invented" windows, the mouse, navigational aids, and interactive graphic displays. Today's cognitive engineers anchor their design recommendations in what we know about human information processing; what we "know" is ideally determined by theoretically-inspired experimentation that yields empirically verifiable results.

Presented below is a set of findings from the cognitive sciences representative of the results that can be obtained through rigorous research. The findings are organized around the two exemplar generic domains of inference- and decision-making (Andriole and Adelman, 1989):

Inference-Making Findings

- Inference-making performance is task dependent;
- Human information processors (HIPS) have poor insight into their own inferential processes;
- "Cognitive feedback" can significantly improve task performance;
- HIPS are susceptible to non-linear-based deception;
- HIPS use simple and complex "causal" schemas to explain behavior and make inferences;
- HIPS often reason from experiences, analogies or "cases";
- HIPS use "plausible scenario generation" to test and generate hypotheses;
- HIPS rely upon "cues-to-causality" to explain current and future behavior;
- HIPS process less information than they think they use to make inferences;
- Experts are as susceptible to poor insight about their own inferential processes as novices;
- HIPS are unsure about how to optimize data, information and knowledge in inference-making;
- HIPS are prone to "cognitive reductionism" to reduce workload and overall mental effort;
- The perception of information is not comprehensive, but selective;
- HIPS tend to under emphasize "base-rate" information;
- HIPS weight the importance of information on the basis of its perceived causal meaningfulness not based upon its statistical diagnosticity;
- HIPS are susceptible to the "availability bias," or the tendency to recall recent or highly publicized events;
- HIPS selectively perceive data, information and knowledge on the basis of experience;
- HIPS are susceptible to confirmation biases that filter incoming information according to preconceived ideas and views;
- The way information is presented often determines how it is perceived; and
- HIPS tend to be conservative inference-makers . . .

Decision-Making Findings

- Human decision-makers (HUDMS) tend to simplify decision problems by setting outcome aspirations;
- HUDMS often choose the first decision alternative that satisfies the outcome aspiration(s);

- HUDMS often simplify decision problems by only considering a small number of alternatives;
- HUDMS use analogies to generate and compare options;
- HUDMS weight criteria to rank-order decision alternatives;
- HUDMS selectively perceive data, information and knowledge, focus on confirming (versus disconfirming) information, and tend to anchor their judgments; and
- HUDMS tend to attribute decision outcomes to chance or the complexity of the problem (not their own decisionmaking deficiencies) . . .

This short list of findings suggests how the cognitive sciences can inform the systems engineering process. User-computer interaction (UCI) routines, displays, dialogue and other communications/presentation techniques can be used to narrow the gap between the way humans make inferences and decisions and the way systems support them. When combined with the power of advanced information technology, new system concepts can be developed and tested (see below).

Advanced Information Technologies for C^2

Substantive and display requirements, and findings from cognitive science, beg the technology challenge. Can advanced technology help? How? Where? What kinds of technologies? How do we know that they will work?

The working premise here is that there are a variety of advanced (and not-so-advanced) information technologies that can greatly aid the inference- and decision-maker.

There are a variety of technological solutions available to the C^2 systems engineer. Some appear below. These (and other) ideas need to be tested in the context of some specific scenarios as working prototypes before they become elements of command center capabilities or operational procedure. The list below contains many ideas that simply could not have been proposed just five years ago; ten years ago they would have been judged unrealistic or prohibitively expensive. It is essential that we exploit advanced information technology; at the same time, it is essential that our technological solutions be requirements-driven (and not unduly "pushed" by the technologies themselves) and informed by what we know about human information processing, decision making and problem solving.

Advanced Information Technology-Based Ideas for C^2

- Adaptive real-time/on-line message processing/routing systems;

- Advanced message typography systems;
- Analogical (case-based) data/knowledge base analyses;
- Spatial/iconic-based search techniques;
- Hypertext-based ultra-rapid reading and information/knowledge processing;
- Multimedia (text, photographic, sound, video . . .)-based situation displays;
- On-line historical data/knowledge bases of previous, pertinent planning and decision making episodes;
- Direct manipulation graphic displays for "what-if" analyses and other functions;
- Knowledge-based option generators;
- Real-time simulation/animation of decision option outcomes;
- Shared data/knowledge bases for group decision making...among other ideas

These ideas—among countless others—are, in principle, viable solutions to complex analytical problems. Are they any good? Will they enhance human performance in C^2? How can they be married to cognitive science to yield the greatest leverage?

Prototype Development

In order to determine the feasibility and applied potential of cognitive systems engineering, we undertook a requirements analysis in the area of strategic air defense intelligence and operations (Andriole and Adelman, 1989). We then designed a number of system concepts intended to demonstrate the extent to which advanced information technology and findings from cognitive science could be merged to yield user-computer interaction routines likely—due to their level of cognitive compatibility with the operators—enhance human inference- and decision-making performance. We designed a suite of interactive "storyboard" prototypes (Andriole, 1989). Storyboard prototypes are simulations of future system's capabilities. They are comprised of a set of screen displays that represent the capabilities of the intended system, though from the user's perspective the system looks and acts "real."

Selected displays from the prototypes appear below. Note the description of the storyboards and their linkage to cognitive science and information technology.

Conclusions

The displays suggest that there is indeed great potential in the marriage of cognitive science and information technology, and hopeful prospects for cognitive systems engineering. While careful evaluations should be conducted to determine where and how the marriage contributes to enhanced human performance (Adelman, 1990), the initial feasibility conditions have been satisfied.

References

Adelman, L. (1990) *Evaluating Decision Support Systems.* Wellesley, MA: QED Information Sciences, Inc.

Andriole, SJ (1989) *Storyboard Prototyping: A New Approach to User Requirements Analysis.* Wellesley, MA: QED Information Sciences, Inc.

__________ and Adelman, A. (1989) *Cognitive Engineering of Advanced Information Technology for Air Force Systems Design and Development: Prototypes for Air Defense Intelligence and Operations.* Marshall, VA: International Information Systems, Inc.

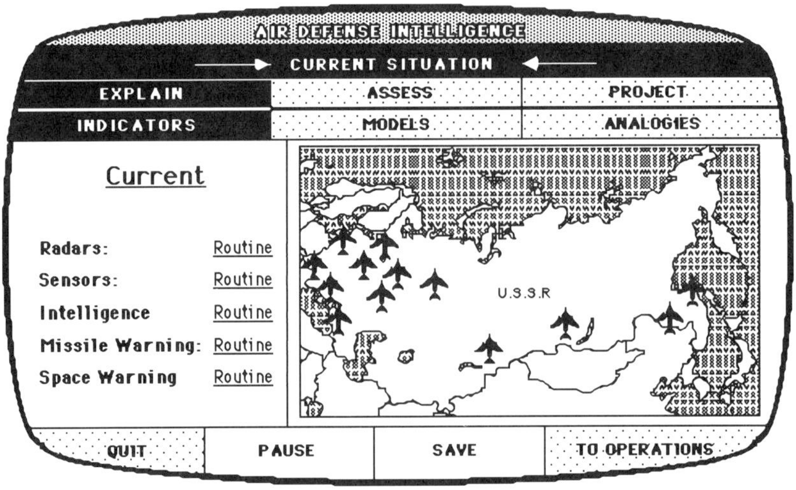
AIR DEFENSE INTELLIGENCE
CURRENT SITUATION
EXPLAIN
ASSESS
PROJECT
INDICATORS
MODELS
ANALOGIES
Current
Radars: Routine
Sensors: Routine
Intelligence Routine
Missile Warning: Routine
Space Warning Routine
U.S.S.R
QUIT
PAUSE
SAVE
TO OPERATIONS

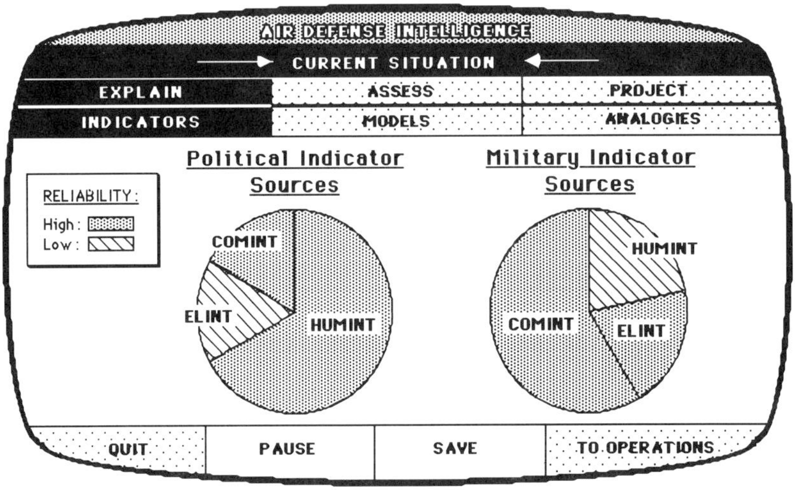
AIR DEFENSE INTELLIGENCE
CURRENT SITUATION
EXPLAIN
ASSESS
PROJECT
INDICATORS
MODELS
ANALOGIES
RELIABILITY:
High:
Low:
Political Indicator Sources
COMINT
ELINT
HUMINT
Military Indicator Sources
HUMINT
COMINT
ELINT
QUIT
PAUSE
SAVE
TO OPERATIONS

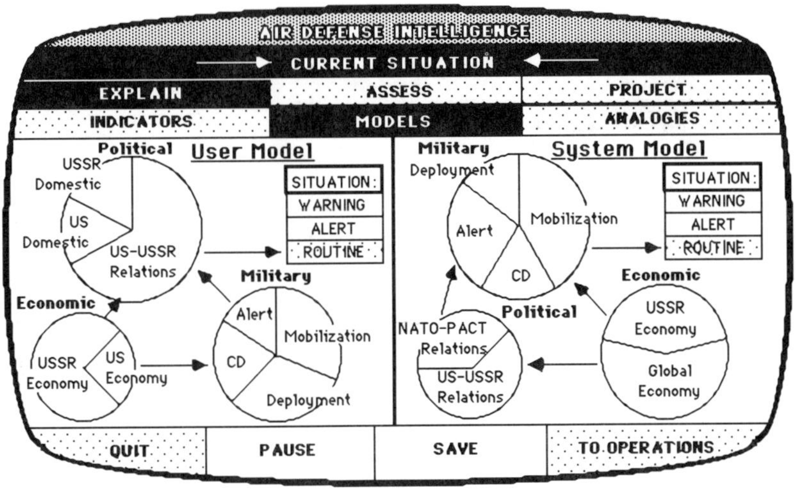
AIR DEFENSE INTELLIGENCE
CURRENT SITUATION
EXPLAIN
ASSESS
PROJECT
INDICATORS
MODELS
ANALOGIES
User Model
Political
USSR Domestic
US Domestic
US-USSR Relations
SITUATION:
WARNING
ALERT
ROUTINE
Military
Alert
Mobilization
CD
Deployment
Economic
USSR Economy
US Economy
System Model
Military
Deployment
Alert
Mobilization
CD
SITUATION:
WARNING
ALERT
ROUTINE
Economic
USSR Economy
Global Economy
Political
NATO-PACT Relations
US-USSR Relations
QUIT
PAUSE
SAVE
TO OPERATIONS

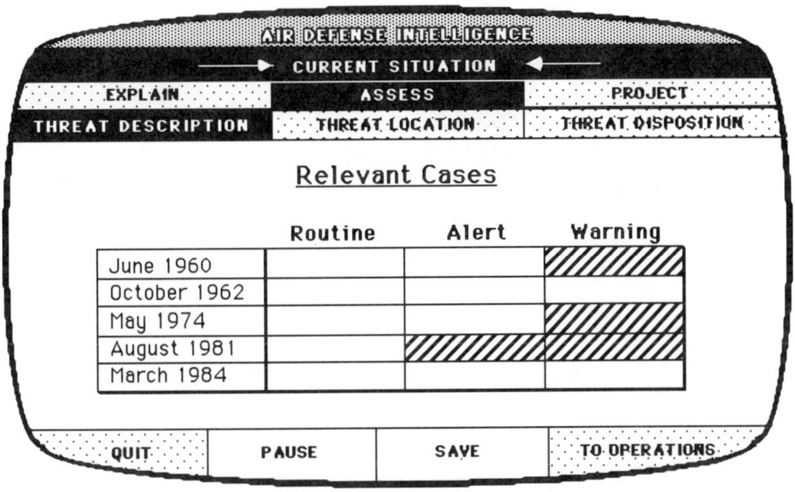
AIR DEFENSE INTELLIGENCE
CURRENT SITUATION
EXPLAIN
ASSESS
PROJECT
THREAT DESCRIPTION
THREAT LOCATION
THREAT DISPOSITION
Relevant Cases
Routine
Alert
Warning
June 1960
October 1962
May 1974
August 1981
March 1984
QUIT
PAUSE
SAVE
TO OPERATIONS

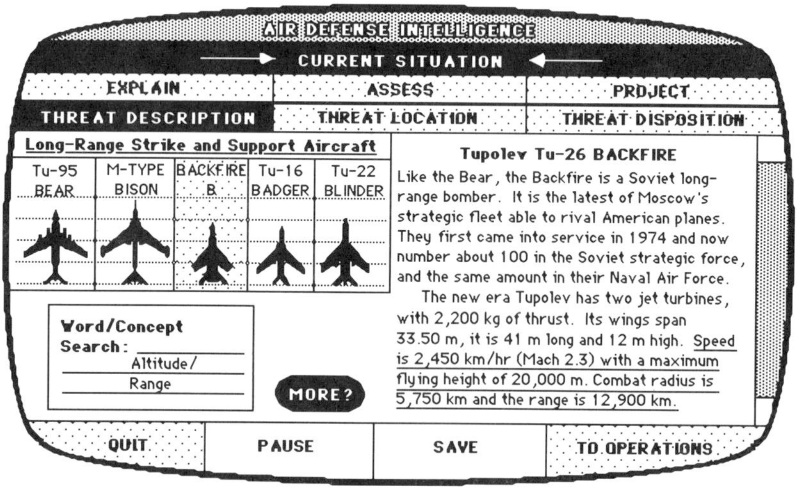
AIR DEFENSE INTELLIGENCE
CURRENT SITUATION
EXPLAIN
ASSESS
PROJECT
THREAT DESCRIPTION
THREAT LOCATION
THREAT DISPOSITION
Long-Range Strike and Support Aircraft
Tu-95 BEAR
M-TYPE BISON
BACKFIRE B
Tu-16 BADGER
Tu-22 BLINDER
Word/Concept Search:
Altitude/
Range
MORE?
Tupolev Tu-26 BACKFIRE
Like the Bear, the Backfire is a Soviet long-range bomber. It is the latest of Moscow's strategic fleet able to rival American planes. They first came into service in 1974 and now number about 100 in the Soviet strategic force, and the same amount in their Naval Air Force.
The new era Tupolev has two jet turbines, with 2,200 kg of thrust. Its wings span 33.50 m, it is 41 m long and 12 m high. Speed is 2,450 km/hr (Mach 2.3) with a maximum flying height of 20,000 m. Combat radius is 5,750 km and the range is 12,900 km.
QUIT
PAUSE
SAVE
TO OPERATIONS

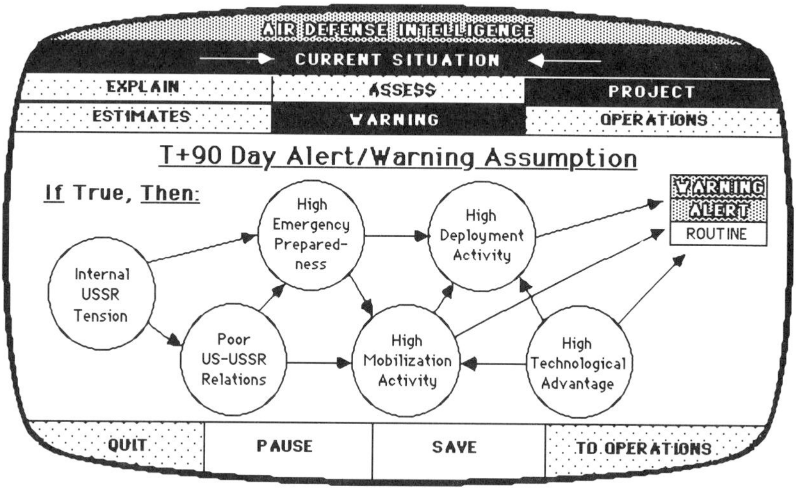
AIR DEFENSE INTELLIGENCE
CURRENT SITUATION
EXPLAIN
ASSESS
PROJECT
ESTIMATES
WARNING
OPERATIONS
T+90 Day Alert/Warning Assumption
If True, Then:
Internal USSR Tension
Poor US-USSR Relations
High Emergency Prepared-ness
High Mobilization Activity
High Deployment Activity
High Technological Advantage
WARNING
ALERT
ROUTINE
QUIT
PAUSE
SAVE
TO OPERATIONS

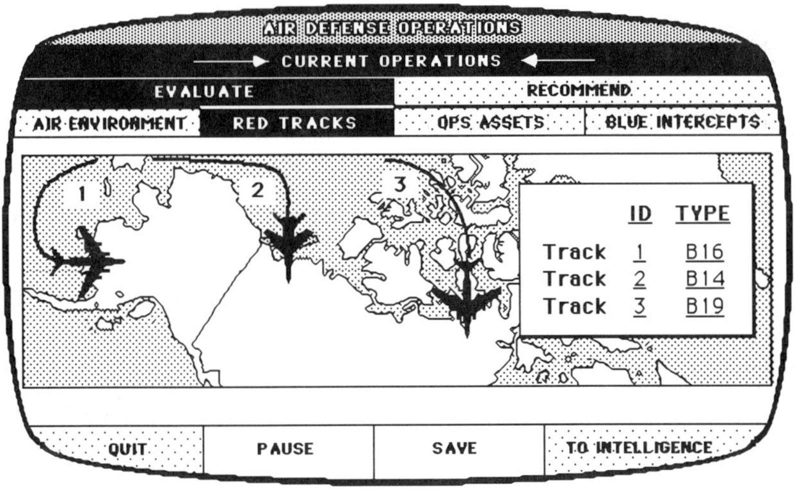
AIR DEFENSE OPERATIONS
CURRENT OPERATIONS
EVALUATE
RECOMMEND
AIR ENVIRONMENT
RED TRACKS
OPS ASSETS
BLUE INTERCEPTS
1
2
3
ID TYPE
Track 1 B16
Track 2 B14
Track 3 B19
QUIT
PAUSE
SAVE
TO INTELLIGENCE

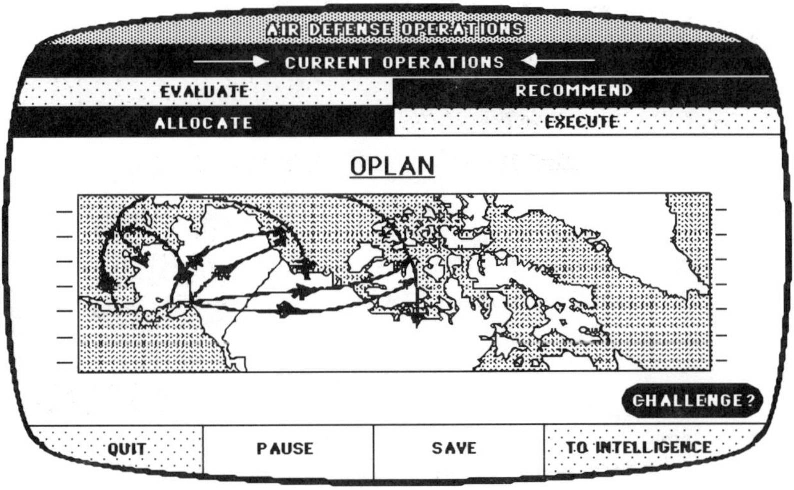
AIR DEFENSE OPERATIONS
CURRENT OPERATIONS
EVALUATE
RECOMMEND
ALLOCATE
EXECUTE
OPLAN
CHALLENGE?
QUIT
PAUSE
SAVE
TO INTELLIGENCE

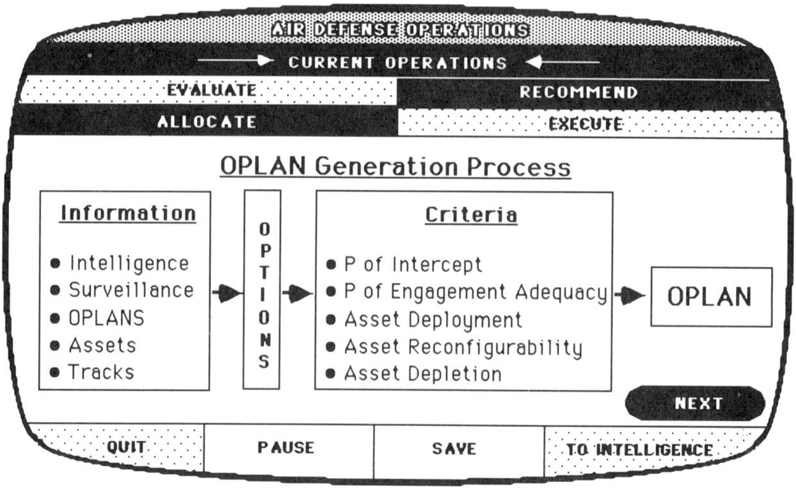
AIR DEFENSE OPERATIONS
CURRENT OPERATIONS
EVALUATE
RECOMMEND
ALLOCATE
EXECUTE
OPLAN Generation Process
Information
• Intelligence
• Surveillance
• OPLANS
• Assets
• Tracks
OPTIONS
Criteria
• P of Intercept
• P of Engagement Adequacy
• Asset Deployment
• Asset Reconfigurability
• Asset Depletion
OPLAN
NEXT
QUIT
PAUSE
SAVE
TO INTELLIGENCE

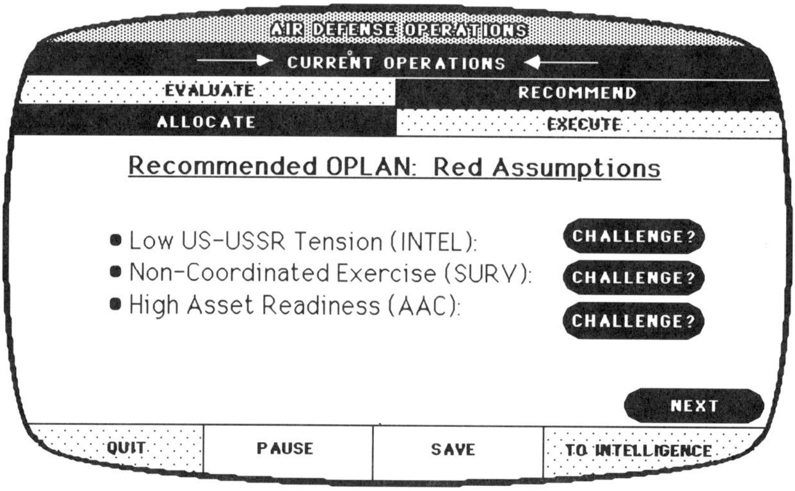
AIR DEFENSE OPERATIONS
CURRENT OPERATIONS
EVALUATE
RECOMMEND
ALLOCATE
EXECUTE
Recommended OPLAN: Red Assumptions
• Low US-USSR Tension (INTEL):
CHALLENGE?
• Non-Coordinated Exercise (SURV):
CHALLENGE?
• High Asset Readiness (AAC):
CHALLENGE?
NEXT
QUIT
PAUSE
SAVE
TO INTELLIGENCE

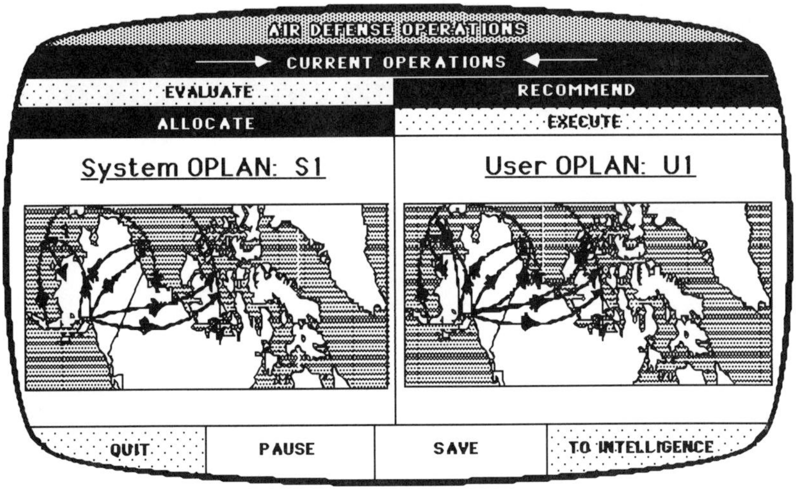
AIR DEFENSE OPERATIONS
CURRENT OPERATIONS
EVALUATE
RECOMMEND
ALLOCATE
EXECUTE
System OPLAN: S1
User OPLAN: U1
QUIT
PAUSE
SAVE
TO INTELLIGENCE

Multidisciplinary Systems Engineering Training and Education: Design and Challenges

Gerald W. Hopple

Definitions of systems engineering (SE) abound, but there is clear consensus on the fact that SE is both inter- and *multidisciplinary* in nature. This in and of itself poses quite vexing challenges to the formidable tasks of SE training and education curriculum design and implementation. This chapter will provide a brief tour of the contemporary SE education design landscape and delineate the central challenges which confront the SE education community. The focus will be on SE, of command and control systems (C^2), but the implications are much broader and can be generalized to the entire field.

As the Chambers (1980) SE bibliography demonstrates, the well-educated systems engineer must master an incredibly diverse range of topics—from analysis, complexity (at various system and subsystem levels), decision theory, design, effectiveness (cost/system), evaluation (cost/performance), functional allocation, and general readings (including the need for dialogues with other engineers, "laws" of system development and problem solving, system management) to human engineering, life cycle costs, logistics support, models/modeling, production engineering, reliability and maintainability, requirements, risk, specifications, systems engineering (general), and systems theory. The latter two subdomains encompass such complex topics as the philosophy and historical development of SE as a concept, the SE process, and alternative SE methods (SE/general) and the more abstract perspective of systems theory as a paradigm for the discipline.

All of this presumes a solid foundation in one or more SE process models and immerses the serious student in the study of (applied) philosophy of (SE) science, research design (the neglected sibling in SE education), and epistemology (a term which is unknown to all too many systems engineers) in addition to the topics delineated above. In addition, contemporary C^2 SE training and education templates increasingly emphasize the *cognitive science*-based concerns of artificial intelligence (AI), psychology, linguistics, and other fields relevant to such *advanced computer-based cognitive problem solving systems* as knowledge-based and decision support systems. Thus, SE today embraces a host of engineering, scientific, mathematical, computer, behavioral, psychological, and management sciences.

Research on the subject of "how advanced information technology has been and will continue to be applied to complicated decision making, evaluation, forecasting, and information management problems as they occur within the context of command and control" (Andriole and Halpin, 1986: 762) provides the foundation for the cognitive science component of the SE curriculum. Such a curriculum would feature information processing models for C^2 (comprised of multiple models and methods of decision making, ranging from cognitive systems engineering and AI tools to optimization and gaming/game theory, among many others), C^2 information systems (design and development issues, command and control as a process, models, evaluation, cognitive engineering and human-machine interface issues), and C^2 information technology issues and challenges (the optimum role of the user, the limits of evolutionary development, dangers emanating from the aiding process, user acceptance, requirements analysis, evaluation method[s]). These modules of the decision making-centered model of C^2 are addressed in detail in the Special Issue of the IEEE *Transactions on Systems, Man, and Cybernetics* devoted to Information Technology for Command and Control (November/December, 1986).

The Evolution of System Engineering into a Multidisciplinary Field

An SE training/education template presumes a definition of systems engineering (and, of course, a definition of C^2). The classical definition of SE highlights the process-based nature of the discipline—that is, SE as a series of discrete but interrelated steps or phases (with feedback and iteration as necessary and desirable). More explicitly, SE can be defined as a " . . . logical sequence of activities and decisions transforming an operational need into a description of system performance parameters and a preferred system configuration [Department of Defense, 1974]." Decisions invariably inject into the discussion the perspectives of behavioral science/psychology and cognitive science, but it should be emphasized at the outset that the succinct definition above obscures a great deal by neglecting the complex process which intervenes between an identified operational need and the description of performance parameters and the selected system configuration.

Complex process models in SE attempt to profile these elements and capture the myriad flows and interrelationships, but it is even more critical to recognize that SE is above all a mindset, a way of viewing the world unique in the engineering sciences. In other words, despite the proliferation of computerized tools and techniques and their incarnation in the acronym CASE (for computer-assisted systems engineering), SE is neither necessarily nor invariably computer-based. Nor is SE reducible to a particular methodology or set of methodologies; SE does not even necessarily entail the use of numbers. Imparting this SE mindset to students is particularly crucial—much more crucial than requiring students to memorize process models or be able to apply a particular analytical tool.

Definitions of SE are perhaps as numerous as systems engineers themselves; each practitioner might provide an at least slightly different definition of the field. However, common themes can be identified. SE features at least nine distinct defining characteristics:

1. Activities that extend over the entire "life cycle" of systems constitute the scope of the discipline; this cycle begins with requirements definition or specification and includes conceptual or functional design, integration, testing, and enhancement.

2. The systems engineer's viewpoint is appreciably different from that of a product engineer, software designer, or technology developer; while the product engineer, for example, deals frequently with detail, the systems engineer takes a "top down" viewpoint, dealing with detail only as necessary.

3. The product engineer deals with internals; the systems engineer, in contrast, deals more directly with the external view of the system—through the system's interfaces to other systems and to its human users, repairers, and managers.

4. The focus of SE is architectural definition and performance evaluation; these perspectives serve as fulcrums for addressing functional integration, maintainability, reliability, and (perhaps most importantly) the proper interfaces that will insure system design for successful human use of or interaction with the completed system.

5. SE activities increasingly center on the design, development, and use of what have previously been referred to as advanced or sophisticated cognitive problem solving systems—knowledge-based systems and processes that support human activities in planning, design, decision making, and associated resource allocation (along with the uncertainty and other complex cognitive concomitants of these activities).

6. Also increasingly, SE emphasizes behavioral factor concerns associated with such subjects as human-machine interaction, the design of decision support and expert systems, command and control systems design, software systems engineering, and system evaluation.

7. SE education should be based on the quantitative skills used in traditional engineering *as well as* additional quantitative *and* qualitative skills from

applied mathematics, psychology, and other disciplines relevant to knowledge organization and design.

8. SE is intrinsically multidisciplinary, integrative, iterative, synthetic, and applied.

9. Systems engineers are professionals with backgrounds and experience in the social, behavioral, management, computer, mathematical, engineering, and physical sciences.

Design

Conventionally, as Rouse (1982) points out, academic disciplines can be profiled on the basis of two key dimensions: the nature of the world of interest (that is, data-driven or real world versus model-based views of reality) and the nature of the empirical phenomena of interest (that is, internal or within-person versus external phenomena). Internal phenomena include the vast realms of human decision making and problem solving. External phenomena refer to entities and processes which exist (for example, the motions of planets). The second dimension differentiates between the study of the real world via inductive reasoning (to reach general conclusions on the basis of particular observed facts) or deductive reasoning (arriving at specific logical conclusions on the basis of general principles).

The cross-classification of the two dimensions yields a disciplinary topology in which we can identify "pure" or corner disciplines. Philosophy represents the model/internal pole; psychology arises at the intersection of the real world/internal perspectives; physical science is also grounded in real world data but assumes an external viewpoint; pure or basic mathematics reflects the model/external pole. In addition to the corner disciplines, a variety of "edge" disciplines can be identified, including such behavioral science fields as history and sociology (between psychology and philosophy) and such engineering or applied mathematics disciplines as operations research, electrical engineering, and mechanical engineering (between physical science and pure mathematics).

Where does SE education fit? As Figure 1 suggests, SE is the synthesis of various multidisciplinary problem areas which fall at neither the corners nor the edges—but draw on the core disciplines in an indirect sense. Thus, SE is comprised of such multidisciplinary concerns as management science, human factors, man-machine systems, biological engineering, cybernetics, and systems science.

The Rouse template and the previously delineated definitions of SE provide a foundation for curriculum design in which the incredibly broad scope of the field is clearly highlighted. The following discussion will look at the "prototypical" curriculum for training systems engineers.

A unique emphasis of some programs is the dimension of *human interaction* with systems (particularly complex and large scale systems). This human interaction may involve human supervisory control of physical processes, such as the robots used in automated manufacturing. It may also involve cognitive tasks at the operational levels of fault diagnosis, detection, and correction or at the level of strategic planning.

Education in SE accordingly stresses behavioral factor concerns associated with such subjects as human-machine interaction, the design of decision support and expert or knowledge-based systems, command and control system design, software engineering, and system evaluation. A principal thrust of this focus is, as noted, the design, development, and application of large, complex systems grounded in knowledge and information bases and processes that support human activities in planning, design, decision making, and associated resource allocation.

Consequently, human information processing principles for command and control emerge as a core facet of the curriculum. Sage (1987) provides a comprehensive and lucid discussion of the domains of human information processing concerns and the enhancement of such processing capabilities via knowledge support systems. Human information processing in C^2 is particularly problematic and susceptible to biases and the inappropriate use of cognitive strategies because of the fact that available information (the basis for inference and choice) is quite typically of questionable validity and reliability, incomplete, inconsistent, associated with severe time constraints, and must be aggregated.

Furthermore, as Sage argues, C^2 decision making consists of a variety of relevant decision/problem solving activities; the C^2 analysis/decision process includes such stages as recognition/assessment of a decision situation, formulation of the decision problem, analysis of alternative decisions, interpretation and selection, specification of decisions, plans, and architectural structures, identification of operational priorities, and forecasting, explaining, and describing the decision to others. Each step of the C^2 process imposes requirements for different types of information and support.

Methodologically, undergraduate and graduate programs require the student to become familiar with the entire spectrum of tools and techniques for applied problem solving. This marries the quantitative skills used in traditional engineering to additional quantitative and qualitative skills from operations research and statistics, applied mathematics, psychology, management, computer science, electrical engineering, data communications, and other disciplines which support knowledge and information organization and systems design.

In addition to C^2, opportunities for specialization in SE include:

- Information systems engineering;
- Expert and decision support systems design/development;
- Telecommunications systems;
- Operations research;
- Software systems engineering;
- Computer systems engineering;
- System acquisition and procurement.

At the undergraduate level, the degree program is designed to prepare students for direct entry into professional practice or for entry into graduate study. In addition to the general degree requirements at the university level, there are specific requirements for the B.S. in SE. The distribution of these courses must be such as to satisfy Accreditation Board for Engineering and Technology requirements for a degree in engineering.

Required non-engineering courses include mathematics (Analytic Geometry and Calculus I, Calculus II, Calculus III, Differential Equations), computer science (Introduction to Computer Science, Data Structures, Analysis of Algorithms), Contemporary Economic Principles I and II, Basic Concepts in Psychology, four physics courses (Energy and Environment, University Physics I, University Physics II, University Physics II-Lab), and English Literature and Composition. In addition, courses are required in Electric Circuit Analysis (I and II), Engineering (Statics), and basic statistics and operations research (a total of four courses). Several humanities, engineering science, and technical electives are also specified.

Interestingly, the curriculum outlined above contains no SE courses and is pursued primarily in the student's first two years of study; it is designed to provide the student with a solid inter- and multidisciplinary background for the junior and senior years. During these years, the student completes elective course requirements as well as a number of SE and information system courses.

The core SE courses in a "prototypical" undergraduate curriculum are SYST 301 (Systems Methodology and Design I) and SYST 401 (Systems Methodology and Design II). SYST 301 introduces the student to morphologies and methodologies of SE, system design, design for functionality, and information requirements determination. SYST 401 (for which SYST 301 is the prerequisite) focuses on the system life cycle, design for functionality, system management, specification of system architecture, evaluation of alternatives, software system engineering, and operational test and evaluation.

Eight courses round out the current SE offerings. One is System Organization, Design, and Development, which is an introduction to systems management that emphasizes information systems definition, organization, design, and development, operational feasibility determination, and design for verification and validation; case studies from software systems engineering are employed. Systems Engineering Laboratory I provides integrated discussions and laboratory exercises concerning the engineering of large systems; experiments focus on information systems design, information requirements determination, systems level architecture specification, and operational test and evaluation. Knowledge Support Systems Design is an introduction to the design of expert systems and decision support systems. The course Behavioral Factors in System Design looks at human information processing, inferential analysis, biases and heuristics in human information processing, support systems to aid in human information processing, human-system interaction, and software systems engineering considerations.

There are two Senior Design project courses. The first portion of I includes familiarization with the knowledge support systems design laboratory software packages at GMU and individual experiments with the packages while the latter portion concerns formulation of a design project. In Senior Design Project II the design project plans formulated in I are reviewed and modified. The systems design effort proposed is completed and a report prepared and evaluated. This "capstone" course in the SE program has as its primary goal the design and implementation of a software package to assist in the performance of a systems engineering function.

A Special Topics course offers in-depth studies of contemporary developments in SE, including software systems engineering, systems project management, systems integration, systems acquisition and procurement and related topics. Independent Study also allows the student to pursue such SE topics as software systems engineering, systems project management, decision support systems, expert systems, economic systems analysis, and related areas of investigation under the supervision of a faculty member.

The graduate program leading to a Master of Science in Systems Engineering is designed to equip students for research and professional practice associated with problem formulation, issue analysis, and evaluation of the consequences of alternative courses of action. The program emphasizes the analytical and behavioral aspects of the engineering of complex systems of large scale and scope. Students are expected to become proficient in operations research and applied statistics, which supplies a key quantitative basis for systems analysis. Cognitive aspects of systems engineering, systems design methodology, and systems management are also essential studies.

To achieve this objective, the program includes four core courses, electives selected by the student with the aid of a faculty advisor (in which the student assembles a program which meets his or her unique interests), and a thesis or systems engineering project. To obtain the Master of Science degree, students must complete an approved plan of study; the plan must contain a minimum of 30 semester hours of graduate level course work.

It is assumed that the student will meet foundation requirements comprised of graduate level courses in Statistical Foundations for Decision Making and Economic Analysis and undergraduate mathematics courses in Analytic Geometry and Calculus II and Elementary Differential Equations.

Degree requirements include required courses, constituting 18 credits. Six semester hours are devoted to two courses in systems engineering concepts and methods; the first of these core SE courses—the foundation for the rest of the program of study—deals with issues of methodology, analysis, and design, while the second is concerned with economic systems analysis, design tradeoffs, and human factors. Also required are two courses in operations research and applied statistics.

Rounding out the required part of the curriculum are the project or thesis options. Three alternatives are offered. First, the student may elect to pursue the Master's SE Project option, designed to permit him or her to carry out systems engineering analysis and design, usually within a multidisciplinary group, under the guidance of faculty members. The emphasis here is on bringing a range of skills to bear on a complex, realistic system problem. The final product of the work is a professional technical report dealing with a predefined portion of the project activity.

Second, the student may conduct Master's level independent research, which involves a Master's thesis as the final result of a significant independent research effort. This is the more conventional "academic" track within the program. The work is conducted under the guidance of a faculty advisor; a faculty committee made up of three members approves the final written thesis.

Third, the individual project is directed toward students whose employment is in SE work; while the first two options are for six credits, this option is for three. Typically, a project objective is selected with the approval of the faculty advisor and may involve analysis of system requirements, development of a prospective system architecture, or use and evaluation of SE methodologies. A Project Report is submitted to meet the requirements of this option.

Grafting C^2 or command, control, communications, and intelligence (C^3I) topics onto this template involves the incorporation of a course (or preferably, a sequence of several courses) which focuses on principles of command and control and a course on the fundamentals of C^2 systems engineering.

The principles course(s) would deal with the theory of warfare and review alternative quantitative models for force effectiveness. Models of the command and control process would also be covered. The concepts of architectures, mission-oriented assessment, and measures of effectiveness would be introduced. Also featured would be a description of the national security structure, the technology needed for each component of the C^2 system, the command decision making process and decision making theories, and quantitative evaluation models and analysis tools.

The C^2 SE course identifies, describes, and assesses the methods, approaches, and tools from the system, behavioral, mathematical, computer, engineering, and management sciences used to design, develop, test, and field systems. The entire panoply of potentially relevant families of methods and tools is examined and assessed—including operations research, control theoretic, artificial intelligence, and decision theoretic (as well as other) methods and analytical approaches. The course highlights the overarching goal of optimal systems design.

Other courses in a C^2 program would include (at a minimum) C^2 systems modeling, simulation, and wargaming, sensor data fusion, decision making in command and control, requirements analysis and prototyping for command and control systems engineering, systems evaluation methodology for C^2, and simulation and training. The student's program would of course also feature courses in systems design methodology, computer architecture, communications theory, and the various ancillary courses profiled in the earlier part of this section of the chapter on generic SE.

Challenges

The undergraduate and graduate programs profiled above are quite ambitious. Not surprisingly, at least several challenges must be confronted and overcome in order to develop and implement successful SE training/education programs of instruction.

The first challenge may be referred to as the need to integrate incommensurable paradigms. Paradigm is a philosophy of science concept which refers to the theories central to a discipline. However, paradigms embrace a wider range of phenomena and include unique discipline-grounded beliefs and views of the world—in other words, a discipline's gestalt. Paradigms vary across disciplines and even within a single discipline across time (as one paradigm replaces another). As Rouse (1982) notes, the concept of "culture" may be associated with an academic discipline; he compares the disciplines of physics and psychology, providing a dramatic illustration of strikingly different paradigmatic frameworks.

Discipline-based problem solving differs particularly with respect to *measures*, *models*, and *methods* employed. Measures refer to the value or utility functions appropriate for determining the degree to which alternative solutions achieve the desired objectives. Assuming implicit or explicit measures, model-based problem solving moves on to modeling the impacts of alternative problem solutions in terms of these measures; models, it should be emphasized, may also be either implicit or explicit. Methods are the procedures by which the researcher manipulates or experimentally determines model parameters for the purpose of assessing the degree to which variations of alternative problem solutions—as represented by explicit or implicit models—can be discovered that improve achievement of objectives in terms of the previously developed measures. Models and methods interact and the modeler's disciplinary affiliation exerts an undeniable influence on this interaction; a disciplinary paradigm provides consensus with respect to what can be modeled, acceptable methods of validation, and the relationships of models to "truth," all of which shape the explicit or implicit choices of models or methods.

Rouse (1982), from whom the discussion above on measures, models, and methods is drawn, notes that conflicts in multidisciplinary problem solving can arise in one of several ways. Differences concerning ends and means (that is, measures and their interpretation and methods, respectively) may be readily apparent but are often difficult to resolve. Model differences are, in contrast, often difficult to even detect (particularly if the model is implicit).

All of this poses enormous problems to SE as a "discipline." In addition to being a young discipline (compared, for example, to a discipline such as physics, which is hundreds of years old), SE constitutes a synthetic and applied science which spans a number of behavioral, physical and mathematical sciences—in striking contrast to such conventional academic disciplines as physics, mechanical engineering, and statistics.

Because SE is synthetic, it does not really have a paradigm. Figure 1 illustrates the multidisciplinary nature of the field. This problem is exacerbated because SE is applied and therefore little thought is even accorded to philosophy of science considerations; there is no coherent body of literature on SE epistemology and philosophy.

Rouse (1982: 608) captures the essence of this challenge in his observation that solutions to SE problems "may [consequently] reflect a mixture of possibly incompatible points of view rather than an *integrated perspective* [emphasis added]." From the vantage point of SE training and education design and implementation, this means that it is exceedingly difficult to design a curriculum which does not reflect the biases of the designers. Further, even an integrated design may be translated quite differently in the classroom if, for example, the instructor has a background in management science or electrical engineering or focuses on information versus communications systems engineering in his or her own applications; an analogy can be made to the realm of decision making, where two individuals may implement the same decision in quite different ways.

A second challenge is the need to avoid the engineering bias. Andriole and Hopple (1982) discuss this issue from a C^2 perspective and link the prevailing hardware/human research imbalance in command and control research and training programs to a view of the domain in which C^2 systems design and development is assumed to be concerned about weapons control and unit maneuvering via the timely, rapid, survivable, and secure operation of data processing, navigation, and sensor systems.

This definition, while far from incorrect, completely neglects the underlying processes by which command decisions are formulated and implemented (see also Sage, 1987); simultaneously, this engineering bias leads to an almost total disregard for the cognitive operations which decision makers perform when confronted with typical C^2 problems.

Extensive empirical research in psychology demonstrates rather convincingly that the human analyst or decision maker brings to the problem solving arena not only a knowledge base (whether it is referred to as a schema, template, cognitive map, mental model, or some other similar entity) but a set of heuristics and biases for performing mental operations generally. These findings can be generalized from the psychological research laboratory to the real world of C^2 decision making.

For example, Wohl et al. (1988) report on an investigation of cognitive activities in the areas of situation assessment and resource management within the domain of antisubmarine warfare. Operationally, the setting is one in which the antisubmarine warfare commander (ASWC) is receiving multiple source contact reports from a set of geographically distributed acoustic sensors; the function is to manage the process of multiobject/multisensor data tracking, correlation, and fusion. Specifically, the task is to estimate enemy submarine tracks (position and velocity) and likelihoods in a passive sonar convergence zone environment.

Initially, a normative (optimal) mathematical model—based on estimation and control theory—is developed; this model transforms imperfect contact data on multiple submarines from distributed sensors into coherent state estimates of submarines (positions and velocities). Juxtaposed against this normative model is a set of descriptive generalizations which incorporate known cognitive limitations and biases. These fall into three areas: the tendency to consider a small number of hypotheses concurrently; the tendency to view the environment as less certain and fuzzier than it is in reality; the tendency to overemphasize the value of sensor-based information (via a recency bias in which subjects tend to overlook the uncertainty associated with measurements). The three factors are introduced into the model via sensitivity analysis on parameter values, using the criterion of minimizing the indices of model-subject performance mismatch.

What does the inclusion of descriptive elements (resulting in a normative-descriptive model) add? There is a significantly better description of subject performance (versus the normative model); the results can be used as the basis for the identification of descriptive elements (biases/heuristics) to incorporate into the overall model. Second, insights can be extracted that are not necessarily at all obvious (for example, the fact that all subjects in the study exhibited limited working memory). Finally, there is a potentially quite useful prescriptive function related to the fact that biases identified can be used for training and/or decision aid design requirements to decrease biases and shift performance toward the optimal.

The emphasis on behavioral factor concerns, human-machine interaction, and related topics in SE curricula reflects this awareness of the need to avoid the engineering bias. The third challenge constitutes the mirror image of this bias—the avoidance of psychocentric overkill. This challenge is embedded in the more general notion of avoiding methodolatry—the apotheosis of a particular method, tool, or disciplinary perspective in SE, a theme of Hopple's (1986) discussion of the "law of the hammer," the idea that a child (or systems engineer) is often tempted to use a hammer on everything encountered. Methodological fads often sweep systems engineering; AI was once invoked as a potential solution-in-search-of-a-problem (any problem, according to the most devoted accolytes of this fetish), while neural network modeling has already emerged as the current favorite fad.

Analogously, cognitive science should be seen as only one discipline among many in the SE armamentarium. Advanced cognitive problem solving systems are particularly in need of psychological science-based tools and principles and command and control is preeminently a cognitive activity. However, certain technical engineering systems have little or no need of such tools and techniques. Planning support systems (again, a cognitive undertaking) can thus be contrasted with robotic systems (although, it should be noted, robotic systems typically draw on planning models and must deal with environmental uncertainty, underlining the partial relevance of cognitive modeling and related constructs).

Furthermore, the emphasis on user friendly *and* user oriented system design and development (where the latter term refers to systems that reflect and use the user's mental model of a domain) provides additional ammunition to those who advocate the incorporation of the cognitive science "paradigm" into SE education. Again, however, it must be stressed that the user is not invariably correct (and may not even be able to specify his or her requirements accurately and reliably at the front end of the SE process, requiring the use of other tools for requirements analysis/knowledge acquisition). This realization leads to what can be labelled the paradox of the user-centric approach, the fact that the user is not always right (and user reactions cannot be the sole basis for system evaluation and modification) and that there arises as a consequence a need to devise system transfer strategies in which users can be convinced to accept new and alien system configurations.

Finally, the fourth challenge revolves around the demonstration of the *value added* by SE training or education programs. To what extent can we attribute improvements in C^2 processes and decision outputs to the formal training received? In a larger sense, this issue surfaces in criticisms that systems engineers are either "born, not made" or can only become genuinely proficient after a fairly long apprenticeship "in the trenches" of system design and development.

A considerable amount of research is now accruing on the question of the value added by decision support and knowledge-based systems. Theoretically, a system of this genre could improve (add value), detract, exert no effects, or have a mixed impact. Unfortunately, empirical work indicates that a system could be neutral, positive, or even negative in its consequences.

Analogously, the question of the value added by education in SE must be raised. However, this is extraordinarily difficult to determine. Value defined as subjective perceptions (such as, student evaluations) is of course only one facet of the overall evaluation task. Student assessments have generally been quite favorable (well above the mean for courses offered by other departments in the relevant school at George Mason) but value also connotes a longer term and more objective dimension. The corresponding concept in the literature on the evaluation of management information and decision support systems is *value of information*. What is the value of education/training in this sense? Unfortunately, value of information can typically be measured theoretically but not empirically, suggesting the difficulties confronting this kind of assessment of performance of any long-term and causally complex process embedded in a tangled set of real world determinants of (multiple) outcomes (because the dependent variable or effect is also complex, subtle, and multidimensional). The point here is to raise this issue and to underline the fact that value added considerations are crucial albeit elusive; efforts to measure value added to the systems engineering process by formal education programs should appear on any research agenda for the future.

Other challenges could undoubtedly be delineated, but these four are clearly among the central issues facing the systems engineering educational community. Future work should attempt to address these and related concerns directly and systematically, thereby providing the basis for improved programs of instruction in the multidisciplinary world of systems engineering.

References

Andriole, S. J. and S. M. Halpin (1986) "Information Technology for Command and Control." *IEEE Transactions on Systems, Man, and Cybernetics*, vol. SMC-16, no. 6 (November/December): 762-765.

Andriole, S. J. and G. W. Hopple (1982) "They're Only Human: Decision Makers in Command and Control." *SIGNAL*, vol. 36, no. 7 (March):45-49.

Chambers, G. J. (1980) "The System Engineering Process: A Technical Bibliography."*IEEE Transactions on Systems, Man, and Cybernetics*, vol. SMC-16, no. 5 (September/October): 712-723.

Department of Defense (1974) *Engineering Management.* MIL-STD-499A (USAF), May 1, pp. 11-18.

Hopple, G. W. (1986) "The Law of the Hammer and Other Maxims." *IEEE Transactions on Systems, Man, and Cybernetics*, vol SMC-16, no. 6 (November/December) : 948-964.

Rouse, W. B. (1982) "Editorial, On Models and Modelers: N Cultures." *IEEE Transactions on Systems, Man, and Cybernetics*, vol. SMC-12, no. 5 (September/October) : 605-610.

Sage, A. P. (1987) "Human Information Processing Principles for Command and Control," pp. 54-74 in J. L. Boyes and S. J. Andriole (eds.), *Principles of Command and Control*. Washington, DC: AFCEA International Press.

Wohl, J., et al. (1988) "Human Cognitive Performance in Antisubmarine Warfare: Situation Assessment and Data Fusion." *IEEE Transactions on Systems, Man, and Cybernetics*, vol. SMC-18, no. 5 (September/October) : 777-786.

ABOUT THE AUTHORS

Glenn Abrett. Senior Scientist with BBN Systems and Technologies Corp. Degree in Psychology from the State University of New York.

Dr. Leonard Adelman. Associate professor, Department of Information Systems and Systems Engineering, George Mason University.

Peter Aiken. Department of Information Systems and Systems Engineering, School of Information Technology and Engineering, George Mason University. Ph.D. George Mason University.

Stephen J. Andriole. Professor, College of Information Studies, Drexel University, Philadelphia, Pennsylvania. Formerly with the Department of Information Systems and Systems Engineering, School of Information Technology and Engineering, George Mason University. He is also the President and Chief Scientist of International Information Systems, Inc. Dr. Andriole is formerly the Director of the Defense Advanced Research Projects Agency's Cybernetics Technology Office, where he was also a Program Manager. He is the author, editor, co-editor and co-author of over twenty books, including Applications in Artificial Intelligence (Petrocelli Books, Inc., 1986), Principles of Command & Control (AFCEA International Press, 1988), and Storyboard Prototyping: A New Approach to User Requirements Analysis (QED Information Sciences, Inc., 1989). He has recently published articles in Large Scale Systems, the IEEE Transactions on Systems, Man & Cybernetics, and Defense Analysis. He is a frequent contributor to edited volumes in information systems design, decision support systems technology, and analytical methodology.

Dr. Eva L. Baker. Professor of Educational Psychology and Director of the UCLA Center for the Study of Evaluation. Ed.D. in Educational Psychology from UCLA.

Dr. Christopher Barrett. Los Alamos National Laboratory, Los Alamos, New Mexico. Ph.D. in Engineering Science/Bioinformation Systems, from Cal Tech.

D. F. Bender. Project Manager, Army Combat Services Support Control System Project, TRW.

Murray R. Berkowitz. Faculty of the Department of Information Systems and Systems Engineering, School of Information Techology and Engineering, George Mason University. Currently studying toward the Ph.D in Information Technology (Command, Control, Communications (C3) Systems Engineering.)

Dr. Barbara A. Black. Research psychologist and team leader at the Army Research Institute for the Behavioral and Social Sciences, Fort Knox Field Unit. Ph.D. in Experimental Psychology from Baylor University.

Christine L. Braun. Manager of the Software Reuse Project, Contel Technology Center. BA in Mathematics from Brown University, MS in Computer Science from the University of Toronto.

Dr. Alan Davis. Acting Chairman of the Department of Computer Science, George Mason University. Ph.D. in Computer Science from the University of Illinois.

Stephen Deutsch. Senior Scientist in the Advanced Simulation Division of BBN Science and Technology Corporation. B.S. in Mathematics from the University of Wisconsin.

Dr. Michael L. Donnell. Associate Professor of Administrative Sciences, George Washington University. Also, President and Chief Scientist of Donnell and Associates, Inc.

Dr. Stephen Downes-Martin. The Analytic Sciences Corp., Reading, Ma. Ph.D. in Theoretical Physics from London University.

Lee S. Ehrhart. Instructor in the Department of Information Systems & Systems Engineering, George Mason University.

Robert G. Eisenhardt. President of the Institute for the Study of Intelligent Systems, Ann Arbor, Michigan. Bachelor's Degree in Mechanical Engineering from the University of Pennsylvania.

Dr. Howard Eisner. Distinguished Research Professor and Professor of Engineering, George Washington University. He was formerly President, C3I Division, ARC Professional Services Group. B.E.E., M.S. and Doctor of Science from The City College of New York, Columbia University and the George Washington University, respectively.

Michael L. Fineberg. Research staff at the Institute for Defense Analyses and a licensed psychologist in private practice. Ph.D. in applied experimental psychology from Catholic University.

Dr. Richard E. Hayes. President of Evidence Based Research, Inc, Vienna, Virginia. Ph.D. in Political Science from Indiana University.

W. A. Holdrege. Senior staff engineer, TRW Electronics Systems Laboratory.

Dr. Gerald Hopple. Associate Professor of Information Systems and Systems Engineering, George Mason University. Ph.D. from the University of Maryland.

Susan M. Horton. Evidence Based Research, Inc., Vienna, Virginia. MA in Political Science, University of California, Santa Barbara.

Colonel Robert F. Gibson. Project manager, Space and Naval Warfare Systems Command.

Frederick Kulik. Director of Advanced Program Development at Bolt, Beranek and Newman Communications, Cambridge, Mass. BS from U.S. Military Academy and MS in Physics from Purdue.

Dr. Paul Lehner. Associate Professor in the Department of Information Systems and Systems Engineering, George Mason University. Ph.D. from University of Michigan.

Dr. Martin A. Leonardo. Senior Scientist, METRON, Inc. Ph. D. and M. S. degrees in Mathematics from Carnegie-Mellon University.

Dr. Robert Linn. Co-Director, Center for Student Testing, Evaluation and Standards and Professor of Education, University of Colorado at Boulder. Ph.D. in Psychological Measurement, University of Illinois.

Dr. Bruce Loatman. Director, AI Development, PRC, Government Information Systems. Ph.D. in Mathematics from Fordham University.

Walter W. Matyskiela. Command Support Division, TRW Defense Systems Group.

Morton L. Metersky. Naval Air Development Center. B. S. degree in aeronautical engineering from the Georgia Institute of Technology. M. S. degree in applied statistics from Villanova University and M. S. degree in management science from Lehigh University.

Dr. Harold F. O'Neil, Jr. Professor of Educational Psychology, University of Southern California. Ph.D. in Psychology, Florida State University.

Dr. James D. Palmer. The BDM International Professor of Information Technology at George Mason University. B.S. and M.S. degrees from the University of California and Ph.D. from the University of Oklahoma.

Dr. John F. Patterson. Research scientist with Bell Communcations Research Inc. Ph.D. in Experimental Psychology, University of Michigan.

C. Pizzutelli, Jr. Interoperability Engineer for the Program Executive Office for Communications Systems, Fort Monmouth, New Jersey.

Stephen D. Post. Director, AI Applications, PRC, Government Information Systems.

R. J. Rechter. Marketing manager for Systems Engineering with TRW Defense Systems Group.

G.A. Redding. Formerly assigned to the Defense Audiovisual Policy Office, American Forces Information Service, OASD(Public Affairs).

Dr. Joan Ryder. Principal Scientist at CHI Systems Inc., Spring House, Pennsylvania. A.B. in Psychology from Brown University and Ph.D. in Experimental Psychology from Brandeis University .

Dr. Alan B. Salisbury. President of the Contel Technology Center and former commander of the Army Information Systems Engineering Command. BS from the U. S. Military Academy and MS and Ph.D. from Standord University in Electrical Engineering and Computer Science.

William T. Scherer. Assistant professor at the University of Virginia. Ph.D. from the University of Virginia.

Bradley S. Stewart. Senior Scientist at the Institute for Parallel Computation in the School of Engineering and Applied Science at the University of Virginia. Ph.D. from the Systems Engineering Department, University of Virginia.

Richard H. Stottler, III. Senior Partner, Stottler Henke Associates, Belmont, California. MS in Computer Science, Stanford University.

John W. Sutherland. Senior Professor of Information Systems, School of Business, Virginia Commonwealth, Richmond, Virginia. Former Dean of the Stillman School of Business, Seton Hall University.

Scott Swarts. Advanced Simulation Division of BBN Systems and Technologies Corporation. BS in Electrical Engineering from MIT.

Edward A. Sykes. Assistant professor, Department of Systems Engineering, University of Virginia. Ph.D. from the University of Virginia.

Dr. Harry L. Van Trees is Distinguished Professor of Information Technology, Electrical and Systems Engineering, George Mason University. He was President, M/A-COM Government Systems, Inc; Assistant Secretary of Defense (C3I); and Chief Scientist U. S. Air Force. Graduate of the U.S. Military Academy. M.Sc.E.E. degree from the University of Maryland and Sc.D.E.E. degree from MIT.

Dr. Stephen T. Vinter. Siemens Nixdorf Information Systems, Cambridge, Mass. Ph.D. from University of Massachusetts, Amherst.

J. Walbeck, Major, USA. Program Executive Office for Command and Control Systems, Fort Monmouth, New Jersey.

Chelsea C. White. Professor and Chairman of Systems Engineering at the University of Michigan. Ph.D. from the University of Michigan in the Computer, Information, and Control Engineering Program.